STP 1115

Plants for Toxicity Assessment: Second Volume

Joseph W. Gorsuch, William R. Lower, Michael A. Lewis, and Wuncheng Wang, Editors

ASTM Publication Code Number (PCN)
04-011150-16

ASTM
1916 Race Street
Philadelphia, PA 19103

Library of Congress Cataloging in Publication Data

ASTM Publication Code Number (PCN): 04-011150-16
ISBN: 0-8031-1422-2
ISSN: 1058-1189

QK750
.P56
1990
Vol. 2

Peer Review Policy

Each paper published in this volume was evaluated by three peer reviewers. The authors addressed all of the reviewers' comments to the satisfaction of both the technical editor(s) and the ASTM Committee on Publications.

The quality of the papers in this publication reflects not only the obvious efforts of the authors and the technical editor(s), but also the work of these peer reviewers. The ASTM Committee on Publications acknowledges with appreciation their dedication and contribution of time and effort on behalf of ASTM.

Printed in Ann Arbor, MI
1991

Foreword

This publication, *Plants for Toxicity Assessment: Second Volume,* contains papers presented at the 2nd Symposium on Use of Plants for Toxicity Assessment, which was held in San Francisco, California on 23–24 April 1990. The symposium was sponsored by Committee E-47 on Biological Effects and Environmental Fate and Subcommittee E-47.11 on Plant Toxicity. Symposium chairman was Joseph W. Gorsuch, Eastman Kodak Company. Co-chairmen were Michael Lewis, Proctor and Gamble, William R. Lower, University of Missouri, Shahbeg S. Sandu, U.S. Environmental Protection Agency, and Wuncheng Wang, Illinois State Water Survey.

Contents

Overview   1

REGULATORY PERSPECTIVES

Keynote Speech: A Brief Overview of Plant Toxicity Testing—JOHN FLETCHER   5

Role of Plant Bioassays in FDA Review: Scenarios for Terrestrial Exposures—
MICHAEL C. HARRASS, CHARLES E. EIRKSON III, AND LISA H. NOWELL   12

The Use of Soil-Based Plant Tests Based on the OECD Guidelines—
ANTHONY J. WINDEATT, JOHN F. TAPP, AND RICHARD D. STANLEY   29

An Inter- and Intra-Agency Survey of the Use of Plants for Toxicity Assessment—
BARBARA M. SMITH   41

An Overview of Experimental Design—MERRILEE RITTER   60

Higher Plants (Common Duckweed, Lettuce, and Rice) for Effluent Toxicity
Assessment—WUNCHENG WANG   68

Testing for Pesticide Toxicity to Aquatic Plants: Recommendations for Test
Species—STELLA M. SWANSON, COLLEEN P. RICKARD, KATHRYN E.
FREEMARK, AND PATRICIA MacQUARRIE·   77

APPLICATIONS OF PLANT BIOASSAYS/PHOTOSYNTHESIS

Peroxidase Activity as an Indicator of Sublethal Stress in the Aquatic Plant
Hydrilla Verticillata (Royle)—THOMAS D. BYL AND STEVEN J. KLAINE   101

Toxicity Testing Using a Chemostat-Grown Green Alga, *Selenastrum
Capricornutum*—HANS G. PETERSON   107

The Effect of Dissolved H_2S and CO_2 on Short-Term Photosynthesis of
Skeletonema Costatum, a Marine Diatom—RONALD J. BRETELER,
ROSANNA L. BUHL, AND ALAN W. MAKI   118

Evaluating Direct Toxicity and Food-Chain Effects in Aquatic Systems Using Natural Periphyton Communities—HARRY L. BOSTON, WALTER R. HILL, AND ARTHUR J. STEWART   126

A Seedling Chlorophyll Fluorescence Toxicity Assay—BARBARA M. JUDY, WILLIAM R. LOWER, FRANK A. IRELAND, AND GARRY F. KRAUSE   146

XENOBIOTIC UPTAKE BY PLANTS

Detecting Ambient Cadmium Toxicity in an Ecosystem— SANDRA M. STEWART-PINKHAM   161

A Plant Bioassay for Assessing Plant Uptake of Contaminants from Freshwater Soils or Dredged Material—BOBBY L. FOLSOM, JR. AND RICHARD A. PRICE   172

A Comparative Study of the Bioconcentration and Toxicity of Chlorinated Hydrocarbons in Aquatic Macrophytes and Fish—FRANK A. P. C. GOBAS, LESLEY LOVETT-DOUST, AND G. DOUGLAS HAFFNER   178

GENERAL PHYTOTOXICOLOGY

Track-Sprayer and Glasshouse Techniques for Terrestrial Plant Bioassays with Pesticides—RICHARD A. BROWN AND DEBORAH FARMER   197

Photoinduced Toxicity of Polycyclic Aromatic Hydrocarbons to the Higher Plant Lemna Gibba L. G-3—XIAO-DONG HUANG, D. GEORGE DIXON, AND BRUCE M. GREENBERG   209

Characterization of Eight Soil Types Using the Selenastrum Capricornutum Bioassay—FRANK A. IRELAND, BARBARA M. JUDY, WILLIAM R. LOWER, MARK W. THOMAS, GARY F. KRAUSE, AMHA ASFAW, AND WILLIAM W. SUTTON   217

The Influence of Test Length and Bacteria on the Results of Algal Bioassays with Monophenolic Acids—LOREN J. LARSON   230

Ecotoxicological Assessment of Toluene and Cadmium Using Plant Cell Cultures— MINOCHER REPORTER, MERLINE ROBIDEAUX, PAUL WICKSTER, JOHN WAGNER, AND LAWRENCE KAPUSTKA   240

Tissue Culture as a Method for Evaluating the Biotransformation of Xenobiotics by Plants—CARLOS WICKLIFF AND JOHN S. FLETCHER   250

Responses of Spruce Trees (Picea Abies. L. KARST) to Fumigation with Halone 1211—First Results of a Pilot Study—PETER SCHRÖDER AND REINHARD DEBUS   258

Response of Sago Pondweed, a Submerged Aquatic Macrophyte, to
   Herbicidese in Three Laboratory Culture Systems—W. JAMES FLEMING,
   M. STEPHEN AILSTOCK, JEFFREY J. MOMOT, AND C. MICHAEL NORMAN   267

X-Ray Microanalysis and Ultrastructure of Lichens from Polluted and Unpolluted
   Areas—MARIA A. FAVALI, MARIA G. CORRADI, AND FABRIZIA FOSSATI   276

BIOCHEMICAL AND GENETIC APPLICATIONS

The Biochemical Basis of the Activation of Promutagens by Plant Cell Systems—
   MICHAEL J. PLEWA   287

^{32}P-Postlabeling for DNA Adduct Determination in Plants—WILLIAM R. LOWER,
   FRANK A. IRELAND, AND BARABARA M. JUDY   297

Application of a Plant Test System in the Identification of Potential Genetic
   Hazards at Chemical Waste Sites—BALJIT S. GILL, SHAHBEG S. SANDHU,
   LORRAINE C. BACKER, AND BRUCE C. CASTO   309

Activation of Aflatoxin B1 and Benzo(a)pyrene by Tobacco Cells in the Plant Cell/
   Microbe Coincubation Assay—JAMES M. GENTILE, PETER JOHNSON, AND
   SUSAN ROBBINS   318

Identification of Aneuploidy-Inducing Agents Using the Wheat Seedling Assay—
   SHAHBEG S. SANDHU, JAGTAR S. DHESI, AND BALJIT S. GILL   326

NEW APPROACHES

A Method for On-Site Evaluation of Phytotoxicity at Hazardous Waste Sites—
   JULIUS U. NWOSU, HILMAN C. RATSCH, AND LARRY A. KAPUSTKA   333

Use of Marsh Plants for Toxicity Testing of Water and Sediment—
   GERALD E. WALSH, DAVID E. WEBER, TASHA L. SIMON,
   LINDA K. BRASHERS, AND JAMES C. MOORE   341

A Short-Term Bioassay for Whole Plant Toxicity—THOMAS PFLEEGER,
   J. CRAIG MC FARLANE, RACINDA SHERMAN, AND GAYLE VOLK   355

A New Plant Life-Cycle Bioassay for Assessment of the Effects of Toxic Chemicals
   Using Rapid Cycling Brassica—R. A. SHIMABUKU, HILMAN C. RATSCH, C. M.
   WISE, JULIUS U. NWOSU, AND LARRY A. KAPUSTKA   365

Use of Plants for Toxicity Assessment of Heavy Metal Rich Industrial Effluents—
   R. K. SOMASHEKAR AND SIDDARAMAIAH   376

Cyanophage/Host Assay for Toxicity Assessment in Water, Wastewater, Sludges,
   and Composts—MARJORIE P. KRAUS   383

Indexes   393

Overview

The use of plants to assess environmental risk is a budding science that has only recently gained the attention of scientists. Although terrestrial plant studies have been used for decades to evaluate the efficacy of potential herbicides, it was not until 1982 that the Office of Pesticide Programs, U.S. Environmental Protection Agency (EPA) published a guideline to evaluate the hazard of pesticides on nontarget plants as part of a Federal Insecticide, Fungicide, and Rodenticide Act (FIFRA) regulation. After nearly a decade of use, these guidelines are still being refined, and steps to conduct studies under field conditions have been initiated. In 1987 the U.S. Federal Drug Administration (FDA) published guidelines to assess environmental risk including two plant bioassay guidelines. In 1985 the U.S. EPA Toxic Substance Control Act (TSCA) seedling test guidelines were published in the *Federal Register*, and in 1984 the current Organization for Economic Cooperation and Development (OECD) terrestrial plant guidelines were published. As can be seen from these recently published guidelines for testing plants, there are numerous differences, often requiring a laboratory to repeat studies to satisfy two or more agencies. This symposium, the Second Symposium on Use of Plants for Toxicity Assessment, focused on some of these issues, in addition to many others relative to the value of using plants to assess environmental risk.

As did the first symposium (Atlanta, Georgia, 1989), the second symposium provided a forum to promote (1) the gathering and dissemination of information, (2) the development of standard practices to assess the impact of chemicals and other xenobiotics upon the plant communities, and (3) the use of plants for toxicity assessment.

The Second Symposium on Use of Plants for Toxicity Assessment was sponsored by ASTM Committee E-47 on Biological Effects and Environmental Fate and its Subcommittee E-47.11 on Plant Toxicology and was held in San Francisco, California, on 23–24 April 1990. The symposium attracted more than 100 researchers from Canada, Denmark, France, Germany, India, the United Kingdom, and the United States. Scientists from academia, industry, consulting laboratories, and governments were represented. The attendees' backgrounds included ecotoxicology, biochemistry, statistics, ecology, plant physiology, and genetics, to name only a few. The platform and poster sessions included a wide range of topics on the use of higher and lower plants for assessing pollution in freshwater, marine environments, air, soil, and water. Themes covered many aspects of ecotoxicology as well as genotoxicology. The plant miniworkshop held toward the end of the second symposium focused on the FDA and OECD seedling test guidelines and some of the associated difficulties in performing these studies and interpreting the results.

The papers indicated the need to eliminate differences and to focus on standardizing bioassays to evaluate the impact of new and existing chemicals, and to evaluate the impact of fugitive emissions (e.g., cadmium and Halone 1211) on forests and trees.

This ASTM special technical publication is an outgrowth of the symposium. It contains 35 refereed papers divided into six groups: Regulatory Perspectives, Applications of Plant Bioassays/Photosynthesis, Xenobiotic Uptake by Plants, General Phytotoxicology, Biochemical and Genetic Applications, and New Approaches. John Fletcher of the University of Oklahoma presented the keynote speech, describing the need for more plant testing in an effort to curtail and avoid the further decline of plant life in threatened habitats such as the forests of central Europe. He also noted the need to review the current test guidelines to determine if they are economically and scientifically sound, and if the test results are being interpreted accurately for the purpose of environmental protection. The use of plants to assess the potential toxicity of chemicals in the air, water, and soil ecosystems has a promising future. In late November 1990, Environmental Research Laboratory (Corvallis, Oregon) of the EPA sponsored a three-day workshop to evaluate the current test guidelines. Several of the recommendations and priority research were concluded, including (1) designing and implementing field experiments and (2) improving the efficiency and validity of test protocols. Scientists are convinced that plants are an important component of our ecosystem and must be protected. Therefore, standardization of methods and participation by scientists worldwide are necessary to accomplish this task. As summed up in the EPA workshop, "One important observation was the need to eliminate differences in the test procedures between different regulatory authorities (e.g., OECD, FIFRA, TSCA, FDA, and CERCLA)." ASTM members certainly have a critical challenge to assist in this process of eliminating differences in guidelines. Another challenge facing many of the ASTM members is learning about and implementing the good laboratory practice standards (GLPs) that are required by these various government agencies. In addition, plant toxicology is evolving and new discoveries are continuously being made. Researchers in plant toxicology are strongly encouraged to continue their work in order to reach new heights.

The symposium committee thanks Tom Doane, Chairman of Committee E-47, for his support of this project; the authors, who contributed and shared their findings with the attendees; and the reviewers, whose time-consuming efforts and constructive comments resulted in much-improved papers. ASTM staff members are acknowledged for their assistance in organizing this symposium as well as their efforts in producing this publication. They include Dorothy Savini, Kathy Greene, Monica Siperko, and Rita Hippensteel.

The symposium committee would also like to acknowledge the efforts of Kenneth R. St. John, who acted as the representative from the ASTM Committee on Publications.

Joseph W. Gorsuch
Symposium Chairman and Editor
Eastman Kodak Company
Rochester, NY   14652-3617

William R. Lower
Symposium Co-Chairman and Editor
University of Missouri
Columbia, MO   65203

Michael A. Lewis
Symposium Co-Chairman
Battelle-Columbus Division
Columbus, Ohio   43201-2693

Shahbeg Sandhu
Symposium Co-Chairman
U.S. Environmental Protection Agency
Research Triangle Park, NC   27711

Wuncheng (Woodrow) Wang
Symposium Co-Chairman
Illinois State Water Survey
Peoria, IL   61652

Regulatory Perspectives

John Fletcher[1]

Keynote Speech: A Brief Overview of Plant Toxicity Testing

REFERENCE: Fletcher, J., **"Keynote Speech: A Brief Overview of Plant Toxicity Testing,"** *Plants for Toxicity Assessment: Second Volume, ASTM STP 1115,* J. W. Gorsuch, W. R. Lower, W. Wang, and M. A. Lewis, Eds., American Society for Testing and Materials, Philadelphia, 1991, pp. 5–11.

ABSTRACT: In 1962 Rachel Carson's book *Silent Spring* challenged the world to reevaluate the use and disposal of manmade chemicals. The initial reaction to this classic book was outrage by industry, bewilderment from government, and deep concern on the part of the general public. In the intervening years, attitudes of outrage and bewilderment have disappeared and a growing concern for protecting the environment has emerged among all factions of our society. Accompanying this concern has been an ever increasing trend towards cooperation and responsible action by involved parties. Fundamental to this trend is an awareness of potential hazard posed by manmade chemicals to the environment and agreement on measures which must be taken to avert deterioration of the environment.

Laboratory testing of new chemicals on selected biota has become the primary means for evaluating the potential threat of chemicals to the environment. Although the amount of plant testing has been minimal in the past, it is almost certain that more plant testing will be conducted in the future in an effort to curtail and avoid the decline of plant life in threatened habitats. Thus, it is very timely to examine plant tests in current use in an effort to ascertain if they are economically and scientifically sound and whether or not the laboratory data from these tests are being interpreted accurately for the purpose of environmental protection.

KEY WORDS: vascular plants, phytotoxicity, toxicity testing

Publication of *Silent Spring* by Rachel Carson in 1962 has proven to be a landmark event in arousing public concern for the environment. This is a concern that has continued and that is indirectly responsible for the regulatory policies which exist today. The 30 years intervening between the appearance of *Silent Spring* and today have been marked by four phases of reaction to the book: (1) attempts in the 1960s to discredit Carson and her book; (2) realignment in the 1970s of government agencies and their regulatory responsibilities; (3) writing of government regulations, (test protocols, etc.) in the 1980s; and now (4) enforcement of government regulations.

The initial phase was short-lived, lasting about two years, but it was colorful and rather intense at times. The flavor of this period is well represented by an article appearing in the July 22, 1962 issue of the *New York Times*. The article was entitled, "Silent Spring Is Now Noisy Summer." The article, written by John M. Lee, begins with the statement, "The $300,000,000 pesticides industry has been highly irritated by a quiet woman author whose previous works on science have been praised for the beauty and precision of the writing." Controversy over the book raged for several years, but after various accusations had been made by numerous parties in all possible places, and counter arguments had been provided by Carson's staunch defenders, I believe it is safe to say that *Silent Spring* and the environment won. Never before had so much attention and concern been focused on the environ-

---

[1] Department of Botany and Microbiology, University of Oklahoma, Norman, OK 73019-0245.

ment, and I believe that the focus brought about by Carson is largely responsible for our current efforts to protect the environment. Thanks, Rachel.

Perhaps the true impact of Carson's book was not what the book said but what people said about the book. In this context, and as a plant scientist, I will take one more swipe at "Silent Spring." It is disappointing that only one chapter in the book pays much attention to vegetation. Plants, the most abundant life form and primary producer for all other life forms, always seem to get the least recognition and finish last in environmental matters. Constant reminders of this last place position come in the form of low numbers of plant scientists employed by EPA and FDA, minimal involvement of federal labs in plant/environment research, and inadequate funding in general for plant/environment research. But perhaps most disconcerting is the misconception held by many that the threat of environmental pollutants to plants is not a major problem. As an example, at a meeting I attended last year, one speaker made the bold statement that the influence of xenobiotics on plant growth and survival was not a problem, that the real problem was the role played by plants in food-chain contamination. The inaccuracy of this statement can be illustrated by citing many environmental disasters occurring around the world, but to me the most dramatic is the death of forests in central Europe. The toxicity of a chemical to plants or the accumulation of a toxicant by plants (food chain contamination) are both major issues of environmental importance, and safeguards must be taken to protect the environment from both of these potential hazards. Protection most certainly must include an early warning system relying on plant tests which are carefully designed, conducted, and interpreted.

During the 1970s and early 1980s numerous plant tests were developed to help safeguard the environment from the chemical threats which Rachel Carson dramatized in *Silent Spring*. The arsenal of plant tests now available or still under development are categorized in Table 1. Five classes of tests exist: (1) biotransformation; (2) food chain uptake; (3) phytotoxicity; (4) sentinel; and (5) surrogate. Of these five classes, the phytotoxicity and sentinel tests have received the most attention and are both in current use by regulatory agencies. The other three are less well developed and have not been adopted by regulatory agencies as required or recommended tests. However, it is very likely that with increased awareness among environmentalists and continued efforts on the part of researchers, all of the various plant tests will eventually gain use. Since more research has been done on phytotoxicity testing, and it is currently required by several agencies, the evolution of this type of testing

TABLE 1—*Classes of plant-chemical tests and their use in environmental assessment analyses.*

Class	Purpose	Example Endpoints
Biotransformation	Determine influence of plants on chemical fate of environmental pollutants	Change in chemical concentration
Foodchain uptake	Establish the amounts and concentrations of toxic chemicals which enter foodchains via plant uptake	Chemical concentration
Phytotoxicity	Evaluate the toxicity and hazard posed by environmental pollutants to the growth and survival of plants	Death, discoloration, reduced growth
Sentinel	Monitor the presence and concentration of toxic chemicals in the environment by observing toxicity symptoms displayed by plants	Death, discoloration, reduced growth
Surrogate	Use inexpensive, socially acceptable plant tests as a substitute for an animal or human assay	Chromosome aberrations

serves as a model for the development and future adoption of other kinds of plant testing. For this reason, I will focus attention in this talk on phytotoxicity testing.

Phytotoxicity testing of new chemical products is now required by U.S. EPA, U.S. FDA, and OECD. The three tests required by these agencies are seed germination, root elongation, and seedling growth (Table 2). Protocols for these tests have been prepared semi-independently by each agency or community and provided to prospective users. Although the protocols have been available in some form for several years, only recently have agencies required that the tests be conducted in fairly large numbers (Table 2). A rough estimate of required testing at present would be 25 and 50 chemicals per year by FDA and EPA, respectively. The majority of the required FDA testing comes from the Veterinarian Science Division, and all of the EPA testing has been at the request of The Office of Pesticides as required under Subdivision J for nontarget plants [4].

The development and use of phytotoxicity tests have progressed to the point where several key questions need to be addressed: (1) Are plant tests needed? (2) Are we using the right tests and the right protocols? (3) Are we interpreting test results accurately? (4) Do current tests do the job? Of these four questions, there is only one that I can answer without any reservations. In response to the first question, plant testing is definitely needed. We must take measures in the United States to avoid what has happened to the European forests. Recognizing the intensity of herbicide use in certain regions of the United States, a constant vigil must be kept to ensure that a new product does not devastate the environment. The other questions I have posed can only be answered with questions. These are questions which I feel must be addressed and resolved to the satisfaction of involved parties in order to accomplish the goal of plant testing, which is the protection of vegetation within the environment.

Concerning the question of whether or not the right tests are being used, at present only three of the seven classes of tests listed in Table 2 are used: (1) seed germination; (2) root elongation, and (3) seedling growth. If all new chemical products were to be subjected to a plant test, then a strong argument could be made for a tiered system wherein the first

TABLE 2—*Plant phytotoxicity tests, their description and use.*

Test	Response Measured to Chemical Treatment	Agencies Requiring Limited Use
Enzyme assay	Enzyme activity	None
Process measurement	Magnitude of a process; photosynthesis respiration, etc.	None
Tissue culture growth	Changes in fresh or dry weight	None
Seed germination (soil solution)	Percent of seeds which germinate	EPA (1985)[a] (50)[b] FDA (1981) (25) OECD
Root elongation (soil solution)	Length of root growth during fixed time period	EPA (1985) (50) FDA (1981) (25) OECD
Seedling growth (soil solution or foliar application)	Changes in height, fresh weight and dry weight	EPA (1985) (50) FDA (1981) (25) OECD
Life cycle	Changes in height, fresh weight, dry weight, flower number, and seed number	None

[a]Year when the agency started requiring plant testing as a concern for potential hazard posed by chemicals to nontarget vegetation.

[b]A rough approximation of the current annual number of agency-requested data sets used for hazard assessment of nontarget vegetation.

round of testing would include a quick-inexpensive enzyme or cell culture test. However, since plant testing is usually required of only those chemical products to which plants may gain high exposure, it seems reasonable to use intact plant material. If it were certain that each of the intact plant assays would always show a response to every and any plant toxicant, which we know is not true, and if we knew the relative sensitivities of the various intact plant assays to different classes of chemical structure, which we don't, then it would be appropriate to use only one kind of intact plant assay. Because of the uncertainties associated with any single plant assay, it is appropriate to examine potential toxicants with a battery of tests, which is the current policy of regulatory agencies. However, omission of a life-cycle test from this battery, also the current policy, may lead to serious consequences in the future since none of the currently required tests (Table 2) are geared towards detection of toxicity towards plant reproduction (flowering, seed set, etc.). A second issue of grave concern is the absence of any air or fog deposition tests from existing protocols. The increasing awareness of both the large quantities and wide diversity of organic chemicals released into the atmosphere [1] make air deposition testing an issue urgently in need of research and development. Of perhaps lesser importance, but certainly a source of aggravation and expense to persons performing required tests, has been the inability of regulatory agencies to adopt uniform or identical test protocols, something which an ASTM subcommittee hopefully can rectify.

The interpretation of laboratory tests raises many questions, most of which center around the issue of how well lab tests reflect what actually happens in nature. Specific questions often voiced are: How does the sensitivity of a plant in the greenhouse compare with that in the field? Can results collected from experiments on one species be extrapolated to another? How does the response of a seedling compare to that of a mature plant? What bearing does the inhibition of seedling growth have on the survival and eventual growth (yield) of a plant? These are all legitimate questions in view of the fact that the primary goal of plant testing is to protect natural plant communities made up of a mixture of taxa, which is comprised of young, old, annual, and perennial plants. The fundamental questions listed above remain subjects of debate and argument because comprehensive studies have rarely if ever been conducted to resolve them, presumably because funds have not been available.

In an effort to give a better perspective to some of these questions, we have recently published a study [2] based on data taken from the PHYTOTOX database. Concerning the question of lab versus field testing, we showed that the variability between laboratory and field data was 1.8 with a confidence interval of +0.4. In contrast to this, when data from PHYTOTOX were used to examine the question of taxonomic influence on chemical toxicity, it was shown that the mean sensitivity ratio between the least and most sensitive species was 10.5 with a confidence interval of 3.5 [2]. In this same study, when PHYTOTOX was used to compare the response of taxa at different taxonomic levels, it was shown that plants within the same genus have similar sensitivities to chemicals, but this is not true within families or orders, contrary to what has been reported for fish [3]. Thus, taxonomic differences among plants appear to have a much greater influence on plant response to chemical treatment than does the testing condition (laboratory versus field); consequently there is a definite need for additional laboratory and field research to confirm these database findings. Until such data are available, there is no alternative but to base judgements upon the patchwork of information which can be drawn from the literature such as found in PHYTOTOX.

Based on our current understanding of how taxonomic differences influence plant response to chemical treatment, it can be concluded that great care must be exercised in extrapolating test results from one species to another unless they belong to the same genus. These findings support the policy of regulatory agencies, which requires chemicals to be tested on more than one species and family. An EPA guideline concerning hazard evaluation of nontarget plants [4] requires testing of ten species in at least six families, while OECD requires three species from three families [5], and FDA ten species at least from two families [6]. Although multifamily testing of plants as required by these agencies is a commendable policy, the suitability of current lists of recommended plants (Table 3) is questionable. Of the 300 families in the plant kingdom [7], eight are represented on EPA's list, seven on FDA's list, and four on OECD's list. Review of these lists (Table 3) shows that no native species have been included. Of great concern is that some major families of native plants in the United States that are abundant, widely distributed, and economically important have been ignored. Most noticeable is the absence of any representatives of the Fagaceae (oaks, beeches,

TABLE 3—*Test species recommended by EPA, OECD, and FDA.*

Common Name	Latin Name	Family
	EPA	
Lettuce	*Lactuca sativa*	Asteraceae
Cabbage	*Brassica oleracea*	Brassicaceae
Cucumber	*Cucumis sativus*	Cucurbitaceae
Soybean	*Glycine max*	Leguminoseae
Pinto bean	*Phaseolus vulgaris*	Leguminoseae
Onion	*Allium cepa*	Liliaceae
Corn	*Zea mays*	Poaceae
Oat	*Avena sativa*	Poaceae
Ryegrass	*Lolium perenne*	Poaceae
Tomato	*Lycopersicon esculentum*	Solanaceae
Carrot	*Daucus carota*	Umbelliferae
	OECD	
Lettuce	*Lactuca sativa*	Asteraceae
Chinese cabbage	*Brassica campestris*	Brassicaceae
Cress	*Lepidium sativum*	Brassicaceae
Mustard	*Brassica alba*	Brassicaceae
Radish	*Raphanus sativus*	Brassicaceae
Rape	*Brassica napus*	Brassicaceae
Turnip	*Brassica rapa*	Brassicaceae
Fenugreek	*Trifolium ornithopodioides*	Leguminoseae
Mungbean	*Phaseolus aureus*	Leguminoseae
Red clover	*Trifolium pratense*	Leguminoseae
Vetch	*Vicia sativa*	Leguminoseae
Oat	*Avena sativa*	Poaceae
Rice	*Oryza sativa*	Poaceae
Ryegrass	*Lolium perenne*	Poaceae
Sorghum	*Sorghum bicolor*	Poaceae
Wheat	*Triticum aestivum*	Poaceae
	FDA	
Lettuce	*Lactuca sativa*	Asteraceae
Cabbage	*Brassica oleracea*	Brassicaceae
Cucumber	*Cucumis sativus*	Cucurbitaceae
Soybean	*Glycine max*	Leguminoseae
Pinto bean	*Phaseolus vulgaris*	Leguminoseae
Wheat	*Triticum aestivus* L.	Poaceae
Corn	*Zea mays*	Poaceae
Oat	*Avena sativa*	Poaceae
Ryegrass	*Lolium perenne*	Poaceae
Tomato	*Lycopersicon esculentum*	Solanaceae
Carrot	*Daucus carota*	Umbelliferae

chestnuts) and Pinaceae (pines, spruces, fir) families. These families are principally located in the southeast, northeast, and south central United States where agroecosystems may be described as a patchwork of cultivated crops and native wood lots often found in close proximity to industrial centers. Under circumstances where agriculture, industry, and native vegetation coexist, the question arises as to whether or not currently recommended surrogate species provide an adequate safeguard for environmental protection against potential organic pollutants. The diebacks of certain species in mature stands of tree communities in the Northeast and northern Midwest [8,9], along with a growing concern for drift of pesticides in cloud formations [10–13], are both compelling reasons to revaluate the taxonomic adequacy of currently recommended lists of surrogate species. Because of the inadequate understanding of chemical toxicity to most native plants [14], such a reevaluation would have to include phytotoxicity testing on numerous families of native plants for which we currently have virtually no toxicology data.

The question of how a laboratory test conducted on a seedling relates to growth and development of a mature plant community is an extremely difficult question to answer. While the question is one which is of great importance, it has been virtually ignored by environmental toxicologists, presumably again for lack of funds. The importance of this question centers around the fact that there are two features of plant growth, regeneration of organs and open-ended growth, which favor plant recovery from temporary setbacks. Because of these features, a reduced growth or partial death of a seedling does not always influence the growth and reproduction of the mature plant [15,16]. This aspect of plant growth makes it very difficult to know how inhibition of seedling growth relates to hazards posed to established plant communities.

In examining plant toxicity testing as it stands today, it becomes apparent that although there exists a definite need for well-described assays whose results can be accurately interpreted, it is questionable that we have a finished product. There are many questions which need to be addressed and resolved before we can be comfortable with the decisions made from current plant test data. These questions, almost without exception, require a better understanding of biological aspects of plant testing and interpretation. Until these questions are resolved by additional research, I predict that major controversy will hover over the interpretation of plant test data.

References

[1] Poje, J., Dean, N. J., and Burke, R. J., "Danger Downwind," Environmental Quality Division, National Wildlife Federation, Washington, DC, 1989.
[2] Fletcher, J. S., Johnson, F. L., and McFarlane, J. C., *Environmental Toxicology and Chemistry,* Vol. 9, 1990, pp. 769–776.
[3] Suter, G. W. II, Vaughan, D. S., and Gardner, R. H., *Environmental Toxicology and Chemistry,* Vol. 2, 1983, pp. 369–378.
[4] "Pesticides Assessment Guidelines Subdivision J. Hazard Evaluation: Nontarget Plants," EPA 540/9-8-020, Office of Pesticides and Toxic Substances, Washington, DC, 1982.
[5] "OECD Guideline For Testing Chemicals," Organization for Economic Cooperation and Development, Paris, France, 1981.
[6] "Environmental Assessment Technical Handbook," Food and Drug Administration, Washington, DC, 1987.
[7] Heywood, V. H., *Flowering Plants of the World,* Mayflower Books, New York, 1978.
[8] Evans, L. S., *Annual Review of Phytopathology,* Vol. 22, 1985, pp. 397–420.
[9] Evans, L. S., *Botanical Reviews,* Vol. 50, 1985, pp. 449–490.
[10] Faulstich, H. and Stournaras, *Nature,* Vol. 317, 1985, pp. 714–715.
[11] Glotfelty, D. E., Seiber, J. N., and Liljedahl, L. A., *Nature,* Vol. 325, 1987, pp. 602–605.

[*12*] Richards, R. P., Kramer, J. W., Baker, D. B., and Krieger, K. A., *Nature,* Vol. 327, 1987, pp. 129–131.
[*13*] Gaffney, G. S., Streit, G. E., Spall, W. D., and Hall, J. H., *Environmental Science Technology,* Vol. 21, 1987, pp. 519–524.
[*14*] Fletcher, J. S., Johnson, F. L., and McFarlane, J. C., *Environmental Toxicology and Chemistry,* Vol. 7, 1988, pp. 615–622.
[*15*] Parochetti, J. V., *Proceedings,* Northeast Weed Society, Vol. 29, 1975, pp. 28–35.
[*16*] Waldrop, D. and Banks, P. A., *Weed Science,* Vol. 31, 1983, pp. 730–734.

Michael C. Harrass,[1] Charles E. Eirkson III,[2] and Lisa H. Nowell[1]

Role of Plant Bioassays in FDA Review: Scenarios for Terrestrial Exposures

REFERENCE: Harrass, M. C., Eirkson, C. E. III, and Nowell, L. H., **"Role of Plant Bioassays in FDA Review: Scenarios for Terrestrial Exposures,"** *Plants for Toxicity Assessment: Second Volume, ASTM STP 1115,* J. W. Gorsuch, W. R. Lower, W. Wang, and M. A. Lewis, Eds., American Society for Testing and Materials, Philadelphia, 1991, pp. 12–28.

ABSTRACT: Plant bioassays are used to predict the effects of chemicals on terrestrial systems. However, in tiered-testing schemes, plants are tested only when introduction and fate information suggest that terrestrial exposure is probable. In this paper we discuss how information on chemical introduction and fate may trigger the need for plant toxicity tests, and how terrestrial exposure estimates are obtained and used at the U.S. Food and Drug Administration (FDA). FDA regulates chemicals used for food additives, food packaging and processing, and animal drugs and feed additives. These materials may be introduced into the environment as a result of their manufacture, use, or disposal. Information on introduction rates, environmental partitioning, and transformation is used to estimate the potential for terrestrial exposure. Described in detail are two scenarios for terrestrial exposure from application of wastewater treatment plant sludges to soils and one scenario for direct introductions of animal wastes to soils.

KEY WORDS: terrestrial effects, environmental assessment, plant toxicity, agricultural soil amendment, sludge, food additives, animal drug wastes

Premarket review of chemicals is an accepted part of consumer and environmental protection laws in the United States. Testing organisms representative of the terrestrial environment, particularly plants, is a relatively recent component of such review, however. The purpose of this paper is to describe how the premarketing review of products regulated by the Food and Drug Administration (FDA) may require and incorporate terrestrial toxicity testing.

FDA has regulatory authority over a wide variety of consumer products and has been described as the oldest federal agency whose primary function is consumer protection, dating back to 1906. Products regulated by FDA, which amount to about 25% of consumer purchases, include foods, food additives, human and animal drugs, human biological products, cosmetics, medical devices, and radiological products. Under the Federal Food, Drug, and Cosmetic Act (FFDCA) and the Public Health Service Act, the safety and effectiveness and proper labeling of certain products must be established before they can be marketed. A company that wishes to market a new chemical for these products must prepare and submit a petition or application for FDA approval.

[1] Center for Food Safety and Applied Nutrition, U.S. Food and Drug Administration, Washington, DC 20204.

[2] Center for Veterinary Medicine, U.S. Food and Drug Administration, Rockville, MD 20857-1706.

FDA's Environmental Review Requirements

FDA actions are subject to the requirements of the National Environmental Policy Act (NEPA), which requires federal agencies to consider the environmental impact of their actions. Consequently, environmental review has been incorporated into FDA procedures and regulations, including premarketing reviews [1]. If an action will have significant environmental impacts, then FDA must prepare an environmental impact statement to describe the impacts of the action and its alternatives. Although NEPA does not supersede FDA's other statutory duties or require FDA decisions to favor environmental protection over other relevant factors, NEPA does provide FDA with supplementary authority to base substantive decisions on environmental considerations. NEPA requires FDA to involve the public in its environmental deliberations and to make environmental documents publicly available.

Under FDA's environmental regulations, the industry sponsor of an application or petition may be required to prepare an environmental assessment of the proposed action. To support the assessment, appropriate testing of the environmental fate and effects of chemicals entering the environment may be required. To assist petitioners and applicants, FDA has described appropriate environmental test methods in some detail [2]. Tests of chemicals on seed germination, seedling growth, and earthworm survival are among the test methods described.

FDA has no routine requirement for specific environmental tests; that is, no base set of data is required for approval of FDA-regulated chemicals. Instead, the need for testing is determined by evaluation of the potential environmental exposure and the toxicity information available for a given chemical. Terrestrial hazard assessment typically involves comparing predicted environmental concentrations with effects tests showing impacts on representative species or systems [3]. FDA's approach is no exception: environmental introduction and fate information is ultimately compared with toxicity information to evaluate anticipated environmental safety. No base set of ecological data is required. Instead, FDA uses a tiered approach in requiring ecological fate and effects data. Available data and structure-activity relationships are considered before additional testing, including terrestrial ecotoxicity testing, is required. Effects tests may be required if substantial exposures are predicted or if the chemical appears to be toxic. Although we have no explicit decision rules, we can describe scenarios that would lead to a requirement for terrestrial plant testing.

Routes of Environmental Introduction

FDA's premarket review considers potential environmental introductions during the entire life cycle of a regulated product, that is, manufacture, use, and disposal. Industry sponsors often think only in terms of emissions from manufacturing processes. Where possible, FDA review relies on industry compliance with existing treatment and emissions requirements at sites of manufacture. This avoids duplication of regulation with other federal, state, and local governmental agencies that regulate manufacturing processes and emissions. However, significant introductions do occur through the use and disposal of products. Several patterns of use and disposal may lead to terrestrial introductions that could trigger the need for toxicity testing.

Animal Wastes

Chemicals of concern in animal wastes (excreta) are those that are used in the treatment, prevention, and control of diseases and to increase the rate of growth and feed efficiency of swine, cattle (dairy and beef), sheep, chickens, turkeys, and, more recently, fish. Ingested animal drugs and animal feed additives are often incompletely absorbed or metabolized by the target animal. Residual chemicals are then introduced in the animal excreta. Because most excreta is used on agricultural crop fields as fertilizer, residual chemicals are introduced into the environment. Any reduction in crop harvest resulting from residual chemicals may have a significant impact on agricultural crop production.

Sewage Effluent and Sludge

Many chemicals that may become a component of food or otherwise affect the characteristics of food are regulated by FDA as food additives. Food additives may be directly added to food or may be used to make articles that contact food (for example, food packaging), thus indirectly becoming a component of food.

Ingested food additives may be introduced to the environment through sewage. In addition, chemicals used to produce food additives or other regulated materials may be added to wastewater streams at production or processing plants. Materials that are not degraded by wastewater treatment may remain with sewage sludge. Land application of sludge may result in the terrestrial introduction of such materials. In some cases, direct land application of sewage may also result in terrestrial introductions of food additives [4].

Other Routes of Introduction

Terrestrial introductions of FDA-regulated chemicals may occur through other routes. Solid waste disposal in landfills usually precludes direct exposure of chemicals to terrestrial plants. However, composting of mixed solid waste is being considered as an alternative management practice to reduce the total volume of solid waste [5]. Terrestrial exposure to chemicals remaining in compost will occur when the compost is incorporated into soil. Incineration of solid waste containing food packaging may emit chemicals that can affect the terrestrial environment such as hydrochloric acid from incineration of chlorinated plastics.

Some animal drugs may be applied in forms that wash off into the environment. For example, some animal drugs used to control cattle grubs and reduce infestation of lice are applied as pour-on liquids. Rain washing off the backs of animals introduces such animal drugs immediately to the environment. To date, these routes of introduction have not led to an estimated exposure that would justify requiring testing for terrestrial effects.

Estimates of Environmental Concentration

Once routes of introduction are identified, introduction levels, rates of incorporation of sewage sludge or animal wastes into soil, and environmental fate data are used to predict environmental concentrations. Because FDA-regulated chemicals are generally used nationwide, FDA's environmental review uses conservative assumptions and may tend to overestimate exposures in the absence of detailed information. Where appropriate, additional assumptions and detailed models are considered. Initial estimates are intended to protect the environment under the most sensitive conditions.

Introduction Levels

The rate of consumption, application or use, and the metabolism of a chemical are used to estimate the upper-bound amount of a chemical that may be released into the environment.

The introduction level combines information on the expected rate of consumption, application, or use with information on metabolism.

Rate of consumption, application or use—In order to estimate the environmental concentrations of a chemical, some information on levels of use must be available. For food additives that are used nationwide and are consumed in patterns corresponding to national population density, the annual market volume provides an estimate of aggregate use. Projected annual markets are usually provided by industry sponsors or may be estimated from current markets for similar products. For ingested chemicals, the maximum annual market volume establishes an upper-bound introduction level in sewage. EPA estimated that about 74% of the U.S. population was served by wastewater treatment plants (WWTPs) in 1984, with a total flow of 102 600 million L/day (27 095 million gal/day), or 3.7×10^{13} L/year [6]. Boethling [7] has described a method to estimate an upper limit for the levels expected in sewage. Using this approach for an ingested chemical, an estimate of the level in sewage is obtained from Eqs 1 and 2

$$MV_{sewage} = (MV_{total})\,(P_{sewered})$$ (1)

where

MV_{sewage} = annual volume entering sewage (in kg),
MV_{total} = total annual market volume (in kg), and
$P_{sewered}$ = proportion of U.S. population served by sewers (74%).

$$C_{sewage} = \frac{(MV_{sewage})\,(10^6\ mg/kg)}{(3.7 \times 10^{13}\ L\ sewage)}$$ (2)

where

C_{sewage} = the estimated concentration in sewage in mg/L or ppm, and 3.7×10^{13} L is the estimated annual sewage flow.

For example, a product with a national market of 1 000 000 kg could be introduced into sewage at an estimated concentration (C_{sewage}) of about 0.02 ppm, assuming that all of an ingested material is excreted unchanged

$$MV_{sewage} = (1\ 000\ 000\ kg)\,(74\%) = 740\ 000\ kg$$

and

$$C_{sewage} = \frac{(740\ 000\ kg)\,(10^6\ mg/kg)}{(3.7 \times 10^{13}\ L\ sewage)} = 0.02\ ppm$$

Alternative methods of calculating concentrations in sewage may be used that do not use the annual market volume. For food additives, FDA establishes an estimated daily intake (EDI), a level of human exposure estimated for the average consumer based upon how the food additive is used. Combined with wastewater engineering data, the EDI can be used to estimate the amount of material received by a wastewater treatment plant of typical size, that is, serving a typical number of households [8]. This is similar to the method described by Holman [9] for predicting chemical concentrations in streams receiving WWTP effluent. Rapaport has described a revised model for predicting the frequency distribution of concentrations in U.S. streams receiving WWTP effluents [10].

For chemicals used to produce or process food-packaging materials, information is usually available to estimate how much chemical is lost during the production process (versus how much is incorporated into the finished packaging material). Combined with waste stream information, process loss information provides an estimated loading to subsequent treatment facilities or to the environment adjacent to the site of packaging production.

For animal drugs and feed additives, the initial introduction level to the environment depends upon the dose administered. Animal drugs or feed additives may be given continuously or in single or multiple doses for a limited period. The continuous use of a chemical is done usually to increase the growth rate and feed efficiency or to prevent diseases. These continuous uses usually result in higher concentrations in the manure over a longer period than do single or multiple doses applied for treatment or control of a disease.

Animal drugs administered to increase growth rate and feed efficiency or to prevent disease may be administered from the time that the animal enters the growout period, for example, in feedlots, broiler houses, or pasture, until just before slaughter. (An animal drug is usually withdrawn before slaughter to deplete animal tissue concentrations and thus reduce human exposure.) During the administration period, the drug is being excreted at a relatively constant rate in the manure or litter of the animal. The concentration in manure or litter can most accurately be determined by integration of the dose and animal size during the entire growout period. For simplicity, an initial, upper-bound estimate can be obtained using the highest expected animal weight, calculated as in Eq 3

$$\text{Concentration in excreta (ppm)} = \frac{(\text{Dosage (mg/kg)}) \, (\text{Animal mass})}{\text{Mass of excretory product}} \qquad (3)$$

One factor that may alter the concentration in manure or litter is the length of the withdrawal period, because the manure or litter will not be removed from the growout area until the animals have been removed. For example, if the animal drug is discontinued from cattle 30 days before slaughter, then drug-free manure will accumulate with manure containing the drug, and the final animal drug concentration in manure would be reduced. Again, for simplicity, we generally can neglect these variations in our estimate.

For example, an animal drug may be administered in the feed at 5 mg/kg/day from the time the animal is 114 kg (250 lb) until it reaches slaughter weight, approximately 455 kg (1000 lb). To estimate the maximum concentration in manure we would use the daily intake for the largest animal, that is, a 455-kg animal, for a total dose of 2275 mg/day. A 455-kg animal can be expected to excrete approximately 27.2-kg manure (urine and feces) per day [11]. The concentration present in manure (C_{manure}) would be approximately 84 ppm, calculated using Eq 3.

Metabolism—Metabolic processes may remove or transform an ingested material or an applied material that is absorbed. Consequently, available information may be used to modify the introduction estimates. In some cases, metabolic information may be used to establish that the material that will enter the environment is not the material that is produced and consumed. Consequently, any subsequent ecotoxicity testing should include any metabolites.

For example, an animal drug that is ingested may or may not be absorbed by the animal and may be totally, partially, or not at all metabolized. Animal drugs that are extensively absorbed and metabolized are usually expected to have little environmental impact, especially if the metabolic products are less toxic than the parent compound. Animal drugs that are only partially metabolized to several products may be more problematic environmentally, because the fate and effects of each component may need to be tested to assess the potential impact of each on the environment. Evaluating several metabolites could substantially increase the cost, complexity, and possibly the time necessary for completing the environmental assessment. Animal drugs that are not metabolized and are essentially excreted entirely as parent compound have a greater potential for environmental impact, although testing and environmental assessment may be relatively simple. The same analysis would be appropriate for an ingested food additive.

Available metabolism and absorption information can readily be used to modify the initial exposure estimates in Eqs 1 and 3, incorporating the fraction excreted, as shown in Eqs 1A and 3A below

$$MV_{sewage} = (MV_{total}) (P_{sewered}) (1 - f_{metabolized,\ absorbed})$$ (1A)

where

MV_{sewage} = annual volume entering sewage (in kg),
MV_{total} = total annual market volume (in kg),
$P_{sewered}$ = proportion of U.S. population served by sewers (74%), and
$f_{metabolized,\ absorbed}$ = fraction of the additive that is metabolized, or absorbed and not excreted.

Concentration in excreta

$$= \frac{(Dosage\ (mg/kg))\ (Animal\ mass)\ (1 - f_{metabolized,\ absorbed})}{Mass\ of\ excretory\ product}$$ (3A)

The fraction excreted $(1 - f_{metabolized,\ absorbed})$ may be entirely the parent compound or may include metabolic products. It is reasonable to assume that the rate of excretion and the products of metabolism will be constant throughout the period that a food additive is ingested or an animal drug is administered.

Continuing with our previous animal drug example, if the animal drug is excreted 90% as parent compound and 10% completely metabolized, our improved estimate of the concentration present manure (C_{manure}) would be 75 ppm, calculated using Eq 3A.

Rate of Incorporation into Soil

Soil amendments are intended to improve soil structure or to enrich the soil. The rate of incorporation of an amendment, also termed the dilution factor, varies with the properties of the amendment itself and the soil to be amended. Two common soil amendments are sewage sludge and animal excreta. Our calculations here assume that the amendment is incorporated into the top 15 cm (6 in.) of soil and that soil density is 1200 kg/m³, as in Holman [9]. However, a soil density of 1500 kg/m³ is generally used in calculations for animal drugs [12]. (The different soil density values reflect how different centers in FDA developed the estimation procedures. Although using the same value would be preferable, the two values are still used in order to allow comparison with previous analyses made by the different FDA groups.)

Sludge-amended soil—Incorporation of sewage sludge into soil achieves two purposes. First, sludge amendment provides plant nutrients and organic material to enrich the soil [13]. Second, it avoids the expenses and drawbacks of landfill disposal. In light of increasing tipping fees and limited landfill capacity, such alternative methods for sludge disposal are expected to become increasingly attractive [14,15].

Depending on degradation and partitioning in a WWTP, concentrations of an ingested food additive in the sewage effluent might remain unchanged or be reduced by degradation or by partitioning to the sludge. Nationwide, municipal WWTPs produce about 5.9 million metric tons/year (5.9×10^9 kg/year or 6.5 million tons/year) of dry sludge [16]. Using our previous food additive example, if all of the additive were to partition to sewage sludge without degrading, then the product's concentration in dry sludge could be estimated as in Eq 4

$$C_{sludge} = \frac{(7.4 \times 10^5 \text{ kg}) (10^6 \text{ mg/kg})}{5.9 \times 10^9 \text{ kg sludge}} = 125 \text{ ppm} \qquad (4)$$

The application rate of sludge to agricultural land or forest land is generally determined by the nutrient content of the sludge, particularly the nitrogen content.[3,4] Application rates may be further limited by regulations restricting concentrations of metals and other inorganic substances or organic compounds such as polychlorinated biphenyls (PCBs) or dioxins [17,18].[5,6,7,8] EPA expects to limit land application of sludges based on concentrations of aldrin/dieldrin, benzo(a)pyrene, cadmium, chlordane, copper, total DDT/DDE/DDD, heptachlor, hexachlorobenzene, hexachlorobutadiene, lead, lindane, mercury, nickel, PCBs, selenium, toxaphene and zinc [19,20].

The maximum rate of application of sludge to agricultural land or forest land is 4.5 kg/m² (20 tons/acre).[7,9] Typical application rates for WWTP sludge are about 2.2 kg/m² (10 tons/acre) [21] but may range from 1.1 to 3.4 kg/m² (5 to 15 tons/acre).[3,9] Because paper mill sludge is generally lower in nutrient content than WWTP sludge, application rates of paper mill sludge tend to be higher.[3,4]

Heavy metal loadings limit the number of years of application of sludge to the same piece of farmland. In some instances, more than ten years of agricultural application may be permitted. For example, the estimated maximum number of years of permissible agricultural application ranged from 18 years (for copper) to 145 years (for lead) based on metal content of San Diego municipal sludge [22].

Incorporating sludge into the top 15 cm (6 in.) of soil of density 1200 kg/m³ at an application rate of 4.5 kg/m² (20 tons/acre) results in a dilution rate of 2.5%, as shown in Eq 5 [9,12].[9]

$$\text{Dilution rate} = \frac{(4.5 \text{ kg sludge/m}^2 \text{ soil}) (100\%)}{0.15 \text{ m} \times 1200 \text{ kg/m}^3} = 2.5\% \qquad (5)$$

Application of paper mill and municipal sludges in land reclamation is usually at higher rates than to agricultural land [23], for example, 13.5 to 22.5 kg/m² (60 to 100 tons/acre), making a dilution rate of 6 to 10%.[7,8] Paper mill sludges may even be added (100%) as topsoil to reclaimed land [8,10]. Sludge may be applied to reclaimed land one time only[7,8] or in successive years until the desired growth is achieved.[3]

The soil concentration immediately following incorporation of the amendment is calculated as the product of the concentration in the amendment source (C_{source}) and the dilution rate,

[3] Personal communication, Dale Brockway, Bureau of Waste Management, Michigan Department of Natural Resources, Lansing, MI.

[4] Personal communication, Richard Dunn, Division of Water Pollution Control, Massachusetts Executive Office of Environmental Affairs, Boston, MA.

[5] Personal communication, Marc Crooks, Industrial Section, Washington Department of Ecology, Olympia, WA.

[6] Personal communication, Mark Jackson, Water Pollution Control, Massachusetts Executive Office for Environmental Affairs, Boston, MA.

[7] Personal communication, William Ginn, President, Resource Conservation Service, Yarmouth, ME.

[8] Personal communication, Robert Phelps, Central Office, Ohio Environmental Protection Agency, Columbus, OH.

[9] Personal communication, William Jewell, Agricultural Engineering Department, Cornell University, Ithaca, NY.

[10] Personal communication, Steven Rine, Chief, Industrial Wastewater Group, SE District Office, Ohio Environmental Protection Agency, Logan, OH.

as shown in Eq 6

$$C_{soil} = \frac{(C_{source})\ (\text{Dilution rate})}{100\%} \tag{6}$$

Thus, for sewage sludge, the initial soil concentration would be 2.5% of the concentration in the sludge.

Because paper mill sludge is not classified as hazardous waste, it falls under the jurisdiction of the states[11]. FDA contacted several states with 20 or more paper mills (Maine, Massachusetts, Michigan, Ohio, Pennsylvania, Washington, Wisconsin) for information on state concerns and regulations of land application of paper mill sludge. Some states (for example, Ohio, Maine) have taken a conservative approach to regulating dioxins in sludge and have limited the concentrations of such chemicals in sludge used for land application, even though little information exists on actual uptake by (or adverse effects on) crops and terrestrial organisms such as earthworms and rodents.[8,12] Other states (Washington, Massachusetts, Pennsylvania) are currently revising their regulations regarding land application of sludge.[5,6,13]

Soil amended with manure or litter—As with sludge, animal manure and litter are routinely spread on agricultural lands as a fertilizing amendment to soils. Typical application rates for cattle, sheep, and swine manures range from 2.2 to 3.4 kg/m² (10 to 15 ton/acre) dry weight, but massive amendments of 220 to 450 kg/m² of cattle manure (100 to 200 tons/acre) have been reported [22]. Poultry litter, which has a higher nitrogen content, is typically spread at a lower rate of approximately 1.1 kg/m² (5 tons/acre). For manure and litter, the application rate generally is determined by the nutrient content, especially nitrogen, storage conditions, method of application, and pollution control concerns.

If we assume that air-dried manure is spread at 3.4 kg/m² (15 tons/acre) [24,25] and that it is incorporated into the soil by plowing to a depth of 15 cm (6 in.), the initial soil concentration from amendment with manure would be 1.9% of the concentration in manure, calculated using Eqs 5 and 6.

As shown above, introduction levels and rates of incorporation establish an upper-bound value for terrestrial concentration following a single application. In the absence of additional information on the transformation or transport within and between environmental compartments, our evaluation of terrestrial hazard uses this estimate.

Environmental Fate

Processes that affect the transport and transformation of introduced chemicals are considered and, when possible, used to modify the environmental concentration estimate. Some fate information may lead to a conclusion that terrestrial exposure is negligible. Other data may indicate the potential for significant terrestrial exposure. For example, strong sorption to sewage sludge and resistance to biodegradation would suggest that most of an ingested food additive would ultimately be present in sludge used for soil amendment. Fate processes may affect chemical concentrations both before and after incorporation into soil. We are

---

[11] Personal communication, Robert Dellinger, Office of Solid Waste, U.S. Environmental Protection Agency, Washington, DC.

[12] Personal communication, Bryce Sproul, Maine Department of Environmental Protection, Augusta, ME.

[13] Personal communication, Steven Socash, Bureau of Solid Waste Management, Pennsylvania Department of Natural Resources, Harrisburg, PA.

not able to use test data for some fate processes to quantitatively modify the exposure estimate, but rather we apply the data in a qualitative way.

Chemical stability—Abiotic reactions, such as hydrolysis and photolysis, can alter the amount of a chemical present in the environment. In some cases, we can use stability to determine whether a route of introduction is of concern. For example, chemicals that readily hydrolyze might be expected to be present only at reduced levels in wastewater effluent.

Biodegradability—Biological degradation of an animal drug or food additive also can alter the amount of chemical present in the environment. Waste treatment processes, such as sewage treatment, may provide opportunity for biodegradation, or biodegradation may occur *in situ* after the material reaches the terrestrial environment. Appropriate biodegradation test data can reduce environmental concentration estimates. If biodegradation is relatively rapid, that is, occurs within several hours to several days, and is expected to be moderately complete, then exposure will be limited and may not be significant. For example, Gillett [26] considers compounds with a half-life less than 14 days to be of reduced concern. In the case of cattle feedlots, biodegradation may occur in the manure before it is spread on a field. If biodegradation is rapid, then plant bioassays may not be needed. If biodegradation is slow, e.g., the half-life is 30 days or greater, then it becomes important to determine what effects might be expected after introduction of the chemical into the soil environment.

Mobility—Properties such as water solubility, soil sorption and desorption, and volatility will determine the movement and partitioning of a chemical between solid, liquid, and gas media. Consequently, these properties are used to estimate transport of chemicals between sewage, wastewater effluent, and sludge, as well as between soil, interstitial soil-water, air, and biota [27]. Although these properties are recognized as important and can be quantified, their relationship to terrestrial exposure is complex.

Water solubility is a primary factor that will determine whether a chemical remains in soils or in soil-water to eventually affect plants. There is an inverse relationship between water solubility and sorptivity to soil: a highly water soluble chemical could leach downward from the surface soils or be carried away in runoff. For such a chemical, used for feedlot raised beef cattle, leaching from manure in the feedlot may be expected to occur, and little of the chemical would remain to be spread on fields. Such a chemical also might leach rapidly from the field after it is applied.

The *n*-octanol/water partition coefficient, K_{ow}, reflects the lipophilic character of a chemical. Lipophilic chemicals are readily adsorbed by oily and organic materials, including sewage sludge and the organic compounds in soil. K_{ow} is a useful predictor of sorption to soil and bioaccumulation. In general, low water solubility correlates with lipophilicity.

Soil sorption of a chemical is complicated further by the dependence of sorption and desorption processes on the makeup of the soil and on environmental conditions such as soil pH and moisture content. Consequently, tests of soil sorptivity should include tests of different soil types. Potential dissociation of the compound determines whether any necessary testing should be conducted at a single pH or over an environmentally relevant pH range (for example, pH 5, 7, and 9). Dissociation data also indicate whether the chemical may be ionically bound to soil.

Another relevant factor is volatility. Two processes affect the potential for impact of the vapor phase of an animal drug or food additive. First, the chemical must volatilize from the solid (soil) or liquid (sewage or soil-water) to the air. Second, deposition may occur from the air onto the leaf surface of the plant. Ryan et al. [28] suggest that a vapor pressure of 10^{-4} mm Hg may be a reasonable transition point for determining when vapor diffusion becomes important, but soil sorption and water solubility must be considered when assessing any potential for a chemical to leave soils in a vapor phase.

Potential for Plant Toxicity

There is no arbitrary soil concentration that FDA uses to trigger the need for testing plant toxicity. If significant terrestrial exposure is expected, the potential for plant toxicity of the chemical is considered, either as a systemic or nonsystemic toxicant. If available information can provide a basis for concluding that the chemical is unlikely to have plant toxicity, then testing may not be required. Such a determination is a matter of professional judgement, so we are not able to state explicit triggers or decision rules.

Systemic toxicants include those chemicals that are taken up by the plant. Properties such as octanol/water partition coefficient, soil sorption, molecular weight,and electronic charge can be used to predict the potential for plant uptake of the chemical [23,25,28], although precise correlations between a chemical's properties and its movement and accumulation are uncertain [29]. Uptake of organic chemicals occurs from soil solution or from absorption of volatile chemicals by roots and shoots [23], so many of the chemical properties considered in environmental mobility are also useful in considering uptake potential. In general, the potential for toxicity increases with the likelihood of uptake by plants.

The octanol/water partition coefficient (K_{ow}) can be useful in determining the potential effects on plants [23,25,28], as well as in evaluating overall environmental fate (discussed previously). K_{ow} has been correlated to the uptake and transport of chemicals by certain plants. Briggs et al. [30] found that root concentration factors in plants with "normal" lipid content and with nonwoody roots were positively related to the logarithm of the K_{ow}. Plants with relatively high lipid content (cress, parsnips, and carrot) may not show the same relationship [31].

The molecular weight of the compound may also significantly affect the uptake of an animal drug or food additive. For barley, Topp et al. [31] found that the total uptake of a chemical correlated well with the molecular weight and felt that the molecular weight may be more predictive of plant uptake than the K_{ow}.

Soil sorption corrected for organic content, measured as K_{oc}, also may be useful in evaluating potential plant uptake. K_{oc} is generally strongly correlated with K_{ow}. High sorption may limit the availability of the chemical for uptake by plants since only the nonadsorbed portion is considered by some to be available for plant uptake. Graham-Bryce [32] considered compounds with a soil adsorption coefficient, K_d, of greater than 1000 to be inactive as herbicide agents. Topp et al. [31] found a negative correlation between barley root concentration factors for some aromatic compounds and K_{oc}.

Nonsystemic toxicity of a chemical may be caused by surface activity or physical mechanisms. For example, electronic charge of a chemical may limit potential uptake of the chemical by plants,[14] but can also affect surface activity and alter nutrient uptake by plant roots. Interference from applied chemicals may affect the availability of water or nutrients from the soil matrix. Such interference may be chemical, or physical, e.g., water repellency [33]. If a specific mechanism for nonsystemic toxicity seems plausible, plant toxicity testing may be required even though plants are unlikely to take up the chemical directly.

Scenarios for Terrestrial Exposure

The environmental introduction and fate data described above may lead to terrestrial exposures. The following three examples illustrate how analysis of environmental introduction and fate data have led to requests for tests of terrestrial plant toxicity. These examples

[14]Personal communication, John Fletcher, Department of Botany, University of Oklahoma, Norman, OK.

are illustrative only, and the values are selected to demonstrate our approach. These values may be similar to actual data, but they are not intended to represent any particular food additive petition or new animal drug application.

Scenario 1: Soil Amendment with Sludge Containing a Direct Food Additive

Direct food additives include chemicals intentionally added to foods. Figure 1 illustrates how the factors affecting environmental introduction and fate are brought together in making a decision about the need for testing the toxicity of a direct food additive to plants. Some direct food additives appear not to be metabolized in humans and to be only minimally degraded as they pass through WWTPs. Suppose that test data about the environmental fate of a hypothetical direct food additive, summarized in Table 1, led to an estimated concentration of the additive in WWTP sludge of at least 1500 ppm. The half-life of this chemical in soil is about 180 days or two half-lives per year, suggesting that about 25% of the chemical would remain after one year. If we assume that sludge containing this additive were used to amend agricultural soil, and the treatment repeated annually, then the estimated soil concentration can be calculated using Eq 7

$$C_{t+1} = [C_t + (C_{\text{sludge}}) (2.5\%)] (f_{\text{biodegradation}}) \qquad (7)$$

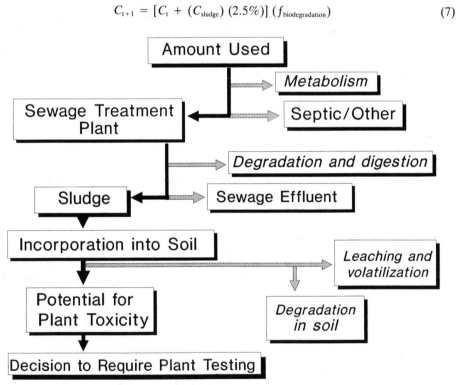

FIG. 1—*Flow chart showing factors considered for food additives. Elements that lead to a decision to require plant toxicity testing are shown for a chemical that is added to food. See discussion of Scenario 1 for a fuller description. Solid arrows show how an upper-bound estimate of the concentration in sludge-amended soil is obtained. Additional test data reflecting various fate processes can be used to reduce the estimate (shaded arrows) For an indirect food additive (Scenario 2), a similar sequence is considered, but modified to reflect where such chemicals might enter sludge and how special sludges (e.g., papermill sludge) are incorporated into soil.*

TABLE 1—*Environmental data profile for a hypothetical direct food additive.*

Metabolic Fate	
Transformation by human digestion or metabolism	Negligible (<5%)
Excretion or nonabsorption	Almost complete (>95%)
Chemical Stability	
Hydrolysis, photolysis	No data; structure appears stable
Thermal stability	Stable
Biodegradability	
Semicontinuous activated sludge	<1% in 48 h
Biodegradation in activated sludge	Half-life >49 days
Biodegradation in soil	Half-life of 180 days
Tendency to Partition and Adsorb	
Molecular weight	Moderate (<1000)
Electronic charge	Neutral at ambient pH
Water solubility	Low (<1 mg/L)
Octanol/water partition coefficient	High (log K_{oc} > 3.5)
Bioaccumulation potential	Low (BCF < 100)
Sorption coefficient	High (K_d > 200 000)
Vapor pressure	No data
Available Toxicity Data	
Bacteria	No effects at >1000 ppm
Selanastrum capricornutum (green alga)	EC_{50} > 1000 ppm
	Growth reduced at 10 ppm, but apparently due to physical effect
Daphnia magna (water flea)	EC_{50} > 1000 ppm
Lepomis macrochirus (bluegill sunfish)	EC_{50} > 1000 ppm

where

C_{t+1} = the soil concentration at the end of the year,

C_t = the soil concentration at the beginning of the year,

(C_{sludge}) (2.5%) = the concentration in the sludge times the dilution factor for the rate of application, and

$f_{biodegradation}$ = the fraction remaining after the annualized loss from biodegradation in the soil, $(0.5)^{(365/180)}$, or 0.25.

Over several iterations, the soil concentration a year after application stabilized at 12.5 ppm for our example. Consequently, terrestrial exposures could range from 50 ppm when additional sludge is incorporated to 12.5 ppm after a year of biodegradation.

Although available information suggested that the material was of low toxicity to aquatic organisms, terrestrial tests would be requested since the analysis showed that terrestrial exposures were relatively high, i.e., 10 to 50 ppm, and were likely to be long-term.

Scenario 2: Soil Amendment with Sludge Containing an Indirect Food Additive

For further illustration, we discuss an additive proposed for use in making paper to be used in contact with food, i.e., an indirect food additive. Numerous chemicals are used in paper processing; wastewater effluents and sludges from processing may contain substantial concentrations of these process chemicals. Suppose that we estimated the concentration of this hypothetical indirect additive in WWTP sludge to be 200 ppm, and test data about the environmental fate of this chemical were available as summarized in Table 2. Concentrations of the additive in WWTP sludge-amended soil would be 5 ppm for agricultural and forest land (assuming a 2.5% dilution rate) and 12 to 20 ppm for reclaimed land (assuming a 6 to

TABLE 2—*Environmental data profile for a hypothetical indirect food additive used for making paper for food-contact use.*

Metabolic Fate	
Transformation by human digestion or metabolism	No data
Excretion or nonabsorption	No data
Chemical Stability	No data
Biodegradability	
Modified semicontinuous activated sludge	>100%[a]
Biodegradation—modified Sturm test	No degradation in 28 days
Biodegradation in soil	No data
Tendency to Partition and Adsorb	
Molecular weight	High (>1000)
Electronic charge	Positive
Water solubility (of alkaline salt)	High (>1 g/L)
Octanol/water partition coefficient	No data
Bioaccumulation potential	No data
Sorption to soil	High (>99%)
Vapor pressure	No data
Toxicity Information	
Bacteria	Bacteriostatic/bactericidal at 10–20 ppm
Selanastrum capricornutum (green alga)	$EC_{50} < 0.05$ ppm
Daphnia magna (water flea)	$EC_{50} < 0.05$ ppm

[a]This procedure does not distinguish between removal by sorption from that by biodegradation. It is likely that removal was entirely by sorption.

10% dilution rate) after the first year of application. When sludge is applied undiluted as topsoil to reclaimed land, the concentration of additive in that topsoil layer is 200 ppm.

Once adsorbed to soil, very little desorption occurs, even though the salt form of the additive is water soluble (Table 2). Since we have no evidence that this additive would biodegrade to any significant extent, the additive may accumulate after repeated application of contaminated sludge to the same tract of land. If the WWTP sludge were applied to the same tract of agricultural land for ten consecutive years, soil concentrations of the additive could reach 50 ppm. Because sludge may be applied to the same tract of forest land only once every 3 to 10 years, accumulation of the additive in forest soil would be slower than for agricultural land. After ten applications of WWTP sludge (which could occur over 30 to 100 years), soil concentrations of the additive could also reach 50 ppm. Where WWTP sludge is applied repeatedly to reclaimed land, the concentrations of the additive in sludge-amended soil may reach 36 to 60 ppm after three applications, and 60 to 100 ppm after five applications.

Overall, then, long-term soil exposures would be expected, with the chemical likely to reach concentrations of about 50 ppm. The observed toxicity, especially to algae, suggest that nonsystemic toxicity, e.g., by surface activity, must be of concern even though the positive electronic charge, high molecular weight, and high soil sorption of this chemical reduce the potential for plant uptake. Therefore, plant toxicity tests would be required.

Scenario 3: Soil Amendment with Animal Wastes Containing an Animal Drug

As a final illustration, we discuss a hypothetical animal drug, proposed to be used in cattle for improved feed efficiency and increased growth rate. Figure 2 illustrates how the factors affecting environmental introduction and fate are brought together in making a decision about the need for testing the toxicity of an animal drug to plants. Test data about the environmental fate of this chemical are in Table 3. This animal drug is to be administered

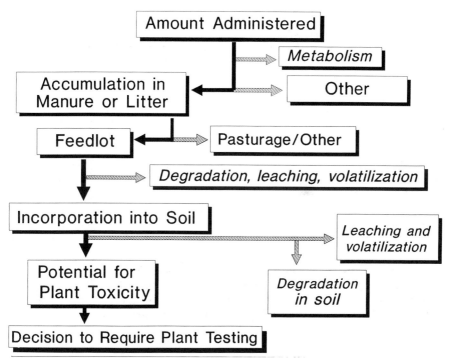

FIG. 2—*Flow chart showing factors considered for an animal drug. Elements that lead to a decision to require plant toxicity testing are shown for a chemical that was used as an animal drug. See discussion of Scenario 3 for a fuller description. Standard animal size, holding period in a feedlot, and rates of incorporation into soil are used for animal drugs. Solid arrows show how an upper-bound estimate of the concentration in manure-amended soil is obtained. Additional test data reflecting various fate processes can be used to reduce the estimate (shaded arrows).*

in the feed from the time the animal is 114 kg (250 lb) until slaughter weight, approximately 455 kg (1000 lb). The animal drug will be administered at a dose of 5 mg/kg per day. Data from a metabolism study indicate that 10% of this animal drug is metabolized to a number of fractions and 90% is excreted in the feces as parent compound. As described above, the estimated concentration in wet manure would be 75 ppm. We are assuming that the dose and excretion rate are the same every day for the duration of the growout period of 136 days. The biodegradation half-life for the chemical is approximately 70 days, so about two half-lives would pass during the growout period for some of the chemical. For simplicity's sake, we will assume that the entire dose of this animal drug is subjected to two half-lives during the growout period. Therefore, we estimate that 25% will remain in the wet manure (19 ppm). If the solubility of this chemical is moderate to low, its soil sorption is moderate and its vapor pressure low (see Table 3), and if the feedlot is not subjected to significant rain, then we would expect most of the animal drug to remain in the manure. We would not expect the chemical to volatilize significantly during this time. Drying the manure during the 136-day period may reduce the initial weight by as much as 60%. Therefore, the estimated concentration of the animal drug in the manure at the end of 136 days is 47 ppm.

If the air-dried manure is spread at 3.3 kg/m² (15 tons/acre) [25] and is incorporated into the soil by plowing to a depth of 15 cm (6 in.), then the dilution rate would be estimated to be 1.9%, using Eq 5. The initial concentration of animal drug in soil would be approximately 0.9 ppm, calculated using Eq 6. Crops might be planted within 30 days of application, at which time the estimated concentration would be about 0.7 ppm. The chemical's moderate

TABLE 3—*Environmental data profile for a hypothetical animal drug used in cattle for improved feed efficiency and increased growth rate.*

Metabolic Fate	
Transformation by animal digestion or metabolism	Low (10%)
Excretion or nonabsorption	High (90%)
Chemical Stability	No data
Biodegradability	
Biodegradation in soil	Half-life of 70 days
Tendency to Partition and Adsorb	
Molecular weight	Moderate (<1000)
Electronic charge	Neutral
Water solubility	Moderate (100 ppm)
Octanol/water partition coefficient	No data
Bioaccumulation potential	No data
Sorption to soil	Moderate (K_{oc} 500)
Vapor pressure	Low (10^{-6} mm Hg)
Toxicity Information	No data

water solubility and moderate soil sorption indicate that a significant amount of the chemical is likely to be present and available for plant uptake at the time of planting.

Because of the anticipated exposure to a biologically active chemical, the likelihood of plant uptake, and the lack of data on environmental effects, plant toxicity testing would be required.

Use of Plant Toxicity Data in the Overall Environmental Analysis

Once the need for plant toxicity testing is established from the analysis of introduction and fate, the appropriate test endpoints should be obtained. For terrestrial exposures, those effects that need to be determined include seed germination, root elongation, and seedling growth (shoot length and weight and root weight). Once the test results are available, the test endpoints are compared with predicted exposure concentrations. If this analysis concludes that no adverse impacts on plants are expected (and assuming all other environmental issues are also resolved), then the agency may find that no significant impacts of the proposed action are expected, and the environmental review can be concluded. If the analysis does not lead to this conclusion, then additional steps must be taken to address the potential impacts, that is, the analysis is an iterative process. These steps may include obtaining additional test results to clarify ambiguities or to test the assumptions made, developing measures to mitigate the potential impact, and preparing an environmental impact statement to describe the anticipated effects and the alternative actions that may be taken.

To illustrate this process, we describe the analysis using the animal drug example from above. We assume that seed germination of six representative plant species was not affected by this animal drug, even at the maximum test concentration of 1000 ppm. In the preliminary bioassay of seedling growth, effects were observed on soybean and cucumber at 100 ppm and in corn at 1000 ppm. Definitive bioassays of soybean and cucumber established a no-observed-effect level (NOEL) of 10 ppm; significant effects were observed at 40 ppm. A NOEL for corn was found to be 200 ppm; significant effects were observed at 400 ppm.

FDA's regulations state that a substance is considered toxic in the environment if the maximum environmental concentration exceeds the concentration that causes any adverse effect in a test organism species (minimum effect level) or exceeds 1% of an LC_{50}, whichever is less [1]. Concentrations at or below the NOEL would not exceed the minimum effect level. Given the expected introductory environmental concentration of this animal drug of 0.9 ppm, and the lowest NOEL of 10 ppm, an eleven-fold margin of safety exists for this animal drug. If manure is not land-applied more frequently than the half-life, the concentration of this drug will decrease, increasing the margin of safety. Therefore, the plant bioassays have indicated that significant effects on plants are not expected to occur in fields where the animal drug is present at low concentrations.

Conclusions

Premarketing evaluation of the environmental impact of products regulated by FDA is required by NEPA. FDA's hazard assessment involves comparing ecotoxicity information with environmental concentration estimates developed from information on introduction and fate. However, ecotoxicity testing is usually required only if the introduction and fate analysis suggests substantial exposure potential, that is, no base set of ecotoxicity data is required. Consequently, tests for plant bioassays are required only in cases where significant terrestrial exposures are expected or when available information indicates likely toxicity.

To determine the significance of terrestrial exposure, information on the rates of consumption, application, or use of a regulated product must be available. In some circumstances, annual market volume and the rate of incorporation into soil provide an upper-bound estimate of potential environmental concentrations. Information on the chemical stability, biodegradability, and mobility is used to determine whether a regulated product may persist in the environment and what type of environment may be exposed to the product.

Incorporation of municipal sewage sludge into soil represents one situation where persistent, water-insoluble products may be introduced into the terrestrial environment. We have required plant bioassay tests for both direct food additives (products added directly to food) and indirect food additives (products used in contact with food).

Incorporation of animal wastes into soil represents another situation where FDA-regulated products may be introduced into the terrestrial environment. We have required plant bioassay tests for animal drugs that are used in the treatment, prevention, and control of diseases or to increase the rate of growth and feed efficiency of livestock.

Plant bioassay data are ultimately used in the overall evaluation of the potential for adverse environmental impacts of FDA regulation. FDA encourages the development of suitable methods for terrestrial plant toxicity testing and hazard assessment methods for incorporating the data from such tests into the environmental review process.

References

[1] Title 21, Part 25, *Code of Federal Regulations*, U.S. Government Printing Office, Washington, DC, 1989.
[2] Food and Drug Administration, *Environmental Assessment Technical Handbook*, PB 87-175345/ AS, National Technical Information Service, Springfield, VA.
[3] Organisation for Economic Co-operation and Development, in *Report of the OECD Workshop on Ecological Effects Assessment*, OECD Environmental Monographs No. 26, OECD, Paris, 1989, Chapter IV, pp. 20–31.
[4] Metcalfe and Eddy, Inc., *Wastewater Engineering*, McGraw-Hill, New York, 1979, Chapter 13, pp. 760–828.
[5] Razvi, A. G., O'Leary, P. R., and Walsh, P., *Waste Age*, Vol. 20, No. 7, 1990, pp. 142–148.
[6] U.S. Environmental Protection Agency, "1984 Needs Survey; Report to Congress: Assessment of Needed Publicly Owned Wastewater Treatment Facilities in the U.S." EPA 430-9-84-011, U.S. Environmental Protection Agency, Washington, DC, 1984.
[7] Boethling, R. S., *Water Research*, Vol. 18, No. 9, 1985, pp. 1061–1076.
[8] Metcalfe and Eddy, Inc., *Wastewater Engineering*, McGraw-Hill, New York, 1979, Chapter 2, pp. 11–55.
[9] Holman, W. F. in *Aquatic Toxicology and Hazard Assessment: Fourth Conference, ASTM STP 737*, D. R. Branson and K. L. Dickson, Eds., American Society for Testing and Materials, Philadelphia, 1981, pp. 159–182.

[10] Rapaport, R. A., *Environmental Toxicology and Chemistry,* Vol. 7, 1988, pp. 107–115.

[11] American Society of Agricultural Engineers, *1977 Agricultural Engineers Yearbook,* American Society of Agricultural Engineers, St. Joseph, MO, 1977, p. 503.

[12] Jackson, M. C., *Soil Chemical Analysis,* Prentice Hall, Inc., Englewood, NJ, 1958, p. 498.

[13] Page, A. L., Logan, T. J., and Ryan, J. A., *Land Application of Sludge: Food Chain Implications,* Lewis Publishers, Inc., Chelsea, MI, 1987, Introduction, pp. 1–4.

[14] Office of Solid Waste, "The Solid Waste Dilemma: An Agenda for Action," EPA/530-SW-89-019, U.S. Environmental Protection Agency, Washington, DC, 1989.

[15] Kellog, S. R., *Pollution Engineering,* November 1989, pp. 50–68.

[16] Council on Environmental Quality, *Fifteenth Annual Report,* U.S. Printing Office, Washington, DC, 1984, p. 242.

[17] Williams, D. E., Vlamis, J., Pukite, A. H., and Corey, J. E., *Soil Science,* Vol. 143, 1987, pp. 124–131.

[18] Bernhart, A. P. and Salvatori, F. G., *Proceedings, Annual Aquatic Toxicity Workshop, Canadian Technical Report, Fisheries and Aquatic Science,* Vol. 1462, 1986, pp. 20–34.

[19] Office of Water, Office of Water Enforcement and Permits, "Sewage Sludge Interim Permitting Strategy," U.S. Environmental Protection Agency, Washington, DC, September 1989.

[20] U.S. Environmental Protection Agency, Office of Water Regulations and Standards, "Technical Support Document: Land Application and Distribution and Marketing of Sewage Sludge," PB 89-136576, National Technical Information Service, Springfield, VA, 1989.

[21] U.S. Environmental Protection Agency, "Sludge Treatment and Disposal," Vol. 2, EPA 625/4-78-012, U.S. Environmental Protection Agency, Washington, DC, October 1978.

[22] McCampbell, R. E., Richard, D., and Crites, R. W., *Biocycle,* Vol. 29, 1988, pp. 26–29.

[23] Jacobs, J. W., O'Connor, G. A., Overcash, M. A., Zabik, M. J., and Rygiewicz, P. in *Land Application of Sludge Food Chain Implications,* A. L. Page, T. J. Logan, and J. A. Ryan, Eds., Lewis Pubishing, Inc., Chelsea, MI, 1987, pp. 101–143.

[24] Fairbanks, W. C. in *The Feedlot,* G. B. Thompson and C. C. O'Mays, Eds., Lea and Febiger, Philadelphia, 1983, pp. 197–212.

[25] Bell, R. M., Sferra, P. R., Ryan, J. R., and Vitello, M. P. in *Contaminated Soil '88,* K. Wolf, W. J. van den Brink, and F. J. Colon, Eds., Kluwer Academic Publishers, Norwell, MA, 1988, pp. 451–458.

[26] Gillett, J. W., *Environmental Toxicology and Chemistry,* Vol. 2, 1983, pp. 463–476.

[27] Jacobs, L. W., O'Connor, G. A., Overcash, M. A., Zabik, M. J., and Rygiewicz, P. in *Land Application of Sludge,* A. L. Page, T. G. Logan, and J. A. Ryan, Eds., Lewis Publishers, Inc., Chelsea, MI 1987, Chapter 6, pp. 101–144.

[28] Ryan, J. A., Bell, R. M., Davidson, J. M., and O'Connor, G. A., *Chemosphere,* Vol. 17, No. 12, 1988, pp. 2299–2323.

[29] Boersma, L., Lindstrom, F. T., McFarlane, C., and McCoy, E. L. *Soil Science,* Vol. 146, 1988, pp. 403–417.

[30] Briggs, G. G., Bromilow, R. H., and Evans, A. A., *Pesticide Science,* Vol. 13, 1982, pp. 495–504.

[31] Topp, E., Scheunert, I., Attar, A., and Korte, F., *Ecotoxicology and Environmental Safety,* Vol. 11, 1986, pp. 219–228.

[32] Graham-Bryce, I. J. in *Pesticide Synthesis Through Rational Approaches, ACS Symposium Series 255,* P. S. Magee, G. K. Kohn, and J. J. Menn, Eds., American Chemical Society, Washington, DC, 1984, pp. 185–207.

[33] Glauser, R., Doner, H. E., and Paul, E. A., *Soil Science,* Vol. 146, 1988, pp. 37–43.

Anthony J. Windeatt,[1] John F. Tapp,[1] and Richard D. Stanley[1]

The Use of Soil-Based Plant Tests Based on the OECD Guidelines

REFERENCE: Windeatt, A. J., Tapp, J. F., and Stanley, R. D., **"The Use of Soil-Based Plant Tests Based on the OECD Guidelines,"** *Plants for Toxicity Assessment: Second Volume, ASTM STP 1115,* J. W. Gorsuch, W. R. Lower, W. Wang, and M. A. Lewis, Eds., American Society for Testing and Materials, Philadelphia, 1991, pp. 29–40.

ABSTRACT: The European Community (EC) has been developing a plant bioassay within the Organization for Economic Cooperation and Development (OECD), Terrestrial Plants, Growth Test guidelines. The aim has been to obtain a relatively simple, reproducible test that is both flexible and economically viable. Over the past ten years, European government and contract laboratories, including the ICI Brixham Laboratory, have examined the suitability of a range of test species and pot sizes for use in plant tests. Other parameters assessed include the use of pregerminated seeds and the value of randomized experimental design. During this period, these establishments have taken part in two Ring Tests (interlaboratory comparative trials), and this has resulted in several refinements to the test. ICI Brixham Laboratory has adopted the basic guideline of this plant test and has used it, with slight modification, to assess the effect of a range of organic and inorganic chemicals on emergence and early plant growth of a number of plant species. The method is also suitable for adaption to assess the effect of volatile chemicals and of soils amended with sludges from treatment processes on plant growth.

KEY WORDS: ecotoxicology, phytotoxicity, germination, plant emergence, early growth, EC_{50}, NOEC, OECD guideline, method

In Europe, the European Community (EC) has been developing a plant test that complies with the Organization for Economic Cooperation and Development (OECD), Terrestrial Plants, Growth Test, Guideline 208 [1]. The aim of this work has been to produce a relatively simple, reproducible test that is both flexible and economically viable. Since the early 1980s, European governments and contract laboratories, including the ICI Brixham Laboratory, have taken part in Ring Tests (interlaboratory comparative trials) which have resulted in refinements to the test. Early work tested the draft proposals for a growth test with higher plants by the OECD Chemical Testing Programme Ecotoxicology Group [2]. ICI Brixham carried out several studies on the practical implications of the protocols. These included the suitability of test species and pot sizes, use of pregerminated or dormant seeds, and randomization of pot layout.

Method Development

Test Species

Plant tests usually specify at least one monocotyledon and one dicotyledon, and the latter is sometimes required to be a legume. A study was carried out by Brixham Laboratory in

---

[1] Project Officer, senior project officer, and functional service head, respectively, ICI Brixham Laboratory, Freshwater Quarry, Brixham, Devon, TQ5 8BA United Kingdom.

1982 on the percentage germination and emergence time of 21 plant species and the optimum plant pot size for each of these species under test conditions. This study showed that some of the plants listed in some protocols were unsuitable for a three-week test for one reason or another [3]. The species considered ideal for testing were those that had similar emergence periods, preferably less than seven days, and that produced a sturdy plant shape at three weeks. Pot size did have an effect on the size of plant produced, but in order to reduce the handling of a large number of different size pots, a compromise between pot and plant size was suggested, and one pot size 10 cm in diameter was chosen for all species.

Treatment Array

It was considered that growth differences could occur in glasshouses, in relation to pot position, due to gradients in external conditions. A study to assess the effect of pot distribution on the growth of plants was done in 1982 [4]. This work was carried out in a very large greenhouse of approximate dimensions 12 by 9 by 7 m (length by width by height). Foliage weights were determined from several plant pots randomly positioned around the greenhouse. One-way analysis of variance showed no difference in foliar growth produced, irrespective of pot position. However, subsequent work has often shown there to be an "edge" or "window" effect producing relatively smaller plants in pots by windows of greenhouses, although the reduction was mainly in plant height. Plant weight was normally affected to a lesser extent. The effect of this problem was reduced by positioning the pots as shown in Fig. 1. This allowed all pots of the same concentrations and of the controls to be kept together and meant that adjacent pots did not differ in concentration by more than one concentration step. It also saved considerable time when assessing emergence and final foliage weight produced. The advantages of this form of design were found to outweigh the disadvantages of the edge effects.

Pregerminated Versus Dormant Seeds

One of the objectives of a Ring Test carried out in August 1982 was to assess the advantages and disadvantages of the use of pregerminated or dormant seeds in plant tests [5]. The use

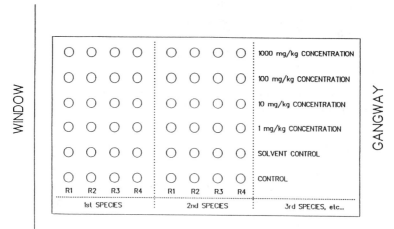

R1, R2, ... etc = Replicate No.

FIG. 1—*Layout of plant pots in greenhouse.*

of pregerminated seeds guaranteed the correct number of plants at the start of the test, although it was time consuming and seedlings were fragile, especially if they were not transplanted within 24 h of germination. With the use of dormant seeds, it was probable that some pots would not produce the correct number of plants. This could be resolved by sowing a larger number of seeds at the start of the test and thinning to a fixed number after a set period of time.

Dormant seeds of different species may have different emergence times requiring staggered sowing times. This increases the operator time and hence the cost of the test, but choosing species with similar emergence times can overcome this problem. The use of dormant seeds also allows the effect of a chemical on emergence to be assessed during the same test as that used to determine the effect on the early growth. The use of dormant seeds, followed by thinning to a predetermined number of seedlings after emergence, has been adopted as the preferred technique.

ICI Brixham Laboratory has adopted the basic OECD guideline and, with the additional information obtained from the studies referred to above, has modified it slightly to assess the effect of several organic and inorganic chemicals on the effect of emergence and early plant growth of a number of plant species. The laboratory has used the test, in adapted forms, to assess the effect of both a volatile chemical [6] and sludges from an industrial effluent treatment works on emergence and early plant growth. The results of this work are not reported here.

The modified OECD test as used by the Brixham Laboratory is described below.

Method

Test Species

Three species of plants, a monocotyledon and two dicotyledons, one of which was a legume, were usually selected for testing. These were, *Sorghum bicolor* (crop sorghum), *Helianthus annuus* (sunflower), and *Glycine max* (soy bean). Soya was replaced by *Phaseolus aureus* (mung bean) in later tests. In some of the earlier studies, *Avena sativa* (oat), *Brassica rapa* (turnip), and *Lycopersicum esculentum* (tomato) were also tested. All seeds were obtained from reputable suppliers.

Plant Pots

Plant pots were new (unused), 10-cm-diameter plastic pots. Each held approximately 600 g of growing medium.

Seeds and Study Arrangement

Each plant pot was sown with eight seeds of one of the three plant species. The test concentrations were usually 1, 10, 100, and 1000 mg chemical/kg dry soil plus control (and solvent control if warranted). There were four replicates for each of the treated soils and controls for each species.

Soil

The soil used was a commercially available compost consisting, by volume, of 50% sifted sterilized Mendip loam and 50% Cornish grit. It contained trace elements and superphosphate. The mix, called WC-B, was specifically designed as a greenhouse potting medium with a low organic content (<3% w/w).

Growth Medium and Incorporation of Test Material

The test material was mixed with silver sand, and this mixture was incorporated into the WC-B soil to produce the growing medium as follows. The silver sand was first washed and sieved through a 1-mm mesh. Sand of <1-mm particle diameter was dried and used for the test. The quantity of silver sand used was in the ratio 1:9, sand:soil. The method of mixing the test material with the sand varied according to whether the material was (*a*) a solid chemical or (*b*) a liquid chemical or chemical soluble in water or volatile solvent such as acetone, methanol, or diethyl ether.

For a solid test material, the growth medium was prepared by adding the test material to a small amount of sand (approximately 100 g) in a mortar and pestle and grinding the two together until the test material was finely and evenly dispersed. This mixture was then added to the remaining sand in a polythene bag and shaken for 30 s.

Liquid test materials, or those that were soluble in water or a volatile solvent, were prepared by weighing the appropriate amount of test material into a glass vial. 100 mL of solvent were then poured into a measuring cylinder, and, using several small quantities (approximately 20 mL) of the solvent, the chemical was rinsed into a 1-L, wide-mouth jar. Any remaining solvent was poured into the 1-L jar. The jar was capped and swirled to dissolve any remaining material.

A small amount of silver sand (10% of the weight of the final soil/sand mix required for the test) was poured into the 1-L jar containing the solvent/test material mixture. At intervals during the addition of the sand (at least twice), the jar was capped and the contents shaken thoroughly. On completion of the addition of the sand, the contents of the glass jar were shaken for a further 30 s. The contents were then emptied into a large shallow aluminum tray and the solvent allowed to evaporate until the sand was completely dry. Similarly, any sand remaining on the inside of the glass jar was also dried. The sand/test material mixture was then returned to the jar until required for mixing with the WC-B soil.

Where water was used as the solvent, a much longer drying period was required because of the slow rate of evaporation. This process was aided by placing the aluminum trays in a warm atmosphere (approximately 25°C). When a solvent other than water was used, a control solvent/sand mixture was also prepared in a similar manner.

The final sand/test material mixture was then added to the appropriate quantity of WC-B soil and shaken in a heavy gauge polythene sack for at least half a minute. Labelled plant pots were then filled with this growth medium.

A control water/sand mixture and, where applicable, a solvent control/sand mixture was prepared using clean sand.

Layout of Plant Pots

After filling the plant pots with growing medium, they were placed on capillary matting on staging in a glasshouse. Pots were arranged (approximately 5 cm apart) such that no concentration was placed next to another that differed from it by more than one order of magnitude (Fig. 1).

Growing Conditions

All tests were carried out in a glasshouse at a nominal temperature of 20°C, although it was accepted that, due to the nature of glasshouses, wide variations from the nominal temperature would be experienced during extreme weather conditions. A daily record of maximum and minimum temperatures was maintained from Mondays to Fridays during the test.

Water was supplied primarily by subirrigation through capillary matting which was kept wet throughout the studies. Pots were examined twice daily, and, if necessary, further water was supplied by top-down watering. No additional chemical fertilizers were added.

During periods of poor natural daylight (less than 5000 lx), mercury vapor lamps provided a minimum day length of 16 h.

A representative sample of WC-B soil was analyzed for particle size, pH, and moisture content. No chemical analysis of the soil or plant foliage was carried out during any of the tests.

Procedure for Sowing Seeds and Subsequent Plant Growth Assessment

On Day minus 7, the treated soils were prepared and put into pots. Eight seeds were sown into each pot and the pots placed onto capillary matting on a bench in the glasshouse.

On Day 0, a first assessment of emergence was made and the numbers recorded. After a further seven days, a second assessment of emergence was made and any growth effects noted. On this day the plants were thinned to four seedlings per pot, aiming to leave plants of uniform size spaced as evenly as possible within the pot.

On Day 21 (14 days after thinning the plants) the phytotoxicity of the test materials was assessed. The plants from each pot were cropped at soil level and the four plants weighed together to obtain a total fresh weight per pot. Weighing was done within 30 s of cropping as the shoots lose weight rapidly after being cut (as described in the OECD guideline [*1*] and in the EC protocol [*7,8*]). Any effects to the plants in the treated soils which distinguished them from those in the controls (for example, stunting, chlorosis, and necrosis) were noted.

Statistical Analysis of Results

Data from the plant weights were entered into a program designed by Brixham Laboratory called "PLANT" [*9*]. Subsequently, PROBIT analysis was used to determine the EC_{50} (growth, 21 days) values. The concentration (calculated) of the test chemical that reduced the emergence of plants through the soil surface by 50% (i.e., EC_{50} emergence) was determined from emergence data recorded on Day 7 of the tests. The highest no-effect concentrations (NOEC) which caused no growth reduction at the 95% significance level were determined by one-way analysis of variance.

Results

The results of the EC_{50} values of the 43 chemicals are shown in Table 1. The highest no-effect concentrations (NOEC), where calculated, are shown in Table 2. In some of the early studies, the NOECs were not determined.

Discussion

The method has been used to assess the potential toxic effects of 43 chemicals on emergence and early growth of several species over a period of about eight years. During this period, experience has shown where improvements to the method could be made. The choice of species was found to be an important factor. In the earlier studies, much of the developmental work was done on oats, turnips, and occasionally tomatoes. Later studies standardized on

TABLE 1—Summary of the results of the OECD soil-based test on 43 chemicals, EC₅₀ values (mg/kg dry soil).

Chemical	Reference	Sorghum	Sunflower	Soya	Mung Bean	Oats	Turnip	Tomato
1 Allylthiourea	[14]	174	127	131				
2 Ammonium metavanadate	[11]					216	13	
	[11]					516	212	
3 Anthracene	[18]	>1000	>1000					
4 Antimony trioxide	[2]				>1000			
5 Biphenyl	[16]	>1000	>1000	>1000		>1000	>1000	
6 Cadmium-based pigment	[2]					>1000	>1000	
7 Cadmium chloride	[2]					329	98	
8 Cetyl trimethyl Ammonium bromide	[14]	>1000	>1000	>1000				
9 Chloral hydrate	[18]	6	496		115			
	[18]	<1	17		4			
10 Chloroacetic acid	[18]	183	248		176			
11 4-chloroaniline	[18]	63	204		95			
12 Chloro-2,4-dinitrobenzene	[18]	18	15		17			
13 Chlorohexidene acetate	[14]	804	326					
14 5-chloro-2-hydroxyaniline	[18]	104	>1000	>1000				
15 4-chlorophenol	[16]	222	354	782	475			
16 2-chloro-ortho-toluidene	[18]	59	113		60			
17 Copper(II) nitrate	[14]	>1000	>1000	>1000		408	156	
18 Cumene(iso-propylbenzene)	[12]	>1000	>1000		>1000	529	70	
	[12]							
19 2,4-dichloroaniline	[17]					51		42
20 ortho-dichlorobenzene	[18]	>1000	>1000		>1000			
21 2,5-dichloronitrobenzene	[18]	26	58		54			
22 3,5-dichlorophenol	[14]	46	172	207				
23 Di-2-ethyl hexyl phthalate	[2]					>1000	>1000	
24 Distearyl dimethyl ammonium chloride	[2]					>1000	>1000	
25 Hexachlorobenzene	[18]	>1000	>1000		>1000			
26 Hexachloroethane	[18]	59	312		172			
27 2-hydroxybiphenyl	[13]	>200						200
28 4-hydroxybiphenyl	[13]	>200						5

No. Compound	Ref							
29 Lead(II) nitrate	[12]	>1000	>1000	>1000		>1000	>1000	
	[12]	>1000				>1000	>1000	
30 Naphthalene	[15]					301	93	
31 Nickel(II) nitrate	[11]					318	167	
32 4-nitrophenol	[15]	140	230	88				
33 Nonyl phenol	[15]	>1000	>1000	>1000				
34 Nonyl phenol 2-ethoxylate	[15]	>1000	>1000	>1000				6
35 Pentachlorophenol	[13]	17	47	113				
	[14]	75	33	73				
	[15]	37	39	96				
	[16]	51						12
	[17]					18		
	[18]	4	5	89	6			
36 Potassium chlorate	[2]	121	95	412				
	[12]	260	130	>1000				
	[14]	135	2	154				
	[15]	78	24	152				
	[16]	30	17		17			
	[16]	21	14		14			
	[18]	29	22		21	120	14	
	[18]	70	5		15	50	11	
37 Potassium dichromate	[19]	56				107	90	18
	[11]					201	14	
	[11]					98		
38 Sodium Tetraborate	[13]	>1000	659	489		301	115	32
	[2]					301		
39 Tin(II) Chloride	[14]	270	519	670	45	>1000	>1000	130
40 Trichloroacetic acid	[11]	49	158			9		
41 1,2,4-trichlorobenzene	[17]	44						
42 2,4,5-trichlorophenol	[16]							
	[13]							
43 Zinc chloride	[18]					>1000	159	15
	[11]					>1000	>1000	
	[11]							

TABLE 2—*Summary of the results of the OECD soil-based test on 43 chemicals, no observed effect concentration (NOEC) (mg/kg dry soil).*

Chemical	Reference	Sorghum	Sunflower	Soya	Mung	Oats	Turnip	Tomato
1 Allylthiourea	[14]	10	10	1				
2 Ammonium metavanadate	[11]					···	···	
3 Anthracene	[11]					···	···	
	[18]	>1000	10			>1000	>1000	
4 Antimony trioxide	[2]				10			
5 Biphenyl	[16]	100	>1000	100		>1000	>1000	
6 Cadmium-based pigment	[2]					>1000	>1000	
7 Cadmium chloride	[2]					···	···	
8 Cetyl trimethyl Ammonium bromide	[14]	>1000	100	100				
9 Chloral hydrate	[18]	<1	<1		<1			
	[18]		10					
10 Chloroacetic acid	[18]	100	100		100			
11 4-chloroaniline	[18]	<1	100		10			
12 Chloro-2,4-dinitrobenzene	[18]	10	1		10			
13 Chlorohexidene acetate	[14]	1	1					
14 5-chloro-2-hydroxyaniline	[18]	10	100	1	10			
15 4-chlorophenol	[16]	100	100	100	10			
16 2-chloro-ortho-toluidene	[18]	1	10		10			
17 Copper(II) nitrate	[14]	>1000	>1000	>1000				
	[12]					···	···	
	[12]					···	···	
18 Cumene(iso-propylbenzene)	[18]	>1000	>1000		>1000			10
19 2,4-dichloroaniline	[17]					10		
20 ortho-dichlorobenzene	[18]	>1000	>1000		>1000			
21 2,5-dichloronitrobenzene	[18]	10	10		10			
22 3,5-dichlorophenol	[14]	10	<1	<1				
23 Di-2-ethyl hexyl phthalate	[2]							
24 Distearyl dimethyl ammonium chloride	[2]							
25 Hexachlorobenzene	[18]	>1000	>1000		>1000	>1000	>1000	
26 Hexachloroethane	[18]	10	100		100	>1000	>1000	
27 2-hydroxybiphenyl	[13]	200(+)						200
28 4-hydroxybiphenyl	[13]	200(+)						
29 Lead(II) Nitrate	[12]	>1000				>1000	>1000	
	[12]	>1000				>1000	>1000	

No. Compound	Ref							
30 Naphthalene	[15]	100	>1000	100				
31 Nickel(II) nitrate	[11]				1	1	–	–
32 4-nitrophenol	[11]	10	100	1			–	–
33 Nonyl phenol	[15]	100	100	100			–	–
34 Nonyl phenol 2-ethoxylate	[15]	10	100	10				
35 Pentachlorophenol	[15]	2	10	10				2
	[13]	10	1	10				
	[14]	10	10	10				
	[15]	10	10	10				
	[16]							
	[17]	>1	>1		>1	1	–	1
36 Potassium Chlorate	[18]							
	[2]	10	10	100		10	–	–
	[12]	10	10	1		10	–	–
	[14]	100	<1	10		10		
	[15]					1		
	[16]	–	–	–		10		
	[16]	10	10	10				
	[18]	10	10					
	[18]	10	1					
	[18]	10						
	[18]	10						
37 Potassium dichromate	[19]	20					–	3
	[11]						–	
38 Sodium tetraborate	[11]	100	100	10				2
	[13]							
39 Tin(II) chloride	[2]	100	100	100			–	
40 Trichloroacetic acid	[14]	20				1		10
41 1,2,4-trichlorobenzene	[11]	10						
42 2,4,5-trichlorophenol	[17]							2
	[16]							
43 Zinc chloride	[13]		10	100	1		–	–
	[18]							
	[11]							
	[11]							

– Not determined
+ Highest concentration tested

three main species: sorghum, sunflower, and soya. This selection fulfilled the requirement to test a monocotyledon and two dicotyledons including one legume. In several studies on soya, the calculated EC_{50} (shoot weight) values were not as low as would be expected from visual inspection. The thick, waxy cotyledons which were often produced contributed considerably to the overall weight of the plant, especially when growth was otherwise severely reduced. The weight of these cotyledons was high compared to that of the secondary leaves produced in a normal healthy plant. For this reason, soya was replaced by mung bean in the most recent studies.

For potassium chlorate, where a comparison between the results with soya and mung bean can be made, the EC_{50} was much lower for mung bean than soya, on average by an order of magnitude (Table 1).

Some species have been found to be unsuitable for use in short-term plant tests for various reasons. For example, the onion group had poor germination and produced small plants, vetch had a trailing habit, and rape tended to produce "floppy" plants which were not self-supporting. Some species, such as beet and tomato, which produced suitable plants, showed relatively poor and variable germination success. These species were considered unsuitable for this reason.

The first EC ring test, carried out in 1984, involved the participation of 19 laboratories. Widely differing results were obtained from several of the laboratories, possibly because there was too much flexibility in the guideline protocol [7]. The main problem was considered to be maintaining sufficient plants in the control pots. For this reason "thinning" was introduced (see method development). Price [7] concluded that the test was simple and well suited as a general indicator of the phytotoxicity of a chemical. EC_{50} values calculated in tests with concentrations an order of magnitude apart would probably be accurate enough for this purpose. This is especially the case with general chemicals (nonagrochemicals) for which it is normally adequate to determine which one of the four following classes the EC_{50} falls into:

1. >1000 mg/kg (nonphytotoxic).
2. 100 to 1000 mg/kg (practically nonphytotoxic).
3. 10 to 100 mg/kg (moderately phytotoxic).
4. <10 mg/kg (highly phytotoxic).
(The class names have been arbitrarily assigned by the Brixham Laboratory.)

Price also suggested that it was unnecessary to stipulate the species for testing, and this is generally supported by the results reported here (Table 1) [2,10–19].

Where repeat tests have been carried out using the same test species on one chemical, the results normally showed good reproducibility. However, on occasion significantly different results were obtained. In the fourth test with pentachlorophenol, for example, the results for sorghum, sunflower, and soya were all about one order of magnitude lower than previously recorded. This is good evidence to support the use of a reference test material with each definitive test. The OECD guideline allows for this option, and potassium chlorate or pentachlorophenol are considered suitable.

Results reported by Adema and Henzen [19] also supported the view that most plant species responded to most chemicals in a similar way. Their result for potassium chlorate on oats was also very similar to the results reported here (Table 1).

Conclusions

The ICI Brixham Laboratory's experience with the OECD plant test has shown it to be simple, easy to run, and amenable to GLP standards. It can be used to generate order of magnitude type results. Such results are adequate for use on material safety data sheets (MSDS) or for the notification of a new chemical under EC legislation. The method does not require the use of expensive growth chambers, although if available they could be used. The test is cost-effective and reasonably reproducible, and the two Ring Tests which have been carried out have resulted in a large database of comparative results. Finally, it is a very flexible test that can be modified easily to allow volatiles to be tested. The effect of 1,1,1-trichloroethane on sorghum and rape was tested in the laboratory using this modification [7]. The guideline has also been used by the Brixham Laboratory to assess the effect of industrial sludges on plant growth.

Acknowledgment

The authors would like to thank the staff of the ecology and weed science sections at Jealott's Hill Research Station, ICI Agrochemicals, Bracknell, Berkshire, England for their assistance and for provision and maintenance of the facilities used in the studies reported here.

References

[1] "OECD Guidelines for Testing of Chemicals, No. 208, Terrestrial Plants, Growth Test," OECD Publications, Paris, France, April 1984.
[2] Stanley, R. D. and Tapp, J. F., "An Assessment of Ecotoxicological Test Methods, Part VIII," ICI Brixham Laboratory Report BL/A/2164, Brixham, Devon, UK, 1982.
[3] Stanley, R. D. and Tapp, J. F., "An Assessment of Ecotoxicological Test Methods, Part XIII and XIV," ICI Brixham Laboratory Report BL/A/2197, Brixham, Devon, UK, 1982.
[4] Stanley, R. D. and Tapp, J. F., "An Assessment of Ecotoxicological Test Methods, Part XII," ICI Brixham Laboratory Report BL/S/0135, Brixham, Devon, UK, 1982.
[5] Stanley, R. D. and Tapp, J. F., "An Assessment of Ecotoxicological Test Methods, Part XV," ICI Brixham Laboratory Report BL/A/2207, Brixham, Devon, UK, 1982.
[6] Thompson, R. S. and Carmichael, N. G., "1,1,1-trichloroethane: Medium-Term Toxicity to Carp, Daphnids, and Higher Plants," *Ecotoxicology and Environmental Safety,* Vol. 17, No. 2, April 1989, pp. 172–181.
[7] Price, C. E., "Higher Plant Test Guideline Ring Test," EEC final report. Contract XI AL(83)647 N (290)," A.E.R. Consultants, Mud Lane, Eversley, Hants, UK.
[8] Price, C. E., "Report on EEC Contract to Analyse the Data from the Higher Plant Ring Test," Contract XI/351/90, A.E.R. Consultants, Mud Lane, Eversley, Hants, UK, 1990.
[9] Jones, M. A., "Computer Program 'PLANT' for Calculation of EC50 by Probit Analysis of Fresh Shoot Weight Data," SOP TX07. ICI Brixham Laboratory, Brixham, Devon, UK, 1982.
[10] Stanley, R. D. and Tapp, J. F., "An Assessment of Ecotoxicological Test Methods, Part V," ICI Brixham Laboratory Report BL/A/2109, Brixham, Devon, UK. 1981.
[11] Stanley, R. D. and Tapp, J. F., "An Assessment of Ecotoxicological Test Methods, Part XVI," ICI Brixham Laboratory Report BL/A/2505, Brixham, Devon, UK, 1984.
[12] Stanley, R. D. and Tapp, J. F., "An Assessment of Ecotoxicological Test Methods, Part XI," ICI Brixham Laboratory Report BL/A/2315, Brixham, Devon, UK, 1984.
[13] Stanley, R. D. and Tapp, J. F., "An Assessment of Ecotoxicological Test Methods, Part XVII," ICI Brixham Laboratory Report BL/A/2528, Brixham, Devon, UK, 1984.
[14] Windeatt, A. J. and Tapp, J. F., "An Assessment of Ecotoxicological Test Methods, Part XVIII," ICI Brixham Laboratory Report BL/A/2742, Brixham, Devon, UK, 1985.
[15] Windeatt, A. J. and Tapp, J. F., "An Assessment of Ecotoxicological Test Methods, Part XIX," ICI Brixham Laboratory Report BL/A/2836, Brixham, Devon, UK, 1986.

[16] Windeatt, A. J. and Tapp, J. F., "An Assessment of Ecotoxicological Test Methods, Part XX," ICI Brixham Laboratory Report BL/A/2959, Brixham, Devon, UK, 1987.

[17] Windeatt, A. J. and Tapp, J. F., "An Assessment of Ecotoxicological Test Methods, Part XXI," ICI Brixham Laboratory Report BL/A/3152, Brixham, Devon, UK, 1987.

[18] Windeatt, A. J. and Tapp, J. F., "An Assessment of Ecotoxicological Test Methods, Part XXII," ICI Brixham Laboratory Report BL/3759/A, Brixham, Devon, UK, 1990.

[19] Adema, D. M. M. and Henzen, L., "A Comparison of Plant Toxicities of Some Industrial Chemicals in Soil Culture and Soilless Culture," *Ecotoxicology and Environmental Safety*, Vol. 18, 1989, pp. 219–229.

Barbara M. Smith[1]

An Inter- and Intra-Agency Survey of the Use of Plants for Toxicity Assessment

REFERENCE: Smith, B. M., "An Inter- and Intra-Agency Survey of the Use of Plants for Toxicity Assessment," *Plants for Toxicity Assessment: Second Volume, ASTM STP 1115,* J. W. Gorsuch, W. R. Lower, W. Wang, and M. A. Lewis, Eds., American Society for Testing and Materials, Philadelphia, 1991, pp. 41–59.

ABSTRACT: The purpose of this project was to perform an inter- and intraagency survey of the use of plants as test systems for ecological risk assessment and remediation for the United States Environmental Protection Agency Office of Research Development (USEPA/ORD). Four broad classes of plant applications were identified: (1) bioassays; (2) bioindicators; (3) bioremediation; and (4) biotechnology. Improvement to research methodologies would include: (1) developing fluorescence bioassays for terrestrial and aquatic plants and algae and combining plant and animal tests to form regionally differentiated suites of bioassays; (2) development of "natural community" databases and small-scale maps for evaluation of changes in plant community structure to determine the "biotic integrity" of specific habitat types; (3) expand work to develop plants and fungi as biologic remediative tools; and (4) development of guidelines for field trials and eventual full-scale agricultural use of pesticide-producing genetically engineered plants (GEPs).

KEY WORDS: plants, algae, bioassay, bioindicator, bioremediation, biotechnology, genetically engineered plants, environmental toxicity assessment, aquatic plants, terrestrial plants, fluorescence

Plants constitute an important, arguably the *most* important, component of the world's ecosystem because of their unique ability to capture the sun's energy and transform it into chemical energy [adenosine triphosphate (ATP) and reducing equivalents], while absorbing carbon dioxide and releasing oxygen into the environment [1]. But, historically, plants (including algae) have received only a modicum of attention in the process of trying to develop environmental risk assessment parameters. This situation is ironic since studies have shown that plants, in general, and algae, in particular, are often more sensitive to toxicants than the animals used in toxicity testing [2–4]. In recognizing the value of plants, recent recommendations to the United States Environmental Protection Agency (USEPA) have stressed the need for greater consideration of plant systems in environmental risk assessment [5].

The purpose of this study, conducted during the summer of 1989 for the USEPA Office of Research Development (ORD), was threefold. The first goal was to give a status report of the existing state of the art of toxicity assessment using plant systems in the USEPA laboratories and as reflected in USEPA-sponsored research. This goal was accomplished by interviewing staff scientists and reading journal articles and internal USEPA documents. The second goal was to present various other techniques and approaches that are currently being used in academia or industry, but may not have been applied to risk assessment using

---

[1] San Francisco Bay Regional Water Quality Control Board, Oakland, CA 94612.

plant systems. This goal was accomplished by reviewing the literature and interviewing various leading scientists, including those in the Department of Energy laboratories. The third goal was to develop a prioritized list of current techniques using plants systems that should be considered for future development by USEPA/ORD into environmental risk assessment tools. Plant and algal applications in four broad categories of research were investigated: (1) bioassays; (2) bioindicators; (3) bioremediation; and (4) biotechnology. The last category of plant applications is not currently under research in the USEPA/ORD, although research using genetically engineered microorganisms is being undertaken.

An informal (nonstatistical) survey of the number of research projects (in USEPA laboratories) using plants as test systems was performed. Data were obtained by reviewing fiscal year 1989 quarterly reports of research from the USEPA regional laboratories and interviewing staff scientists. Of 219 research efforts (this tally may include "double hits" where more than one type of organism or various aspects of the response of that organism were investigated), 22% included work on plants. Of that number, about 7% of the projects dealt with algae and nearly 14% of the projects with forest and crop plants. The remaining 1% of the projects used seagrasses and noncrop vascular plants as test systems. Although nearly 19% of all projects dealt with bacteria or genetically engineered microorganisms (GEMs), no research effort was allotted to issues of risk assessment or regulation of transgenic or genetically engineered plants (GEPs).

Plants Used as Bioassays—Present Applications

The mandate of the USEPA, in its broadest interpretation, is to protect the environment [6]. To accomplish this goal, early regulatory standards were set forth that allowed fixed concentrations of potentially hazardous chemicals to be released into the environment (technology-based standards). This strategy has proved infeasible [7], and research has been undertaken to develop tests that quantified the responses of living organisms, plants, and animals to these chemicals (environmentally based standards) [8]. The term "bioassay" refers to types of laboratory tests in which organisms are exposed to toxicants for defined periods of time and a biological endpoint (e.g., growth or death) is measured. The major focus of plant research in the USEPA laboratories has been in the development of single-species algal toxicity test protocols [9]. The publication record in this area is fairly extensive, and the following more recent examples will serve to demonstrate the types of technologies involved.

Bioassays Using Algae

A study of the response of the marine diatom *Skeletonema costatum* to several pesticides was conducted [10]. The endpoints of this test were the number of cells at the beginning and end of the exposure period (cell count) and the number of dead cells (evaluated by using a differential staining response). From this information, the EC_{50} for cell division ("effective concentration" at which 50% of the test organisms exhibited the response) for 2, 3, and 4-day exposures to the pesticides was estimated. Similar types of studies have included testing the growth responses of two marine diatoms to organotin compounds [11], three marine unicellular algae to brominated organic compounds [12], and two marine diatom species to 19 toxicants [13].

Algal toxicity tests currently under development by the USEPA that have reproductive rather than growth or biomass endpoints include the red algal (*Champia parvula*) and the brown algal (*Laminaria saccharina*) tests. Other types of algal toxicity tests developed for agencies other than USEPA include a germination efficacy test using spores from the giant kelp *Macrocystis pyrifera* [16], a *Macrocystis* gametophyte growth bioassay [17], and numerous other independent research projects in which single-species algal toxicity testing has been reported.

Bioassays Using Vascular Plants

The development of vascular plant bioassay testing by the USEPA, particularly terrestrial plants, has lagged behind that for algae. This reflects a focus on the aquatic environment due perhaps to the realization that ocean and aquatic habitats have been the recipient of the majority of commercial chemical exposures [14]. Only three vascular plant toxicity tests developed by the USEPA are currently approved for laboratory use: (1) the duckweed (*Lemna*) acute toxicity test; (2) the seed germination/root elongation toxicity test; and (3) the early seedling growth toxicity test [5]. In one study, growth reduction in response to exposure to pentachlorophenol (PCP) in the aquatic vascular plant *Elodea sp.* was observed, but not in *Lemna minor*, even at the highest experimental exposure level [15]. A similar type of toxicity test using vascular (marsh) plants was reported in an internal report prepared by the EPA-Gulf Breeze laboratory in which two species of freshwater marsh plants were exposed to liquid waste from a sewage treatment plant at several dilutions in natural and simulated sediments. Results showed that germination, early development, survival, and growth of the plants were inhibited by the liquid waste.

Toxicity tests with reproductive endpoints for vascular plants developed by the USEPA are very few and have centered on genotoxic assays using *Zea mays* (induction of mutant pollen grains) and *Tradescantia paludosa* (induction of micronuclei) [18]. These bioassays are now seriously outdated despite continued support for their application [19]. Plant molecular biologic techniques have made such tremendous strides (see below) that reevaluation and perhaps "mothballing" of these bioassays must be seriously considered.

Bioassays Using Microcosms

With the realization that "(s)ingle species toxicology data is inadequate to predict population effects in ecological systems" [20], microcosm toxicity tests have been developed by the USEPA. This type of multispecies testing approach grew out of the realization that poor correlations between single-species toxicity tests and *in situ* toxicity tests have been documented by both environmentalist and industrialist camps [7]. Microcosm, or multispecies tests, fall into two major categories that include both algal and vascular plant test systems. The first involves testing of "natural" communities, and the second involves testing defined groups of organisms or "standardized" microcosms. For example, in a study of the effects of PCP on the algal component of the natural periphyton community in outdoor experimental streams, the investigators determined biomass [ash free dry weight, chlorophyll (Chl) *a* and pheophytin (Pheo) *a* concentrations] and changes in community structure (presence/absence data, species diversity index) for three concentrations of toxicant during short-term (2-week) and long-term (9-week) exposures over two experimental seasons. The responsevariables were quantified using cell counts, taxonomic identification to genus, and fluorometric measurement of pigment concentrations [21]. In a similar study of a natural seagrass community, positive and negative controls were used to distinguish physical from chemical effects of drilling fluids. Effects on *Thalassia testudinum*, the seagrass component, and its epiphytic algal (diatom) community were distinguished using the growth endpoints, ash-free dry weight (AFDW) per unit leaf area and Chl *a* per unit leaf area, as well as Chl *a* per gram AFDW of the diatom epiphyte tissue. Benthic invertebrates were identified to species, and effects were based on enumeration data of the mean number of organisms per microcosm for the ten most abundant invertebrate species. Results showed significant differences in the control and treated mesocosms with respect to biomass and community structure [22]. Additional results of the above experiment presented pigment analyses (Chls a, b, and c) and uptake of radio-labelled sodium carbonate data to distinguish effects on *Thalassia* and the epiphytic diatom assemblage. Photosynthesis, as carbon fixation, was the most sensitive response, showing significant reduction in response to both clay (positive control) and drilling fluid exposures [23].

In an effort to develop multispecies toxicity tests to improve the correlation between laboratory-based and *in situ* toxicity testing, a series of experiments were conducted that culminated in an interlaboratory effort to test a defined standard aquatic microcosm (SAM) protocol [24,25]. Four laboratories participated in the experiment: University of Washington; EPA-Duluth; U.S. Army, Aberdeen, MD; Marine Bioassay Laboratory, Carlsbad, CA. The standard "community" consisted of ten species of algae and five species of zooplankters with well-defined culture characteristics. Algal response was determined by estimating algal biovolume, defined as total number of cells of each algal species multiplied by a predetermined mean cell volume for each algal species. Responses of zooplankters were determined by enumeration of each species. Results showed that, although the precise timing of events among experimental trials differed, the outcome and conclusions related to the effects of copper exposure on the SAMs were the same. Experiments provided similar statistical differences between control and treated microcosms within the same experiment and gave the same rank order of the "day-weighted variable," a statistic designed to characterize the "center of gravity" (mean) of a variable with an unknown probability distribution [25]. The development of terrestrial vascular plant microcosms, "natural" or "standardized," has not received attention. Techniques developed using photosynthetic parameters [26,27] and using mycorrhizal associations [28–30] would be appropriate bases from which to develop terrestrial vascular plant microcosm bioassay protocols.

Photosynthesis for Plant Toxicity Evaluation

The use of enumeration techniques to determine growth of unicellular plants is a simple, however time-consuming technique. The use of more rapid, less labor-intensive technologies has distinct advantages, especially when testing rapidly changing effluents such as those of municipal and/or industrial waste outfalls. Growth bioassays for plants have been adopted due to the difficulty of determining when a plant cell is "dead." Further explanation may be had by analogy drawn with animal bioassays. If the biological endpoint of, e.g., a fish bioassay was death, the researcher could safely assume the mortality of the test organism if it stopped breathing for some extended period of time, rather than having to wait to see if the fish was able to grow or reproduce. The goal of a more rapid algal bioassay, then, is to evaluate the plant's "breathing." This can be accomplished by monitoring photosynthesis as carbon dioxide uptake, oxygen evolution, or fluorescence emission characteristics.

Carbon Dioxide Uptake Techniques for Plant Toxicity Evaluation

Carbon dioxide uptake techniques have been used successfully in vascular plant and algal studies, and there exists a vast body of literature that supports their use. However, those techniques that require the use of radioactive material are not recommended for broad use because of the level of expertise necessary to perform the tests and the specific safety needs for radioactive waste handling and disposal. Other techniques, such as the use of infrared gas analyzers (IRGA) to monitor photosynthesis in land plants, have been used extensively and effectively in both laboratory and field experiments [31–35]. This technique can evaluate sublethal effects of many toxicants on plants before such obvious effects as chlorosis or necrosis of leaves occur [26,27].

Oxygen Evolution Techniques for Plant Toxicity Evaluation

Oxygen evolution is another measure of photosynthetic activity. This technique is used on aquatic plants and algae. Although oxygen electrodes are relatively inexpensive and simple to operate, they must be used in the laboratory, are fairly insensitive, and are subject to significant within-experiment variability. Using a flow-through system for exposure of seagrass to pollutants, oxygen evolution was used to evaluate the effect of PCP and atrazine on photosynthesis and respiration of the seagrass *Thalassia testudinum* [2]. The endpoints of this test were oxygen evolution by whole plants or leaves, measured after 40 or 88-h exposures to the toxicants using an oxygen electrode. Results, expressed as oxygen concentration of the water (rather than on the more accepted per weight plant tissue or per Chl bases) and as changes in the photosynthesis to respiration (P/R) ratio, showed a depression of photosynthesis in response to the pollutants [2].

Fluorescence as an Index of Photosynthesis

Another technique to monitor photosynthesis that is inexpensive, technically simple to master, and that can be performed in the laboratory or in the field exploits an intrinsic quality of the photosynthetic apparatus of all plants, terrestrial and aquatic—fluorescence. Fluorescence occurs in plants when excess light, absorbed by Chl and other light-harvesting pigments, that cannot be used for photochemistry is reradiated at a lower, specific energy level (wavelength). *In vivo* (protein-bound) Chl fluorescence of chloroplast membranes and algal cells is emitted at room temperature in the 650 to 800-nm (red) region of the spectrum. At physiologic temperatures, the vast majority of Chl fluorescence arises from Photosystem II (PSII) [36]. The total amount of Chl *a* fluorescence has been correlated with content of Chl *a*. This technique is used as an indirect method for estimating biomass in algae. The kinetics of the light absorption and reradiation phenomenon in plants and algae are quite complex (Fig. 1) and are dependent on a number of physical, chemical, and biological factors, including, but not limited to: the type of chlorophylls and accessory light-harvesting pigments in the photosynthetic apparatus (i.e., is the organism a red, brown, or green alga, see Fig. 2); the content of the pigments that quench excitation (e.g., carotenoids); biophysical "non-photochemical quenching" phenomena; the quality and intensity of light used to excite the photosynthetic apparatus; and especially stress factors, including nutrient status, osmotic balance, heavy metals, toxins, and herbicides.

The value of fluorescence as an estimate of Chl *a* content in aquatic systems has been recognized by oceanographers for many years [37,38]. The use of fluorescence techniques in the USEPA research effort has been limited to estimations of the Chl concentration of algal cells as an indirect measure of biomass accumulation [23,39,40]. While this is obviously an important use of the tool, the potential for application of this technique is much broader and can give information about the ongoing process of photosynthesis, not just about its end result, growth. Several examples to follow will serve to demonstrate the use of fluorescence as a vital indicator of the "health" of the photosynthetic apparatus in test plants, both terrestrial and aquatic.

The majority of research using fluorescence as an index of photosynthetic capacity has focussed on terrestrial plant systems [41–43]. This work has a long research history dating

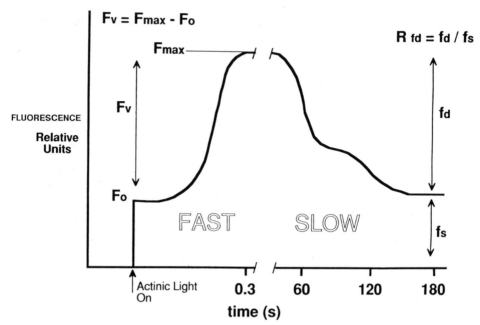

FIG. 1—*Typical fluorescence induction curve of a terrestrial green plant. The figure shows the time-dependent fluorescence induction kinetics of dark-adapted chloroplasts or whole leaves. The shape of the curve is also typical of green algae. When the actinic light is turned on, the fluorescence rises immediately to the level F_0, then more slowly to a maximum level, F_{max}. The difference in fluorescence level between F_0 and F_{max} is termed the variable fluorescence, F_v. The fast fluorescence rise to F_{max} is one indicator of the health of the plant. After several minutes in the light, the level of fluorescence declines to a steady state level, referred to as fs. A so-called viability index, Rfd, is calculated as the fluorescence difference fd (usually greater than, and not equivalent to, F_v), divided by the fs value. This viability index has been used successfully to study the effects of acid rain on coniferous trees [43]. The shape of the fluorescence curve can also be used to identify (Photosystem II) herbicide-resistant plants [45–48].*

to the early 1960s. While some controversy still exists as to the exact biophysical mechanisms by which fluorescence is emitted and from which pigment system(s) it derives, its use is widespread, technically simple, economical, and well accepted in the scientific community. Most studies of the fluorescence characteristics of plants have been centered on understanding the intricacies of the process of photosynthesis in isolated chloroplast membranes, but practical application of the technique as a noninvasive tool for evaluating photosynthesis in laboratory and field studies has been rigorously undertaken [43].

In one study, simultaneous determinations of the fluorescence induction kinetics and CO_2 uptake (IRGA) were performed to test the effects of H_2S in air on photosynthesis in laboratory-grown spinach plants [44]. Results showed that prolonged exposure to H_2S, a common anthropogenic air pollutant, reduced CO_2 fixation as well as several fluorescence parameters that were monitored, although dark respiration was apparently not affected [44]. Fluorescence induction kinetics can also be used to screen for herbicide resistance in plants and algae because the shape of the fluorescence curve differs between these classes of resistant and sensitive organisms [45,46]. This characteristic has been used successfully in the laboratory to identify mutants for development of herbicide-resistant lines of crop species [47]. Future application at chemical waste sites may permit field identification of resistant or remediative biotypes [48].

Evaluation of fluorescence parameters has also been used successfully with algal systems

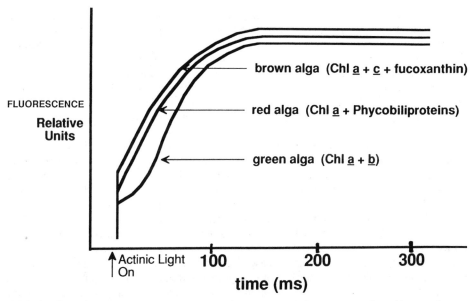

FIG. 2—*Typical fluorescence induction curves of three types of algae: green, red, and brown. Species differences can be identified from the shape of the fluorescence curves. Red and brown algae have a single exponential component to the initial portion of their fluorescence curves, while green algae and terrestrial plants exhibit a biphasic curve in the millisecond time range* [50,53–55].

in determining effects on photosynthesis. Photoinhibition (reduction in photosynthetic capacity due to photodestruction of the reaction center of PSII) in naturally occurring mixed populations of phytoplankton in Lake Titicaca has been documented using a field fluorometer [49,50]. Fluorescence emission of DCMU-poisoned cells measured at various depths in the photic zone of the lake showed that the subpopulations of microalgae at the surface had impaired photosynthesis when compared to algae collected at deeper sampling stations [49]. The effect of copper on photosynthesis in the marine diatom *Skeletonema costatum* in laboratory experiments (and field verification using naturally occurring phytoplankton populations) was tested by monitoring *in vivo* fluorescence and the uptake of ^{14}C bicarbonate in light and dark bottles. Results showed that effects of copper on photosynthesis were reflected in changes in fluorescence parameters in both the laboratory and field experiments [51]. In what could be classified as a single species toxicity test, variable fluorescence was used as the endpoint to test the effect of exposure to various concentrations of $CuSO_4$ on the green microalga *Dunaliella tertiolecta* [52]. Reduction in variable fluorescence was well correlated with reduction in photosynthesis, as measured by oxygen evolution. Spectrophotometric techniques, including fluorescence, have been used to characterize the photosynthetic apparatus of such ecologically important algae as the giant kelp *Macrocystis pyrifera*, the diatom *Cylindrotheca fusiformis*, as well as an economically important green algal genus *Dunaliella* [53–55].

Recommendations for Improved Algal and Terrestrial Plant Bioassay Techniques

The future development of additional single-species algal toxicity tests using growth or cell division as endpoints may not be necessary except where the intent is to isolate and characterize particularly sensitive species to use as test organisms to reduce environmental risk [56]. By taking advantage of the willingness of the scientific community to share such discoveries, USEPA may benefit without expenditure of large amounts of resources. Measurements of variable fluorescence parameters as indices of photosynthesis using commercially available fluorometers are technically simple, rapid, and inexpensive. In many instances, aquatic effluents are complex and change composition rapidly. Algal growth bioassays may be good long-term (3 to 5-day) evaluative techniques, but in highly dynamic systems such as wastewater outfalls short-term (2 to 3-h) diagnostic testing is desirable. Under these circumstances, an aquatic algal fluorescence bioassay would be particularly appropriate.

The use of regionally differentiated suites of bioassays to regulate discharges into the aquatic environment, including both plant and animal bioassays, could be highly beneficial in the maintenance or attainment of water quality standards [57]. This approach has been successfully attempted in several instances. In 1986, the California State Water Resources Control Board Marine Bioassay Project was instituted to develop in house bioassay protocols for the Orange County Sanitation District to monitor municipal waste toxicity [14]. In addition, the EPA-Narragansett Laboratory has prepared a document containing bioassay protocols intended to respond to the need for testing programs that address the change from technology-based to environmentally based water quality standards. A suite of bioassays (algae, *Daphnia*, Microtox, lettuce root elongation, and earthworm) was used to evaluate the toxicity of soil from the Rocky Mountain Arsenal and aid in focusing clean-up operations of this hazardous waste site [58]. A "User's Guide" was developed by EPA-Corvallis to assist decision makers in the use of aquatic and terrestrial acute toxicity bioassays to evaluate chemical waste sites. Included were case studies, similar to that cited above, that demonstrated the efficacy of bioassays as evaluative and predictive tools in developing hazardous waste site sampling strategies [59].

The development of fluorescence bioassays for terrestrial plants is also strongly recommended. Where waste storage, accidental spills, and intended applications of chemicals occur on land, it is important to have toxicity tests that can evaluate the environmental risk to terrestrial plants in the locations of interest. Fluorescence bioassays permit adequate evaluation of the risk associated with exposure to toxicants of terrestrial plants without the necessity of referring to an algal model that may be inappropriate for the specific circumstances surrounding the regulatory question. As discussed above, regionally differentiated suites of bioassays would most adequately protect the environment. Improved bioassay testing was the subject of a workshop that dealt with the issues of the limited groups of organisms used for bioassay test systems [5]. The author strongly supports the recommendations of the workshop participants who recognized the need to develop bioassays using ecologically important groups of terrestrial and aquatic plants such as long-lived and perennial species from a variety of habitats including: forests, deserts, prairies, lakes, freshwater and saltwater wetlands and marshes, streams, and agricultural regions [5].

Plants as Bioindicators

The term "bioindicator" refers to species of plants (including algae) that may, by their presence or absence, be indicators of environmental "ecoregions" [60], pollution or environmental degradation (sentinel species). The concept of bioindicator is also intended to include the use of plants as monitors or accumulators of toxic substances, such as heavy metals or organic compounds (see also below, bioremediation). In contrast to the current understanding of the need for in-stream biological monitoring and biocriteria using fish populations [61], no similar proposals have been formally advanced to permit development of similar "biocriteria" for algal or terrestrial plant populations. Although there are significant data describing (disturbed and undisturbed) terrestrial and aquatic floras of various regions of the United States, no database nor "index of biotic integrity" [62] exists from which to draw inferences about the relative quality of the environment with respect to the plant communities that inhabit it. This may be due in part to the comparative difficulty in convincing funding agencies that noncrop plants and trees may be as useful as fish and other wildlife in performing environmental risk assessment [63].

Algae and Terrestrial Plants as Biomonitors

There has been much information gathered about the use of algae and plants as monitors of heavy metal pollution by virtue of their ability to bioaccumulate this class of toxins to levels several thousand times the concentrations measured in water [64,65]. Results showed the value of plants and algae as indicators and integrators of ambient metal concentrations and changes in community structure that reflected tolerant and nontolerant algal and terrestrial plant groups [65,66].

Another quality of algal communities that has proven to be a powerful monitoring tool is the quality of species diversity. The value of phytoplankton species diversity as a tool for evaluating and monitoring acidity in lakes has been demonstrated by several authors [67,68]. Using sediment cores as an *in situ* historical record of algal (diatom and chrysophyte) distribution and abundance, acidity of lake water has been shown to affect the relative distribution of these two classes of phytoplankton and can be used to predict lake acidity based on cell counts [69]. The USEPA at present has no experimental research designed to investigate the use of terrestrial plants as biomonitors. However, much work has been done in other laboratories to address these issues, and one effective role for the USEPA to play may be in the accumulation and organization of this information.

PHYTOTOX and Plant Biomonitor Databases

Development and updating of databases that catalog the responses of terrestrial and aquatic plants and algae to various toxicants is an important tool to aid in responding to the USEPA's mandate to regulate the use of chemicals in the environment. Although it has not been updated since 1984, PHYTOTOX has been developed to respond to one aspect of this need [70]. The database deals with the direct effects of exogenously supplied organic chemicals on terrestrial vascular plants. PHYTOTOX consists of a "reference file" containing 9700 bibliographic records and an "effects file" containing 78 000 dose-response records [71]. The database is heavily weighted toward agricultural species and emphasizes the paucity of information concerning the influence of organic chemicals on noncrop plants, especially those growing in natural habitats. Continued updating of this database is strongly urged, and several areas of expansion are suggested. Inclusion of literature on submergent and emergent aquatic vascular plant responses to chemical exposure would improve understanding and could help predict responses of lake, stream, and wetland ecosystems to chemical insult. Expansion of the database to include terrestrial and aquatic plant responses to inorganic chemicals (especially heavy metals) could help predict responses of terrestrial ecosystems to mining and manufacturing processes and might indicate species for future investigation as remediative tools. Unfortunately, no similar database resource exists for algal toxicity information despite the fact that there exists a considerable body of research in this area. In developing this much-needed algal database, PHYTOTOX could be used as a model in recording algal responses, first to organic, then to inorganic compounds.

The importance of having adequate information on which to base management decisions is obvious. One such area of need is in knowing the components of the "natural" community so that evaluations can be made about whether changes may have occurred. The development of the Ecoregion Map [60] and the subsequent testing of its applicability in evaluating fish populations [72] has demonstrated the need for similar information in terrestrial and aquatic ecosystems for evaluating water quality and terrestrial habitat quality with respect to plant communities. Historical data can be acquired, and changes in the community structure of the "hard body" members of the phytoplankton (e.g., diatoms and chrysophytes) can be evaluated using sediment cores [67–69]. Much information exists about phytoplankton community structure, distribution, and abundance that could be used in setting attainable water quality standards. Using historical and modern studies to develop a similar type of database, the USEPA could develop "biocriteria" for aquatic ecosystems [73].

Recommendations For Improved Use of Plants as Biomonitors

The author recommends the development of a terrestrial plant ecology database that would include herbaceous plants and wetlands and marsh ecosystems to further refine the "potential natural vegetation" component of the "ecoregion" mapping technique [60]. Initial steps would include determination of the amount of existing data and an assessment of its applicability and quality. Data may already exist that would provide much of the missing information about local and regional changes in terrestrial plant communities. Such a database would permit the determination of "indicator" or "sentinel" species and a better understanding of the "natural" plant communities that could be used to monitor environmental quality. Inclusion of information about metal-tolerant and pollution-tolerant species would help to document changes in community structure associated with environmental degradation of terrestrial ecosystems. In conjunction with such existing databases as PHYTOTOX, the quality of environmental risk assessment would improve because the uncertainty about "what is out there" would decrease. As stated above, better management decisions can be made when the components of both the ambient and historic, or natural, community are known.

Plants as Bioremediators

The use of plants as bioremediative tools can take many forms, from disposal/recycling of municipal sewage sludge [74,75] to ecological restoration as a strategy for conserving biological diversity [76]. To date, the major focus of research in the USEPA, with respect to biologic solutions to waste site remediation, has centered on soil microbes. Because biological remediation is, on the whole, more economical than technological solutions, especially when dealing with large areas of low-level contamination, emphasis should also be placed on the plant and algal components of bioremediative strategies [77,78].

Wetlands Ecosystems and Bioremediation

Increased concern that the degradation of the nation's wetland resources were not being adequately assessed, increased interest in the use of wetlands as low-cost alternative waste water treatment solutions, and the realization that many Superfund sites were contiguous to or contained within wetland regions were all factors that contributed to the development of a wetlands research program by the USEPA. The Wetlands Research Plan, submitted in 1985 [79] and implemented in 1989 [80], identifies several research goals including: (1) development of water quality criteria for wetland function; (2) determination of the status of wetlands with respect to their community structure and quantification of changes, over time, in a subset of monitored wetlands; and (3) determination of the waste assimilation limits of wetlands. The USEPA has additional research in the area of constructed wetlands [81].

Plants Used in Other Bioremediative Technologies

Focusing first on the terrestrial environment, soil contamination may be broadly classified into two categories: (1) inorganic substances such as heavy metals; and (2) organic wastes, e.g., oils, solvents, pesticides, PCBs, etc. One approach to soil decontamination is to remove the toxic element. For example, metal-tolerant plants are known to exhibit at least two types of strategies when grown on contaminated soil, exclusion or detoxification and storage of the metals. Although metal hyperaccumulation has evolved in a wide range of apparently unrelated plant taxa, one unifying feature of the ecology of these plants is their restricted field distribution and high degree of endemism. Preliminary field experiments, in which soil contaminated with radioactive cobalt was continuously cropped with a Zairian plant, *Haumaniastrum katangese*, were found to have reduced the toxicity of the soil [65]. Biorecovery procedures of metal contaminants present one drawback, however, in that the metals are not metabolized but merely accumulated by the plants, and thus these technologies may be considered as only a part of a remediative plan. Such a plan might also include subsequent removal of the metal-laden plants and reintroduction of naturally occurring species. Disposition of the hyperaccumulator species could possibly include recovery of the metal for industrial reuse.

A more desirable outcome of bioremediation is the detoxification and degradation of wastes to nonhazardous substances. This goal is more easily achieved with organic wastes. Although the major focus of USEPA research has been to investigate microorganisms (including bacteria and fungi) to serve this purpose, plants may also be of value. An exhaustive literature search prepared as an internal document for the USEPA addressed the subject of accumulation of organic pollutants from soils and the potential use of plants as bioremediative tools [82]. Included in the results of this search was the formulation of a model demonstrating a linear correlation between the organic matter sorption coefficient of a soil (K_{oc}) and the octanol/water partition coefficient (K_{ov}) of the pollutant and showing the relationship between transpiration rate and root uptake in plants of specific pollutants. A thorough review of the interaction among the various compartments of the above-proposed model of the plant uptake of nonionic organic chemicals from soils has been performed [83].

Research by the USEPA has not focused on terrestrial plants as bioremediative tools, but rather on experiments that would assist modeling efforts in the area of the transport of organic chemicals into the food chain through terrestrial plant systems. In work to evaluate the effect, uptake, and disposition of nitrobenzene in terrestrial plants, the responses of three herbaceous crops (soybean, barley, and lettuce) and five species of woody perennials (recommended for the revegetation of abandoned waste sites) were tested [84,85]. Endpoints included: photosynthesis (CO_2 uptake); uptake of radiolabeled nitrobenzene; transpiration of labeled parent compound; and partitioning of metabolic breakdown products into polar, nonpolar, and insoluble fractions. Results showed that volatilization of nitrobenzene was a major route of chemical loss from the hydroponic reservoir used in the experiment; but, in general, passive uptake was proportional to the rate of water flux through the plant. In addition, effects on photosynthesis, partitioning within plant tissues, and metabolism or biotransformation of nitrobenzene varied with species. These results suggest that plants may play a significant role in governing the fate of organic chemicals at waste sites.

The current technologies for using algae as bioremediative tools are limited. Although some marine algae are effective bioaccumulators of heavy metals [64,65,86], their use as metal scavengers from polluted waters has been proposed but not practiced. Research to develop algal systems that can metabolize organic wastes would be of particular interest since aquatic environments are the major recipients of organic wastes [16].

Recommendations for Improved Use of Plants as Bioremediative Tools

Recommendations for the improvement of research in developing plants as bioremediative tools would include: (1) the development of a database that catalogs plants with known bioremediative properties; (2) financial support for the "Wetlands Research Plan"; and (3) support of research to evaluate the ability of plants and plant-mycorrhyzal associations to break down organic contaminants in soils.

Plants in Biotechnology

Genetically engineered organisms, including plants, have been developed in the past few years and are now at the stage of small-scale field testing. Transgenic or genetically engineered plants (GEPs) are plants in which genetic material has been added or deleted via recombinant DNA or other related techniques [87,88]. To date, the primary focus of work in the area of transgenic plants has been in the development of agricultural species with "improved" characteristics including, but not limited to, herbicide resistance, disease resistance, insect resistance, drought and salt tolerance, and biochemical composition modification (e.g., protein content, quality, or compositional alteration). Future applications may include development of plants that metabolize hazardous wastes, improve biomass production for use in nonpetroleum-fuel technologies, or produce a myriad of algal products (especially "boutique" polysaccharides such as agar and carrageenan) [89–94]. Uncertainties are presently large with respect to the long-term effects on man and the environment of widespread agricultural application of this new and exciting technology. It is highly unlikely that "the plant that took over the world" will be accidentally released, but the concept of "genetically-engineered superfund sites" is enough to make even the most cavalier observer cringe. With careful attention to the risks associated with the release of these organisms, a reasonable and responsible management strategy will be developed. A thoughtful presentation of some of the environmental issues surrounding the planned introduction of genetically engineered organisms has been published [95]. Particularly useful were a number of tables, listing attributes of organisms, that were designed to assist decision makers in risk evaluation. In addition, a number of recommendations were put forth to guide development of "a scientifically based regulatory policy that encourages innovation without compromising sound environmental management" [95].

Field Trials of GEPs

By the end of 1989, nearly 50 field trials of transgenic plants will have been initiated or completed. The approval for the majority of the field trials has been obtained through the regulatory arm of the United States Department of Agriculture, the Animal and Plant Health Inspection Service (USDA-APHIS) [94]. Environmental risk associated with the plants field-tested has been minimal. As to future field tests, the question of where to draw the line between environmental safety and risk remains [96]. Transgenic plants of 22 herbaceous dicots, 3 woody dicots, and 5 monocot species have been reported to be in some stage of development [97]. Among these plants, several taxa contain weedy relatives or genetic alterations or both that could broaden their growth range. The release of these types of plants into the environment are of particular concern [95]. The industry perspective asserts that the regulation of transgenic plants should be "based on scientific principles that (i) meet the general public's need for a safe and reasonably priced food supply and (ii) recognize the inherent low risk of gene transfer technology and the benefits afforded by genetically engineered crops to growers, food processors, and consumers" [97]. The emphasis, understandably, is on the safety of the food supply to humans and domestic animals; but the government must also protect the environment in which man and animals live. The development of a balanced program of regulation of, and research on, transgenic plants is essential to the development of a viable biotechnology industry in the United States. The benefits of promised effective and efficient agricultural and industrial applications of biotechnology must be weighed, however, with the risks associated with their application. With more certainty about what those risks are, better decisions will be made.

Recommendations on Research and Regulation of Genetically Engineered Plants

Several recommendations are presented below that outline the focus of research to support the regulatory needs in the area of GEPs. Coordination of research efforts between USEPA and USDA is essential to the success of this program. With respect to specific research objectives, the major focus is to address the questions of movement of genetic information in plants (genetically engineered traits) within the natural community. Using currently available techniques, surrogate marker genes can be traced from the test plot into natural and weedy populations and within populations of those engineered species with natural and weedy relatives (e.g., oats, clover, sorghum) to determine rates of transfer and document changes in community structure or other ecological effects about which information is poor but concern is great. An essential part of the research effort is to develop an ecological database to geographically pinpoint areas that are inhabited by potentially receptive non-target plants. This database would assist decision makers in evaluation of the appropriateness of proposed sitings of large- and small-scale field tests.

The rapid development of guidelines of the release of transgenic plants is essential to the development of a strong biotechnology industry in the United States. Draft guidelines for field testing of GEPs proposed by USDA [98] provide an excellent basis for regulatory development. The concern over environmental safety need not impede the timely testing of these plants. In order to respond to these needs, however, action must be taken soon, since industry expects to have products ready for large-scale testing in one year and production in two to five years (1990 to 1994). For this reason, a balanced and coordinated program of regulation and research needs to be developed immediately. Although field testing issues have not yet been raised for algal biotechnology, this industry is taking advantage of the newest molecular techniques to develop many industrial products not usually considered for regulation [90]. Releases of "superalgae" into the marine environment pose many of the same environmental risks as GEPs in the terrestrial environment.

To deal with the specific issues of the ecological effects of transgenic plants, it is necessary to have expertise in molecular plant biology, molecular ecology, weed genetics, and terrestrial plant population genetics. USEPA laboratories currently do not have sufficient expertise to appropriately address these research questions. For this reason it is recommended that the USEPA should sponsor, through grants and contracts with universities and private research groups, short-term research projects that begin to address field research issues with those already trained in these technologies. Concomitantly, the Agency should establish a long-term GEP research program to support the development of guidelines and to determine long-term ecological effects of GEPs at one of the research laboratories that has molecular biologists and related technical expertise. Assistance in planning and developing the program could be obtained from the Science Advisory Board, the Biotechnology

Science Advisory Committee, the National Academy of Science, and other advisory entities. The development of an ecological database to support both the regulatory and research sides of the environmental effects questions is of primary concern. Adequate funds should be allotted such that data acquisition and storage and development of the ecological database are considered integral parts of the research effort. Present efforts should concentrate on terrestrial plant issues and, as the algal biotechnology industry approaches field testing, expand to answer research and regulatory questions in this area. Recognizing the future increases in pressure to regulate a larger number of field trials, it is important to have adequate and expert staffing on the regulatory side. Cooperation between USEPA headquarters (OTS, OPP, ORD) and USDA (APHIS, ARS) is essential, but USEPA needs additional staffing to respond to these anticipated needs. As above, outside advisory panels may assist in identifying the needed expertise to improve the regulatory capability of the USEPA.

Conclusion

The recommendations that derived from this study were intended to respond to the need expressed by USEPA/ORD for improved methodologies for the use of plants in risk assessment and remediation. Many technologies that were formerly considered esoteric, experimental, or too expensive have been shown to have practical and economical uses in both laboratory and field applications. With respect to the development of a prioritized list of recommendations, suggestions have been outlined for improvements in research concerning each of the categories of plant applications Prioritization of the four most pressing needs for the improvement of plant science in the USEPA/ORD would place the development of research to support a regulatory responsibility in the area of plant biotechnology as the first goal. The second area of need is to develop photosynthesis (fluorescence) bioassays for algae and vascular plants. The third area of need is to develop regionally differentiated suites of bioassays to use as regulatory tools for aquatic and terrestrial systems. The fourth major area of need, and one that is overarching in all categories of plant applications, is the development of databases for determining "what's out there." It is hoped that this "status report" will aid in the development of a greater appreciation for the value of plants as risk assessment tools as well as an understanding of their role as a major focus of biotechnology.

Acknowledgments

The author would like to thank the many helpful persons who contributed to this work, including Bob Frederick and Joanne Sulac (EPA/ORD/OPER), Fred Betz (EPA/OPPTS), the staff scientists at the USEPA Laboratories, Morris Levin, the OPP work group, the OTS work group, Claudia Sturges (AAAS), and the other AAAS/EPA Fellows. This manuscript was developed from a draft report prepared during the author's tenure as a Fellow under the American Association for the Advancement of Science/U.S. Environmental Protection Agency Fellowship Program during the summer of 1989. The views expressed herein are entirely those of the author and do not represent official policy of the USEPA or AAAS. Mention of trade names or commercial products does not constitute endorsement or recommendation.

References

[*1*] Arnon, D., "Sunlight, Earth, Life, the Grand Design of Photosynthesis," *The Sciences*, October 1982, pp. 22–27.

[2] Walsh, G., Hansen, D., and Lawrence, D., "A Flow-through System for Exposure of Seagrass to Pollutants," *Marine Environmental Research*, Vol. 7, 1982, pp. 1–11.

[3] Miller, W., Peterson, S., Green, J., and Callahan, C., "Comparative Toxicology of Laboratory Organisms for Assessing Hazardous Waste Sites," *Journal of Environmental Quality*, Vol. 14, 1985, pp. 569–574.

[4] Benenati, F., "Plants—Keystone to Environmental Risk Assessment," seminar presented on 19 April 1989, U.S. Environmental Protection Agency, Office of Toxic Substances, Washington, DC, pp. 1–30.

[5] Eastern Research Group, Inc., "Workshop on the Use of Biological Mechanisms in Ecotoxicity Testing Decisions," prepared for U.S. Environmental Protection Agency, San Diego, CA, 10–12 March 1987.

[6] USEPA, "Short-Term Methods For Estimating Chronic Toxicity of Effluents and Receiving Waters to Marine and Estuarine Organisms," EPA 600/4-87/028, USEPA Environmental Research Laboratory, Cincinnati, OH, 1988.

[7] Hellawell, J., "Toxic Substances in Rivers and Streams," *Environmental Pollution*, Vol. 50, 1988, pp. 61–85.

[8] Cairns, J., "Regulating Hazardous Chemicals in Aquatic Environments," *Boston College Environmental Affairs Law Review*, Vol. 11, 1983, pp. 1–10.

[9] Walsh, G., "Principles of Toxicity Testing with Marine Unicellular Algae," *Environmental Toxicology and Chemistry*, Vol. 7, 1988, pp. 979–987.

[10] Walsh, G., "Cell Death and Inhibition of Population Growth of Marine Unicellular Algae by Pesticides," *Aquatic Toxicology*, Vol. 3, 1983, pp. 290–214.

[11] Walsh, G., McLaughlin, L., Lores, E., Louie, M., and Deans, C., "Effects of Organotins on Growth and Survival of Two Marine Diatoms, *Skeletonema costatum* and *Thalassiosira pseudonana*," *Chemosphere*, Vol. 14, 1985, pp. 383–392.

[12] Walsh, G., Yoder, M., McLaughlin, L., and Lores, E., "Responses of Marine Unicellular Algae to Brominated Organic Compounds in Six Growth Media," *Ecotoxicology and Environmental Safety*, Vol. 14, 1987, pp. 215–222.

[13] Walsh, G., "Methods for Toxicity Tests of Single Substances and Liquid Complex Wastes with Marine Unicellular Algae," EPA/600/8-87/043, U.S. Environmental Protection Agency, Washington, DC, 1988.

[14] Anderson, B. and Hunt, J., "Bioassay Methods for Evaluating the Toxicity of Heavy Metals, Biocides and Sewage Effluent Using Microscopic Stages of Giant Kelp," *Marine Environmental Research*, Vol. 26, 1988, pp. 113–134.

[15] Smith, B. and Harrison, F., "Sensitivity of *Macrocystis* Gametophytes to Copper," UCRL-52481 or NUREG/CR-0684, Lawrence Livermore National Laboratory, Livermore CA, 1978.

[16] U.S. Congress, Office of Technology Assessment, "Wastes in Marine Environments," OTA-O-344, U.S. Government Printing Office, Washington, DC, April 1987.

[17] Hedtke, S., West, C., Allen, K., Norberg-King, T., and Mount, D., "Toxicity of Pentachlorophenol to Aquatic Organisms Under Naturally Varying and Controlled Environmental Conditions," *Environmental Toxicology and Chemistry*, Vol. 5, 1986, pp. 531–542.

[18] Plewa, M., "Plant Genetic Assays to Evaluate Complex Environmental Mixtures," Fourth Symposium on the Application of Short-term Bioassays in the Analysis of Complex Environmental Mixtures, *Environmental Science Research*, Vol. 32, 1986, pp. 45–64.

[19] Sandu, S. and Acedo, G., "Assessment of Genotoxicity of Hazardous Wastes with the *Arabidopsis* Embryo Assay," EPA-600/D-87/272, U.S. Environmental Protection Agency, Washington, DC, 1987, pp. 1–21.

[20] Taub, F., "Standardized Aquatic Microcosm—Development and Testing," in *Aquatic Ecotoxicology: Fundamental Concepts and Methodologies, Vol. II,* CRC Press, Florida, 1989, pp. 47–91.

[21] Yount, J. and Richter, J., "Effects of Pentachlorophenol on Periphyton Communities in Outdoor Experimental Streams," *Archives Environmental Contamination Toxicology*, Vol. 15, 1986, pp. 51–60.

[22] Morton, R., Duke, T., Macauley, J., Clark, J., Price, W., Hendricks, S., Owsley-Montgomery, S., and Plaia, G., "Impact of Drilling Fluids on Seagrasses: an Experimental Community Approach Community Toxicity Testing," in *ASTM STP 920,* American Society for Testing and Materials, Philadelphia, 1986, pp. 199–212.

[23] Price, A., Macauley, J., and Clark, J., "Effects of Drilling Fluids on *Thalassia testudinum* and its Epiphytic Algae," *Environmental and Experimental Botany*, Vol. 26, 1986, pp. 321–330.

[24] Taub, F., Kindig, A., and Conquest, F, "Interlaboratory Testing of a Standardized Aquatic Microcosm," in *Aquatic Toxicology and Hazard Assessment: 10th Volume, ASTM STP 971,* American Society for Testing and Materials, Philadelphia, 1988, pp. 384–405.

[25] Taub, F., Kindig, A., Conquest, F. and Meador, J., "Results of Interlaboratory Testing of the Standardized Aquatic Microcosm Protocol," in *Aquatic Toxicology and Environmental Fate: Eleventh Volume, ASTM STP 1007*, American Society for Testing and Materials, Philadelphia, 1989, pp. 368–394.

[26] Surano, K., Daley, P., Houpis, J., Shinn, J., Helms, J., Palassou, J., and Costella, M., "Growth and Physiological Responses of *Pinus ponderosa* Dougl ex P. Laws. to Long-term Elevated CO_2 Concentrations," *Tree Physiology*, Vol. 2, 1987, pp. 243–259.

[27] Houpis, J., Surano, K., Cowles, S., and Shinn, J., "Chlorophyll and Carotenoid Concentrations in Two Varieties of *Pinus ponderosa* Seedlings Subjected to Long-term Elevated Carbon Dioxide," *Tree Physiology*, Vol. 4, 1988, pp. 187–193.

[28] McFarlane, C., Pfleeger, T., and Fletcher, J., "Transpiration Effect on the Uptake and Distribution of Bromacil, Nitrobenzene, and Phenol in Soybean Plants," *Journal of Environmental Quality*, Vol. 16, 1987, pp. 372–376.

[29] Rygiewicz, P. and Castellano, M., "Effects of Atrazine, Hexazinone and Picloram on Mycorrhizal Douglas-fir and Ponderosa Pine Seedlings. I. Mycorrhizal Formation. *In Vitro* Growth, and Seedling Germination," submitted to *Canadian Journal of Forestry Research*, 1989.

[30] Walton, B. and Anderson, T., "Microbial Degradation in the Rhizosphere: Potential Application to Biological Remediation," *Environmental Applied Microbiology*, Vol. 56, 1990, pp. 1012–1016.

[31] Kaplan, A. and Bjorkman, O., "Ratio of CO_2 (Carbon Dioxide) Uptake to O_2 (Oxygen) Evolution During Photosynthesis in Higher Plants (*Encelia californica, Atriplex rosea*)," *Zeitschrift Pflanenphysiologisches*, Vol. 96, 1980, pp. 185–188.

[32] Powles, S. and Bjorkman, O., "High Light and Water Stress Effects on Photosynthesis in *Nerium Oliander*," Carnegie Institute of Washington Year Book, 1982, pp. 74–77.

[33] Nobel, P. and Hartsock, D., "Physiological Responses of *Opuntia ficus-india* to Growth Temperature (Nocturnal Carbon Dioxide Uptake)," *Physiology Plantarum*, Vol. 60, 1984, pp. 98–105.

[34] Nobel, P. and Long, S., "Canopy Structure and Light Interception," in *Techniques in Bioproductivity and Photosynthesis*, 1985, pp. 41–49.

[35] Geller, G. and Nobel, P., "Comparative Cactus Architecture and PAR Interception," *American Journal of Botany*, Vol. 74, 1987, pp. 998–1005.

[36] Butler, W. and Katijima, T., "A Tripartite Model for Chloroplast Fluorescence," in *Proceedings, Third International Congress on Photosynthesis*, Elsevier Press, Amsterdam, 1975, pp. 13–24.

[37] Whitledge, T. and Wirick, C., 1986. "Development of a Moored In Situ Fluorometer for Phytoplankton Studies," in *Lecture Notes on Coastal and Estuarine Studies, Vol. 17: Tidal Mixing and Plankton Dynamics*, Springer-Verlag, Berlin Heidelberg, 1986, pp. 449–462.

[38] Cowles, T., Moum, J., Desidero, R. and Angel, M., "*In Situ* Monitoring of Ocean Chlorophyll Via Laser-induced Fluorescence Backscattering Through an Optical Fiber," *Applied Optics*, Vol. 28, 1989, pp. 595–600.

[39] Bristow, M., Bundy, D., Edmonds, C., Ponto, P., Frey, B. and Small, L., "Airborne Laser Fluorosensor Survey of the Columbia and Snake Rivers: Simultaneous Measurements of Chlorophyll, Dissolved Organics and Optical Attenuation," *International Journal of Remote Sensing*, Vol. 6, 1985, pp. 1707–1734.

[40] Bristow, M. and Zimmerman, R., "Remote Water Quality Monitoring With an Airborne Laser Fluorosensor," *Proceedings, 7th International Conference on Chemistry for the Protection of the Environment*, Lublin, Poland, 4–7 Sept. 1989.

[41] Melis, A., Manodori, A., Glick, R., Ghirardi, M., McCauley, S., and Neale, P., "The Mechanism of Photosynthetic Membrane Adaptation to Environmental Stress Conditions: a Hypothesis," *Physiology Vegetale*, Vol. 23, 1985, pp. 757–765.

[42] Anderson, J., "Photoregulation of the Composition, Function, and Structure of Thylakoid Membranes," *Annual Review of Plant Physiology*, Vol. 37, 1986, pp. 93–136.

[43] Lichtenthaler, H. and Rinderele, U., "The Role of Chlorophyll Fluorescence in the Detection of Stress Conditions in Plants," *CRC Critical Reviews in Analytical Chemistry*, Vol. 19, 1988, pp. 29–85.

[44] Maas, F., van Loo, E. and van Hasselt, P., "Effect of Long-term H_2S Fumigation on Photosynthesis in Spinach. Correlation Between CO_2 Fixation and Chlorophyll *a* Fluorescence," *Physiology Plantarum*, Vol. 72, 1988, pp. 77–83.

[45] Voss, M., Renger, G., Kotter, C. and Graber, P., "Fluorometric Detection of Photosystem II Herbicide Penetration and Detoxification in Whole Leaves," *Weed Science*, Vol. 32, 1984, pp. 675–680.

[46] Gressel, J., "The Molecular Anatomy of Resistance to Photosystem II Herbicides," *Oxford Survey of Plant & Cell Biology*, Vol. 2, 1985, pp. 321–328.

[47] Stowe, A. and DiTomasso, J., "Evidence for Increased Herbicide Detoxification in Triazine-

tolerant Velvetleaf," *Proceedings of the Annual Meeting of the Northeastern Weed Science Society,* Vol. 43, 1989, pp. 31–35.

[48] Solymosi, P., Lehoczki, E. and Laskay, G., "Difference in Herbicide Resistance to Various Taxonomic Populations of Common Lambsquarters (*Chenopodium album*) and Late-flowering Goosefoot (*Chenopodium strictum*) in Hungary," *Weed Science,* Vol. 34, 1986, pp. 175–180.

[49] Neale, P. and Richerson, P., "Photoinhibition and the Diurnal Variation of Phytoplankton Photosynthesis. I. Development of a Photosynthesis-Irradiance Model from Studies of *In Situ* Responses," *Journal of Plankton Research,* Vol. 9, 1987, pp. 167–193.

[50] Cleland, R., Melis, A., and Neale, P., "Mechanism of Photoinhibition: Photochemical Reaction Center Inactivation in System II of Chloroplasts," *Photosynthesis Research,* Vol. 9, 1986, pp. 79–88.

[51] Cullen, J., Zhu, M., and Pierson, D., "A Technique to Assess the Harmful Effects of Sampling and Containment from Determination of Primary Production," *Limnology and Oceanography,* Vol. 31, 1986, pp. 1364–1373.

[52] Samson, G., Morissette, J., and Popovic, R., "Copper Quenching of the Variable Fluorescence in *Dunaliella tertiolecta.* Evidence for a Copper Inhibition Effect on the Oxidizing Side of PSII," *Photochemistry and Photobiology,* Vol. 48, 1988, pp. 329–332.

[53] Smith, B. and Melis, A., "Photosystem Stoichiometry and Excitation Distribution in Chloroplasts From Surface and Minus 20 Meter Blades of *Macrocystis pyrifera,* the Giant Kelp," *Plant Physiology,* Vol. 84, 1987, pp. 1325–1330.

[54] Smith, B. and Melis, A., "Photochemical Apparatus Organization in the Diatom *Cylindrotheca fusiformis:* Photosystem Stoichiometry and Excitation Distribution in Cells Grown Under High and Low Irradiance," *Plant and Cell Physiology,* Vol. 29, 1988, pp. 761–769.

[55] Smith, B., Morrissey, P., Guenther, J., Nemson, J., Harrison, M., Allen, M., and Melis, A., "Photosynthetic Apparatus Response to Irradiance Stress," *Plant Physiology,* Vol. 93, 1990, pp. 1433–1440.

[56] Walsh, G., McLaughlin, L., Yoder, M., Moody, P., Lores, E., Forester, J., and Wessinger-Duvall, P., "*Minutocellus polymorphus:* a New Marine Diatom for Use in Algal Toxicity Tests," *Environmental and Toxicological Chemistry,* Vol. 7, 1988, pp. 925–929.

[57] Eastern Research Group, Inc., "Biomonitoring for Control of Toxicity in Effluent Discharges to the Marine Environment," prepared for Center for Environmental Research Information,USEPA, Cincinnati, OH and ERL-Narragansett, USEPA, Narragansett, RI, 1988, pp. 1–143.

[58] Thomas, J. Skalski, J., Cline, J., McShane, M., Simpson, J., Miller, W., Peterson, S., Callahan, C., and Green, J., "Characterization of Chemical Waste Site Contamination and Determination of Its Extent Using Bioassays," *Environmental Toxicology and Chemistry,* Vol. 5, 1986, pp. 487–501.

[59] Athey, L., Thomas, J., Skalski, J., and Miller, W. "Role of Acute Toxicity Bioassays in the Remedial Action as Hazardous Waste Sites," *User's Guide,* Battele PNWL, Richland, WA and USEPA, ERL-Corvallis, Corvallis, OR, 1987, pp. 1–119.

[60] Omernik, J., "Ecoregions of the Coterminous United States," *Annual Association of American Geographers,* Vol. 77, 1987, pp. 118–125.

[61] Whittier, T., Hughes, R., and Larsen, D., "Correspondence Between Ecoregions and Spatial Patterns in Stream Ecosystems in Oregon," *Canadian Journal of Fisheries and Aquatic Sciences,* Vol. 45, 1988, pp. 1264–1278.

[62] Miller, D., Leonard, P., Hughes, R., Karr, J., Moyle, P., Schrader, L., Thompson, B., Daniels, R., Fausch, K., Fitzhugh, G., Gammon, J., Halliwell, D., Angermeier, P., and Orth, D., "Regional Applications of an Index of Biotic Integrity for Use in Water Resource Management," *Fisheries,* Vol. 13, 1988, pp. 12–20.

[63] USEPA, "Report of the Research Strategies Committee of the Science Advisory Board, Appendix C: Strategies for Ecological Effects Research," SAB-EC-88-040C, 1988, pp. 1–35.

[64] Bryan, G. and Hummerstone, L., "Indicators of Heavy-metal Contamination in the Looe Estuary (Cornwall) With Particular Regard to Silver and Lead," *Journal of the Marine Biological Association of the U.K.,* Vol. 57, 1973, pp. 75–92.

[65] Baker, A. and Brooks, R., "Terrestrial Higher Plants Which Hyperaccumulate Metallic Elements—a Review of Their Distribution, Ecology and Phytochemistry," *Biorecovery,* Vol. 1, 1989, pp. 81–126.

[66] Foster, P., "Concentrations and Concentration Factors of Heavy Metals in Brown Algae," *Environmental Pollution,* Vol. 10, 1976, pp. 45–53.

[67] Thornton, K., Payne, F., Ford, J., and Landers, D., "The Concept of Time: Temporally Integrated Monitoring of Ecosystems," Corvallis Environmental Research Laboratory, Corvallis, OR, 1987.

[68] Payne, F. and Ford, J., "The Concept of TIME (Temporally Integrated Monitoring of Ecosys-

tems)," supplement prepared for U.S. Environmental Protection Agency, Corvallis, OR, 1988, pp. 1–70.

[69] Dixit, S., Dixit, A., and Evans, R., "Scaled Chrysophytes (Chrysophyceae) as Indicators of pH in Sudbury, Ontario, Lakes," *Canadian Journal of Fisheries and Aquatic Sciences*, Vol. 45, 1988, pp. 1411–1421.

[70] Royce, C., Fletcher, J., Risser, P., McFarlane, J., and Benenati, F., "PHYTOTOX: A Database Dealing With the Effects of Organic Chemicals on Terrestrial Vascular Plants," *Journal of Chemical Information and Computer Science*, Vol. 24, 1984, pp. 7–10.

[71] Fletcher, J., Johnson, F., and McFarlane, J., "Database Assessment of Phytotoxicity Data Published on Terrestrial Vascular Plants," *Environmental Toxicology and Chemistry*, Vol. 7, 1988, pp. 615–622.

[72] Larsen, D., Omernik, J., Hughes, R., Rohm, C., Whittier, T., Kinney, A., Gallant, A., and Dudley, D., "Correspondence Between Spatial Patterns in Fish Assemblages in Ohio Streams and Aquatic Ecoregions," *Environmental Management*, Vol. 10, 1986, pp. 815–828.

[73] Davis, W., Luey, J., Simon, T., Maxted, J., Plafkin, J., Hughes, R., and Whittier, T., "Draft Report of the National Biocriteria Workshop, Lincolnwood, Illinois, December 2–4, 1987," draft report prepared for U.S. Environmental Protection Agency, Corvallis, OR, 1987, pp. 1–31.

[74] Page, A., Logan, T., and Ryan, J., Eds., *Land Application of Sludge*, Lewis Publishers, Inc., Chelsea, MI, 1987, pp. 1–168.

[75] Vriens, L., Nihoul, R., and Verachtert, H., "Activated Sludges as Animal Feed: a Review," *Biological Wastes*, Vol. 27, 1989, pp. 161–207.

[76] Jordan, W., Peters, R., and Allen, E., "Ecological Restoration as a Strategy for Conserving Biological Diversity," *Environmental Management*, Vol. 12, 1988, pp. 55–72.

[77] U.S. Congress, Office of Technology Assessment, "Superfund Strategy," OTA-ITE-252, U.S. Government Printing Office, Washington, DC, April 1985, pp. 1–282.

[78] Richards, D., "Remedy Selection at Superfund Sites: an Analysis of Bioremediation," *Reports*, American Association for the Advancement of Science/Environmental Science and Engineering Fellows Program, AAAS Publication No. 90-055, 1989, pp. 48–50.

[79] Zedler, J., Kentula, M., and Kibby, H., "Wetlands Research Plan, November, 1985," submitted to Environmental Research Laboratory, Office of Research and Development, U.S. Environmental Protection Agency, Washington, DC, 1985.

[80] Adamus, P. and Preston, E., "Wetlands and Water Quality: EPA's Research and Monitoring Implementation Plan for the Years 1989–1994," Environmental Research Laboratory, Office of Research and Development, U.S. Environmental Protection Agency, Corvallis, OR, 1989.

[81] Forde, L., "A Performance Evaluation of Natural Systems for Wastewater Treatment," *Reports*, 1989, American Association for the Advancement of Science/Environmental Science and Engineering Fellows Program, AAAS Publication No. 90-055, 1989, pp. 14–15.

[82] Sferra, P., "Higher Plant Accumulation of Organic Pollutants From Soils," EPA/University of Liverpool Cooperative Research Programme, draft report, P. Sferra, Project Officer, U.S. Environmental Protection Agency, Hazardous Waste Engineering Research Laboratory, Cincinnati, OH, Hazardous Waste Engineering Research Laboratory, Office of Research and Development, U.S. Environmental Protection Agency, Cincinnati, OH. Environmental Advisory Unit, University of Liverpool, Merseyside Innovation Centre, 131 Mount Pleasant, Liverpool, UK, 1988.

[83] Ryan, J., Bell, R., Davidson, J., and O'Connor, G., "Plant Uptake of Non-ionic Organic Chemicals From Soils," *Chemosphere*, Vol. 17, 1988, pp. 2299–2323.

[84] McFarlane, C. and Pfleeger, T., "Plant Exposure Chambers for Study of Toxic Chemical-Plant Interactions," *Journal of Environmental Quality*, Vol. 16, 1987, pp. 361–371.

[85] McFarlane, C., Nolt, C., Wickliff, C., Pfleeger, T., Shimabuku, R., and McDowell, M., "The Uptake, Distribution and Metabolism of Four Organic Chemicals by Soybean Plants and Barley Roots," *Environmental Toxicology and Chemistry*, Vol. 6, 1987, pp. 847–856.

[86] Forsberg, A., Soderland, S., Frank, A., Petersson, L., and Pedersen, M., "Studies on Metal Content in the Brown Seaweed, *Fucus vesiculosus* from Archipelago of Stockholm," *Environmental Pollution*, Vol. 49, 1988, pp. 245–263.

[87] Weising, K., Schell, J. and Kahl, G., "Foreign Genes in Plants: Transfer, Structure, Expression, and Applications," *Annual Review of Genetics*, Vol. 22, 1988, pp. 421–477.

[88] Benfey, P. and Chua, N., "Regulated Genes in Transgenic Plants," *Science*, Vol. 244, 1989, pp. 174–181.

[89] Beneman, J. and Weismann, J., "Chemicals From Microalgae," *Bioconversion Systems*, CRC Press, pp. 59–70.

[90] De La Noue, J. and de Pauw, N., "The Potential of Microalgal Biotechnology: a Review of Production and Uses of Microalgae," *Biotechnology Advances*, Vol. 6, 1988, pp. 725–770.

[*91*] Grohmann, K. and Villet, R., "Fuels and Chemicals from Biomass: a Role for Gene-splicing Technology," *Bioconversion Systems,* CRC Press, Boca Raton, FL, 1984, pp. 1–16.

[*92*] Ryther, J., "Technology for the Commercial Production of Macroalgae," in *Energy Applications of Biomass,* Elsevier, London, 1985, pp. 177–188.

[*93*] McIntosh, R., "Intensive Microalgae Culture for Production of Lipids for Fuel," in *Energy Applications of Biomass,* Elsevier, London, 1985, pp. 169–175.

[*94*] U.S. Congress, Office of Technology Assessment, "New Developments in Biotechnology—Field-Testing Engineered Organisms: Genetic and Ecological Issues," OTA-BA-350, U.S. Government Printing Office, Washington, DC, May 1988.

[*95*] Tiedje, J., Colwell, R., Grossman, Y., Hodson, R., Lenski, R., Mack, R., and Regal, P., "The Planned Introduction of Genetically Engineered Organisms: Ecological Considerations and Recommendations," *Ecology,* Vol. 70, 1989, pp. 298–315.

[*96*] Smith, B., "Should EPA Have a Role in Regulating Transgenic Plants?," in preparation.

[*97*] Gasser, C. and Fraley, R., "Genetically Engineering Plants for Crop Improvement," *Science,* Vol. 224, 1989, pp. 1293–1299.

[*98*] USDA, "Draft Guidelines for Research with Genetically Modified Organisms Outside Contained Facilities," Office of Agricultural Biotechnology, Room 321-A, Administration Building, 14th and Independence Avenue, S.W. Washington, DC, 1 May 1989, pp. 1–24.

Merrilee Ritter[1]

An Overview of Experimental Design

REFERENCE: Ritter, M., **"An Overview of Experimental Design,"** *Plants for Toxicity Assessment: Second Volume, ASTM STP 1115,* J. W. Gorsuch, W. R. Lower, W. Wang, and M. A. Lewis, Eds., American Society for Testing and Materials, Philadelphia, 1991, pp. 60–67.

ABSTRACT: In experiments that examine the effects of chemicals on plant behavior, good design can overcome the difficulties imposed by limited resources. An appropriate design is one that obtains information about the interactions between and among factors, as well as the curvature, if any, in factor behavior. This paper outlines the chief concepts inherent in effective experimental design: purpose, objective, hypothesis, statistical power, treatment level determination, response variables, design structure, and replication.

KEY WORDS: experimental design, plants, chemical effects, statistics, factorial design, blocking, randomization

A well-designed experiment enables the investigator to draw conclusions about the effects of one or more variables on other variables. Good design can overcome the difficulties imposed by limited resources. Given the resources available, an appropriately designed experiment takes these limitations into account, yielding more information than the investigator could obtain from an experiment that examines the effect of each variable in turn or from a study of historical data alone. Two examples of the kinds of information obtainable with effective experimental design are: (1) interactions between and among the factors; and (2) curvature in factor behavior.

This paper outlines the chief concepts inherent in good experimental design, with particular emphasis on studies of chemical effects on plant behavior.

Concepts

Purpose

An investigator states the purpose of an experiment in general terms that describe what the experiment is intended to discover. Some examples of experimental purposes are:

1. Screening of chemicals for their effects on plant behavior.
2. Effects on plant behavior caused by single or multiple chemical exposures over time.
3. Effects of chemical exposure on plant behavior within a range of exposure levels.
4. Detecting the effects of multiple chemical interactions on plant behavior.

Objective

An investigator states the objective of an experiment in precise, specific terms that define what the experiment is expected to accomplish. As the basis of the experiment, the objective

---

[1] Biostatistician, Eastman Kodak Co., Rochester, NY 14652-3615.

challenges the investigator to design the treatment structures and select the methodology so that the experimental data answer the questions posed by the objective. An example is:

Objective: To determine the toxicity of Chemical X on plant development when given once in aqueous solution by top watering to seedlings of corn, lettuce, and soybeans, with an observation period of 21 days.

Hypothesis [1]

The null hypothesis (H_0) is an assumption of interest made about the population under consideration. The alternative hypothesis, H_1, reflects the actual state of nature when the null hypothesis is not true. Some examples are:

1. H_0: treatment mean 1 = treatment mean 2   means are equal
   H_1: treatment mean 1 $\neq$ treatment mean 2   means are not equal
                                    or
   treatment mean 1 > treatment mean 2   mean 1 greater than mean 2
                                    or
   treatment mean 1 < treatment mean 2   mean 1 less than mean 2

	HO TRUE	HO FALSE
ACCEPT HO	CORRECT DECISION	BETA
REJECT HO	ALPHA	CORRECT DECISION

HO: NULL HYPOTHESIS:

ON THE AVERAGE, THE TREATMENT MEAN IS NOT DIFFERENT FROM THE CONTROL MEAN.

H1: ALTERNATIVE HYPOTHESIS:

ON THE AVERAGE, THE TREATMENT MEAN IS DIFFERENT FROM THE CONTROL MEAN.

ALPHA RISK -- THE CHANCE OF FINDING A TREATMENT EFFECT WHERE NONE EXISTS -- FALSE POSITIVE.

BETA RISK -- THE CHANCE OF NOT FINDING A TREATMENT WHERE ONE EXISTS -- FALSE NEGATIVE.

FIG. 1—*Risk levels.*

2. H_0: mean 1 = mean 2 = mean 3 = mean 4   all means are equal

   H_1: mean 1 < mean 2 < mean 3 < mean 4   dose response, trend

Statistical Power [2]

Statistical power (Fig. 1) is the ability of an experiment to detect statistically significant results of a chemical effect on plants, which are of interest to the investigator. The factors that control the statistical power of the experiment are the alpha risk, the sample size, the critical difference, the noise in the response variable, and the statistical tests used to analyze the data.

Alpha and beta designate the risks an investigator assumes for drawing false positive or false negative conclusions, respectively. For example, the investigator may decide there is a chemical effect, when actually there is not (false positive), or the investigator may decide that there is no chemical effect when actually there is (false negative).

Statistical power is calculated as 1-beta. Before executing the study, the investigator should determine the calculated power of the experiment; the magnitude chosen for each risk level depends on the investigator's willingness to assume that much risk of making the corresponding error.

The critical difference is the minimum difference in the response variable, which is judged by the investigator to be a change of interest or importance. An example is a 10% reduction in the growth of treated plants when compared to the growth of untreated (control) plants.

When the investigator has established the alpha risk, the critical difference, a measure of noise in the response variable, and a desired sample size, the investigator can determine the calculated power of the experiment. If the calculated power is insufficient (i.e., a level the investigator is uncomfortable with), then compromises must be made. The alpha risk, the sample size, and the critical difference can be juggled until an agreeable power is obtained within available resources. If it's impossible to establish a desirable level of power given the available resources, then there is no useful purpose in conducting the experiment.

ADEQUATE DESIGN

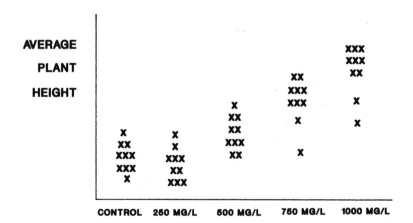

FIG. 2—*Single factor design.*

PLANT RESPONSES ARE MEASURED REPEATEDLY THROUGHOUT THE PERIOD OF EXPOSURE AND/OR OBSERVATION.

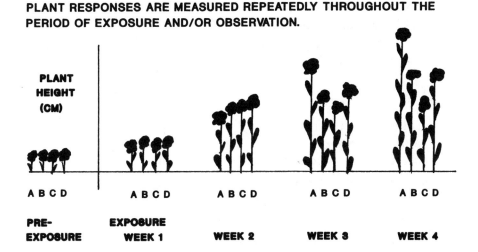

A • LOW DOSE OF WEED CONTROL CHEMICAL
B • MEDIUM DOSE OF WEED CONTROL CHEMICAL
C • HIGH DOSE OF WEED CONTROL CHEMICAL
D • CONTROL

FIG. 3—*Repeated measures design.*

When the experiment is complete, the investigator should be sure that the statistical analyses meet their assumptions in order to bring the actual power of the experiment as close to the calculated power as possible.

Actual Designs [1,3–5]

Below are some examples of designs (Figs. 2–4). In each case, in order to meet the objectives of the experiment, the investigator determines the number of factors, the length of time needed to study the effects of the factors on the plant behaviors, and the number and frequency of observations required.

1. To determine the effect of Chemical X on plant growth after a single application:

One factor, Chemical X, has one treatment level (or more, evenly spaced), plus control. The investigator chooses one time interval and makes observations at the end of the interval.

2. To determine the dose response of plant growth to chemical X after a single application:

One factor, Chemical X, has a minimum of three treatment levels plus control, with the treatment levels spaced evenly or in a geometric series, to cover the desired exposure range. Usually, the investigator administers one application of the chemical, then makes observations on the plants at the end of a specific time interval.

3. To determine the effects of Chemical X on plants over time, with repeated exposures:

One factor, Chemical X, has one or more treatment levels, usually evenly spaced, with exposures at specified time intervals. The investigator observes the plants for a specific length of time, repeatedly measuring their behavioral responses.

- **FACTORIAL DESIGNS, INVOLVING TWO OR MORE
  VARIABLES, ARE USED TO DETECT INTERACTIONS
  AMONG VARIABLES AND THEIR EFFECTS ON
  THE RESPONSE.**

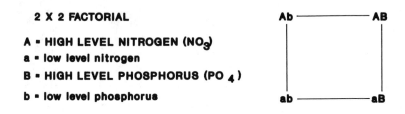

2 X 2 FACTORIAL

A = HIGH LEVEL NITROGEN (NO$_3$)
a = low level nitrogen
B = HIGH LEVEL PHOSPHORUS (PO$_4$)
b = low level phosphorus

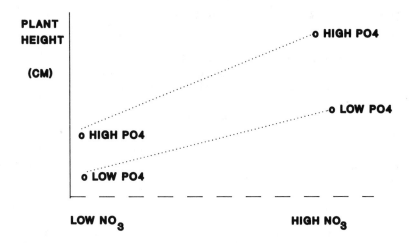

FIG. 4—*Variable interactions.*

4. To determine the effects of Chemicals A and B on plant behavior:

In this case, the factors have at least two treatment levels each; the experiment may be a factorial or a fractional factorial design.

The factorial design tests all possible combinations of the factors' treatment levels, i.e., three levels of A times two levels of B equals six treatment level combinations. These combinations define the factorial space or the part of the total environment under study.

The fractional factorial design uses a fraction of the possible treatment combinations to cover the factorial space. Confounding of factor effects occurs when the effects cannot be separated from each other. In a fractional factorial design, some confounding of effects is inevitable because the experiment does not test all the treatment combinations. The investigator selects the combinations so as to avoid confounding the most important effects and interactions and to be able to estimate them using the statistical model.

Treatment Level Determination

In plant studies, if the effects are expected to be linear in nature, then two treatment levels are sufficient to collect the needed information. In order to detect curvature in the response data, if any, at least three levels of the treatment plus a vehicle control are needed. The investigator can add more levels as resources permit. Government regulations, available resources, or end use of a chemical product can determine the range of exposure to be studied; treatment levels, evenly spaced if possible, should cover this range. If a suitable arithmetic series is not available, the investigator may use a geometric one. The nature of the phenomena under study may also dictate a geometric series of dose levels in order to cover all likelihoods of exposure. Plant growth, when the plants are being exposed to chemicals, is an example of a phenomenon usually tested with a geometric series of exposures. The number of control groups is determined by the treatment entity and its ease of application to the plants. For example, a solvent control may be needed in addition to the vehicle control, and a positively reacting control may be required to verify the correctness of the experimental system.

Response Variables (Endpoints)

The investigator interprets changes in the response variable data to determine whether the chemical had an effect on the plants. For plant studies, some endpoints are: number of

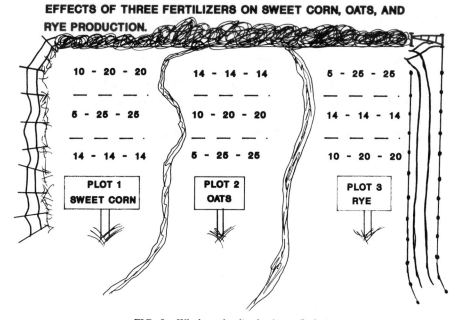

FIG. 5—*Whole and split plot (nested) designs.*

seeds germinated, time to emergence of seeds, plant height, root length, dry weight of roots and shoots, and clinical observations of the condition of the plants.

Design Structure [1,3,5]

The physical setup of an experiment is called the design structure (Fig. 5). It attempts to detect a particular factor effect by separating it from the variation or noise present in the environment. Randomization and blocking are two of the techniques of design structure available to the investigator.

Randomization is the allocation of plants to the experiment such that each one has equal opportunity to experience the noise of the experimental environment. Blocking cordons off the experimental environment so that the noise within each block is as small as possible, giving the chemical effect a better chance of being detected. If blocking is part of the experimental design, the investigator sets up the blocks first, then randomizes the treatments within each block.

An example is several agricultural fields divided into more uniform subfields. The treatments are then randomized on each subfield independently. Each whole field is confounded with the levels of one factor, but the other factors are not confounded inside the subfields.

Replication

Replication (Fig. 6) is an "exact" repeat of the experimental conditions on new plants at a particular sampling time. Replication allows the investigator to measure variation caused by the individuality of the plants.

Duplication is additional sampling on the same plant at a particular sampling time. It measures the variation of the sampling technique.

Depending on the intent of the investigator, many replicates may be needed to achieve the experimental objective or duplication may satisfy the objective.

REPLICATION -- AN APPLICATION OF THE TREATMENT TO AN INDEPENDENT PLANT IN THE EXPERIMENT, WITH SAMPLING DONE ONLY ONCE ON THE PLANT AT ANY PARTICULAR SAMPLING TIME.

DUPLICATION -- AN APPLICATION OF THE TREATMENT TO AN INDEPENDENT PLANT IN THE EXPERIMENT, WITH SAMPLING DONE REPEATEDLY ON THE PLANT AT ANY PARTICULAR SAMPLING TIME.

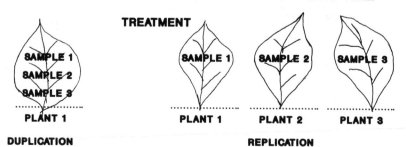

FIG. 6—*Replication versus duplication.*

Conclusion

An experiment designed in accordance with the concepts outlined above allows an investigator to separate factor effects readily from the noise in the experimental data and to make judgments about the importance of the effects. Based on the sound evidence of a well-designed experiment, the investigator can proceed to conduct further experimental work with confidence in the choice of direction of the research.

References

[1] Winer, B. J., *Statistical Principles in Experimental Design,* 2nd ed. McGraw-Hill Book Co., New York, NY, 1971.

[2] Lipsey, M. W., *Design Sensitivity,* Sage Publications, Newbury Park, CA, 1990.

[3] Box, G. E. P., Hunter, W. G., and Hunter, J. S., *Statistics for Experimenters,* John Wiley and Sons, Inc., New York, NY, 1978.

[4] Hicks, C. R., *Fundamental Concepts in the Design of Experiments,* 2nd ed., Holt, Rinehart and Winston, New York, NY, 1973.

[5] Milliken, G. A. and Johnson, D. E., *Analysis of Messy Data, Volume I: Designed Experiments,* Van Nostrand Reinhold Co., New York, NY, 1984.

[6] Gomez, K. A. and Gomez, A. A., *Statistical Procedures for Agricultural Research,* 2nd ed., John Wiley and Sons, Inc., New York, NY, 1984.

Wuncheng Wang[1]

Higher Plants (Common Duckweed, Lettuce, and Rice) for Effluent Toxicity Assessment

REFERENCE: Wang, W., "**Higher Plants (Common Duckweed, Lettuce, and Rice) for Effluent Toxicity Assessment,**" *Plants for Toxicity Assessment: Second Volume, ASTM STP 1115,* J. W. Gorsuch, W. R. Lower, W. Wang, and M. A. Lewis, Eds., American Society for Testing and Materials, Philadelphia, 1991, pp. 68–76.

ABSTRACT: The objective of this study was to evaluate effluent toxicity using phytotoxicity tests. Samples including receiving water, treated effluent, and pretreated industrial wastewaters were collected from three Illinoisian sanitary districts. These samples were tested as is, using duckweed, lettuce, and rice. The treated effluents from the three sources were not toxic in all tests. A pretreated industrial sample from a metal processing plant showed 98% inhibitory effect on duckweed. The same sample was not toxic to lettuce and rice germination. A sample from a transportation industry was toxic using duckweed, lettuce, and rice tests, having inhibitory effects of 45, 13, and 39%, in that order. A sample from a dairy plant was extremely toxic. The sample as received caused inhibitory effects of 100, 87, and 100%, in the same order of three species. The IC_{50} value and 95% confidence limit for this sample was 50% and 47 to 53% effluent concentration, respectively, using the duckweed toxicity test method.

KEY WORDS: effluent toxicity, phytotoxicity, toxicity assessment, duckweed, toxicity test, industrial effluent, municipal effluent, rice, lettuce

In general, plant toxicity tests can be described as simple, cost-effective, and having low or no maintenance requirements. They are uniquely suitable for testing herbicides or herbicide-like compounds. In addition, the tests can be used for general toxicity assessment. Wang [1] conducted a comparative toxicology study using cucumber, lettuce, and millet species. The results indicated that lettuce and millet were sensitive to inorganic and organic compounds, respectively. Many other studies using plant tests have also been reported [2–5].

The use of plant tests for toxicity assessment of complex effluents is especially attractive. Many industrial and municipal wastewaters are discolored and/or turbid, making the use of algal tests for toxicity assessment of the samples difficult. Other samples may be unstable during a standard 96-h test period. The methods commonly used to test this type of sample are flow-through or static-renewal test methods; algal tests are not applicable. Some samples contain low dissolved oxygen or high biochemical oxygen demand. These samples cannot be tested using fish or daphnids without aeration, a step that violates sample integrity. Plant tests, on the contrary, can be used with little or no difficulty for all these complex effluent samples [6–10].

In a recent study, Wang and Keturi [11] found that rice was the most promising among ten plant species for toxicity testing of a complex effluent. Rice was found to be highly sensitive to this sample, to have high germination rate, and to have long shelf life.

---

[1] Professional scientist, Water Quality Section, Illinois State Water Survey, Peoria, IL 61652.

The objective of this study was the toxicity assessment of industrial and municipal effluents using higher plants. The treated effluent and pretreated industrial samples were collected from three Illinoisian sanitary districts (I, II, and III). Under the National Pollutant Discharge Elimination System permit requirements, Sanitary Districts II and III had performed six-month biomonitoring in 1989, while Sanitary District I has not done so so far. Three plant toxicity tests were performed using common duckweed (*Lemna minor*), lettuce (*Lactuca sativa*, var. Buttercrunch), and rice (*Oryza sativa*).

Methods

Test Samples

Five or six samples were collected on January 23 and 24, 1989, from each of the three sanitary districts. Samples included a receiving water upstream from the effluent discharge point, a treated effluent prior to chlorination, and pretreated industrial samples at the point of entering into the sewer system. The industries in these districts were diverse, including auto parts, food processing, dairy, and metal processing. The plant size varied from small to large. The industries were required to meet federal pretreatment standards using chemical parameters [12]. The stream water was taken using the grab method, while the other samples were collected using the 24-h composite method [13].

The collected samples were shipped immediately to the laboratory in ice. They were analyzed according to Standard Methods [13]. These samples were tested as is for toxicity.

Duckweed Growth Tests

Duckweed stock culture has been maintained in the laboratory since 1982. Water and plant nutrients were added weekly. Light was maintained continuously using cool-white fluorescent light. Duckweed tests were modified from those of previous studies [2]. Only duckweed plants having two fronds per colony were used in this study. The test conditions are summarized in Table 1. In this study, the duckweed growth medium was ten-fold the

TABLE 1—*Summary of test conditions.*

	Duckweed Growth Test	Lettuce and Rice Seed Germination Test
Test species	*Lemna minor*	*Lactuca sativa* (Var. Buttercrunch) *Oryza sativa*
Test type	Static	Static
Temperature	25–28°C	25–25.2°C
Light quality	Cool-white fluorescent	Dark
Light intensity	86 µEi/m^2/s	NA
Photoperiod	Continuous	NA
Test vesel	100 by 15 mm petric dish	100 by 15 mm petric dish plus Whatman No. 1 filter paper
Test solution volume	35 mL	5 mL (lettuce) 7 mL (rice)
Specimens per vessel	30 fronds (15 colonies)	20 seeds
Replicates	2	4
Water control and dilution water	Duckweed growth medium	Hard standard water
Test duration	120 h	120 h
End point	Frond increase	Germination

strength of algal growth medium as given in Standard Methods [13]. The test end point was frond number increase; any visible, protruding frond was counted.

Two types of experiments were performed: screening tests and definitive tests. In screening tests, 100% effluent samples and a control sample were used. Observations were made on duckweed plants showing visible symptoms such as chlorosis (loss of pigment), necrosis (localized dead tissue), lesions (breakup of colony structure), and loss of buoyancy.

Two samples were noted for extreme toxicity: Industry Sample D from Sanitary District I and Industry Sample C from Sanitary District III. Definitive tests were performed on these two samples. They were diluted using duckweed growth medium to 100, 60, 36, 22, 13, and 7.8% effluent concentration. They were tested along with a control sample in the same manner as the screening test. The exposure period was 120 h.

Seed Germination Tests Using Lettuce and Rice

The same bulk of seeds indicated in other studies [1] were used in this study. They were kept at $-10°C$. The test conditions are summarized in Table 1. The seeds were pretreated with hypochlorite solution (3.33 g OCl^-/L) for 20 min and rinsed repeatedly with deionized water. Both screening and definitive tests were performed. Each culture dish contained 5 or 7-mL test solution for lettuce or rice, respectively. After incubation in the dark at 25 to 25.2°C for 120 h, each seed was observed to determine whether it had germinated; a seed was defined to have germinated if it developed a primary root 5 mm or longer [14]. Industry Sample C from the Sanitary District III was also tested, using the definitive test. The sample was diluted using hard, standard water [13] to 100, 60, 36, 22, 13, and 7.8% effluent concentration. The other test conditions are the same as those of the screening test.

Statistical Tests

Student t-tests were used to determine if there was a significant difference ($p < 0.05$) in duckweed frond increase (or seed germination) between the water control and the test samples. IC_{50} values (the concentration that caused 50% inhibition relative to the control sample) and 95% confidence limits were calculated using the moving average method [15].

Results

Test Samples

Selected water quality characteristics of test samples are given in Table 2. Several industrial samples contained strong odor, as shown by Footnote b in the table. The pungent or moth-ball-like odor suggested the presence of volatile compounds which might or might not be toxic. Two samples were oily: Samples B and D from Sanitary District II. All pH values were in the range of 7 to 9. A high alkalinity value was noted in Sample C, Sanitary District III. This sample had low hardness (51 mg/L as $CaCO_3$) and high NH_4-N value (37.8 mg/L). Two other samples having high NH_4-N values (35.3 and 445 mg/L) were Samples B and D from Sanitary District I.

Control Tests

One measure of test reproducibility is the results of control tests. Because control sample tests were performed each time a test was conducted, the overall control test results can be used as an indication of precision of the toxicity tests. The duckweed tests were conducted

TABLE 2—*Selected water quality of test samples.*

Industry		pH	Alkalinity[a], mg/L	Hardness[a], mg/L	NH₄-N, mg/L	Turbidity, NTU
		SANITARY DISTRICT I				
Receiving water		8.00	163	255	0.15	36
Finished effluent		7.59	101	202	1.68	18
Industry A[b]	Food	7.21	159	216	0.11	60
B	Metal	7.29	175	330	35.3	105
C	Metal	7.60	103	180	2.31	190
D	Metal	8.11	303	176	44.5	68
		SANITARY DISTRICT II				
Receiving water		8.05	279	415	0.38	6
Finished effluent		7.71	213	300	1.65	12
Industry A	Appliance	7.90	261	232	7.39	32
B[b]	Transportation	7.35	621	NT	NT	526
C[b]	Auto parts	7.80	223	260	0.22	85
D[b]	Wood treatment	7.69	167	205	0.09	358
		SANITARY DISTRICT III				
Receiving water		8.21	219	336	0.05	6
Finished effluent		7.99	143	196	0.08	3
Industry A	Soybean	8.05	235	702	0.01	18
B[b]	Food	8.20	382	65	2.29	1200
C[b]	Dairy	8.80	2927	51	37.8	518

NT = Not tested; NTU = nephelometric turbidity unit.
[a] Expressed as CaCO₃.
[b] Sample odorous.

in duplicates. The mean values (and standard deviations) for all duckweed tests were 97 (5), 82 (2), 89 (8), 93 (1), and 84 (5), all expressed in frond increase per dish. The mean coefficient of variation was 5%. The values varied in a relatively narrow range, 82 to 97 frond counts, indicating consistency of control tests. The initial inoculation was 30 fronds per dish, and the net frond increase in the 5-day incubation period was averaged 2.9-fold. For comparison, Wang and Williams [*10*] reported that *Lemna minor* frond increase in a 5-day period was 2.7-fold, whereas the increase in a 4-day period was 2.2-fold [*16*].

Several control tests using seed germination were conducted. Each test vessel contained 20 seeds. The seed germination tests were conducted in quadruplicates. The means (and standard deviations) for lettuce germination counts per dish were 19 (0.8), 19 (1.4), 19 (1.0), 19.8 (0.5), 19.3 (0.5), and 19.5 (0.6). The mean coefficient of variation was 4%. The means (and standard deviations) for rice germination counts per dish were 18 (0.8), 18.3 (1.0), 18.8 (0.5), and 18 (0). The mean coefficient of variation was 3%. Lettuce and rice both had high germination rates, which was considered an important criterion for plant toxicity testing.

Toxicity Testing

The results of toxicity tests are given in Table 3. Finished effluents from all these districts were not significantly toxic in comparison with the control tests, using a 95% confidence level.

Among six samples from Sanitary District I, only one sample from metal processing plant (Industry D) showed an inhibitory effect indicated by the duckweed test, 98% inhibition. In 24-h exposure, duckweed exhibited chlorosis, necrosis, and lesion in 14 out of 15 initial colonies. Lettuce and rice seed germination tests did not show a significant effect.

The receiving water in the Sanitary District II was found to be significantly toxic, $p <$

TABLE 3—*Phytotoxicity test results. Duckweed results are expressed in increased frond count (30 fronds initially) per dish, while lettuce and rice results are expressed in seeds germinated (20 seeds initially) per dish. The other conditions refer to Table 1.*

	Duckweed			Lettuce			Rice		
	Mean	SD	% Inhibition	Mean	SD	% Inhibition	Mean	SD	% Inhibition
	SANITARY DISTRICT I								
Control	88.5	7.8		19.3	1.0		18.8	0.5	
Receiving water	64.5	3.5	27	19.5	0.6		18.3	1.0	
Finished effluent	100	14.1		19.3	1.0		18.5	1.0	
Industry A	73	18.4	18	19.8	0.5		18.5	1.0	
B	63.5	3.5	28	19.3	0.5		17.3	2.1	
C	96	2.8		19.5	0.6		17.5	1.9	
D	2	1.4	98[a]	18.5	1.3		19	1.4	
	SANITARY DISTRICT II								
Control	96.5	4.9		19	0.8		18	0.8	
Receiving water	71	1.4	26[a]	19.5	0.6		19.5	0.6	
Finished effluent	76	2.8	21	19.5	0.6		17.3	0.5	
Industry A	105	15.6		19.3	1.0		18.5	0.6	
B	53	5.7	45[a]	16.5	3.1	13[a]	11	2.9	39[a]
C	74	7.1	23[a]	20	0		18.5	0.6	
D	109.5	5.0		18.8	0.5		17	0.8	
	SANITARY DISTRICT III								
Control	81.5	2.1		19	1.4		18.3	1.0	
Receiving water	84.5	2.1		19.5	0.6		18.8	1.0	
Finished effluent	120.5	2.1		19.3	0.5		17.5	0.6	
Industry A	93	2.8		19	1.4		18.8	1.3	
B	110	15.6		18	1.4		17.8	1.7	
C	0	0	100[a]	2.5	1	87[a]	0	0	100[a]

[a] Significance at $p = 0.05$

0.05. The inhibitory effect on duckweed growth was 26%. One sample from a transportation industry (Industry B) was oily and toxic, having a 45% inhibitory effect on duckweed growth. This sample also caused 13 and 39% inhibitory effects on lettuce and rice germination, respectively.

One sample from a dairy industry (Industry C) from Sanitary District III was extremely toxic. In 24-h exposure, the duckweed plants showed signs of chlorosis. In 48 h, the plants died completely. There was no increase of frond count from 60 fronds initially inoculated, attesting extreme toxicity. The colony structure, unlike that in Sample D, remained intact in 90% of the specimens after 5-day exposure. The inhibitory effects in the 5-day test period on duckweed, lettuce, and rice were 100, 87, and 100%, in that order.

Definitive tests were performed on two samples: Industry D from Sanitary District I and Industry C from Sanitary District III. The results in Fig. 1 indicated that Sample C (District III) was more toxic than Sample D (District I). In 60% effluent concentration, Sample C caused 96% inhibition, while Sample D caused 27%. In 36% effluent, both samples did not show significant toxicity. The IC_{50} value and 95% confidence limit for Sample C was 50% and 47 to 53%, respectively. The same values for Sample D were 70% and 67 to 75%, respectively.

Sample C of Sanitary District III was also extremely toxic to lettuce and rice. The toxic effects as shown in Fig. 2 were less dramatic than in Fig. 1. For example, 60% effluent concentration of Sample C caused 24 and 17% inhibition on lettuce and rice, respectively, while the same concentration caused 96% inhibition on duckweed. The IC_{50} value and 95%

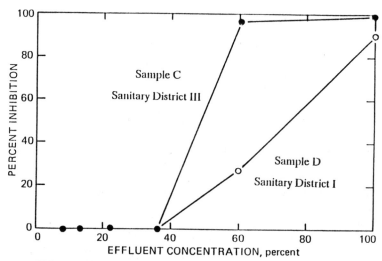

FIG. 1—*Inhibition of duckweed frond increase in selected test samples.*

confidence limit for lettuce tests were 65% and 61 to 71%, respectively. The same values for rice tests were 70% and 67 to 73%, respectively.

Discussion

Three standard tests are used by various regulatory agencies and industries as a part of the National Pollutant Discharge Elimination System for biomonitoring of effluent toxicity. The test species are fish, daphnids, and algae [15]. Higher plants are seldom used. Many recent reports indicated that higher plants could become a viable part of ecotoxicity tests [1–3,5,10,11]. The author recently compiled two extensive literature reviews, one on use

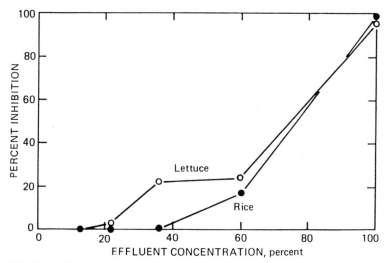

FIG. 2—*Inhibition of seed germination in Sample C from Sanitary District III.*

of plants for toxicity assessment (to be published) and the other on duckweed toxicity testing [16].

The phytotoxicity tests using higher plants have three important features. One, higher plants are uniquely suitable for phytotoxic compounds—herbicide pollution in the environment is little understood and the herbicide toxicity cannot be detected by means other than plant tests. More phytotoxicity tests are required to shed some light in this little known area. Two, tests using higher plants are relatively simple and reliable. Three, complex effluent samples can be tested without pretreatment. These features make plant tests a valuable part of a test battery for ecotoxicology.

In this study, results of duckweed, lettuce, and rice tests all indicated that the finished effluents from these sanitary districts were not toxic. It is of interest to compare the results of phytotoxicity tests with those of the standard tests; the comparative study is in progress. The results will be informative as to the relative sensitivity between phytotoxicity tests and standard tests.

The results in Table 3 indicated that industrial samples from three districts ranged from nontoxic to extremely toxic. It is possible that some toxicity reduction of these industrial wastewaters might have been achieved during the wastewater treatment processes. The conventional treatment processes, however, are designed to remove biodegradable materials. Many organic (e.g., aromatic compounds) and inorganic toxic substances (e.g., metals) cannot be removed effectively in the treatment plants. The lack of toxicity in the treated effluents as shown in Table 3 is partly due to the dilution effect of the toxic industrial samples by other nontoxic industrial effluents and domestic sewage.

Another possibility, especially relevant to duckweed tests, is the luxuriant plant nutrients present in the treated effluents. Seto et al. [18] reported that duckweed cultured in a high nutrient solution tended to be more resistant to environmental toxicity. This is perhaps the reason that duckweed growth in the finished effluents from Sanitary Districts I and III was greater than the respective controls. This possibility, nevertheless, should be balanced with the fact that seed germination tests on the effluent samples did not detect any toxic effect (Table 3). In the seed germination stage, plant nutrients had little effect on the germination rate [19].

The results of this study and others [10] generally indicated that duckweed was more sensitive to these effluents than lettuce and rice, making an impression that duckweed test is far superior than using other macrophytes. However, there are two important aspects deserved to be mentioned. In a recently published study, Wang [20] tested 23 wastewater samples from different industries. The results showed that rice was far more sensitive to certain industrial wastewaters than duckweed, suggesting that plant toxic response may be wastewater specific. Furthermore, lettuce and rice tests have two interesting and valuable characteristics. The tests require no light and thus are especially useful for photosensitive compounds. The tests require no nutrient supplements so that the potential interference between the supplements and toxic substances is nonexistent.

Publicly owned treatment works regulate industrial wastewaters using the federal pretreatment standards based on chemical parameters [12]. Compliance with the standards, however, does not guarantee that the effluents are not toxic. Here is a case in point [21]. The wastewater from a metal engraving plant entering into an Illinoisian sanitary district was pretreated. Using lime precipitation the plant effectively removed Zn from 1207 to 4.24 mg/L, although still surpassing the pretreatment standard of 2.61 mg/L. Another toxic metal (Pb) at 0.906 mg/L also violated the standard of 0.69 mg/L. All other parameters were within the standards. The toxicity test showed that the effluent sample was extremely toxic, causing 100% inhibition of seed germination in both 100 and 80% effluent concentrations. This extreme toxicity was unlikely due to elevated Zn and Pb levels by comparing with literature values [2]. More likely, it was due to interaction and matrix effects among complex constituents in the sample.

The toxicity test results of industrial samples are important information and can be used in several ways. For example, the federal pretreatment standard can be amended to include both chemical quality criteria [12] and biological quality criteria. Furthermore, industries that fail to meet these criteria may be (1) required to upgrade pretreatment processes or (2) levied treatment charges based on a formula consisting of the quantity (volume) and the quality (toxicity) of effluents.

Acknowledgment

I thank Tim Bachman, Jim Browning, and Mike Callahan for their full cooperation on this project.

References

[1] Wang, W., "Root Elongation Method for Toxicity Testing of Organic and Inorganic Pollutants," *Environmental Toxicology Chemistry*, Vol. 6, 1987, pp. 409–414.
[2] Wang, W., "Toxicity Tests of Aquatic Pollutants by Using Common Duckweed," *Environmental Pollution*, Series B, Vol. 11, 1986, pp. 1–14.
[3] Wong, M. and Bradshaw, A. D., "A Comparison of the Toxicity of Heavy Metals Using Root Elongation of Ryegrass, *Lolium Perenne*," *New Phytology*, Vol. 91, 1982, pp. 255–261.
[4] Ratsch, N. C., "Interlaboratory Root Elongation Testing of Toxic Substances on Selected Plant Species," EPA-600/S3-83-051, U.S. Environmental Protection Agency, Washington, DC, 1983.
[5] Miller, W. E., Peterson, S. A., Greene, J. C., and Callahan, C. A., "Comparative Toxicology of Laboratory Organisms for Assessing Hazardous Waste Sites," *Journal of Environmental Quality*, Vol. 14, 1985, pp. 569–574.
[6] Perez, J. D., Esteban, E., Gomez, M., and Gallardo-Lara, F., "Effects of Wastewater from Olive Processing on Seed Germination and Early Plant Growth of Different Vegetable Species," *Journal of Environmental Science and Health*, B21, 1986, pp. 349–357.
[7] Srivastava, N. and Sahai, R., "Effects of Distillery Waste on Performance of *Cicer arietinum* L.," *Environmental Pollution*, Vol. 43, 1987, pp. 91–102.
[8] Behera, B. K. and Misra, B. N., "Analysis of the Effect of Industrial Effluent on Growth and Development of Rice Seedlings," *Environmental Research*, Vol. 28, 1982, pp. 10–20.
[9] Thomas, J. M., Skalske, J. P., Cline, J. F., McShane, K. C., Simpson, J. C., Miller, W. E., Peterson, S. A., Callahan, C. A., and Greene, J. C., "Characterization of Chemical Waste Site Contamination and Determination of Its Extent Using Bioassay," *Environmental Toxicology and Chemistry*, Vol. 5, 1986, pp. 487–501.
[10] Wang, W. and Williams, J. M., "Screening and Monitoring of Industrial Effluents Using Phytotoxicity Tests," *Environmental Toxicology and Chemistry*, Vol. 7, 1988, pp. 645–652.
[11] Wang, W. and Keturi, P., "Comparative Seed Germination Tests Using Ten Species for Assessing Effluent Toxicity from a Metal Engraving Sample," *Water, Air, and Soil Pollution*, in press.
[12] *Federal Register*, 1983, *Rules and Regulations*, Section 433.15: Pretreatment Standards for Existing Sources, Vol. 48, pp. 32487.
[13] American Public Health Association, American Water Works Association, and Water Pollution Control Federation, "Standard Methods for Examination of Water and Wastewater," Washington, DC, 1985.
[14] U.S. Environmental Protection Agency, "Seed Germination/Root Elongation Toxicity Tests," EG-12, Office of Toxic Substances, Washington, DC, 1982.
[15] Peltier, W. H. and Weber, C. I., Eds., "Methods for Measuring the Acute Toxicity of Effluents to Freshwater and Marine Organisms," 3rd ed., EPA/600/4-85/O13, U.S. Environmental Protection Agency, Cincinnati, OH, 1985.
[16] Wang, W., "Chromate Ion as a Reference Toxicant for Aquatic Phytotoxicity Tests," *Environmental Toxicology and Chemistry*, Vol. 6, 1987, pp. 953–960.
[17] Wang, W., "Literature Review on Duckweed Toxicity Testing," *Environmental Research*, Vol. 51, 1990, pp. 7–22.

[*18*] Seto, M., Takahashi, Y., Ushijima, T., and Tazaki, T., "Chlorotic Death of *Lemna gibba* by Cadmium in Different Concentrations of Nutritional Minerals," *Japanese Journal of Limnology*, Vol. 40, 1979, pp. 61–65.

[*19*] Wang, W., "The Use of Plant Seeds in Toxicity Tests of Phenolic Compounds," *Environmental International*, Vol. 11, 1985, pp. 49–55.

[*20*] Wang, W., "Toxicity Assessment of Pretreated Industrial Effluents Using Higher Plants," *Research Journal of Water Pollution Control Federation*, Vol. 62, 1990, pp. 853–860.

[*21*] Wang, W., "Characterization of Phytotoxicity of Metal Engraving Effluent Samples," *Environmental Monitoring and Assessment*, Vol. 14, 1990, pp. 45–50.

Stella M. Swanson,[1] *Colleen P. Rickard,*[1] *Kathryn E. Freemark,*[2]
and Patricia MacQuarrie[2]

Testing for Pesticide Toxicity to Aquatic Plants: Recommendations for Test Species

REFERENCE: Swanson, S. M., Rickard, C. P., Freemark, K. E., and MacQuarrie, P.,
"Testing for Pesticide Toxicity to Aquatic Plants: Recommendations for Test Species," *Plants
for Toxicity Assessment: Second Volume, ASTM STP 1115,* J. W. Gorsuch, W. R. Lower, W.
Wang, and M. A. Lewis, Eds., American Society for Testing and Materials, Philadelphia,
1991, pp. 77–97.

ABSTRACT: A literature review of pesticide toxicity to aquatic plants was conducted in order
to arrive at a set of species for use in preregistration testing. Criteria used in the selection
process were: (1) existing database; (2) interspecies variability in response to pesticides; (3)
availability of well-characterized plant cultures; and (4) ease of culture. Freshwater and marine
algae and freshwater and estuarine macrophytes were included in the study.

Results showed that the three practical criteria [1,3,4] all resulted in the selection of unicellular green algae (Chlorophyceae) for the freshwater algae tests. However, variability among
freshwater species and classes was so great and so unpredictable that it became obvious that
a species battery approach was required. The battery should consist of representatives of each
of the major algal classes. Research into culture methods and alternative endpoints is required
before species other than unicellular green algae can be routinely incorporated into species
battery tests.

The studies in the marine algal database had much more consistent methodology; therefore,
it was possible to include relative sensitivity and variability as well as practical considerations
in the selection of marine tests species. The golden-brown algae (Chrysophyceae) were generally the most sensitive. However, because variability in response was still high, we recommend a species battery for marine algae as well. Four species of golden brown alga, two
diatoms (Bacillariophyceae) and two green algae, were chosen based on the four criteria.

Macrophytes should be part of preregistration testing because they can be more sensitive
than algae and because of their ecological importance. The database showed a great variety
of test methods and very few laboratory methods. No species emerged as the most consistently
sensitive among either freshwater or estuarine macrophytes. *Lemna gibba* and *Lemna minor*
are fast growing, easy to culture, and available commercially; they are logical test species for
effects of pesticide drift and surface films. However, they are not always very sensitive. Rooted
macrophytes belong in a comprehensive test program; however, test methods have to be
developed.

One criterion not included in this study was ecological importance. We recommend that
ecological importance become part of any selection process in order to facilitate extrapolation
to the field during hazard assessments.

KEY WORDS: algae, aquatic macrophytes, pesticides, preregistration testing, test species,
toxicity

[1] Research scientist and research technician, respectively, Saskatchewan Research Council, Saskatoon, Saskatchewan, Canada. (Both are now at Beak Associates Consulting, Ltd., Saskatoon, Saskatchewan, Canada.)
[2] Songbird ecologist and pesticide evaluation officer, respectively, Environment Canada, Conservation and Protection, Ottawa, Ontario, Canada.

Scientific evidence is needed to support selection of aquatic plant species for preregistration testing of pesticides [1]. In the past, there has not been any "profound rationale" behind selection of test species. Species which have already been used extensively in toxicity testing have been favored [1,2]. For example, the aquatic plant test species recommended for testing under the Federal Insecticide, Fungicide and Rodenticide Act (FIFRA) by the Environmental Protection Agency (EPA) in the United States were selected on the basis of extensive testing by chemical companies, economic significance, existence of good data on growth parameters, and availability of specific strains [3]. Test species recommended by the Organization for Economic Cooperation and Development (OECD) were selected on the basis of consensus among member countries. A particularly important criterion which has not generally been assessed is variability in species responses to pesticides [4,5].

In order to address several questions regarding the selection of test species, Environment Canada contracted with the Saskatchewan Research Council to conduct a review of the scientific literature to evaluate the suitability of algal and aquatic vascular plant species for pesticide toxicity testing [6]. Results were to be organized by plant type and habitat (i.e., freshwater algae, marine algae, and freshwater and estuarine macrophytes). Marine algae and estuarine macrophytes were included in this study because pesticides in runoff from coastal areas is a legitimate environmental concern.

Species suitability was to be assessed on the following criteria:

1. *Extent of the existing database*—The pesticide test record for each species has to be large enough to provide a basis for comparison among different pesticides and test methods.
2. *Variability in species response to pesticides*—The variability in response among test species is central to species selection because when variability is small, testing of a single sensitive species is appropriate. However, if variability is great, a battery of species need to be tested in order that those hazardous to the environment can be adequately assessed.
3. *Availability of stock cultures*—The availability in culture collections of species with good taxonomic characterization and small genotypic and phenotypic variability [7] is a necessary practical consideration.
4. *Ease of culture*—Ease of culture in defined media, well-known nutrient requirements, growth habit that allows rapid efficient determination of growth, and a growth form and rate which are suited to the use of well-established endpoints are necessary practical considerations [7,8]. For algae, growth should be rapid enough to allow estimation of density within three days after inoculation.

It was recognized that there are other criteria that can be used when choosing test species, notably ecological importance. These were not within the scope of this project but are discussed briefly.

Methods

Literature searches were conducted in order to obtain as much information as possible on the toxicity of pesticides to algae and aquatic vascular plants. Searches focused on pesticides, not toxic chemicals in general.

The Aquatic Sciences and Fisheries Abstracts (1978 to present), Chemistry Abstracts (1972 to present), Enviroline (1971 to present), Environmental Bibliography (1973 to present), NTIS (1964 to present), and Pollution Abstracts (1970 to present) databases of the DIALOG System were searched. In addition, the AQUIRE (1970–1985) and PHYTOTOX (1980–1984) databases of the Chemical Information System (CIS) were searched. The search strategy was based on sets of key words as follows: (1) general terms for pesticides (e.g., herbicide, algicide, carbamate) and common names and CAS numbers for 32 pesticides that are either currently widely used or still persist in the environment (e.g., atrazine, 2,4-D, DDT); (2) general terms for aquatic plants (e.g., alga, phytoplankton, macrophyte, vascular) and scientific names of algal and aquatic vascular plant genera (e.g., *Lemna, Potamogeton, Myriophyllum*); and (3) toxicity terms (e.g., EC_{50}, inhibition, effect).

The CIS output and relevant scientific papers and reports from each search were examined for plant species, pesticides, methods, endpoints used, and study reliability. The reliability of CIS data is judged on the basis of methodology (acceptability of procedures and quality assurance), control performance, measurement of toxicant concentration, use of solvent controls, reporting of temperature, pH and oxygen, and the number of toxicant concentrations tested [9]. We eliminated all CIS records with a reliability code greater than 2 (Table 1). Studies with codes greater than 2 had unsatisfactory controls, weaknesses in methods, insufficient description of methods, or were conducted improperly [9]. We attempted to follow some of the same criteria for the literature not covered by CIS.

The criteria used for examination of suitability of species for pesticide toxicity are given in the introduction. Only the data for the most commonly tested pesticides and species were used to determine the record of species use and to examine variability in species response because other data were too limited to allow meaningful comparisons. We conducted empirical analyses of the data only. Statistical analyses were not possible because of incompatibility of the CIS output with commonly available statistical software and because the datasets were very small once data had been stratified by similarity in methodology and endpoint.

Throughout this paper, algal classes follow Prescott [10]. The classes used are Chlorophyceae (green algae), Bacillariophyceae (diatoms), Chrysophyceae (golden brown algae), Dinophyceae (dinoflagellates), Rhodophyceae (red algae), Cryptophyceae, and the Xanthophyceae (yellow-green algae). The blue-green algae are now more correctly designated as the Cyanobacteria because of their procaryotic nature. Because "blue-green algae" is still in common usage, this term is used as the common name for the Cyanobacteria in this paper. The classification of aquatic vascular plants follows Scoggan [11,12].

Results and Discussion

Freshwater Algae

Existing Database—Our searches of the open literature yielded test results for 97 pesticides and 68 algal species. A lack of consistency with respect to species tested resulted in poor

TABLE 1—*Summary of reliability assessment of papers accessed through the AQUIRE database.*

Code	Study Reliability Description	No. Citations[a]
1	Study meets all criteria [b] for reliability	0
2	Study meets some criteria for reliability	151
3	Study does not meet criteria for reliability	90
4	No review of study quality	50
		Total 291

[a] Number of database records and published papers.
[b] See Appendix I for criteria definition.

databases for individual species. The database for each species was even smaller when data with unsatisfactory reliability were eliminated.

The most frequently used species in pesticide tests in the open literature were *Scenedesmus quadricauda* and *Chlorella vulgaris* (Table 2). *Scenedesmus quadricauda, Chlorella pyrenoidosa*, and *Selenastrum capricornutum* have been tested on the greatest number of pesticides. All of these species belong to the Chlorophyceae (green algae). Many of the references to *S. quadricauda* may have actually been tests on *Scenedesmus subspicatus* [13]. The open literature contained more references which used *Scenedesmus* despite *Selenastrum* being the species recommended by the EPA for preregistration testing. Since this report is based upon the published literature only, the relative importance of *Selenastrum* cannot be thoroughly assessed. A complete evaluation of *Selenastrum* would require access to proprietary data. Two blue-green algae (Cyanobacteria) were frequently tested, *Anabaena flos-aquae* and *Microcystis aeruginosa*. Only one diatom (Bacillariophyceae), *Navicula* sp., was frequently tested. The dinoflagellates (Dinophyceae), yellow-green algae (Xanthophyceae) and golden-brown algae (Chrysophyceae), were rarely tested.

A wide variety of endpoints have been used in algal testing (Table 2). For example, seven different endpoints were used in tests of 15 different pesticides using *Scenedesmus quadricauda*. The most common endpoint used was "growth," usually determined by cell counts or measurements of light absorbance. However, in many cases, the methods of determining endpoints were inadequately described, making comparison among the different studies difficult. Test duration also varied greatly. There were no cases where test results on the same pesticide using the same test duration and endpoints can be compared for several freshwater algal species among studies.

If the extent of the existing database is the only criterion used for selection, *Scenedesmus quadricauda, Chlorella pyrenoidosa*, and *Chlorella vulgaris* would be candidate species for pesticide testing. *Selenastrum capricornutum* could also be included because it is the species required by the EPA and has been widely used in tests with other chemicals and effluents.

TABLE 2—*Most common freshwater algae species used in pesticide toxicity tests reported in the open literature.*

Species	No. of Citations	No. of Pesticides	Endpoints Used
Green Algae (Chlorophyceae)			
Ankistrodesmus falcatus	9	9	BCF, ABD, PGR, GRO, CUP
Chlamydomonas reinhardtii or sp.	8	8	GRO, CLR, CUP
Chlorella vulgaris	11	7	GRO, CLR, CUP, ABD
Chlorella pyrenoidosa	9	17	GRO, CLR, PGR
Oedogonium cardiacum	9	8	BCF, MOR, LET
Scenedesmus quadricauda	14	15	PGR, PSE, CLR, ABD, MOR, GRO, BCF
Scenedesmus obliquus	7	7	GRO
Selenastrum capricornutum	5	16	GRO, CUP, O$_2$P
Blue-Green Algae (Cyanobacteria)			
Anabaena flos-aquae or *cylindrica*	8	6	GRO, LET, CUP
Microcystis aeruginosa	8	6	GRO
Diatoms (Bacillariophyceae)			
Navicula pelliculosa or *Navicula* sp.	5	6	GRO, PGR

NOTE: ABD = abundance   LET = lethality
BCF = bioaccumulation factor   MOR = mortality
CLR = chlorophyll   O$_2$P = oxygen production
CUP = C^{14} uptake   PGR = population growth
GRO = growth   PSE = photosynthesis

TABLE 3—*Test record for atrazine effects on aquatic plants reported in the open literature.*

	No. of Records	No. of Genera/Spp.	No. of Test Durations	No. of Endpoints	No. of Citations
Freshwater Algae					
Green	83	19	18	8	14
Blue-Green	52	6	5	3	5
Diatoms	3	1	1	1	1
Dinoflagellates	3	3	1	2	1
Yellow-Green	3	3	2	2	2
Marine Algae					
Green	18	7	5	1	3
Diatoms	23	11	4	6	6
Dinoflagellates	2	1	2	1	2
Golden-Brown	12	3	4	1	2
Freshwater Macrophytes	42	10	12	11	13
Estuarine Macrophytes	18	3	6	10	7

Variability in Species Response—This criterion proved to be very difficult to evaluate because of the lack of uniformity in test methodology and the paucity of studies in which a variety of species were tested on the same pesticide using comparable methods. The herbicide, atrazine, was by far the most common pesticide tested in the literature (Table 3). Yet even with atrazine, no two studies used the same combination of test duration and endpoint.

Test duration often had a major impact on the reported concentrations at which a toxic response was observed. For the insecticide DDT (which is no longer registered for use in Canada or the United States), test duration varied between 1 h and two months among species and in many cases was not reported at all. Even within a species, test duration can have a major effect on the toxic response observed (Table 4). *Scenedesmus quadricauda*

TABLE 4—*Effect of test duration on toxic response of freshwater algae (all data sourced through CIS).*

Pesticide	Test Duration	Toxic Concentration, µg/L	Endpoint
	SCENEDESMUS QUADRICAUDA		
DDT	48 h	100	Photosynthesis
	96 h	100	Photosynthesis
	6 d	1000	Photosynthesis
	8 d	1000	Photosynthesis
Carbaryl	2 d	100	Photosynthesis
	96 h	1000	Photosynthesis
	6 d	1000	Photosynthesis
2,4-D	48 h	1000	Photosynthesis
	96 h	1000	Photosynthesis
	6 d	100	Photosynthesis
	8 d	100	Photosynthesis
	10 d	1000	Photosynthesis
	CHLORELLA PYRENOIDOSA		
Carbaryl	4 d	100	Growth
	7 d	100 000	Growth

exhibited orders of magnitude difference in its response to the insecticides DDT and carbaryl, or the herbicide 2,4-D, depending upon test duration. The toxic response of *Chlorella pyrenoidosa* to carbaryl also varied with test duration. Differences in response with time can be due to changes in pH caused by photosynthesis and nutrient uptake, by breakdown of the pesticides, and/or by recovery of the algae over the duration of the test [13,14]. Differences can also be caused by the use of static versus renewal test protocols or by variability in toxicity of technical grade versus formulated pesticides.

The wide variety of endpoints used for the most commonly tested pesticides further complicated comparisons among species. More research is needed on the relationships among different endpoints and, in turn, the relevance of the endpoints to ecological effects in the field. At present, growth (measured by cell numbers) is the endpoint most commonly recommended in pesticide testing guidelines [1]. Measurement of cell numbers is not suitable for filamentous species or species that clump together. Alternative endpoints need to be developed, such as fluorescence and turbidity as indicators of growth or phosphorus uptake and ^{14}C uptake as indicators of algal activity.

The only useful information on variation in species response came from individual studies within which several species were tested under uniform conditions. Many of these studies have tested atrazine (Table 5).

None of the studies summarized in Table 5 revealed either freshwater algal classes or species that were consistently the most or the least sensitive to pesticides. In some cases, there was a distinct difference among algal classes, in others not. Variation among endpoints within species or among strains within species was significant in some studies.

In a review of the effect of contaminants on aquatic algae, Boyle [15] stated that "tests on single species of algae are . . . of limited applicability in assessing the effects of contaminants on algal communities that are composed of an array of species with different sensitivities." Because of the uncertainty regarding sensitivity among algal species and the difficulty in identifying any particular species or group of species that appear to be more sensitive than others, the OECD has recommended that their algae growth inhibition test be changed from a single species to a species battery test, with several species being tested in parallel [24]. They did not recommend a particular battery of species and commented that the basic concept of species battery testing requires validation through experimental work. Nyholm and Kallqvist [13] stated that because the sensitivity of algae may vary greatly among species, a test battery approach that represents different functional groups (e.g., green algae, diatoms, blue-greens) is required.

Our review of the published literature supports the species battery approach. The approach is necessary if the range in algal sensitivity to pesticides is to be covered. Taxonomic diversity within the battery is needed to account for the great variability both within and among algal classes. There is some evidence that, at the least, representatives of the Chlorococcales and the Cyanobacteria should be included. Interestingly, the data also showed that, in some cases, algae were as sensitive to insecticides as herbicides (Table 6). Toxicity testing for nontarget plants should be required for insecticides as well as herbicides.

Availability of Cultures—Species recommended for testing must be readily available from reputable suppliers. The species should be well characterized taxonomically and the origin of the culture known. Several culture collections in Canada, the United States, or Europe meet these criteria. These suppliers have all of the commonly tested species (Table 2), most of which are green algae. The culture collections also have representatives of other algal classes which should be part of a species battery. For example, species of blue-greens, diatoms, yellow-greens, and dinoflagellates can be obtained from the University of Toronto culture collection [25]. Therefore, it would be possible to include a wide taxonomic range in a test battery if availability was the only criterion.

TABLE 5—*Summary of results from studies which tested several species.*

Reference	Pesticide Tested	Range of Species Tested	Endpoints	Findings
16	Atrazine	3 greens 2 yellow-greens 1 blue-green	Chlorophyll a	All similar except one of the greens
	Malathion	3 greens 2 yellow-greens 1 blue-green	Chlorophyll a	All similar except one of the greens
17	Atrazine	2 greens (3 blue-green)	Photosynthesis, yield, growth rate	More variation among endpoints than among species; e.g. 100–500 mg/L among species versus 500–3600 mg/L among endpoints in one species
18	Atrazine	6 greens 1 blue-green	C^{14} uptake	No distinct separation between greens and blue-green
19	Atrazine	2 *Chlorella* strains 2 *Chlamydomonas reinhardtii* strains	Growth	Variation between strains as great as between species
20	Diquat	1 green 2 blue-greens 1 diatom	Growth	Two orders of magnitude difference among classes; blue-greens most sensitive
21	Diquat	7 greens	Growth	Two orders of magnitude difference among species
4 5	19 chemicals including pesticides	13 species	Growth	No generally sensitive or insensitive species; 3 orders of magnitude variation among species
22	19 chemicals including pesticides	13 species	Growth	Two groups with significantly different sensitivities—the Chlorococcales and the Cyanobacteria; other taxa did not have consistent sensitivity patterns
23	15 pesticides	1 freshwater green 1 freshwater blue-green 1 freshwater diatom 1 marine diatom Duckweed	Growth	No one species consistently the most or least sensitive

TABLE 6—*Some comparisons of algae sensitivity to insecticides and herbicides.*

Reference	Algal Classes Tested	Pesticide	Toxic Concentration, (mg/L)
16	Greens	Malathion	1.0
	Yellow-greens	Atrazine	1.0
	Blue-green		
CIS[a]	Golden-brown	Chlorpyrifos	1.0–200
	Green		
	Dinoflagellates		
CIS[a]	Green	DDT	100–1000
		Carbaryl	100–1000
		2,4-D	100–1000

[a] Chemical Information System.

Ease of Culture—Ease of culture is one of the major factors influencing the choice of test species. All of the most commonly tested species (Table 2) except *Navicula* are easy to grow using standard inorganic artificial growth media [26]. *Navicula* is not always easy to grow; it may require a complex medium with vitamins (ibid.). Some researchers have had success culturing *Navicula* in organic media with silica enrichments [27]. There are some other diatom species that have been used in testing. These are *Cyclotella* spp., *Nitzschia palea*, and *Rhizosolenia* spp. These diatoms are fairly easy to culture and have rapid growth rates [26]. However, because they tend to clump or attach to the side of test containers, the use of endpoints based on cell counts is impractical. The dinoflagellates *Dinobyron* and *Ceratium* grow well in culture [26]. However, there is very little experience with using these species in toxicity tests. The yellow-green algae *Vaucheria germinata* and *Tribonema* spp. have been used in longer duration (seven days) tests [16]. Whether these species are suitable for short-term tests is unknown. There is a definite need for more research into culturing techniques for diatoms and more test experience with dinoflagellates, yellow-green and golden-brown algae.

The species selected for testing should grow rapidly to avoid the problems associated with longer running tests. These problems center around the changes in pH that occur over time due to the utilization of carbon dioxide and bicarbonate and uptake of nitrogen from either ammonia or nitrate [13,14]. As the pH changes, metabolic processes of algae change, chemical bahavior of toxicants change, and interactions between metabolism and the mode of action of toxicants changes. Therefore, it is best to avoid such changes by keeping the duration of the test to a minimum. Some current test protocols [the OECD and International Standards Organization (ISO) algae tests] now recommend a 72-h duration [1,2]. These protocols also recommend keeping the initial inoculum to 10 000 cells. Thus, test species must have rapid growth; growth rate in controls should be higher than 0.9 divisions per day [2]. Because many of the algal species common in the wild do not grow rapidly in the laboratory (notably the diatoms), most test species are rapidly growing unicellular green algae such as *Chlamydomonas*, *Chlorella*, and *Selenastrum*.

The morphology of algal species also directly affects their usefulness in routine toxicity testing. Most standard protocols are based upon endpoints involving the counting of individual cells with electronic particle counters (EPA, OECD, ISO) [2]. This method is rapid and accurate for species with uniformly shaped, unicellular forms such as *Chlamydomonas*, *Selenastrum*, *Chlorella*, and *Ankistrodesmus*. However, problems arise with colonial or filamentous species (such as *Anabaena* and *Ulothrix*) and species that clump (such as *Oed-*

ogonium, Stigeoclonium, and many diatoms). These species cannot be accurately counted on electronic particle counters, making costly and time-consuming manual counting necessary. Manual counting is also not very accurate [28]. Sometimes growth habits of colonial species can be altered by using particular growth media. *Scenedesmus* and *Microcystis* cultures can contain up to 95% single-celled forms under laboratory conditions [26]. The tendency to grow unicellular forms can vary with strains within species [29]. These species may therefore be used in standard bioassays providing estimations of the percentage of colonies is carried out by manual microscopic counting and providing the strain is one that will form unicells in laboratory culture.

Another solution to the morphology problem is to use endpoints that do not rely on manual or machine-aided cell counts. These may include optical density, turbidity, fluorescence, chlorophyll, particulate carbon, carbohydrate carbon, nitrogen, or protein content [1,2]. All of these endpoints are substitutes for direct estimation of biomass. They should be calibrated against direct growth determinations because in most cases growth in the presence of toxicants will not be uniform in that the relationship between these endpoints and growth will be variable. Some endpoints, such as optical density, may not be sensitive enough to detect changes in growth at the low inocula recommended for these tests. Other endpoints include ATP formation/degradation, movement (for motile species), nutrient uptake, oxygen evolution, photosynthetic inhibition, or pH change of the medium [1,2]. These endpoints may not always be related to growth. At present, fluorescence appears to be an appropriate substitute for cell counting [13]. However, more research into alternative endpoints is required.

An alternative to improvements in culture techniques or development of different endpoints that do not rely on rapid growth or unicellular growth habit may be to include longer-term tests. It has been argued that these tests would encompass all stages of the logarithmic growth profile of algae and would represent the large diurnal pH changes encountered by algal populations growing in the wild [13]. Physical and chemical parameters would have to be monitored closely throughout the test to aid the interpretation of results. These longer-term tests may allow the inclusion of a greater variety of species. However, standard methodology has not been developed. Furthermore, results from such tests will be difficult to reproduce and difficult to use for predicting effects at specific pH conditions [13]. In the meantime, we are much closer to developing alternative culture techniques and endpoints that will also expand the list of test species.

Selection of Freshwater Algal Test Species—When the four selection criteria are combined, it is evident that no particular set of freshwater algal species can be recommended at this time, primarily because of lack of data on relative sensitivity and inconsistency in the data that do exist. However, it is clear that a species battery approach is preferable to the selection of a single "indicator" species. We recommend that a species battery that includes a representative of each of the major classes be developed (Table 7).

Candidate species for this battery are listed in Appendix II. Each species has been tested for its response to a pesticide at least once. Species are arranged in approximate order of decreasing frequency of use and ease of culture, with single-celled, commonly used species appearing first followed by colonial or filamentous species that have not been frequently tested. They could not be rated for relative sensitivity because there was not enough consistency in methods and species used in the data extracted to allow such an evaluation. Species were annotated when there were limitations with respect to use in testing, culturing, growth form, or commercial availability.

Marine Algae

Existing Database—The problem with inconsistency of test methodology was not as pronounced with marine algae as it was with freshwater algae. The endpoints and test durations

TABLE 7—*Candidate freshwater species/algal class for a species battery, based on species used at least once for pesticide tests (details by species in Appendix II).*

Algal Class		No. Candidate Species
Common Name	Latin Name	
Green	*Chlorophyceae*	29
Blue-green	*Cyanobacteria*	11
Diatoms	*Bacillariophyceae*	6
Yellow-green	*Xanthophyceae*	4
Golden Brown	*Chrysophyceae*	1
Dinoflagellates	*Dinophyceae*	2
···	*Cryptophyceae*	2
···	*Rhodophyceae*	1
Total		56

used were less variable (Table 8). The most common endpoint used was growth as determined by electronic cell counting. Occasionally, chlorophyll, photosynthetic activity, carbohydrate metabolism, or bioconcentration were used as endpoints. The duration of many tests was either 48 or 96 h with 72 h becoming more common in later years [30]. Lately, many tests use two durations on each pesticide, 2 h and 240 h [31].

Variability in Species Response—Sensitivity among the commonly tested species can be compared for marine algae because of the higher quality database. Generally, *Isochrysis galbana* (a golden brown alga) was the most sensitive species when tests using the same endpoint, duration, and pesticide were compared (Table 9). *Dunaliella tertiolecta* (a green alga) was the least sensitive. In some cases, species of green algae and diatoms can be the most sensitive groups.

The range in response among different marine algae can be as great as for freshwater species. Hollister and Walsh [32] found a seven-fold difference in response to atrazine among

TABLE 8—*Most common marine algae species used in pesticide toxicity tests reported in the open literature.*

Species	No. of Citations	No. of Pesticides	Endpoints Used
Diatoms (Bacillariophyceae)			
Skeletonema costatum	10	26	GRO
Thalassiosira fluviatilis	6	6	CLR, ABD, BCF, PSE, GRO
Nitzschia spp., *N. closterium*	8	9	CLR, ABD, GRO
Thalassiosira pseudonana	8	10	GRO
Golden-Brown Algae (Chrysophyceae)			
Isochrysis galbana	5	22	CAR, GRO
Monochrysis lutheri	5	8	CAR, GRO
Phaeodactylum tricornutum	5	20	GRO
Green Algae (Chlorophyceae)			
Chlorococcum sp.	4	17	GRO
Dunaliella tertiolecta	9	26	CAR, GRO

NOTE: ABD = abundance (decrease)   CLR = chlorophyll
      BCF = bioaccumulation factor   GRO = growth
      CAR = carbohydrate inhibition   PSE = photosynthesis effect

TABLE 9—*Relative sensitivity of marine algae.*

Species	Number of Tests Conducted	Cases When Most Sensitive	Cases When Least Sensitive
Diatoms (Bacillariophyceae)			
Skeletonema costatum	35	2	1
Thalassiosira pseudonana	15	0	2
Nitzschia spp.	24	0	3
Golden-Brown Algae (Chrysophyceae)			
Isochrysis galbana	42	13	0
Phaeodactylum tricornutum	38	4	0
Monochrysis lutheri	13	2	0
Green Algae (Chlorophyceae)			
Chlorococcum sp.	37	1	2
Dunaliella tertiolecta	46	0	11

17 algal species. Four marine species exposed to dieldrin showed responses ranging over five orders of magnitude; the diatom *Cyclotella nana* did not grow when exposed to 0.1 mg/L while the green alga *Dunaliella tertiolecta* showed no response to 1000 mg/L [*33*]. Walsh [*8*] states that "it is impossible to predict which species, or which clone of a species, will be most sensitive or most resistant to a particular toxicant."

Availability and Ease of Culture—Ease of culture and availability criteria indicated the suitability of the species most commonly tested (Table 8) as well as other candidate species. All of the species most commonly tested (Table 8) are available from commercial culture collections. The golden-brown alga, *Pavlova* (= *Monochrysis*) *lutheri* is also available. *Minutocellus polymorphus* is not commercially available.

Selection of Marine Algal Test Species—Evaluation of relative sensitivity among marine algal species indicated that the golden-brown algae were generally more sensitive. However, testing of a species battery including species of diatoms, green algae, and dinoflagellates as well as golden-brown algae would be prudent given the variability in response among classes.

Test species recommended by other authors include *Skeletonema costatum* because it meets most selection criteria and is generally sensitive to pollutants. Cell numbers cannot be estimated on an electronic particle counter because *Sicostatum* forms long chains [*8*]; however, alternative endpoints can be used. *Thalassiosira pseudonana* is recommended by Bonin et al. [*8*]. Recently, the diatom *Minutocellus polymorphus* has been recommended because of its general sensitivity, countability (due to its unicellular growth habit), and rapid growth allowing a test duration of 48 h [*34*]. Maestrini et al. [*7*] discuss the choice of test species for algae growth potential (yield) bioassays in detail. They chose the golden-brown algae *Emiliania* (= *Coccolithus*) *huxleyi*, *Pavlova* (= *Monochrysis*) *lutheri* and *Phaeodactylum tricornutum*, the green alga *Dunaliella tertiolecta* and two diatoms *Skeletonema costatum*, and *Thalassiosira pseudonana*.

When all our selection criteria are combined, seven marine species from three algal classes appear to be suitable for inclusion in a species battery (Table 10).

Freshwater and Estuarine Macrophytes

It has sometimes been assumed that toxicity tests on algae can be used to predict possible effects on aquatic vascular plants. Given the variability in algal response shown in this paper, this assumption appears to be incorrect even without examining comparative data for algae and macrophytes. The following is a brief summary of the data obtained in this study

TABLE 10—*Candidate marine species by algal class for a species battery.*

Class	Species
Chrysophyceae	*Isochrysis galbana*
	Emiliania (= *Coccolithus*) *huxleyi*
	Phaeodactylum tricornutum
Bacillariophyceae	*Skeletonema costatum*
	Thalassiosira pseudonana
Chlorophyceae	*Chlorococcum*
	Dunaliella tertiolecta

supporting the view that tests on vascular plants must be used to predict effects on vascular plants.

Tests on a range of algae and macrophytes using 2,4-D showed that the vascular plants were more sensitive to this herbicide than either freshwater or marine algae [16,31,35,36,37]. This is no doubt related to the mode of action of 2,4-D. Herbicides affecting growth regulation are targeted specifically towards vascular plants. Atrazine and diquat are also more toxic to macrophytes than to algae. For example, *Vallisneria americana* showed effects of atrazine concentrations as low as 4 µg/L [38]. A range of algal species had their growth reduced by 50% by diquat levels of 100 to 10 000 µg/L [20]. In contrast, *Lemna* showed 50% growth reduction at 2.7 to 4.6 µg/L of diquat (CIS). Some algae exposed to fluridone were affected by 1000 µg/L, while most macrophytes tested were affected by 20 to 30 µg/L (CIS). The fluridone data do show some similarities between some algal tests and macrophyte tests; however, because algal species were not specified, the algal species that would be good indicators of macrophyte response cannot be identified.

Although data comparing algae and macrophyte responses to pesticides are not plentiful, the evidence suggests that it would be prudent to include macrophyte tests in screening for nontarget plant effects. Because many macrophytes can actually become aquatic nuisances, the knowledge about their general biology and life history is quite extensive. Therefore, design of toxicity tests using these plants does not have to wait for basic biological research [39]. Also, many of these plants are ecologically important to wildlife in providing both cover and food (see below). The main question regarding macrophytes should not be whether they should be included in toxicity testing, but rather, which species would be most appropriate.

Existing Database—*Lemna minor, L. gibba, L.* spp., and *Myriophyllum spicatum* or *M.* spp. were commonly tested (Table 11), especially for herbicides used for macrophyte control such as diquat and fluridone. Common test species belong to six different families of macrophytes.

Test methodology for macrophytes was the most variable of all the groups surveyed. Although there are now standard protocols for *Lemna* [1], the open literature does not reflect this. Many of the tests on other species were longer-term field tests, with widely varying test durations, methodologies, and endpoints. Thus, although there is a wide taxonomic range represented in the database, there is an extremely limited amount of information on any one particular species using similar test methods.

Variability in Species Response—Relative sensitivity of species cannot be compared over a wide variety of pesticides because of the great variability in methodology. Atrazine testing has resulted in the most extensive database for macrophytes. Within the atrazine database, there are several studies that used photosynthetic activity as an endpoint [40–45]. These studies show sensitivity to atrazine among *Potamogeton perfoliatus, Potamogeton pectinatus,*

TABLE 11—*Most common macrophyte species used in pesticide toxicity tests reported in the open literature.*

Species	No. of Citations	No. of Pesticides	Endpoints Used
Duckweed (Lemnaceae)			
Lemna minor or *L. gibba* or *L. sp.*	14	15	MOR, CEL, LET, GRO, CLR, PSE, CUP
Pondweed (Najadaceae)			
Potamogeton perfoliatus, or *P. pectinatus* or *P. sp.*	9	2	PSE, length, dry weight, ABD, MOR, LOEL
Milfoil (Haloragidaceae)			
Myriophyllum spicatum or *Myriophyllum sp.*	12	5	PSE, length, MOR
Tapegrass (Hydrocharitaceae)			
Vallisneria americana (estuarine)	5	2	LGB, SRE, OVW, PSE, MOR, length, dry weight
Hornwort (Ceratophyllaceae)			
Ceratophyllum spp.	9	4	MOR, GRO
Waterweed (Hydrocharitaceae)			
Elodea spp.	6	4	length, dry weight, MOR, GRO, PGR

NOTE:
ABD	= abundance (decrease)	LOEL	= least observed effect level
CEL	= cellular effect (membrane permeability)	MOR	= mortality
CLR	= chlorophyll	OVW	= overwintering
CUP	= decreased C^{14} uptake	PGR	= population growth
GRO	= growth	PSE	= photosynthesis effect
LET	= lethality	SRE	= sexual reproduction
LGB	= leaf growth biomass		

and *Myriophyllum spicatum* as well as the estuarine species *Zannichellia palustris* and *Vallisneria americana.* The toxic concentrations for these species varied from 50 to 1310 µg/L. No one freshwater species emerged as the most consistently sensitive. For example, toxic concentrations for *P. perfoliatus* varied from 50 to 1310 µg/L. Among the estuarine species, *Zannichellia palustris* appeared to be more sensitive to atrazine than *Vallisneria americana;* however, the available data are few. Duration of the tests was an important influence on the toxic response observed. *Potamogeton perfoliatus* was affected by 130 to 140 µg/L of atrazine for the first week of an experiment; however, the toxic concentration for the entire test (30 days) was 1100 to 1310 µg/L [*43*].

When growth was used as an endpoint, toxic concentration of atrazine varied from 80 µg/L for *Elodea canadensis* to 1104 µg/L for *Myriophyllum spicatum* [*41,46*]. The database is not large enough to indicate a particularly sensitive freshwater species. The estuarine species *Vallisneria americana* responded to atrazine at concentrations as low as 8 µg/L when leaf growth biomass was the endpoint.

Responses to atrazine using other endpoints were similarly variable. The databases for other endpoints are smaller. In studies using other endpoints, toxic concentrations ranged from 12 to 650 µg/L [*38*].

Fluridone was the only other pesticide tested on a wide variety of macrophytes. Based on CIS data, *Lemna, Ceratophyllum demersum, Vallisneria,* and *Potamogeton pectinatus* showed similar sensitivity to fluridone. *Myriophyllum spicatum* and *Utricularia* were less sensitive than either *Ceratophyllum demersum* or *Elodea* over a 30-day period.

Although the database for macrophytes was relatively large, lack of consistent methodology and inherent variability in response make the choice of a set of test species difficult. For example, there is some evidence that rooted macrophytes can be more sensitive to pesticides than the only existing "standard" macrophyte used for testing, *Lemna*. *Potamogeton, Ruppia, Myriophyllum,* and the estuarine species *Zannichellia* all appear more sensitive to atrazine than *Lemna gibba* [42,47]. *Myriophyllum* was killed by 44 μg/L of 2,4-D after 30 days exposure, while *Lemna* was not killed by 5000 μg/L after more than six days exposure. On the other hand, *Lemna* was found to be an extremely sensitive tool for screening and biomonitoring complex effluent samples [48] and was very sensitive to sulfonylurea herbicides [23]. In addition, because *Lemna* is a floating macrophyte, it provides a good test species for surface active and hydrophobic pesticides that form a film at the air-water interface.

Availability and Ease of Culture—These selection criteria have not been thoroughly examined in the literature for macrophyte species. *Lemna* is fast growing and reproduces rapidly compared with other vascular plants. The species recommended by the EPA, *L. gibba,* is available commercially from the UTEX collection and from Dr. Charles Cleland, U.S. Department of Agriculture, Washington, DC. Recent work using *Hydrilla verticillata* has shown that it is possible to conduct well-controlled tests using defined media over a 14-day period for both water-borne and sediment-borne pesticides [49]. *Potamogeton pectinatus* can be tested using culture in large mason jars over a four-week period for effects on photosynthesis, respiration, growth, and morphology [50].

Selection of Macrophyte Species—Macrophytes belong in nontarget test guidelines. Available data indicate that algae cannot be used to reliably predict effects on aquatic macrophytes.

Lemna is a logical test organism for the effects of pesticide drift and the effects of surface films at the air-water interface. *Lemna* tests are quite widely used, simple, quick, inexpensive, and quite reproducible and are therefore practical components of any comprehensive nontarget plant screening scheme. The use of *Lemna* in nontarget plant screening is recommended by the EPA and was supported by evidence gathered by Swanson and Peterson [1,2]. There are insufficient data to distinguish between the suitability of *Lemna gibba* and *Lemna minor* as test species.

Rooted macrophytes can be more sensitive to pesticides than *Lemna*. They have been used in laboratory testing, although not as extensively as *Lemna*, have been frequently tested in the field, and for some species, the biology is well known. Because of their importance in aquatic habitats, they should be part of any testing scheme where the aim is as accurate an assessment of ecological effects as possible. Some freshwater species such as *Potamogeton pectinatus* have already been tested in the laboratory and would be good candidates for development as test species in nontarget plant test guidelines. *Vallisneria americana* is a relatively common estuarine test species and can be quite sensitive (depending on the endpoint). Rooted macrophyte test methods require much more development.

Ecological Importance as an Additional Criterion

The criteria used for selection of test species in this study were based on practical considerations and variability in species response (which included sensitivity). When the suitability of test species as representatives of larger groups was examined further, it became apparent that the ecological importance of the species in the Canadian environment should also have been used.

The use of ecological importance as a criterion for species selection would make bridging the gap between laboratory testing and hazard assessment easier. One of the most frequent criticisms of toxicity tests is that the species used are not representative of the environment that we are trying to protect [1,2].

There are several obstacles to overcome before the use of such a criterion would be possible. Ecological importance has to be defined. It could include factors such as numerical dominance, productivity, importance in food webs, importance as habitat, importance in determining chemical fate (especially bioaccumulation), or geographical distribution. Secondly, a database has to be assembled for use in selecting the appropriate species. Depending upon the definition used, this task could be relatively straightforward (e.g., if geographical distribution is used) or quite difficult (e.g., if importance in food webs is used).

One initial approach may be to use geographical distribution from which a list of candidate species for each region is derived. The regions could be political (i.e., pacific, prairie, central, maritime) or ecological (i.e., ecozones such as alpine, prairie, parkland, boreal).

Despite the difficulties associated with including ecological importance as a criterion, it is important that it eventually be included as part of the process in choosing candidate species for species battery testing. Otherwise, the battery will always be open to the criticism that it does not adequately represent the "real world."

Research Needed

This study has shown that there are still several areas that require more study before a confident recommendation can be made regarding algae and macrophyte species for pre-registration testing. We have arranged those areas of research in order of priority. However, for the purposes of developing defensible guidelines from both a practical and ecological point of view, the first three are of equal priority and should (if possible) be conducted concurrently.

1. The ecological importance of candidate test species has to be investigated, using a clear definition of ecological importance.
2. Culture methods and test methods for algae other than unicellular, rapidly growing species need development. Methods for species that are currently difficult to culture (especially diatoms) are required. Suitable alternatives to cell counting as an endpoint will allow the use of species that clump or are filamentous. Alternatives to the bottle test, such as the microplate technique, need to be developed for several species to make species battery tests more economical.
3. Tests for rooted aquatic macrophytes require development. Techniques should be investigated for use with macrophytes that are important components of wildlife habitat such as nesting habitat (e.g., *Scirpus*, *Typha*, and *Juncus*) or food (*Potamogeton*, *Ceratophyllum*, *Carex*, *Beckmania*, *Eleocharis*, *Myriophyllum*, *Polygonum*, *Scirpus*, *Sparganium*, *Nuphar*, and *Typha*) [51–54].
4. The sensitivity of *Lemna* must be studied so that its suitability as an indicator for macrophytes in general can be assessed.
5. The applicability of species battery tests to the field should be tested through comparative laboratory and field studies.
6. Once culture methods are available for a wider range of algal species, a standardized series of tests on species batteries should be conducted to isolate sensitive species and species that appear to be consistent in their response to pesticides. This will be a long-term iterative process that requires cooperation among academic, government and industry researchers.

Conclusions and Recommendations

Existing guidelines for the testing of aquatic plants represent "a series of compromises between often conflicting constraints" [7]. For the most part, selection of test species has been based primarily upon a balance among several practical criteria. If significant advances are to be made in increasing the usefulness of nontarget plant tests in hazard assessment, several areas require more research. Without these additional studies, our recommendations have to be based on a set of criteria that are primarily practical. These practical considerations are the same ones currently used by other agencies (notably the EPA and OECD). Therefore, the recommended test species will not differ significantly from the species currently being used. The major difference is that our recommendations include the testing of a species battery rather than one or two indicator species. We believe that although an interim set of test guidelines can be drawn up using the current state of knowledge, it should be made clear that these are indeed interim guidelines that will be refined after further research has been conducted.

APPENDIX I

Criteria for Study Reliability (CIS) [9] (See Table 1 of text)

Definition

Studies assigned a reliability code of 1 meet all of the criteria listed below:

Methodology judged satisfactory (e.g., accepted procedures and quality assurance)
Control mortality or effect satisfactory (e.g., equal to or less than 10% of acute tests)
Toxicant concentration measured
Solvent control used if solvent used
Temperature, pH, and dissolved oxygen reported for organics and nonmetallic inorganics
Temperature, pH, dissolved oxygen, and either alkalinity or hardness reported for metals
     (saltwater tests do not need alkalinity or hardness studies)
Minimum of five organisms per treatment
Minimum of three different toxicant concentrations in addition to the control

Studies assigned a reliability code of 2 meet all the criteria for quality testing; however, the study may be characterized by one or more of the following deficiencies:

Control mortality not reported or control mortality high but accounted for statistically
Toxicant concentration unmeasured
Water chemistry variables not reported
No solvent control when solvent used

Studies assigned a reliability code of 3 do not meet the criteria for quality testing and are generally characterized by:

Control mortality unsatisfactory (e.g., greater than 10% and not accounted for statistically)
Methods section shows weakness in experimental procedures or insufficient description
     to judge quality of experimental design
Static test with unmeasured concentration conducted improperly (e.g., in the presence of
     precipitate or some undissolved chemical or in an unacceptable container)

Studies assigned a reliability code of 4 are either an abstract only or a foreign paper. Alternatively, data may have been extracted from the study prior to the development of the coding scheme.

APPENDIX II

Candidate Freshwater Algae Species for a Species Battery Based on Species Used at Least Once for Pesticide Tests

Species	Limitations (Based on Ease of Culture and Availability Only)
Chlorophyceae (Green Algae)	
Scenedesmus quadricauda	
Scenedesmus obliquus	
Scenedesmus bijuga	
Chlorella pyrenoidosa	
Chlorella vulgaris	
Chlorella emersonii	
Selenastrum capricornutum	
Chlamydomonas reinhardtii	
Chlamydomonas spp.	
Chlamydomonas dysosmos	limited testing experience
Ankistrodesmus spp.	limited testing experience
Tetraedron minimum	limited culturing or testing experience
Schroederia setigera	limited culturing or testing experience
Oedogonium cardiacum	filamentous, needs alternative endpoint
Spirogyra spp.	filamentous, needs alternative endpoint; limited testing experience
Stigeoclonium spp.	filamentous, needs alternative endpoint; limited testing experience
Ulothrix spp.	filamentous, needs alternative endpoint; limited testing experience
Kirchneriella subsolitaria or *contora*	gelatinous colonies, needs alternative endpoint; limited testing experience
Elakatothrix spp.	gelatinous colonies, needs alternative endpoint; limited culturing or testing experience
Chlorococcum hypnospermum	clump, needs alternative endpoint; limited culturing or testing experience
Sphaerocystis schroeteri	colonial, needs alternative endpoint; limited testing experience; only available from European suppliers
Oocystis lacustris	colonial, needs alternative endpoint; limited testing experience; this species available only from European supplier; other *Oocystis* species available from Canadian and U.S. suppliers
Coccomyxa subellipsoidea	may form gelatinous colonies, needs alternative endpoint; limited culturing or testing experience
Stichococcus bacillaris	filamentous, needs alternative endpoint; limited culturing or testing experience

Spongiochloris excentrica filamentous, needs alternative endpoint; limited culturing or testing experience

Golenkinia sp. limited culturing or testing experience

Pediastrum sp. colonial, needs alternative endpoints; limited culturing or testing experience

Klebsormidium marinum limited culturing or testing experience

Raphidonema longiseta limited culturing or testing experience

Monoraphidium pussilum limited culturing or testing experience; not available from North American suppliers

Cyanobacteria (Blue-Green Algae)

Microcystis aeruginosa gelatinous colonies, needs alternative endpoint

Anabaena flos-aquae filamentous, needs alternative endpoint

Anabaena cylindrica filamentous, needs alternative endpoint

Oscillatoria spp. filamentous, needs alternative endpoint; limited culturing or testing experience

Gloeocapsa alpicola forms gelatinous masses, needs alternative endpoint; limited culturing or testing experience

Anacystis nidulans forms gelatinous masses, needs alternative endpoint; limited culturing and testing experience

Aphanizomenon flos-aquae filamentous, needs alternative endpoint, limited testing experience

Cylindrospermum sp. filamentous, needs alternative endpoint; limited culturing or testing experience

Aulosira fertillissima filamentous, needs alternative endpoint; limited culturing or testing experience

Plectonema boryanum filamentous, needs alternative endpoint; limited culturing or testing experience

Synechococcus leopoliensis limited culturing or testing experience

Bacillariophyceae (Diatoms)

Navicula pelliculosa can clump, needs alternative endpoint

Navicula incerta can clump, needs alternative endpoint; may be difficult to culture

Cyclotella meneghiniana & spp. limited culturing and testing experience

Nitzschia palea clumps and attaches to sides of test chambers, needs alternative endpoints

Rhizosolenia sp. clumps, needs alternative endpoints; limited culturing or testing experience; unavailable from suppliers

Diatoma sp. difficult to culture, limited culturing or testing experience

Xanthophyceae (yellow-green algae)

Vaucheria geminata siphonaceaus, forming felt-like masses, needs alternative endpoints; limited culturing or testing experience

Tribonema sp. dilamentous, needs alternative endpoints; limited culturing and testing experience

Bumilleriopsis filiformis	can form short filaments, needs alternative endpoints
Monodus subterraneus	limited culturing or testing experience; not available from North American suppliers
Chrysophyceae (Golden-Brown Algae)	
Dinobyron sp.	colonial, needs alternative endpoints; limited culturing or testing experience
Dinophyceae (Dinoflagellates)	
Glenodinium sp.	limited culturing or testing experience
Gymnodinium sp.	limited culturing or testing experience
Cryptophyceae	
Rhodomonas sp.	limited culturing or testing experience
Cryptomonas sp.	limited culturing or testing experience
Rhodophyceae	
Batrachospermum	colonial, needs alternative endpoints; limited culturing or testing experience

References

[1] Freemark, K., MacQuarrie, P., Swanson, S., and Peterson, H. in *Plants for Toxicity Assessment, ASTM STP 1091,* W. Wang, J. W. Gorsuch and W. R. Lower, Eds., American Society for Testing and Materials, Philadelphia, 1990.

[2] Swanson, S. M. and Peterson, H., "Development of Guidelines for Testing Pesticide Toxicity to Nontarget Plants," Environmental Protection Series, Environment Canada, Ontario, 1990.

[3] Host, R. W. and Ellwanger, T. C., "Pesticide Assessment Guidelines, Subdivision J., Hazard Evaluation: Nontarget Plants," Office of Pesticides and Toxic Substances, U.S. Environmental Protection Agency, Washington, DC, 1982.

[4] Blanck, H., *Ecological Bulletin,* Vol. 36, 1984, pp. 107–119.

[5] Blanck, H., Wallin, G., and Wangberg, S., *Ecotoxicology and Environmental Safety,* Vol. 8, 1984, pp. 339–351.

[6] Swanson, S., "Aquatic Plant Toxicity Testing: Recommendations for Test Species," unpublished manuscript, Environment Canada, Ottawa, Ontario, 1989.

[7] Maestrini, S. Y., Droop, M. R., and Bonin, D. J. in *Algae as Ecological Indicators,* L. E. Shubert, Ed., Academic Press, Toronto, 1984, pp. 135–188.

[8] Walsh, G. E. in *Principles of Toxicity Testing with Marine Unicellular Algae,* Vol. 7, 1988, pp. 979–987.

[9] Pilli, A., Carle, D. O., and Sheedy, B. R., "AQUIRE (Aquatic Information Retrieval), A Technical Support Document," U.S. EPA Environmental Research Laboratory, Duluth, MN, 1988.

[10] Prescott, G. W., "Algae of the Western Great Lakes Area," W. C. Brown Co., Dubuque, IA, 1962.

[11] Scoggan, H. J., "The Flora of Canada," Part 3, National Museum of Natural Sciences Publications in Botany, No. 7(3), Ottawa, Ontario, 1978.

[12] Scoggan, H. J., "The Flora of Canada," Part 4, National Museum of Natural Sciences Publications in Botany, No. 7(4), Ottawa, Ontario, 1978.

[13] Nyholm, N. and Kallqvist, T., *Environmental Toxicology and Chemistry,* Vol. 8, 1989, pp. 689–703.

[14] Soeder, C. J. and Stengel, E., in *Algal Physiology and Biochemistry,* W. D. P. Stewart, Ed., Blackwell Scientific Publishers, Oxford, 1974, pp. 714–740.

[15] Boyle, T. P. in *Algae as Ecological Indicators,* L. E. Shubert, Ed., Academic Press, Toronto, Ontario, 1984, pp. 238–256.

[16] Torres, A. M. R. and O'Flaherty, L. M., *Phycologia,* Vol. 15, No. 1, 1976, pp. 25–36.

[17] Stratton, J. C., Jones, T. W., Kemp, W. M., Boynton, W. R., and Means, J. C., *Archives of Environmental Contamination and Toxicology,* Vol. 13, 1984, pp. 35–42.

[18] Larsen, D. P., deNoyelles, Jr., F., and Shiroyama, T., *Environmental Toxicology and Chemistry,* Vol. 5, 1986, pp. 179–190.

[*19*] Hersh, C. M. and Crumpton, W. G., *Bulletin of Environmental Contamination*, Vol. 39, 1987, pp. 1041–1048.

[*20*] Birmingham, B. C. and Coleman, B., *Water, Air and Soil Pollution*, Vol. 19, 1983, pp. 121–131.

[*21*] Cullimore, D. R., *Weed Research*, Vol. 15, 1975, p. 401.

[*22*] Wangberg, S. A. and Blanck, H., *Ecotoxicology and Environmental Safety*, Vol. 16, 1988, pp. 72–82.

[*23*] Hughes, J. S. and Erb, K., "The Relative Sensitivity of five Nontarget Plant Species to Various Pesticides," paper presented at the 10th Annual Meeting of Society of Environmental Toxicology and Chemistry, Toronto, Ontario, 1989.

[*24*] Organization for Economic Co-operation and Development (OECD), "Report of the OECD Workshop on Ecological Effects Assessment," OECD Environment Monographs No. 26, Paris, France, 1989.

[*25*] Acreman, J. C., University of Toronto Culture Collection: Listing by Taxa, as of 28 March 1989.

[*26*] Peterson, H., personal communication, Saskatchewan Research Council, Saskatoon, Saskatchewan, Canada, 1989.

[*27*] Kent, R., personal communication, Environment Canada, Water Quality Branch, Ottawa, Ontario, 1989.

[*28*] Lund, J. W. G., Kipling, C., and LeCren, E. D., *Hydrobiologia*, Vol. 11, 1958, pp. 143–170.

[*29*] Trainor, F. R. in *Indicator Algal Assays: Laboratory and Field Approaches*, L. E. Shubert, Ed., Academic Press, Toronto, 1984, pp. 3–14.

[*30*] Walsh, G. E., McLaughlin, L. L., Lores, E. M., Louis, M. K., and Deans, C. H., *Chemosphere*, Vol. 14, No. 3/4, 1985, pp. 383–392.

[*31*] Mayer, F. L. "Acute Toxicity Handbook of Chemicals to Estuarine Organisms," U.S. EPA Report EPA/600/8-87/017, 1987.

[*32*] Hollister, T. A. and Walsh, G. E., *Bulletin Environmental Contamination Toxicology*, Vol. 9, 1973, pp. 291–295.

[*33*] Moore, Jr., S. A., "Impact of Pesticides on Phytoplankton in Everglades Estuaries," South Florida Environmental Project: Ecological Report No. DI-SFEP-74-15, National Park Service, Southeastern Region, Atlanta, Georgia, 1973.

[*34*] Walsh, G. E., McLaughlin, L. L., Yoder, M. J., Moody, P. H., Lores, E. M., Forester, J., and Wessinger-Duvall, P. B., *Environmental Toxicology and Chemistry*, Vol. 37, 1988, pp. 925–929.

[*35*] O'Brian, M. C. and Prendeville, G. N., *Weed Research*, Vol. 19, 1979, pp. 331–334.

[*36*] Boyle, T. P., *Environmental Pollution*, Vol. 21, 1980, pp. 35–49.

[*37*] Miller, W. E., Peterson, S. A., Greene, J. C., and Callahan, C. A., *Journal of Environmental Quality*, Vol. 14, No. 4, 1985, pp. 569–574.

[*38*] Cohn, S. L., "An Evaluation of the Toxicity and Sublethal Effects of Atrazine on the Physiology and Growth Phases of the Aquatic Macrophyte *Vallisneria americana* L.," Ph.D. thesis, Faculty of the College of Arts and Science, The American University, Washington, DC, 1985.

[*39*] Sortkjaer, O., *Ecology Bulletin*, Vol. 36, 1984, pp. 75–80.

[*40*] Correll, D. L. and Wu, T. L., *Aquatic Biology*, Vol. 14, 1982, pp. 151–158.

[*41*] Stevenson, J. C., Jones, T. W., Kemp, W. M., Boynton, W. R., and Means, J. C., "An Overview of Atrazine Dynamics in Estuarine Ecosystems," Chesapeake Biological Laboratory, Center for Environmental and Estuarine Studies, University of Maryland, Solomons, Maryland, 1983.

[*42*] Jones, T. W. and Winchell, L., *Journal of Environmental Quality*, Vol. 13, No. 2, 1984, pp. 243–247.

[*43*] Cunningham, J. J., Kemp, W. M., Lewis, M. R., and Stevenson, J. C., *Estuaries*, Vol. 7, pp. 519–530.

[*44*] Kemp, W. M., Boynton, W. R., Cunningham, J. J., Stevenson, J. C., Jones, T. W., and Means, J. C., *Marine Environmental Research*, Vol. 16, 1985, pp. 255–280.

[*45*] Jones, T. W., Kemp, W. M., Estes, P. S., and Stevenson, J. C., *Archives of Environmental Contamination and Toxicology*, Vol. 15, pp. 277–283.

[*46*] Forney, D. R., "Effects of Atrazine on Chesapeake Bay Aquatic Plants," Office of Water Research and Technology, OWRTA-067-ALAC1.

[*47*] Hughes, J. S., Alexander, M. M., and Balu, K. in *Aquatic Toxicology and Hazard Assessment, 10th Volume, ASTM STP 971*, W. J. Adams, G. A. Chapman, and W. G. Landis, Eds., 1988, pp. 531–542.

[*48*] Wang, W. and Williams, J. M., *Environmental Toxicology and Chemistry*, Vol. 7, 1988, pp. 645–652.

[*49*] Hinman, M. L. and Klaine, S. J., paper presented at the Ninth Annual Meeting of the Society of Environmental Toxicology and Chemistry, November 1988.

[50] Fleming, W. J. and Momot, J. J., paper presented at Ninth Annual Meeting of the Society of Environmental Toxicology and Chemistry, November 1988.

[51] Sheehan, P. J., Baril, A., Mineau, P., Smith, D. K., Harfenist, A., and Marshall, W. K., "The Impact of Pesticides on the Ecology of Prairie Nesting Ducks," Canadian Wildlife Service Technical Report Series No. 19, Ottawa, Ontario, Canada, 1987.

[52] Sugden, L., "'Feeding Ecology of Pintail, Gadwall, American Widgeon and Lesser Scaup Ducklings in Southern Alberta," Canadian Wildlife Service Report Series No. 24, Ottawa, Ontario, Canada, 1973.

[53] Dirschl, H. J., Journal of Wildlife Management, Vol. 33, No. 1, 1969, pp. 77–87.

[54] Keith, L. B., Wildlife Monographs, No. 6, 1961, p. 88.

Applications of Plant
Bioassays / Photosynthesis

Thomas D. Byl[1] and S. J. Klaine[1]

Peroxidase Activity as an Indicator of Sublethal Stress in the Aquatic Plant *Hydrilla Verticillata* (Royle)

REFERENCE: Byl, T. D. and Klaine, S. J., **"Peroxidase Activity as an Indicator of Sublethal Stress in the Aquatic Plant** *Hydrilla Verticillata* **(Royle),"** *Plants for Toxicity Assessment: Second Volume, ASTM STP 1115,* J. W. Gorsuch, W. R. Lower, W. Wang, and M. A. Lewis, Eds., American Society for Testing and Materials, Philadelphia, 1991, pp. 101–106.

ABSTRACT: This study examines the utility of plant peroxidase (POD) activity as a biochemical indicator of chemical stress in the aquatic plant *Hydrilla verticillata* (Royle). Plants were raised in 10% Hoagland's solution augmented with 200 mg $NaHCO_3$/L. The plants were exposed to concentrations of exponential increment of Cu^{2+} or the sulfonylurea herbicide, sulfometuron methyl, (methyl 2-[[[[(4,6-dimethyl-2-pyrimidinyl) amino]-carbonyl] amino] sulfonyl] benzoate), also known by the trade name Oust[R]. After seven days of exposure the plants were harvested, macerated, and the POD extracted with salt solution. POD was measured by a spectrophotometric assay. There was a significant increase in POD activity after exposure to sublethal concentrations of Cu^{2+} or Oust. This increase in POD activity was dose dependant and concomitant with a decrease in root and shoot elongation. Plants exposed to 1 mg/L Cu^{2+} resulted in a 1.75-fold increase over the controls in shoot POD activity. Plants treated with 1 mg/L Oust resulted in a 1.5-fold increase in shoot POD activity over the controls, yet no necrosis was visible after seven days. These results indicate that measuring the salt-extractable POD activity has great potential as a biochemical indicator of sublethal stress in the aquatic plant *H. verticillata*.

KEY WORDS: peroxidase, biochemical indicator, sublethal stress, *Hydrilla verticillata*, copper, sulfometuron methyl, toxicity

The study of biochemical responses to sublethal stress in terrestrial and aquatic plants is an important area of increasing interest. One biochemical marker that has shown potential as an indicator of stress in terrestrial plants is the enzyme peroxidase (POD). POD is an enzyme found in large quantities throughout plant cells, especially the cell wall. Several studies have shown a change in extractable POD activity due to various pollutants. For example, soybean plants treated with cadmium (Cd) illicited an increase in POD activity [1]. Lettuce seedlings exposed to copper (Cu) also had an increase in POD [2]. Cotton plants raised in a medium high in manganese (Mn) also displayed greater POD activity than the controls [3]. Conifers exposed to gaseous fluoride (F) from aluminum smelters had enhanced POD activity [4]. Tomato and bean plants exposed to ozone, nitrogen oxide (NO_x), or sulfur dioxide (SO_2) were found to have higher POD activity [5–7]. Even plant growth regulators, such as gibberellic acid, caused a change in POD activity [8]. Despite the apparent response of POD to terrestrial plant stress, little attention has been given to aquatic plant POD activity as a measure of stress.

This research investigated the use of POD activity as an indicator of sublethal stress in

---

[1] Department of Biology, Memphis State University, Memphis, TN 38152.

the aquatic angiosperm *Hydrilla verticillata* (Royle). *H. verticillata* is a rooted, aquatic vascular plant found in the southern United States. It is very prolific and grows well under adverse conditions. The plant is very amenable to culturing in the laboratory and has previously been examined as a bioassay organism for aquatic toxicity bioassays [9–10]. The study done by Hinman [10] used total plant dehydrogenase activity and chlorophyll$_a$ concentration as biochemical measures of chemical stress. This paper presents the response of *H. verticillata* POD to the chemical stresses, copper and a sulfonylurea herbicide, sulfometuron methyl (Oust). The chemical name of the herbicide is methyl 2-[[[[(4,6-dimethyl-2-pyrimidinyl) amino]-carbonyl] amino] sulfonyl] benzoate.

Materials and Methods

The plants were handled and grown using sterile techniques as described by Hinman [10]. Algal-free *H. verticillata* cultures were maintained in autoclaved 1-qt Mason jars with 500 mL of sterile, 10% Hoagland's solution. The Mason jars were covered with inverted petri dishes, and the Hoagland's solution was augmented with 200 mg sodium bicarbonate (NaHCO$_3$)/L of solution [9]. The plants were maintained in a growth chamber with constant temperature (25°) and light (40 to 50 μE m^{-2} s^{-1}). The stock plants were split up and transferred to fresh medium every six to eight weeks.

Jars with plants were set up just prior to initiating each experiment. The length of the roots and shoots were measured prior to placing the plants in the jars. Two plants were used per jar. There were three replicate jars. The experimental design was a randomized complete block. Stock solutions of Oust (technical grade) or CuSO$_4$ · 7H$_2$O were used to spike the media. The nominal concentrations of Oust tested were 0.0001, 0.001, 0.01, 0.1, 1.0, and 10.0 mg active ingredient/L. Nominal concentrations of Cu tested were 0.01, 0.1, 1.0, and 3.0 mg/L. The jars were randomized and placed in a growth chamber for seven days.

After seven days exposure, the plants were harvested. The plants were removed from the jars by forceps and rinsed in a water bath to remove any herbicide or copper residue on the plant surface. The plants were gently dried between paper towels. The final root and shoot length was measured. Then the shoots were removed from the roots and weighed. The excised shoots were placed into a cold porcelain mortar and macerated with a pestle. An amount of cold 0.5 *M* CaCl$_2$ solution equal to five times the wet weight (e.g., 5 mL/g tissue) was added to the slurry and mixed. This slurry was placed into clean, labelled test tubes and stored on ice. When eight samples were macerated, the test tubes were centrifuged for 8 min at 1000 *g*. The supernatant was decanted and saved on ice. The cell wall pellet was resuspended with an amount of CaCl$_2$ equal to 2.5 times the wet weight of the tissue (e.g., 2.5 mL/g tissue). This was centrifuged again and the supernatants collected in the previous containers. The cell wall pellet was resuspended a third time with CaCl$_2$ (a volume equal to 2.5 times the wet weight of the tissue). This slurry was centrifuged again and the supernatant was added to the previous extractions. Thus, the final supernatant extract was 10 mL CaCl$_2$ solution per gram wet tissue weight. This crude extract was stored on ice for 2 h before assaying for POD activity.

POD activity was measured using a modified version of the assay by Maehly and Chance [11]. The following two assay solutions were made. Solution A contained 810 mg phenol and 25 mg 4-aminoantipyrene in 50 mL distilled water. Solution B consisted of 0.3% hydrogen peroxide (H$_2$O$_2$) in 0.01 *M* HEPES buffer (pH 7.1). Just prior to assaying, 1.4 mL of solution A was added to a 3-mL glass cuvette. A 1.5-mL aliquot of Solution B was added to the cuvette. A 400-μL aliquot of the crude extract was the final addition to the assay mixture. The contents of the cuvettes were mixed. The cuvettes were placed in a spectro-photometer, and the change in absorbance was monitored at 510 nm for 3 min. Standards (Sigma Chemical Co.) of known horseradish POD activity were made in 0.5 M CaCl$_2$ and assayed under similar conditions. The resulting standard curve was used to transform the results of the crude extract into units of activity (as defined by Sigma Chemical Co).

Results

The results of the salt-extracted POD assays are shown in Figs. 1a and 2a. The results of the root and shoot growth are shown in Figs. 1b and 2b. Plants exposed to Oust herbicide displayed a dose-response curve for POD activity and growth (Figs. 1a and 1b). Concentrations of Oust equal to or greater than 0.01 mg/L showed a statistically significant (p = 0.05) two-fold increase in POD compared to the controls. Concurrent with the increase in POD activity, there was a 50% decrease in shoot length and a 75% reduction in root length. However, the change in shoot length was not significantly different from the controls except at 10 mg/L. Plants exposed to Oust concentrations of 0.001 to 10 mg/L showed a significant decrease in root length as compared to the control plants. The plants exposed to 10.0 mg/L had no change in root or shoot length and were discolored with some necrosis at the leaf tips.

Plants exposed to Cu^{2+} also displayed a dose-dependent response of POD activity (Fig. 2a). There was no significant change in shoot length at Cu^{2+} concentrations of 0.01 to 1.0 mg/L (Fig. 2b) or in root length at 0.01 to 0.1 mg/L. There was a significant decrease in shoot elongation at 3.0 mg/L and root elongation at 1.0 and 3.0 mg Cu^{2+}/L. Even though there was no statistically significant change in shoot length at 0.01 to 1.0 mg Cu^{2+}/L there was a noticeable, but not statistically significant (c.v. = 28%) increase in extractable POD activity from plants treated with 0.01 mg Cu^{2+}/L (Fig. 2a). The 0.1 mg Cu^{2+}/L dosage

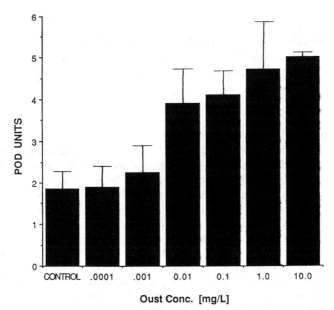

FIG. 1a—*Peroxidase activity extracted from* Hydrilla *shoots after exposure to Oust for seven days. Standard deviations are shown.*

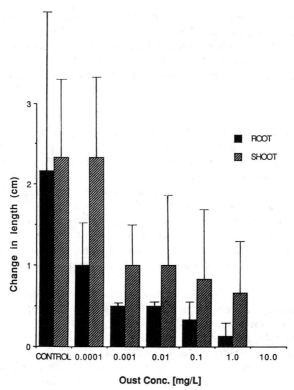

FIG. 1b—*Change in length of* Hydrilla *roots and shoots exposed to Oust for seven days. Standard deviations are shown.*

resulted in a statistically significant two-fold increase in POD activity. The POD activity peaked in plants treated with 1.0 mg Cu^{2+}/L for seven days. The plants treated with 3.0 mg/L started to show visible signs of necrosis in the tissue after seven days.

Discussion

Salt-extractable shoot POD activity from *Hydrilla verticillata* increased with increasing copper or Oust herbicide concentration. The change in shoot POD activity in response to Oust was comparable to the change in root length and more sensitive than that of the shoot. The herbicide Oust has its mode of action by inhibiting branched amino acid synthesis [12]. Thus, at the high concentrations of Oust, protein synthesis would have been interrupted and one might expect the plateau in POD activity (Fig. 1a). The change in shoot POD activity in response to Cu^{2+} was much more sensitive than elongation of either the roots or shoots. This increase in POD activity was probably the result of free radicles formed by the Cu^{2+} in the cell. The POD activity peaked in plants exposed to 1 mg Cu^{2+}/L and decreased in plants exposed to 3 mg Cu^{2+}/L. This reduction in POD activity was due to the advanced necrosis of the tissue. Divalent copper, at 3 mg/L, was lethal to *H. verticillata* under these experimental conditions after approximately two weeks exposure.

As with any biological system, there was variability among the plants within a treatment. This variability is illustrated by the standard deviation bars in Figs. 1 and 2. The coefficient

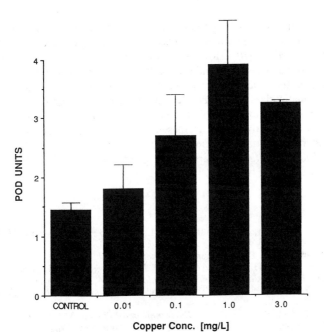

FIG. 2a—*Peroxidase activity extracted from* Hydrilla *shoots after exposure to* Cu^{2+} *for seven days. Standard deviations are shown.*

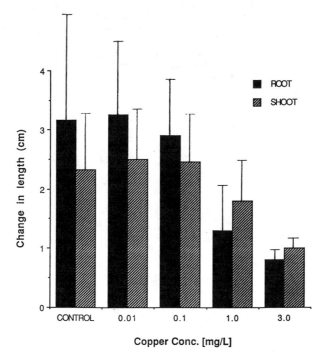

FIG. 2b—*Change in length of* Hydrilla *roots and shoots exposed to* Cu^{2+} *for seven days. Standard deviations are shown.*

of variations ranged from 28% in both low dose treatments to 1.7% in the high Oust concentration. This was good when compared to previous work done with biochemical indicators in *H. verticillata* [*10*]. However, future studies should consider using four replicates, which has proven to reduce standard deviation substantially in our continued research.

In summary, this study was performed to determine the feasibility of using salt extractable POD from the aquatic plant *H. verticillata* as a biochemical indicator of short-term chronic stress. Exposure of plants to sublethal concentrations of Oust or Cu^{2+} resulted in increased extractable shoot POD activity and decreased growth. The increase in POD activity corresponded to increasing concentrations of toxin. Based on these results, salt-extractable POD from *Hydrilla verticillata* appears to be a good indicator of sublethal stress and may be useful in future aquatic bioassays.

References

[*1*] Lee, K. C., Cunningham, B. A., Paulsen, G. M., Liang, G. H., and Moore, R. B., *Physiologia Planta*, Vol. 36, 1976, pp. 4–6.
[2] Mukherji, S. and Gupta, B. D., *Physiologia Planta*, Vol. 27, 1972, pp. 126–129.
[*3*] Morgan, P. W., Joham, H. E., and Amin, J. V., *Plant Physiology*, Vol. 41, 1966, pp. 718–724.
[*4*] Keller, T. H., *European Journal of Forest Pathology*, Vol. 4, 1974, pp. 11–19.
[*5*] Endress, A. G., Suarez, S. J., and Taylor, O. C., *Environmental Pollution, Series A.*, Vol. 22, 1980, pp. 47–58.
[6] Horsman, D. C. and Wellburn, A. R., *Environmental Pollution*, Vol. 8, 1975, pp. 123–134.
[7] Peters, J. L., Castillo, F. J., and Heath, R. L., *Plant Physiology*, Vol. 89, 1989, pp. 159–164.
[8] Fry, S. C. in *Oxford Surveys of Plant Molecular and Cell Biology*, B. J. Milfin, Ed., Vol. 2, 1985, pp. 1–42.
[9] Klaine, S. J. and Ward, C. H., *Annals of Botany*, Vol. 53, 1984, pp. 503–514.
[*10*] Hinman, M. L., "Utility of Rooted Aquatic Vascular Plants for Aquatic Sediment Hazard Evaluation," dissertation, Memphis State University, December 1989, pp. 1–31.
[*11*] Maehly, A. C. and Chance, B. in *Methods of Biochemical Analysis*, Vol. 1, D. Glick, Ed., Interscience, New York, 1954, p. 357.
[*12*] Beyer, Jr., E. M., Duffy, M. J., Hay, J. V., and Schuelter, D. D., "Sulfonylureas," *Herbicides—Chemistry, Degradation and Mode of Action*, P. C. Kearney and D. D. Kaufman, Eds., Vol. 3, Marcel Dekker, Inc., New York, 1987.

Hans G. Peterson[1]

Toxicity Testing Using a Chemostat-Grown Green Alga, *Selenastrum Capricornutum*

REFERENCE: Peterson, H. G., **"Toxicity Testing Using a Chemostat-Grown Green Alga,** *Selenastrum Capricornutum*," *Plants for Toxicity Assessment: Second Volume, ASTM STP 1115,* J. W. Gorsuch, W. R. Lower, W. Wang, and M. A. Lewis, Eds., American Society for Testing and Materials, Philadelphia, 1991, pp. 107–117.

ABSTRACT: Regulatory toxicity testing with algae commonly use 72- or 96-h inhibition of growth by the green algae *Selenastrum, Chlorella,* or *Scenedesmus*. The long incubation period is sufficient for the algae to change the composition of the test medium. Organism-induced changes include the release of organics and the uptake of nutrients with accompanying pH changes. Both the released organics and the variable pH can change the chemical speciation of the toxicants; this may alter their toxicity. In addition, the competitive interactions between hydrogen ions and toxicants can affect toxicity as the pH varies. It is shown here that cadmium (Cd) toxicity to *Selenastrum capricornutum* increases by up to eight times for every unit of pH increase. Copper (Cu) with its different speciation pattern does not vary as much as Cd, and the pattern is different. Because the toxicity of the two metals vary differently with pH, Cd is 500 times less toxic than Cu at pH 6, but Cd is twice as toxic as Cu at pH 10.

The present regulatory protocols call for nutrient-sufficient conditions, but algae are likely to be limited by either nitrogen (N) or phosphorus (P) in nature. Here, *Selenastrum* has been grown in chemostats with both N and P approaching limiting levels. Inhibition of N and P uptake, rather than growth, has been used to assess toxicity in short-term experiments. The advantage of chemostat, rather than batch cultivation (as used in present regulatory tests), is the ability to produce algae of known and well-defined physiological states.

KEY WORDS: *Selenastrum,* toxicity testing, regulatory protocols, test conditions, copper, cadmium, metal speciation, pH

Methods used in regulatory protocols [1–3] should ideally account for chemical and biological factors that may affect toxicity. Chemical factors include speciation of toxicant, amount of toxicant in true solution, and hydrogen ion competition with toxicants on the algal cell surface. Biological factors include the physiological state of the algae and the choice of toxicity indicator (endpoint). Because these factors may affect toxicity, we must address how test conditions and endpoints can be optimized to reflect accurately toxicant effects.

Nyholm and Kallqvist [4] addressed some of these concerns when reviewing the USEPA *Selenastrum* test [1] as well as some other algal bioassays used for regulatory purposes [2,3]. They showed that when the *Selenastrum* test is carried out under standard conditions the pH may change by two units (100-fold change in the hydrogen ion concentration) from the start to the end of the incubation. Peterson et al. [5] and Peterson and Healey [6] showed that a pH shift of one unit may alter metal toxicity to *Scenedesmus* ten-fold. The increase in toxicity with an increase in pH was attributed to decreasing competition from hydrogen

---

[1] Lead scientist, Water Quality Section, Saskatchewan Research Council, Saskatoon, Saskatchewan, Canada S7N 2X8.

ions. Therefore, metal ion access to the surface of algal cells is increased with a decrease in the concentration of hydrogen ions associated with the cell surface. *Selenastrum* may therefore be exposed to toxicity levels that vary 100-fold during the 96-h incubation. Nyholm and Kallqvist [4] made the following comment: "It should be rather obvious that no meaningful use can be made of algal toxicity data as currently reported in the literature unless a critical evaluation of the test systems used to generate the data is made."

The hydrogen ion concentration and dissolved organics released by the algae affect the chemical speciation of both inorganic and organic toxicants. It is now possible to use computer models to predict which form some chemicals will be in by using the general chemical composition of the medium along with other characteristics, such as temperature and pH. One such model is MINTEQA1 [7; updated version MINTEQA2]. Equilibrium between different forms of the toxic substance and other components in the artificial media is usually rapid (<1 min), but may be slow if naturally occurring dissolved organic substances are present (>1 h).

Cadmium toxicity to *Scenedesmus quadricauda* was strongly linked to pH, and copper toxicity was also pH dependent in the range where free hydrated copper ions dominated the copper pool [5,6]. The simplest explanation of the pH-dependent toxicity of copper (Cu) and cadmium (Cd) may be high toxicity for free hydrated metal ions (and positive inorganic complexes) with lower toxicity for uncharged complexes [5,6]. The toxicity for these ions is not constant, but may increase by more than 100 times from pH 5 to 8. Although inorganic complexes dominate the Cu pool at pH 8, the toxicity of Cu is likely due to the low level of free cupric ions which successfully displace hydrogen ions on the cell surface [5]. The pH dependency of organic complexation was not addressed in these studies, but organic complexes have often been shown to be nontoxic [8,9].

Stringent control of pH and the formation of inorganic and organic complexes that may not be toxic should therefore be monitored. It is also important to ensure that precipitation of toxicants or major elements in the nutrient medium is controlled or eliminated. The elements that will potentially precipitate can also be predicted from different computer programs. A systematic evaluation of the importance of speciation in the toxicity of metals and pesticides may show that speciation characteristics must be addressed in regulatory assessments of toxic compounds.

The present regulatory testing requirements rely on nutrient-sufficient cells, i.e., the physiological state of the cells is excellent. Peterson et al. [5] and Peterson and Healey [6] argued that most natural phytoplankton blooms will be limited by either phosphorus (P) or nitrogen (N); any impairment of these nutrient uptake systems is likely to have both short- and long-term effects on the phytoplankton. The most sensitive indicator of toxicity may therefore be inhibition of the N and P uptake systems.

It is possible to produce steady-state populations of nutrient-limited cells in chemostats. Peterson and Healey [6] used inhibition of the nitrate, ammonium, and phosphate uptake systems when determining toxic effects of copper and cadmium in relation to pH. Hall, Healey, and Robinson [10] showed that chemostat cultures of *Chlorella* were four orders of magnitude more sensitive to copper than batch cultures. Both Peterson and Healey [6] and Hall, Healey, and Robinson [11] pointed out differences in sensitivity between the N and P uptake systems.

A further advantage of using inhibition of nutrient uptake systems is the ability to automate the tests. We have adapted nutrient uptake for 96-well microplate assays. The algae are incubated, and the nutrient uptake is determined on the microplate. After incubation, a colorimetric reaction is used to determine nutrient uptake. The color is measured, and the data for a 96-well plate are stored on a floppy disc within 15 s. These data can then be transferred directly into a spreadsheet program such as Lotus-123.

There are two microplate characteristics that limit their usefulness in several applications. First, the test volume per well is usually not larger than 0.25 mL, and longer-term toxicant tests, such as 72 to 96-h growth tests, are difficult to carry out because of evaporation losses from the small wells. The plates have to be shaken to achieve efficient gas transfer (reducing pH drift); this further increases evaporation losses. However, even intense shaking may not provide sufficient gas transfer to avoid pH drift. It is therefore likely that only certain tests are suitable for microplates or the data can be used for preliminary screenings only. Second, microplates are made of plastics and some toxicants adhere to plastics, in which case experiments are better performed in glass.

Present regulatory tests with algae are difficult to automate because cell counting is mostly required. Sometimes the optical density is used, but the cell densities where this is a satisfactory endpoint coincide with changes induced by the test algae to the medium. The need to use lower inoculum or shorter exposure times therefore increases the need for more suitable endpoints.

Improvements in present tests can therefore be accomplished in several different areas. Here, test conditions and endpoints are addressed.

Method

The methods are similar to those described previously [5,6]. Axenic cultures of *Selenastrum capricornutum* obtained from the University of Texas culture collection (UTEX 1648) were grown in chemostats at a growth rate of 0.25 divisions/day. Both nitrogen and phosphorus were set to approach limiting levels at a molar N/P ratio of 20 [267 μM sodium nitrate ($NaNO_3$), 13.3 μM potassium dihydrogen orthophosphate (KH_2PO_4)] (μM = mmol/m^3). Other components of the medium were as in Woods Hole MBL [*12*] except a 50% reduction in ferric chloride ($FeCl_3$) additions and additions of boric acid (H_3BO_3) (16 μM) and the pH buffer HEPES (1 mM). The alga was grown at a light intensity of 60 μmol photons m^{-2} s^{-1} in 200-mL chemostats similar to those described by Healey and Hendzel [*13*]. The chemostats were inoculated with exponentially growing *Selenastrum* into around 30 mL of medium; this was kept at low light until visible growth was apparent. The chemostat tube was then transferred to a constant temperature water bath (25°C), and culture medium was pumped into the tube. Steady state of the alga was determined from optical density readings at 750 nm; at steady state the daily variation in optical density was less than 10%.

The media used for the tests were composed of the following chemical components: calcium chloride ($CaCl_2$) 250 μM, magnesium sulfate ($MgSO_4$) 150 μM, and sodium silicate (Na_2SiO_3) 100 μM. Sodium hydrogen carbonate ($NaHCO_3$) was added in quantities close to those required for atmospheric equilibrium with inorganic carbon at each pH. Hydrochloric acid (HCl) and sodium hydroxide (NaOH) were used to adjust pH. The water used to make up the media was purified using softening followed by carbon filtration; it then passed through Millipore units, first one carbon filter followed by reverse osmosis; this water was fed into a Milli-Q Plus consisting of one carbon filtration unit, two mixed-bed deionizing resins, one Organex-Q unit (organic carbon scavenger), one ultra-filter (to remove low-molecular weight organics including endotoxins), and one Teflon filter (pore size 0.22 μm). The media were prepared in 2-L Pyrex flasks before the experiments. To achieve fast inorganic carbon equilibrium and pH stabilization of the media, they were sparged with organic free air (trace organics removed by an activated carbon filter). The pH of the media before incubation with algae were kept within 0.05 pH units of the target pH as measured with an Orion Research Expandable Ionanalyzer EA 940 with a Ross Orion pH combination electrode model 81-02.

Algae were withdrawn from the chemostat and centrifuged four times at 2500 g and resuspended in new medium. Algae (2×10^6 cells/mL) and toxicants were added to Corning 96-well microplates No. 25855-96 with Corning microplate lids No. 25803. The algae were incubated with toxicants (total volume 240 μL) on a microplate shaker at 650 rpm (SLT Labinstruments, Fisher Scientific) for 1 h. Nutrient was then added, and the algae were then incubated for either 1, 2, or 3 h.

To stop the uptake, 35 μL of 0.625 M NaOH (first reagent for nitrate analysis, pH > 12) was added to the nitrate uptake experiments. Phosphate uptake was stopped by adding 40-μL ammonium-molybdate and sulphuric acid (pH < 3). Phosphate was analyzed by a stannous chloride method [14]. Nitrate was analyzed by adapting USEPA method 353.1 [15].

The optical density of the chemical determinations was read in a VMAX Kinetic Microplate Reader (Molecular Devices Corp., Menlo Park, California) connected to an Epson Equity 1 computer. The optical density data were converted into print files and imported into a Lotus 123 spreadsheet. Nutrient uptake and percent inhibition were calculated and the data imported into a Sigmaplot Vers. 4 graphing package (Jandel Scientific). Percent inhibition of nutrient uptake by the metal was calculated using the control uptake rate as 0% inhibition. Four replicate wells were used for each concentration.

Results

The effects of Cu and Cd were determined on washed, chemostat-grown, N- and P-deficient *Selenastrum*. Percent inhibition of nitrate and phosphate uptake were determined for five different pH levels: pH 6, pH 7, pH 8, pH 9, and pH 10. Initial pH was within 0.05 pH units of the target pH. It was not possible to avoid pH drift even with rapidly shaken microplates; the pH drift averaged 0.3 units. The rate of nutrient uptake at pH 5 was too low to be used as a toxicity endpoint. To compensate for lower nutrient uptake rates at extreme pH levels, the incubation time with the nutrient was increased to up to 3 h. The aim was to have uptake rates around 75% of the added nutrient in control cultures [nitrate additions (NO$_3$) 35 μM, phosphate additions (PO$_4$) 20 μM].

Percent inhibition of the nitrate (Fig. 1) and phosphate (Fig. 2) uptake systems by Cd show saturation-type kinetics. Drastically different concentrations of Cd were used depending on pH to inhibit the nutrient uptake systems 50%. The aim was to have two concentrations on either side of the 50%-inhibition level. The concentrations required to inhibit both phosphate and nitrate uptake were therefore highly pH dependent, and in Fig. 3 the 50% inhibition levels are shown.

Both nutrient uptake systems show a similar pattern with 50% inhibition levels above 10-μM Cd at pH 6; these levels gradually decrease with an increase in pH to as low as 0.07 μM for the nitrate uptake system. At neutral or acidic pH levels the nitrate uptake system shows similar sensitivity to the phosphate uptake system. However, the nitrate uptake system becomes five times more sensitive than the phosphate uptake system at pH 8 and 9 with a further increase to more than 10 at pH 10. At pH 10, the nitrate and phosphate uptake systems were inhibited 50% by 0.07 and 1.03-μM Cd, respectively.

Cu inhibition of the nitrate and phosphate uptake systems also showed saturation-type kinetics. The 50% inhibition of the nitrate and phosphate uptake systems (Fig. 4) show that the nitrate uptake system was more sensitive to Cu; this was especially pronounced at neutral or acidic pH levels. The nitrate uptake system was most sensitive at pH 6, where Cu was 9 times more toxic than at pH 9. At pH 6 the nitrate uptake system was inhibited 25% at a Cu level of 0.013 μM (0.8 μg Cu/L), 50% at a level of 0.02 μM (1.3 μg/L), and 75% at a level of 0.056 μM (3.6 μg/L).

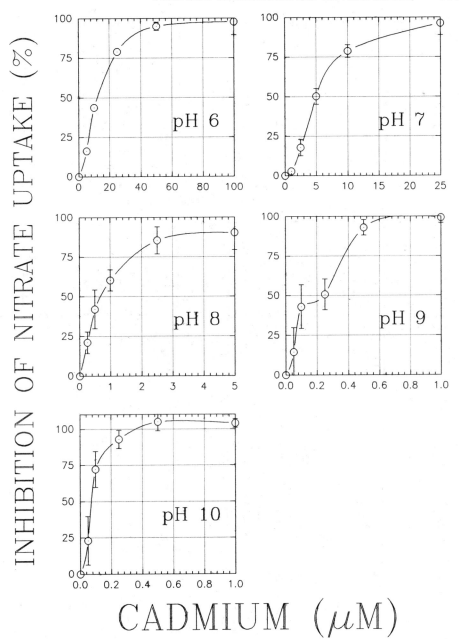

FIG. 1—*Total dissolved Cd inhibition (%) of nitrate uptake relative to controls by* Selenastrum capricornutum *as a function of pH. Each datapoint represents the mean of four replicates with bars representing the standard deviation.*

Discussion

The physical and chemical characteristics of aquatic ecosystems are extremely variable. The availability and toxicity of pesticides and metals to aquatic organisms are linked to this physico-chemical variability. The problem is that the linkages are dependent on the toxicant

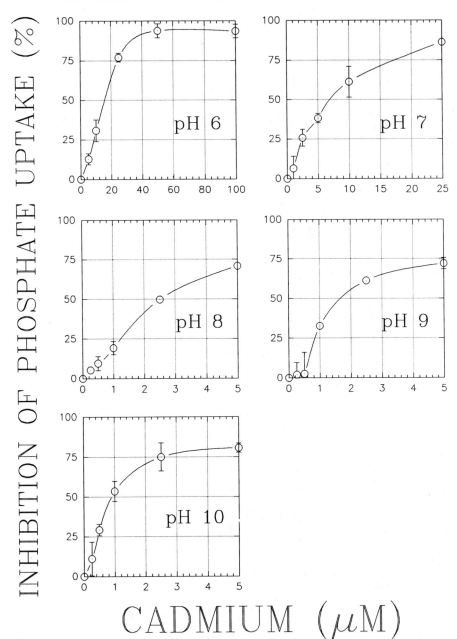

FIG. 2—*Total dissolved Cd inhibition (%) of phosphate uptake relative to controls by* Selenastrum capricornutum *as a function of pH. Each datapoint represents the mean of four replicates with bars representing the standard deviation.*

studied. Without knowledge of these physical and chemical linkages to the toxicity of a compound, it is difficult to predict correctly the impact of the compound in different aquatic systems. Regulatory toxicity testing is not designed to assess these links. Indeed, the USEPA *Selenastrum* test, when carried out according to the test protocol, is a victim of physicochemical linkages.

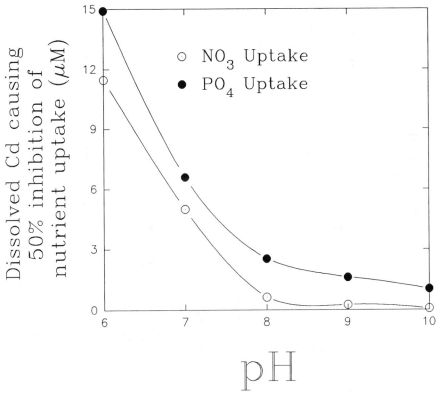

FIG. 3—*Total dissolved Cd required to cause 50% inhibition of nitrate and phosphate uptake (determined graphically from Figs. 1 and 2).*

Nyholm and Kallqvist [4] showed that the pH may increase by two pH units during a standard *Selenastrum* 96-h incubation. Peterson et al. [5] showed that the toxicity of Cd increased by almost 200 times over three pH units for the green alga *Scenedesmus quadricauda*. Here we have also shown for *Selenastrum capricornutum* that pH is an important factor in determining the toxicity of a compound. The toxicity of Cd increased 170 times from pH 6 to pH 10 using inhibition of nitrate uptake as the toxicity endpoint. Using phosphate uptake, toxicity increased 15 times from pH 6 to 10. With Cu, however, the pH dependency was different and not as pronounced. The inhibition of phosphate uptake was five to nine times less sensitive to Cu at pH 6 than at higher pH levels; Cu was most toxic at pH 7 and 8. For nitrate uptake, however, there was a decrease in Cu toxicity with pH with highest toxicity at pH 6 to 8. The different patterns in toxic response to Cu and Cd with pH make comparisons between the toxicity of these two metals difficult.

Comparisons of toxic effects are therefore dependent on pH and the endpoint used. If the discussion is restricted to the most sensitive endpoint in the experiments reported here, the inhibition of nitrate uptake, the comparison can be summarized as follows. Cd is 500 times less toxic than copper at pH 6, but Cd is twice as toxic as Cu at pH 10. The evaluation of toxic effects without any knowledge of such dramatic pH-mediated toxicity alterations is not acceptable.

To gain an understanding of the reasons for differential pH responses, one must evaluate the speciation of the toxicants. The speciation of Cd and Cu in our media was calculated using the MINTEQA2 chemical speciation program [7]. In Fig. 5 the inorganic speciation

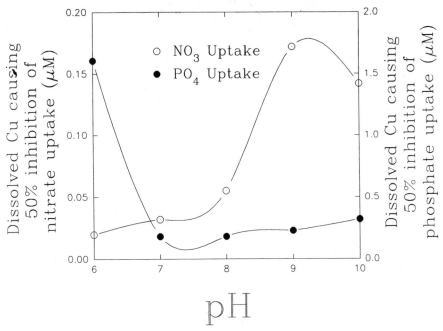

FIG. 4—*Total dissolved Cu required to cause 50% inhibition of nitrate and phosphate uptake (determined graphically from inhibition curves as in Figs. 1 and 2).*

characteristics were separated into four groups: free hydrated ions and positive, negative, and uncharged complexes.

The free metal ions remain high for both Cu and Cd from pH 5 to 6. For Cu the free metal ion levels drop strongly at pH 7, and at pH 8 less than 1.0% of the total Cu pool is present as free metal ions. For Cd, there is a drop at pH 8, but the free hydrated Cd ions are not reduced to less than 1% until pH 10. *Selenastrum* was only tested from pH 6 to 10 because of low nutrient uptake rates at pH 5. However, for Cu the only two pH levels (between 5 and 10) where the free cupric ions were unchanged were pH 5 and 6. These are therefore the only two pH levels where we can study the effects of hydrogen ions without major changes in the speciation of the Cu pool. *Selenastrum* is therefore not a good candidate for toxicity testing when speciation effects of Cu are to be evaluated.

The free hydrated Cd ions remain almost unchanged from pH 5 to 7 with a 30% drop at pH 8; this is followed by a reduction to 2 and 1% at pH 9 and 10, respectively. The amount of Cd required to inhibit both the nitrate and phosphate uptake systems decreases strongly with an increase in pH. A similarly strong pH dependency of Cd toxicity was shown for *Scenedesmus quadricauda* [5,6]. In contrast, Cu toxicity did not conform to this simple pattern [5,6]. Instead, copper toxicity increased from pH 5 to 6 and then either remained unchanged or decreased at high pH. This was accompanied by large changes in copper speciation from pH 6 to 8. Peterson et al. [5] suggested that these results could be most simply interpreted as a protection of the cells by hydrogen ions; positively charged ions or complexes will not have access to uptake sites on the cell surface. The lack of increased toxicity for copper at higher pH levels was explained as lower toxicity of the uncharged inorganic complexes that dominate the copper pool at high pH.

Cu was most toxic to the nitrate uptake system at pH 6 for *Selenastrum*; this was also the

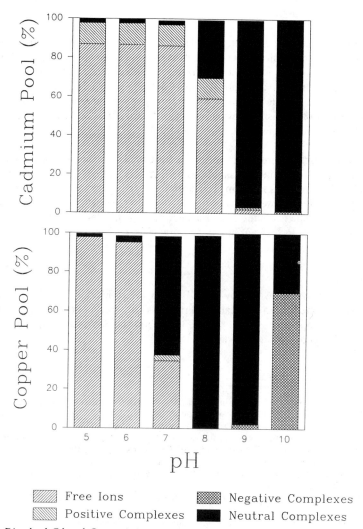

Free Ions   Negative Complexes
Positive Complexes   Neutral Complexes

FIG. 5—*Dissolved Cd and Cu speciation as a function of pH (calculated with MINTEQA2). Four groups of compounds are shown, free hydrated ions, negative, positive, and uncharged inorganic complexes.*

case for *Scenedesmus* [6]. Where the free hydrated Cu ions remained almost constant (pH 5 to pH 6), toxicity increased three-fold for *Scenedesmus* [6]. Our data on *Selenastrum* are therefore quite similar to those of Peterson and Healey [6]. Using inhibition of nitrate uptake as the toxicity indicator, the similarities are evident. The 50% Cu inhibition level for *Scenedesmus* was 0.017 μM at pH 6 (0.020 μM for *Selenastrum* in our study), 0.027 μM at pH 7 (0.032 μM), 0.061 μM at pH 8 (0.055 μM). The Cu levels that were toxic to nitrate uptake by both *Scenedesmus* and *Selenastrum* were lower than standard algal assay test results [e.g., *16,17,18,19*]. The ability to detect low levels of Cu could make nitrate uptake by chemostat-grown algae an ideal candidate for detecting Cu at levels that have been regulated for the protection of the aquatic environment. The Cu levels that have been suggested for the protection of aquatic life range from 0.03 to 0.09 μM (2 to 6 $\mu g/L$) [*20*].

Conclusions

Toxicity testing will become a better regulatory tool if toxicant speciation and its effects can be resolved. Other chemical linkages, such as hydrogen ion competition with toxicants, should also be determined. Rather than striving to determine one key concentration on which to base regulations, it may be more useful to determine the toxic profile of a compound. Toxicity is dependent on the endpoint used, and several endpoints may be selected to establish the toxic profile. Using inhibition of nitrate uptake as an endpoint, it was shown here that Cd was 500 times less toxic than Cu at pH 6, but Cd was twice as toxic as Cu at pH 10.

Algae and cyanobacteria are ideal candidates for establishing such toxic profiles because they grow fast, can be manipulated simply using a variety of methods, and can be exposed to toxicants in a homogenous medium. In addition, these organisms rarely have specialized cells, making toxicity interpretations easier. These characteristics also make them ideal candidates for improvements in the cost efficiency of the tests. Several genera must be tested to establish generalities in toxicant action and organism response.

Acknowledgments

Technical support was provided by Jaret Bogdan, Mary Moody, and Lloyd McKenzie. Discussions with Celine Boutin and Kathryn Freemark (Environment Canada) were essential to this work. The support of Environment Canada and the Saskatchewan Agriculture Development Fund is gratefully acknowledged.

References

[1] Miller, W. E., Greene, J. C., and Shiroyama, T., "The *Selenastrum capricornutum* Printz Algal Assay Bottle Test: Experimental Design, Application and Data Interpretation Protocol," EPA-600/9-78-018, U.S. Environmental Protection Agency, Corvallis, OR, 1978.
[2] Organization for Economic Cooperation and Development (OECD), "Guideline for Testing of Chemicals," No. 201, Algae Growth Inhibition Test, Paris, France, 1984.
[3] International Standards Organization (ISO), "Water Quality—Fresh Water Algal Growth Inhibition Test with *Scenedesmus subspicatus* and *Selenastrum capricornutum*," International Standard ISO/DIS 8692, Geneva, Switzerland, 1989.
[4] Nyholm, N. and Kallqvist, T., *Environmental Toxicology and Chemistry*, Vol. 8, 1989, pp. 689–703.
[5] Peterson, H. G., Healey, F. P., and Wagemann, R., *Canadian Journal of Fisheries and Aquatic Sciences*, Vol. 41, 1984, pp. 974–979.
[6] Peterson, H. G. and Healey, F. P., *Journal of Phycology*, Vol. 21, 1985, pp. 217–222.
[7] Brown, S. D. and Allison, J. D., "Minteqal, an Equilibrium Metal Speciation Model: User's Manual," EPA/600/3-87/012, U.S. Environmental Protection Agency, Environmental Research Laboratory, Athens, GA, 1987.
[8] Sunda, W. G. and Guillard, R. R. L., *Journal of Marine Research*, Vol. 34, 1976, pp. 511–529.
[9] Anderson, D. M. and Morel, F. M. M., *Limnology and Oceanography*, Vol. 23, 1978, pp. 283–295.
[10] Hall, J., Healey, F. P., and Robinson, G. G. C., *Aquatic Toxicology*, Vol. 14, 1989, pp. 15–26.
[11] Hall, J., Healey, F. P., and Robinson, G. G. C., *Aquatic Toxicology*, Vol. 14, 1989, pp. 1–14.
[12] Nichols, H. W. in *Handbook of Phycological Methods: Culture Methods and Growth Measurements*, J. R. Stein, Ed., Cambridge University Press, New York, 1973, pp. 7–24.
[13] Healey, F. P. and Hendzel, L., *Journal of Phycology*, Vol. 11, 1975, pp. 303–309.
[14] *NAQUADAT—Dictionary of Parameter Codes 1988*, Data Systems Section, Water Quality Branch, Environment Canada, Ottawa, Canada.
[15] United States Environmental Protection Agency, 1983, "Methods for Chemical Analysis of Water and Wastes," Environmental Monitoring and Support Laboratories, Cincinnati, OH.
[16] Bartlett, L., Rabe, F. W., and Funk, W. H., *Water Research*, Vol. 8, 1973, pp. 179–185.

[*17*] Blaise, C., Legault, R., Bermingham, N., Coillie, van R., and Vasseur, P., *Toxicity Assessment,* Vol. 1, 1986, pp. 261–281.
[*18*] Christensen, E. R., Scherfig, J., and Dixon, P. S., *Water Research,* Vol. 13, 1979, pp. 79–92.
[*19*] Michnowicz, C. J. and Weaks, T. E., *Hydrobiologia,* Vol. 118, 1984, pp. 299–305.
[*20*] "Canadian Water Quality Guidelines," Water Quality Branch, Environment Canada, Ottawa, Ontario, 1987.

Ronald J. Breteler,[1] *Rosanna L. Buhl,*[2] *and Alan W. Maki*[3]

The Effect of Dissolved H₂S and CO₂ on Short-Term Photosynthesis of *Skeletonema Costatum*, a Marine Diatom

REFERENCE: Breteler, R. J., Buhl, R. L., and Maki, A. W., "**The Effect of Dissolved H₂S and CO₂ on Short-Term Photosynthesis of** *Skeletonema Costatum*, **a Marine Diatom,**" *Plants for Toxicity Assessment: Second Volume, ASTM STP 1115*, J. W. Gorsuch, W. R. Lower, W. Wang, and M. A. Lewis, Eds., American Society for Testing and Materials, Philadelphia, 1991, pp. 118–125.

ABSTRACT: The planktonic diatom *Skeletonema costatum* was exposed during 4-h periods to dissolved forms of H_2S and CO_2. The experimental design permitted evaluation of the acute toxicity of these gases by measuring the reduction in photosynthesis using the ^{14}C assimilation technique. Carbon dioxide gas, measured with a CO_2-specific electrode, demonstrated a EC_{50} value of 0.69-mM molecular CO_2. The inhibition in algal photosynthesis was largely attributed to the decrease in pH with increasing exposure concentrations (from pH 8.45 to 5.73). Hydrogen sulfide, measured colorimetrically, showed no adverse effects up to 24-μM total sulfide corresponding with 1.2-μM molecular H_2S. The calculated EC_{50} value was 3.1-μM H_2S. The toxicity tests yielded results useful for the assessment of potential harmful effects to marine diatoms of discharge of these gases into ocean waters by off-shore natural gas production platforms.

KEY WORDS: *Skeletonema costatum*, H_2S, CO_2 marine ecotoxicology, hazard assessment, bioassays, diatoms

The fixation of carbon by phytoplankton photosynthetic processes depends on the aqueous carbonate system. Yet, excessive dissolved carbon dioxide, such as may occur when waste gases are released below the sea surface during offshore gas production, may cause a drop in pH and thereby alter the natural, buffered environment essential for primary productivity. Similarly, the dissolution of hydrogen sulfide during subsea surface bubbling of off gases is expected to exert acute toxic effects to algae, as this gas is known to be highly toxic to fish and invertebrates [*1*].

The toxic effects to marine phytoplankton of dissolved CO_2 and H_2S have not previously been investigated, perhaps due to the practical difficulty in maintaining stable exposure concentrations of these unstable materials. Dissolved carbon dioxide is readily assimilated by algae, while dissolved sulfide is rapidly converted to sulfate or to intermediate oxidation products [2]. The study reported here was designed to assess the toxicity of these gases, both in total dissolved form and as undissociated molecular gas, to a representative planktonic diatom, *Skeletonema costatum* (Greville). The toxicity test was based on measurements of inhibition of photosynthesis during short-term, 4-h exposures to several concentrations of these gases under controlled laboratory conditions. The purpose of the investigation was

---

[1] Springborn Laboratories, Inc., Wareham, MA 02571.
[2] Battelle Ocean Sciences, Duxbury, MA 02359.
[3] Exxon USA, Western Division, Anchorage, Alaska 99502.

to make predictions concerning the potential deleterious effects of below sea surface sparging of these gases during the production of natural gas.

Procedure

Algal Cultures

An axenic culture of the diatom *Skeletonema costatum* (Greville) clone *Skel* was maintained in an artificial, autoclaved seawater medium of 30 ‰ salinity according to the methods of Walsh and Alexander [3]. The original cultures were provided by G. E. Walsh, EPA, Environmental Research Laboratory, Gulf Breeze, Florida. Cultures were grown under constant temperature (20 ± 2°C and light (4000 lm/m^2 and 10-h light and 14-h dark cycle) conditions and continuously swirled on a gyratory shaker. To obtain cells in an exponential growth phase, sterile seawater medium with ethylenediamine tetraacetate (EDTA) was inoculated with *S. costatum* one week prior to the start of a test. The photosynthetic activity of an aliquot of the culture was measured one day before the test start to ensure that the culture was in a logarithmic growth phase. This was estimated by using the 4-h test technique described below. The culture was found to be acceptable if the inoculum resulted in a carbon assimilation equivalent to at least 2000 counts/min [4].

Test Procedures

Procedures used to measure reduction in photosynthesis were based largely on Giddings et al. [4] with test-specific modifications. Prior to performance of the definitive tests, several range-finding tests were conducted to select the most appropriate series of test material concentrations. Because photosynthetic activity and chemical oxydation were expected to result in a decrease in carbon dioxide and hydrogen sulfide, respectively in at least some of the concentrations during the 4-h test period, sufficient concentrations were included to ensure that both effect and no effect levels were included. Seven CO_2 and twelve H_2S test concentrations were prepared in 125-mL Erlenmeyer flasks by addition of an artificial seawater solution, inoculating with 1 mL of a dense suspension of *S. costatum* (approximately 1.25 x 10^7 cells/mL), and spiking with incremental volumes of sodium bicarbonate or sodium sulfide stock solutions. Concentrated stock solutions of CO_2 and H_2S were prepared by addition of reagent grade $NaHCO_3$ and Na_2S, respectively, to distilled water. Total volume of the medium in each flask was 92 mL. Three standard liquid scintillation vials (25-mL volume) were filled with 23.5 mL of the test solution containing the algae. The remaining solution was used for H_2S or CO_2 analyses and for pH determinations. Blanks containing all chemicals but no algae also were included.

Experiments were carried out at 20°C, and 30‰ salinity. H_2S test solutions were adjusted to pH = 7.9 ± 0.2, and the pH was measured at 0 and 4 h. As pH was a test variable in the CO_2 exposure bioassay, it was not adjusted. The pH was determined at 0, 2, and 4 h.

The vials were placed on a light table lit with two cool white fluorescent lamps with an approximate light intensity of 1400 lm/m^2. After 2 h, each vial was spiked with 40-μL ^{14}C-labeled sodium bicarbonate (New England Nuclear, specific activity = 8.4 mCi/mM, concentration 1.19 mM), a glass bead was added, the solution was thoroughly mixed using a vortex mixer and returned to the light table. After two additional hours, the vials were removed from the light table, mixed again, and 5 mL of solution was removed for analysis of the test material and pH determination. To terminate carbon assimilation, a 0.31-mL mixture of deionized water:formalin:glacial acetic acid (5:4:1, v/v) was added, and the vials were capped and mixed. Residual, inorganic ^{14}C was removed by aeration for 30 to 45 min. Next, 1.0 mL of solution was added to a new vial and mixed with 15 mL of liquid scintillation cocktail (Aquasol-2, New England Nuclear). The vials were radioassayed with a Beckman LS 7500 liquid scintillation system.

In the experiment with CO_2, radiolabeled sodium bicarbonate ($NaH^{14}CO_3$) was spiked to test solutions containing increasing amounts of unlabeled CO_2. Therefore, the measured photosynthetic response (in cpm) was converted to account for the fraction of tagged carbon dioxide of the total CO_2 pool. Following standard limnological techniques, and ignoring possible discrimination of $^{14}CO_2$ from the total inorganic carbon supply used for cell growth [5], the ratio of ^{14}C assimilated and ^{14}C available was assumed to equal the total quantity of carbonate assimilated divided by the total available carbon pool. In our experiment, the fraction of assimilated ^{14}C of the total phytoplankton organic carbon was normalized by multiplying the measured photosynthesis (in cpm, corrected for blank) with the ratio of the molarities of the mean total carbonate measured and the total ^{14}C added to the growth medium (2.026×10^{-6} mol). The resulting normalized photosynthetic response was divided by 1000 for convenience.

Analytical Procedures

Concentrations of the test material were measured at start and termination of the 4-h exposures. CO_2 was measured using a CO_2-specific electrode (HNU, Model ISE-10-22-00). Prior to measurement of CO_2 concentrations, samples were adjusted to a pH of ≤ 4.5, using a 5-M NaH_2PO_4 buffer to ensure that almost all dissolved carbonate was present as molecular CO_2. The electropotential readings were converted to total inorganic CO_2 by linear calibration of mV against log molar bicarbonate standards. The concentrations of molecular, aqueous CO_2 were calculated from each corresponding measured concentration of total inorganic carbonate and from the corresponding pH, salinity, and temperature, and the means and standard deviations of molecular carbon dioxide presented elsewhere in this report were obtained from these individual calculated concentration values. Calculations were based on the two ionization reactions of the aqueous carbonate system [6]. Values of the dissociation constants K_1 and K_2 are salinity and temperature dependent and were obtained from Lyman [7]. The following relationship, incorporating the two ionization reactions between bicarbonate, carbonate, and aqueous carbon dioxide [6], was derived to express aqueous, molecular CO_2 as a function of total dissolved carbonate, C_T

$$\frac{CO_2,aq}{C_T} = \frac{1}{1 + 10^{(pH - pK_1)} + 10^{(2pH - pK_1 - pK_2)}} \tag{1}$$

Concentrations of total dissolved sulfide were measured based on the reaction among *P*-aminodimethylaniline, ferric chloride, and sulfide ion, resulting in the formation of methylene blue, which was measured colorimetrically at 667 nm using a Perkin-Elmer Lambda 3 UV/VIS spectrophotometer. The method was based largely on Standard Methods [8] and incorporated modifications by Goldhaber and Kaplan [9] and Broderius and Smith [10], mostly regarding computation of the fraction molecular, aqueous H_2S present at the pH, temperature, and salinity conditions of the test system. The formula used for calculating concentrations of molecular H_2S from measured total dissolved sulfide, S_T, was

$$\frac{H_2S_{aq}}{S_T} = \frac{1}{1 + 10^{(pH - pK_1)}} \tag{2}$$

The dissociation constant, K_1, was calculated from the relation:

$$pK_1 = 2.527 - 0.1388 \frac{(\text{Salinity}^{1/3} + 1359.96)}{\text{Temperature} + 273} \tag{3}$$

modified from Goldhaber and Kaplan [9]. As the suspension of algal cells would interfere with the absorbance readings for the H_2S measurements, the samples were filtered to remove algae prior to the addition of reagents using a 0.45-μm disposable filter holder. This filtration process resulted in minimal losses of H_2S due to volatilization, which were ignored.

Statistical Procedures

EC_{50} values (the concentrations of test material which reduced culture photosynthesis by 50%) were determined by linear regression analysis using the method of inverse prediction [11]. The biological response, expressed as percent reduction in photosynthesis compared with controls, was plotted against the corresponding logarithm-transformed exposure concentrations. Only values critical to the computation of the EC_{50} computations, i.e., those within the critical regression interval, were included in the calculation.

Results and Discussion

Effects of CO_2

Mean concentrations of total, inorganic CO_2 at the initiation and termination of the 4-h exposure were up to seven-fold higher than control concentrations (Table 1), corresponding to a difference of almost three orders of magnitude in the concentration of aqueous, molecular CO_2. This difference between total and aqueous CO_2 resulted from shifts in the carbonate equilibrium system caused by the difference in pH between the highest and control exposure media. Test concentrations of CO_2 dropped during the 4-h exposure period by 10 to 30% in all but the two highest concentrations, where photosynthesis was severely inhibited. The pH of the test solutions increased mostly during the initial 2 h of exposure in all but

TABLE 1—*Mean concentrations and effect of CO_2 during 4-h exposure. Standard deviations for measured total carbonate and calculated molecular CO_2 are based on 0-h and 4-h observations. Mean and standard deviations of photosynthetic response are based on three replicate samples.*

Total Carbonate, mM		Molecular CO_2, mM		pH		Photosynthetic Response, cpm		
						Uncorrected		
Mean	SD	Mean	SD	Mean	SD	Mean	SD	Corrected[b]
3.2	0.09	2.1	0.03	5.73	0.07	42	13	40
1.7	0.03	0.72	0.11	6.17	0.10	253	87	203
1.1	0.09	0.17	0.17	6.92	0.42	2611	469	1383
0.79	0.13	0.048	0.061	7.78	0.69	2833	882	1098
0.60	0.18	0.012	0.061	8.16	0.58	5105	944	1507
0.54	0.13	0.0043	0.0048	8.37	0.38	5686	173	1511
0.56	0.05	0.0034	0.0033	8.39	0.32	5137	506	1415
0.47[a]	0.47	0.0025	0.0023	8.45	0.29	4324	1541	1000

[a] Control.
[b] Normalized for ^{14}C present as test material versus ^{14}C added to determine photosynthesis.

the highest two concentrations and stabilized thereafter. This increase in pH varied from about 1 pH at the midconcentrations to 0.5-pH unit at the lower exposure concentrations. No increase in pH was observed at the highest two exposure concentrations. Photosynthesis had greatly diminished at CO_2 test concentrations ≥ 1.7 mM total carbonate, explaining the stable pH and CO_2 at these exposure concentrations.

The toxicity test with dissolved CO_2 did not reveal any inhibition of photosynthesis up to mean total carbonate concentrations of 1.1 mM, or 0.17 mM molecular CO_2. By comparison, mean control concentrations were 0.47-mM total carbonate, or 2.5-μM molecular CO_2. Based on the results of this toxicity test in artificial seawater medium, the EC_{50} was 0.69-mM molecular CO_2.

Consideration was given to possible isotope discrimination, as discussed by Nielsen [14]. This author estimated that, during 4-h experiments, unlabeled carbon dioxide is assimilated by plankton at a rate of approximately 10% higher than $^{14}CO_2$. The resulting possible underestimation of the true quantity of assimilated cell carbon in our experiments was considered to be within the experimental uncertainty of the algal bioassays and was therefore ignored.

To place our experimental results in a broader perspective, the buffering capacity of the artificial seawater should be considered in comparison with the buffering capacity of natural seawater. The borate concentration of the artificial medium was 0.20 mM, or 45% that of natural seawater [12], while the inorganic carbon content of the control medium was only about one fourth of measured average concentrations in coastal seawater. Comparison with other toxicity studies conducted in natural seawater [13] indicate that the pH is about 0.3 to 0.4 pH units lower in artificial seawater used during the algal toxicity tests than in natural seawater at comparable concentrations of (calculated) aqueous CO_2 (Fig. 1). Increasing concentrations of CO_2 therefore affected pH more rapidly than would be the case in the natural environment, and the observed reduction in photosynthesis at relatively low total inorganic carbon levels were amplified by the low buffering capacity of the algal seawater

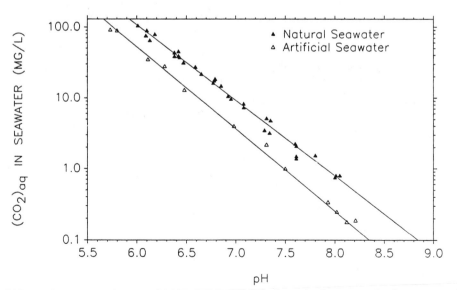

FIG. 1—*Comparison of aqueous (CO_2) concentration-pH relationship between artificial seawater used during the algal toxicity tests and natural seawater.*

medium. Linear regression of the logarithm of the molecular CO_2 concentration (in mM) in the artificial seawater medium versus pH (Table 1) yielded the following equation

$$pH = -0.95 \log [CO_2] + 6.15 \qquad (R^2 = 0.97)$$

Following this equation, the pH at the CO_2 concentration causing 50% inhibition in photosynthesis (0.69 mM) was estimated to be 6.31. Similar linear regression of the molecular CO_2 concentration (in mg/L) in natural seawater versus pH (Fig. 1) yielded

$$\log [CO_2] = -1.066 \, pH + 8.428 \qquad (R^2 = 0.99)$$

Using this equation, and assuming that the inhibition in algal photosynthesis was primarily due to pH, the EC_{50} of molecular CO_2 in natural seawater was estimated to be 52 mg/L, or 1.17 mM CO_2.

Effects of H_2S

Hydrogen sulfide concentrations dropped between 30 and 50% from time 0 exposure concentrations during the 4-h exposure period as a result of chemical oxidation, and the values of total sulfide presented in Table 2 represent means of measured initial and final concentrations.

Algal photosynthesis dropped abruptly at total dissolved sulfide concentrations in excess of 24 μM and stopped completely at ≥ 74 μM total sulfide. The estimated EC_{50} value was 48 μM total HS^-, or 104 $\mu g/L$ molecular H_2S.

During 60-h experiments under continuous illumination using *S. costatum*, Matudaira [15] found that addition of 1-mg/L Na_2S $9H_2O$ (equivalent to 4.2 μM total sulfide) to either natural or synthetic seawater increased the number of cells compared to the control. In our experiments, photosynthetic activity also appeared to have been stimulated by addition of 1 to 5-μM total sulfide.

TABLE 2—*Mean concentrations and effect of H_2S during 4-h exposure. Standard deviations for sulfide and pH measurements are based on 0-h and 4-h observations. Mean and standard deviations of photosynthetic response are based on three replicate samples.*

Total Sulfide, μM		Molecular H_2S, μM		pH		Photosynthetic Response, cpm	
Mean	SD	Mean	SD	Mean	SD	Mean	SD
410	39	24	1.0	7.96	0.06	22	1
330	43	29	10	7.77	0.23	41	20
140	36	14	5.6	7.70	0.33	33	7
74	50	8.6	0.35	7.56	0.35	30	1
50	16	3.6	0.32	7.84	0.11	565	53
24	6.5	1.2	0.74	8.04	0.16	4033	356
15	4.3	0.66	0.41	8.11	0.17	4347	223
4.8	2.6	0.24	0.24	8.12	0.25	6211	263
2.3	1.2	0.11	0.09	8.13	0.24	5240	1293
1.5	0.1	0.065	0.059	8.13	0.30	6913	2007
1.0	0.3	0.038	0.012	8.17	0.29	7007	619
0.91	0.39	0.032	0.012	8.15	0.37	5513	2130
0.73[a]	0.12	0.035	0.002	8.04	0.13	3260	256

[a] Control.

Initially, the pH of all control and test solutions was 7.9 ± 0.2. During the subsequent 4-h period, the pH either increased or diminished by about 4% compared to the initial pH, depending on the presence or absence, respectively, of algal photosynthesis. At the lower exposure concentrations, where the presence of sulfide appeared to have a stimulatory effect, carbon assimilation caused some increase in pH compared to the control. In contrast to the more dramatic shift in pH during the toxicity test with carbon dioxide, increasing sulfide concentrations caused only a minor decrease in pH (Table 2), and pH was unlikely to have been a significant factor in the observed toxicity pattern. However, the slight decrease in pH at the higher concentrations of sulfide caused a relatively larger proportion of the available sulfide pool to be present in its more toxic form, molecular H_2S, and may thus indirectly have enhanced the sulfide toxicity. This may partially explain the observed steep concentration-response relationship.

Conclusions

This investigation, which quantified the toxicity of CO_2 and H_2S to a representative marine diatom, provided useful information with respect to the potential deleterious effects of the dispersion of these off gases in natural seawater around gas production platforms. As the concentrations of dissolved off gases would vary from saturation down to background levels, resulting deleterious effects to algal populations would closely follow along these concentration gradients. Carbon dioxide dissolution would result in a decrease in pH, which was shown to affect photosynthesis. The toxicity of dissolved sulfides, which is strongly pH dependent, would be greatly enhanced by increased carbon dioxide concentrations in the seawater. Such releases would lower the pH and result in a substantial increase in highly toxic molecular H_2S in the seawater. The results of this toxicological investigation have been used to mathematically predict the zone of influence of off gas releases in tropical seas. Due to the extremely large volumes of off gases which may potentially be released into the sea, substantial adverse effects to algal populations should be expected in large areas surrounding the dispersion tubes if this method of waste gas management were practiced.

Acknowledgment

The authors wish to thank Gerald E. Walsh, EPA/ERL, Gulf Breeze, Florida, for providing diatom cultures. This work was sponsored by Exxon Biomedical Sciences, Inc.

References

[1] Smith, R. E., "The Estimation of Phytoplankton Production and Excretion by Carbon-14," *Marine Biology Letters*, Vol. 3, 1982, pp. 325–334.

[2] Richards, T. A., "Anoxic Basins and Fjords," in *Chemical Oceanography*, Vol. 1, J. P. Riley and G. Skirrow, Eds., Academic Press, New York, 1965, pp. 611–646.

[3] Walsh, G. E. and S. V. Alexander, "A Marine Algal Bioassay Method: Results with Pesticides and Industrial Wastes," *Water, Air, and Soil Pollution*, Vol. 13, 1980, pp. 45–55.

[4] Giddings, J. M., Stewart, A. J., O'Neill, R. V., and Gardner, R. H., "An Efficient Algal Bioassay Based on Short-Term Photosythetic Response," *Aquatic Toxicology and Hazard Assessment: Sixth Symposium, ASTM STP 802*, W. E. Bishop, R. D. Cardwell, and B. B. Heidolph, Eds., American Society for Testing and Materials, Philadelphia, 1983, pp. 445–459.

[5] Welschmeyer, N. A. and Lorenzen, C. J., "Carbon-14 Labeling of Phytoplankton Carbon and Chlorophyll *a* Carbon: Determination of Specific Growth Rates," *Limnology and Oceanography*, Vol. 29, 1984, pp. 135–145.

[6] Stumm, W. and Morgan, Y. Y., *Aquatic Chemistry*, Wiley Interscience, New York, p. 120.

[7] Lyman, quoted in *Chemical Oceanography*, Vol. 1, J. P. Riley and G. Skirrow, Eds., Academic Press, New York, 1956, p. 651.

[8] American Public Health Association, Standard Methods for the Examination of Water and Wastewater, 13th ed., Washington, DC, 1971, pp. 551–558.

[9] Goldhaber, M. B. and Kaplan, I. R., "Apparent Dissociation Constants of Hydrogen Sulfide in Chloride Solutions," *Marine Chemistry*, Vol. 3, 1975, pp. 83–104.

[10] Broderius S. J. and Smith, L. L., "Direct Determination and Calculation of Aqueous Hydrogen Sulfide," *Analytical Chemistry*, Vol. 49, 1977, pp. 424–428.

[11] Sokal R. R. and Rohlf, F. J., *Biometry*, W. H. Freeman and Co., San Francisco, CA, 1969, p. 447.

[12] Skirrow, G., "The Dissolved Gases—Carbon Dioxide," in *Chemical Oceanography*, Vol. 1, J. P. Riley and G. Skirrow, Eds., Academic Press, New York, 1965, pp. 227–322.

[13] Breteler, R. J., "Acute and Chronic Effects of Dissolved Gases, H_2S and CO_2, on Marine Species," final Battelle report to Exxon Biomedical Sciences, Inc., 1985.

[14] Nielsen, E. S., "The Use of Radio-Active Carbon (C^{14}) for Measuring Organic Production in the Sea," *Journal of Conservation and International Marine Exploration*, Vol. 18, 1952, pp. 117–140.

[15] Matudaira, T., quoted in *The Chemistry and Fertility of Sea Waters*, W. H. Harvey, Ed., Cambridge at the University Press, 1969, p. 108.

Harry L. Boston,[1] Walter R. Hill,[1] and Arthur J. Stewart[1]

Evaluating Direct Toxicity and Food-Chain Effects in Aquatic Systems Using Natural Periphyton Communities

REFERENCE: Boston, H. L., Hill, W. R., and Stewart, A. J., "**Evaluating Direct Toxicity and Food-Chain Effects in Aquatic Systems Using Natural Periphyton Communities,**" *Plants for Toxicity Assessment: Second Volume, ASTM STP 1115,* J. W. Gorsuch, W. R. Lower, W. Wang, and M. A. Lewis, Eds., American Society for Testing and Materials, Philadelphia, 1991, pp. 126–145.

ABSTRACT: Periphytic algae are a ubiquitous and ecologically important component of many rivers and streams and are useful for assessing toxicity. Biotic communities in streams near Oak Ridge National Laboratory were apparently impacted by industrial activities; however, no one toxicant had been clearly implicated in these streams and consequently no exposure data were available. We used natural periphyton communities in several streams to (1) identify stream reaches with ambient toxicity, (2) characterize conditions for the growth of periphyton and other aquatic biota, and (3) evaluate contaminant movement and effects through food-chain interactions.

Algal biomass (as chlorophyll *a*) and photosynthetic (PS) rates (based on $H^{14}CO_3^-$ incorporation) were measured monthly for two years at 15 sites. Data for periphyton PS were analyzed using analysis of covariance (ANCOVA) with biomass as the covariate. The ANCOVA generated a chlorophyll-adjusted photosynthetic (CAPS) rate for the periphyton at each site that provided a means to track and compare functional aspects (e.g., PS) of periphyton communities among sites where biomass differed. The CAPS rate appears to provide an index of the physiological condition of the periphyton that was (1) sensitive to stress, and (2) robust for different communities and for different light and nutrient levels based on experimental evidence and comparison with data for other biotic communities.

Commonly used periphyton evaluation techniques (e.g., species composition, percent organic content, and accrual rates) were used, determined, and compared with CAPS rates. A 23-day periphyton bioassay suggested that toxicity from chlorine in once-through cooling water could result in low taxonomic diversity and a patchy algal distribution at sites near cooling water discharges. These experiments showed that PS was more sensitive to chlorine stress than was either biomass or species composition.

In systems where toxic stress is intermittent, cumulative, or the result of biotic interactions, integrative measures of natural communities, such as CAPS and biomass accrual, are relatively simple and useful for assessing toxicity. Data from these integrative measures were better correlated with data for other biotic communities than were the results of single species toxicity tests.

Periphyton at sites directly downstream of industrial discharges was enriched with various toxic metals [e.g., cadmium (Cd), chromium (Cr), and mercury (Hg)]. Because several taxa of periphyton-grazing invertebrates were coincidentally absent from these sites, we conducted laboratory experiments to assess the effects of contaminants on invertebrates. Experiments (five-week duration) showed that the growth of snails and grazing caddisflies on periphyton from three sites was positively correlated with algal biomass and was apparently not affected by the metals in the periphyton consumed. Thus, although periphyton may be good indicators of metal loading, these metals may not be rapidly transferred to consumers.

KEY WORDS: algae, bioassay, biomonitoring, periphyton, photosynthesis, stream, toxicity assessment

[1] Research staff member, research associate, and research staff member, respectively. Environmental Sciences Division, Oak Ridge National Laboratory, Oak Ridge, TN 37831-6351.

Freshwater algae are structurally simple compared to terrestrial plants, yet metabolically similar to higher plants. Periphyton is a complex matrix of algae and heterotrophic microbes attached to submersed surfaces in aquatic systems. It is of fundamental importance in most lotic ecosystems, as it provides reduced carbon compounds to higher trophic levels [1,2]. Periphyton may accumulate potentially toxic compounds such as arsenic (As), Cd, Cr, Hg, and selenium (Se) [3–6] that can then be transferred through the aquatic food chain. Thus, periphyton is useful for monitoring pollutant input, and periphyton influences the fate of ecologically important contaminants.

Various studies have explored the use of periphyton community structure [7–9] and function [10–12] to assess water quality. In the context of biomonitoring, photosynthesis (PS), an easily measured, integrative attribute of natural periphyton communities, was useful for evaluating biological impacts of toxic materials [13]. Good dose-response relationships were found between periphyton community PS and exposure to aliphatic amines and waste-water from textile industries [13]. In other studies, short-term PS was used to show the induction of arsenate tolerance in phytoplankton [14,15].

Laboratory tests have shown that algae are about as sensitive as freshwater animals to various toxicants [16–19]. Thus periphyton has three attributes that make it particularly suitable for biomonitoring purposes: (1) it is ubiquitous and is an ecologically important component in aquatic systems most subject to pollutant input; (2) it is moderately sensitive, has relatively high turnover rates, and thus may respond rapidly to stresses integrated by organisms with longer life spans; and (3) it has an easily measured environmentally important process (PS) that responds to toxic chemicals and complex effluents [16].

As part of National Pollution Discharge Elimination System (NPDES) permits for the U.S. Department of Energy's (DOE) Y-12 Plant and Oak Ridge National Laboratory (ORNL), the effects of effluents on biota in two streams near these facilities were evaluated. Although biotic effects were apparent [20–22], no one toxicant had been identified as being responsible. Consequently, no exposure data were available. We used natural periphyton communities and integrative measures of biotic conditions to assess toxicity in selected stream reaches. We regularly determined algal periphyton biomass, PS, and community composition, and used these data to track and compare the physiological condition of the algae at the study sites. These efforts were augmented with several short-term studies addressing factors influencing the algal periphyton and their interaction with other biota. The experiments were: (1) determining rates of algal accrual (colonization and growth); (2) performing algal bioassays of water quality; (3) evaluating the effects of industrial discharges of chlorine on algal taxonomic composition, biomass, and PS; and (4) exploring food-chain effects of contaminants in the periphyton.

Procedure

General Site Characteristics

East Fork Poplar Creek (EFPC) and the White Oak Creek (WOC) system, located near Oak Ridge, Tennessee (35°56′N, 84°18′W), receive effluents from the DOE Y-12 Plant and ORNL, respectively. Studies of the periphyton communities began in March 1986. Sites were selected based upon their distance from Y-12 or ORNL discharges and to coincide with study sites for investigations of other biotic communities. Sites on EFPC were located from 6.3 to 24.4 km upstream from the confluence with Poplar Creek (Fig. 1), corresponding to the riffle/pool region of the stream. The headwaters of EFPC receive discharges from the Y-12 Plant at site E-2 (24.4 km upstream). The City of Oak Ridge Sewage Treatment Plant (OR-STP) also discharges into EFPC about 200 m upstream of site E-5 (about 10.6

ORNL-DWG 90M-8903

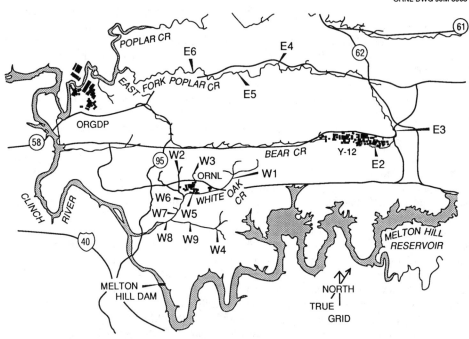

FIG. 1—*Periphyton study sites. Brushy Fork site not shown.*

km upstream of Poplar Creek). A site in nearby Brushy Fork (E-1) received some agricultural inputs but was not impacted by industrial activities.

For streams in the WOC watershed (Fig. 1), five sites were located downstream of ORNL discharges (W-5 through W-9). Four sites located upstream of ORNL inputs (W-1 through W-4) were used as reference sites. ORNL discharges entered WOC upstream of W-5 (about 4 km above White Oak Dam), and the ORNL-STP discharged about 100 m upstream of W-6. The ORNL High Flux Isotope Reactor (HFIR) facility (when operating) discharged about 0.7 km upstream of W-9 on Melton Branch. A more detailed site description can be found in Loar (1989) [20].

We have characterized the habitat (light, canopy cover, etc.) during winter and summer for 50-m reaches at each of the sampling sites and have measured selected water quality parameters monthly [20–22]. The taxonomic composition of the algal assemblages at each site was determined several times each year. At all sites periphyton samples were collected from shallow riffles.

Periphyton Chlorophyll and Carbon Incorporation

Each month, four small (10 to 60-cm²) flat rocks with periphyton were collected from shallow riffle areas (10 to 30 cm deep, with 0.2 to 0.4 m s^{-1} water velocity) at each site. In the laboratory the periphyton was incubated for 2 h in water from the collection site with 10 μCi (3.7×10^5 Bq) sodium bicarbonate (NaH^{14}CO$_3$). The ^{14}C solution contained 20 μCi (7.4×10^5 Bq) per mL and about 200 μg total inorganic carbon (Dupont-New England

Nuclear NSC-086S). The carbon added with the isotope solution contributed negligibly to the inorganic carbon in the incubation chamber. During the incubation, water temperature was maintained within 2°C of ambient stream temperature. Light was supplied at 400 μmol $m^{-2} s^{-1}$ photosynthetically active radiation ($\sim$16% full sun) by a 1000-W metal halide lamp. Photosynthetic light response curves for periphyton from these or similar sites [23] indicate that 400 μmol $m^{-2} s^{-1}$ was sufficient to saturate PS for most of these communities without causing photoinhibition. Dark controls were also run to correct for nonphotosynthetic carbon uptake.

The water in the incubation chambers was circulated by submersible pumps to simulate natural conditions. Less than 10% of the added ^{14}C was taken up during the 2-h incubation, suggesting that the pool of available carbon did not change greatly during incubation.

After incubation the periphyton on the rocks was rinsed twice in distilled water to remove residual inorganic ^{14}C. The periphyton on the rocks was then placed in 30 mL of dimethylsulfoxide (DMSO) and kept in darkness for 24 h to extract organically incorporated ^{14}C and chlorophyll [24,25]. Five millilitres of the DMSO extract was diluted 1:1 with 90% acetone, and Chl a was determined spectrophotometrically [26] and corrected for phaeo pigments [27]. A 500-μL aliquot of the extract was added to 10 mL of Aquasol (scintillation cocktail), and the ^{14}C in the aliquot was determined by liquid scintillation spectrometry to quantify inorganic carbon incorporation.

The surface area of each rock was determined by covering the upper surface with aluminum foil, weighing the foil, and converting weight to surface area based on a known weight per unit area of foil. Chlorophyll a (Chl a) and PS were then expressed on a substratum surface area basis.

Algal Accrual

The rates of periphyton accrual (algal colonization and growth) were measured by quantifying Chl a through time on ceramic tiles attached to bricks placed at each site except E-5, W-4, and W-7. On 22 June 1988, four bricks, each with 24 2.5 by 2.5-cm unglazed ceramic tiles, were placed at the sites. Starting three weeks later and continuing weekly thereafter for six weeks, one tile was randomly selected from each brick and Chl a was measured as described above. On each date, a fifth tile was selected for the determination of algal taxonomic composition.

Chlorine Toxicity Bioassay

To determine if the toxicity from chlorine in cooling water discharges from the Y-12 Plant and ORNL could be responsible for the low taxonomic diversity of algae and patchy periphyton distribution on rocks at sites near discharges (E-2 and W-5), we conducted a 23-day bioassay by using natural periphyton communities. Small rocks with adhering periphyton were collected from riffles at three EFPC sites. One site (E-2) was known to experience occasionally high chlorine concentrations, and two sites further downstream (E-3 and E-4) and were rarely or never exposed to measurable concentrations. The periphyton from each site was exposed to 0, 50, 100, 250, 500, or 1000 μg L^{-1} total residual chlorine (TRC) in six continuous-flow outdoor chambers (volume replaced every 0.5 h). Submersible pumps provided water velocities similar to those found under field conditions. Concentrations of TRC were controlled by mixing tap water (containing about 1 mg L^{-1} TRC) with dechlorinated tap water. On days 0, 5, and 23, periphyton samples were collected from each treatment to determine algal species composition. On day 23, we also measured algal biomass (Chl a) and periphyton PS on four rocks from each treatment for each site.

TABLE 1—*Characteristics of EFPC and WOC sites. Site type: R = upstream or reference; E = receiving effluent; DS = downstream of effluent; FD = far downstream. Location is the distance in km upstream of reference point (e.g., EFK 24.4 is 24.4 km upstream of the confluence with Poplar Creek). SRP = soluble reactive P (μg P L^{-1}, $NO_3 + NH_4$ (mg NL^{-1}). % canopy = cover over stream channel. Daily PAR is daily integrated photosynthetically active radiation (mol m^{-2} day^{-1}) measured during summer using 3 to 5 ozalid light meters [38]. Canopy cover was not estimated for WOC sites; NA = data not available. Data for algal composition and pollution-sensitive species of fish and benthic macroinvertebrates from Loar et al. (1989) and Loar (1989) [22,20].*

Site	Location	Type	SRP, $NO_3 + NH_4$	% Canopy, daily PAR	Algal Community	Fish and Invertebrates
				EAST FORK POPLAR CREEK SITES		
E-1	BFK 7.6	R	12 0.52	93 1.2	Prostrate *Stigeoclonium* small diatoms	Pollution sensitive species present, grazing snails abundant
E-2	EFK 24.4	E	80 3.5	0 22	Small green unicells	Few fish or invertebrates
E-3	EFK 23.2	DS	70 3.5	54 12	Green and bluegreen filaments, small diatoms	Pollution sensitive species absent, algivorus fish abundant
E-4	EFK 13.8	FD	95 2.5	0 31	Green filaments and large diatoms	Pollution-sensitive species present
E-5	EFK 10.6	DS	700 3.5	98 1.2	Small diatoms, green unicells, bluegreen filaments	Pollution-sensitive species absent
E-6	EFK 6.3	FD	700 3.2	95 1.3	Large diatoms, green filaments	Pollution-sensitive species present

WHITE OAK CREEK SITES

Site	Location	Category					Dominant forms	Notes
W-1	WCK 6.8	R	6	0.1	NA	3.8	Prostrate *Stigeoclonium* small diatoms	Pollution-sensitive species present, grazing snails abundant
W-2	FCK 1.0	R	7	0.1	NA	1.0	Prostrate *Stigeoclonium* small diatoms	Pollution-sensitive species present, grazing snails abundant
W-3	FFK 1.1	R	10	0.16	NA	2.4	Large stalked and filamentous diatoms	Pollution-sensitive species present, no grazing snails
W-4	MEK 1.5	R	7	0.1	NA	NA	Prostrate *Stigeoclonium* small diatoms	Pollution-sensitive species present, no grazing snails
W-5	WCK 3.9	E	60	1.1	0	21	Small green unicells	Pollution-sensitive Seldom present
W-6	WCK 3.4	DS	200	1.2	NA	2.1	Green and bluegreen filaments, diatoms, unicells	Pollution-sensitive species absent
W-7	WCK 2.9	DS	200	1.0	NA	2.2	Small diatoms, green and bluegreen filaments	Pollution-sensitive species absent
W-8	WCK 2.4	FD	170	0.9	NA	1.0	Large and small diatoms green & bluegreen filaments	Pollution-sensitive species present
W-9	MEK 0.6	DS[a]	120	0.2	NA	3.0	Large diatoms	Pollution-sensitive species present

[a] Site W-9 was located about 0.7 km downstream of discharge from the HFIR facility; however, that facility shut down early in this study period.

Periphyton Metal Content and Invertebrate Growth Studies

During February and May 1987, periphyton was brushed from three rocks collected from each site in EFPC and WOC. The samples were dried at 65°C for 24 h, digested in nitric acid (HNO_3) + sulfuric acid (H_2SO_4), and analyzed for metals by inductively coupled plasma spectroscopy (ICP). Samples collected on February 10 were also analyzed for total Hg, using cold-vapor atomic absorption. To investigate the effects and trophic transfer of metals, two taxa of macroinvertebrates were fed periphyton with high metal concentrations collected from E-3 (immediately downstream of the Y-12 Plant), periphyton from E-4 (10 km downstream from the Y-12 Plant), and periphyton from reference site W-1. Grazing caddisfly larvae (*Neophylax* sp.) and snails (*Elimia clavaeformis*) collected from W-1 were maintained in eleven miniature laboratory stream chambers [28] from 20 April to 27 May 1989. Four chambers contained rocks from E-3, four contained rocks from E-4, and three contained rocks from W-1. The periphyton-covered rocks were replaced every two days. All chambers contained water from W-1 at ambient temperature (13 to 18°C) that was replaced daily. Individual *Neophylax* were collected from the chambers when they began larval diapause. They were dried at 60°C and weighed. Snails were weighed live at the beginning and at the end of the experiment. For metal analyses, snails were removed from their shells, and the soft tissues were digested (HNO_3 + H_2SO_4) for analysis by ICP. Periphyton was brushed from rocks from the three sites, dried at 60°C, and digested for analysis by ICP.

Results and Discussion

Site Characteristics

Sites were selected based on their location relative to discharges and the availability of ancillary data for other biotic communities. Some sites lacked any riparian vegetation, while others were heavily canopied, thus incident light varied substantially among sites (Table 1). The water at sites downstream of the Y-12 Plant and ORNL was nutrient [nitrogen (N) and phosphorus (P)] rich, while nutrient concentrations were generally low at upstream or reference sites (Table 1). Livestock pastures adjacent to E-4 and the sewage treatment plants upstream of E-5 and W-5 also contributed N and P to these streams. To facilitate the discussion, we have classified sites based on their location relative to ORNL and Y-12 Plant discharges and information for biotic communities (Table 1).

Algal Communities

Diatoms and *Stigeoclonium* sp. basal cells were the dominant algal taxa at most reference sites (Table 1). *Stigeoclonium* sp. was characteristic at sites with low nutrients and high snail grazing pressure (E-1, W-1, and W-2), while, in the absence of snails, stalked or filamentous diatoms dominated (e.g., W-3). At sites occasionally receiving discharges containing chlorine (E-2 and W-5), a peculiar, small (<5μ dia) round, nonmotile unicellular green alga (SNGA) dominated the community. At those sites acute toxicity due to chlorine discharges [29] was evidenced by dramatic monthly changes in alga cover and macroinvertebrate numbers and the scarcity of fish. This SNGA persisted in moderate to high numbers at sites just downstream of these areas (E-3 and W-6) where the community was usually dominated by *Stigeoclonium* sp., bluegreen algal filaments, and diatoms. These sites were at least occasionally exposed to toxicants, as evidenced by the absence of pollution-sensitive fish and macroinvertebrates.

Large diatoms and green algal filaments (e.g., *Cladophora* sp.) became more common with distance downstream of industrial discharges. Small diatoms were common at E-5 (below the OR-STP) during summer; however, filamentous bluegreen algae dominated that community during winter.

Monthly Determination of Algal Periphyton Biomass and PS

The monthly sampling of the periphyton was intended primarily for evaluating environmental conditions at each site, rather than for estimating primary production. At E-2 and W-5, where chlorine concentrations were occasionally high (>250 μg L^{-1} TRC), the periphyton was patchily distributed. It was not uncommon to find single rocks with a thick periphyton covering, while surrounding rocks were bare. At those sites we chose rocks from among those with obvious periphyton to avoid excessive variance. Thus, the data from those sites did not represent average biomass and were used primarily for evaluating the physiological condition of the algae.

Monthly data from each site were pooled to determine the annual average density of Chl *a* per unit rock surface area. Data from EFPC for 1988 and from WOC for 1989 (Table 2) were similar to those found previously [20,22]. Multiple comparisons of means for EFPC and WOC sites showed that average annual Chl *a* was generally lowest at upstream or reference sites (likely due to low nutrients, low light intensities, and many invertebrate grazers). The data for sites near discharges (E-2 and W-5) overestimate the biomass present, as noted above; however, biomass on colonized rocks was usually quite high. Average annual biomass was also high at E-4, a downstream site with no canopy cover (Tables 1 and 2).

Data for periphyton community PS tended to be higher at sites far downstream from discharges (E-4, E-6, W-8, and W-9) but were not highly correlated with algal biomass. Periphyton PS was lowest at reference sites with low biomass and many grazers (e.g., E-1, W-1, and W-2).

TABLE 2—*Mean annual periphyton chlorophyll* a *(μg cm^{-2}) and photosynthesis (mg C m^{-2} h^{-1}) for EFPC sites (October 1987 to September 1988) and WOC sites (1989). Site type R = upstream or reference; E = near effluent; DS = downstream of effluent discharge; and FD = far downstream of discharges. Communities at site W-9 were recovering following the shutdown of the HFIR facility about 1 km upstream. Values are means ±1 S.E. (n = 48). For each stream, values with the same letter are not significantly different based on Tukey's multiple comparison procedure (p < 0.05).*

Site	Type	Chlorophyll *a*		Photosynthesis	
EFPC SYSTEM					
E-1	R	10.1 ± 1.3	C D	27.6 ± 3.1	D
E-2	E	30.9 ± 3.3	A B	61.4 ± 5.7	A B
E-3	DS	19.5 ± 1.7	A B	54.0 ± 3.7	A B
E-4	FD	19.7 ± 1.5	A B	61.3 ± 6.4	A B
E-5	DS	17.4 ± 2.2	B C	37.5 ± 3.3	C
E-6	FD	10.9 ± 1.3	C D	52.1 ± 4.2	B
WOC SYSTEM					
W-1	R	4.3 ± 0.5	C	8.4 ± 0.7	E
W-2	R	4.6 ± 0.3	C	13.6 ± 0.7	D
W-3	R	11.6 ± 0.8	B	39.1 ± 3.0	B
W-4	R	4.3 ± 0.3	C	15.8 ± 1.6	D
W-5	E	21.5 ± 1.9	A	47.1 ± 4.7	A B
W-6	DS	17.5 ± 1.5	A	42.8 ± 4.1	B
W-7	DS	8.7 ± 1.1	B	32.5 ± 4.3	C
W-8	FD	12.7 ± 1.3	B	48.6 ± 5.7	A B
W-9	DS‡	8.3 ± 0.7	B	60.6 ± 5.9	A

Low-nutrient upstream sites with many snails (e.g., W-1) did not provide a good reference condition for sites near discharges where nutrient concentrations were high, because in these streams nutrients and grazing greatly influenced periphyton biomass and PS. Sites far downstream of discharges (with similar nutrient concentrations) allowed a better comparison with sites near discharges (Table 2).

We examined the monthly data for periphyton Chl a and PS per unit area at all sites for evidence of acute toxicity. As noted above, the biomass data for sites near discharges are not representative of natural variability and therefore are not presented. At those sites (E-2 and W-5), periphyton cover typically varied greatly among rocks and among months. The patchy periphyton cover, low species diversity, and the scarcity of fish and macroinvertebrates suggested chronic toxicity. However, because periphyton biomass and PS on colonized rocks was frequently high (Table 2), we suspect that acutely toxic events were apparently short-lived and so were difficult to detect. For these sites most of the evidence for acute toxicity to periphyton was anecdotal (e.g., a sudden decrease in periphyton biomass in conjunction with a loss of macroinvertebrates).

Monthly data for E-5 (downstream of the OR-STP) and E-6 (away from discharges) (Fig. 2) were typical of results from sites in similar locations relative to industrial discharges (e.g.,

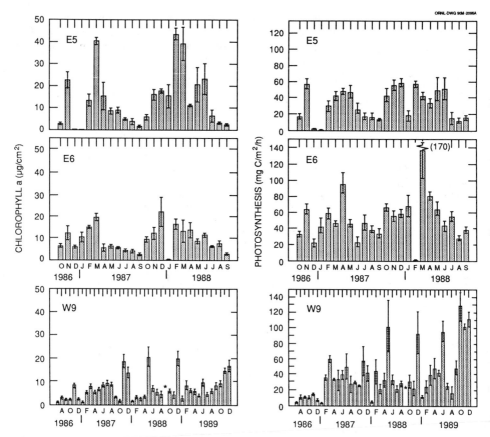

FIG. 2—*Monthly periphyton chlorophyll* a *($\mu g\ cm^{-2}$) and photosynthesis (mg C $m^{-2}h^{-1}$) for sites E-5, E-6, W-1, and W-9. Values are means and 1 standard error (S.E.) (n = 4).*

E-3 and W-8). Variations in biomass and PS at all sites reflected seasonal changes in light and temperature, and discrete occurrences such as scouring during storms. It was particularly difficult to evaluate the effects of storms because these effects varied with initial biomass and other factors. At low-nutrient reference sites, biomass and PS were generally much lower and less variable than at downstream sites.

Overall, month-to-month variability at each site was high enough to make it difficult to determine if acutely toxic events occurred at downstream sites. However, some apparently toxic incidents were identified. For example, data for Chl a and PS from E-5 and E-6 suggest toxicity just downstream of the OR-STP (i.e., at E-5) during December 1986 and January 1987. At this time algal biomass at E-5 was very low compared with relatively high biomass and PS at E-6 (Fig. 2) and sites upstream of the STP (data not shown). On occasion, this type of comparison suggested toxicity rather than storm effects; however, without exposure data such evaluations are subjective. Efforts to detect discrete toxic events would have benefited from a sampling frequency closer to the turnover time of the algae (i.e., days) and exposure information for industrial discharges.

Data from W-9 (Fig. 2) show a recovery following the cessation of toxic stress (after the shutdown of the HFIR facility in November 1986). The increase in periphyton biomass and PS was correlated with some recovery in numbers and diversity of macroinvertebrates and fishes at that site.

Using PS Rates to Evaluate Periphyton Condition

Near discharge points, the effects on periphyton appeared to be subtle or chronic (e.g., influencing species composition more than biomass), with only occasional evidence of acute toxicity. Consequently, it was difficult to characterize and evaluate the frequency of toxic stress. Assessment was further complicated by differences in light, nutrients, and grazing among sites. Here we used community PS to provide an integrative measure of the condition of the periphyton at each site. PS per unit algal biomass (Chl a) should provide an index of periphyton status, with higher values assumed to represent better physiological condition and to reflect environmental conditions [30] (see also the effects of chlorine on PS and Chl a in the results of the chlorine bioassay). However, the comparison of PS/Chl a among communities is complicated because algal biomass and the rate of PS are nonlinearly related. Self-limitation apparently causes carbon incorporation to increase more slowly than biomass as biomass accumulates and the periphyton matrix thickens; thus, PS/Chl a decreases as biomass increases [23,31,32].

To parcel out the effect of algal biomass on biomass-specific PS, we performed an analysis of covariance (ANCOVA) of PS for all sites and times in each system (with Chl a as the covariate) using natural log (ln) transformed data. The model used was

$$Y_{ijk} = \mu + \alpha_i + \delta_j + (\alpha\delta)_{ij} + \beta(X_{ijk} - X \ldots) + \epsilon_{ijk}$$

where

Y_{ijk} = the predicted chlorophyll-adjusted PS (CAPS) rate (ln mg C m^{-2} h^{-1}) for periphyton at site i, on month j, for sample k (of replicate 1 to 4).

μ = the overall average ln PS rate (all sites, all months),

α_i = the effect of site i on the PS rate,

δ_j = the effect of month j on the PS rate,

$\mu\delta_{ij}$ = the effect of the interaction of site i and month j,

β = the common slope of ln PS relative to ln Chl a,

$(X_{ijk} - X \ldots)$ = the adjustment of sample chlorophyll to the average value, or the difference between chlorophyll for sample ijk and the overall average chlorophyll, and

ϵ = the residual error due to sample differences and measurement.

In simpler terms, the ANCOVA analysis determined an average relationship between ln PS and ln Chl a using the data from all sites (i.e., one average slope) and then used the four replicate values for each site-month combination to fit lines with this common slope (β) but differing intercepts (Fig. 3). For each site-month regression, the PS rate at the overall average biomass value is the CAPS rate for periphyton from that site during that month. The assumption of a uniform slope for the relationship of ln PS to ln Chl for all sites and months was required because the four monthly data points available for each site and month were not sufficient to develop reasonable estimates of the PS-chlorophyll relationship for each site-month combination. The assumption was valid statistically, and, although the raw data showed that the slopes varied nonrandomly among communities [33], this variation did not appear to impart unreasonable bias. The ANCOVA therefore is used primarily as a simple computational method to compare PS rates among communities that have different algal biomass.

We used the mean CAPS from the ANCOVA [LSMEANS (least square means) in SAS procedure: general linear model (SAS PROC GLM)] as a measure of the physiological condition of the algal community for each site and each month. Here a higher CAPS rate indicates higher growth and better condition. This CAPS rate reflects both the inherent physiological potential of the algal community and the influence of recent environmental conditions.

Annual average CAPS rates (LSMEANS for each site for twelve months from the AN-COVA) were similar among stream systems (Table 3). CAPS rates were highest for communities far downstream from the Y-12 Plant and ORNL discharges (e.g., E-4, E-6, and

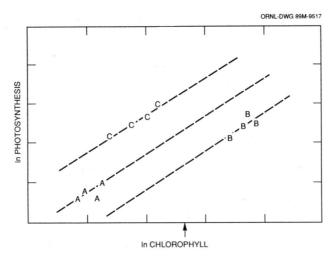

ORNL-DWG 89M-9517

FIG. 3—*Conceptual diagram of ANCOVA generation of chlorophyll-adjusted PS rates. Letters (A, B, and C) are monthly values for ln Chl a cm^{-2} and ln PS cm^{-2} from three sites. The data are fit with lines of identical slope based on the average relationship of ln Chl to ln PS for all sites. For each site-date (i.e., each regression), the model determines a PS rate at average Chl a for all sites (arrow), which is the chlorophyll-adjusted PS for that site-date.*

W-8) where pollution-sensitive species of fish and invertebrates were found (Tables 1 and 3). The average CAPS rates was also high for W-9 following the shutdown of the HFIR facility, although fish and invertebrate communities at that site had not fully recovered. CAPS rates were lower for communities nearer to discharge points (E-2, E-3, E-5, W-5, and W-6) where pollution-sensitive fish and invertebrates were absent. CAPS rates were unexpectedly low for communities at several reference sites (e.g., E-1, W-1, and W-2). Experimental evidence (unpublished data, W. R. Hill) has shown that low-nutrient levels and intense grazing by snails can result in algal communities with slow growth rates compared with communities at low-nutrient sites without snails (e.g., W-3).

The CAPS rates were similar among downstream communities at E-4 and E-6 that had high nutrient concentrations and diverse biotic communities but differed in available light (Tables 1 and 3). In contrast, where conditions of light and nutrients were similar but biotic community diversity differed (e.g., E-5 and E-6), the CAPS rates also differed. This suggests that the periphyton CAPS rates were more sensitive to stress (as evidenced by data for other biotic communities) than to general environmental conditions, at least among sites with similar nutrient status. For communities at some low-nutrient reference sites, low CAPS rates apparently reflect "natural stresses." Thus, when using CAPS rates to evaluate conditions for algal growth (i.e., toxicity assessment), CAPS rates must be compared among sites with similar nutrient levels. This complicates the interpretation of data for CAPS rates and is a potential limitation of this approach. However, appropriate reference sites may be difficult to find for any comparative assessment, and factors such as nutrient concentrations complicate even simple and commonly used measures such as taxonomic comparisons.

TABLE 3—*Average annual chlorophyll-adjusted photosynthetic rates (ln μg C cm^2 h^{-1}) for EFPC sites (1987 and 1988) and WOC sites (1988 and 1989). For each stream system and year, sites with the same letter are not significantly different based on multiple comparison by SAS GLM PDIFF procedure (p < 0.05).*

| | | EFPC System | | | |
| | | 1987 | | 1988 | |
Site	Type	Mean	Comparison	Mean	Comparison
E-1	R	0.91	C	0.99	E
E-2	E	1.2	B	1.2	C D
E-3	DS	1.3	B	1.4	B C
E-4	FD	1.5	A	1.5	A B
E-5	DS	0.66	D	1.1	D E
E-6	FD	1.6	A	1.6	A
		WOC System			
		1988		1989	
Site	Type	Mean	Comparison	Mean	Comparison
W-1	R	0.37	D	0.05	F
W-2	R	0.85	C	0.52	E
W-3	R	1.1	A B	0.93	C
W-4	R	NA		0.56	E
W-5	E	0.97	B C	0.72	D
W-6	DS	0.90	C	0.70	D
W-7	DS	1.2	A	0.69	D
W-8	FD	1.2	A	1.1	B
W-9	DS[a]	1.2	A	1.4	A

[a] Sampled after the HFIR facility shut down.

Despite this limitation, we found the CAPS rates provided an indication of the condition of the periphyton community that (1) agreed with data for fish and invertebrates, and (2) allowed us to compare communities that differed in many respects.

We also used the monthly CAPS rates for each site to track the communities through time. Monthly CAPS for selected WOC sites (1989) and EFPC sites (October 1987 to September 1988) are shown in Fig. 4. Seasonal changes in CAPS at most sites were small

CHLOROPHYLL-ADJUSTED PHOTOSYNTHESIS
EFPC 1987-1988

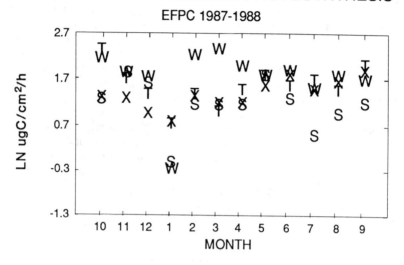

CHLOROPHYLL-ADJUSTED PHOTOSYNTHESIS
WOC 1989

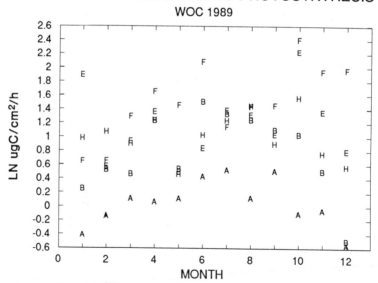

FIG. 4—*Monthly chlorophyll-adjusted photosynthetic rates (ln µg C cm⁻² h⁻¹) based on ANCOVA for EFPC sites (October 1987 through September 1988) and WOC sites (1989). A = W-1, B = W-5, E = W-8, F = W-9, H = W-3, S = E-1, T = E-3, W = E-6, and X = E-2.*

and could be related to temperature. However, high CAPS rates during summer at sites where potentially toxic concentrations of chlorine occasionally occurred (E-2 and W-5) apparently reflected less-frequent or less-intense stress as evidenced by greater periphyton cover and the more-frequent appearance of macroinvertebrates during summer (H. Boston, personal observation). During any month, low CAPS rates may result from acute toxic stresses or natural factors such as storms that remove the upper more-productive layers of the periphyton community (e.g., E-6 for January 1988). The ANCOVA-generated CAPS rates provided a means for comparing PS rates of natural periphyton communities and as such served as a tool for in situ toxicity assessment. Because this is a comparative approach, some subjectivity was involved in choosing appropriate reference conditions (e.g., all high-nutrient sites). However, the technique is promising for systems with subtle or intermittent stress.

Periphyton Accrual Rates

Sites were grouped by location and information for stress on other biotic communities (Table 1) into four categories: upstream sites typified by E-1 and W-1, highly impacted communities (e.g., E-2 and W-5); mildly impacted communities (E-3 and W-6); and sites far from effluents (e.g., E-6 and W-9). Site W-9 was included in the later group due of the partial recovery of biotic communities at that site following the shutdown of the HFIR facility. Algal biomass accumulated rapidly on tiles at mildly impacted and downstream sites (Fig. 5). At one site (E-4), we observed that colonization was rapid; however, it was six to eight weeks before the biomass on the tiles was similar to that on the rocks at the site because the filamentous green alga (*Cladophora*) covering the rocks at that site was slow to colonize the tiles.

PERIPHYTON COLONIZATION STUDY

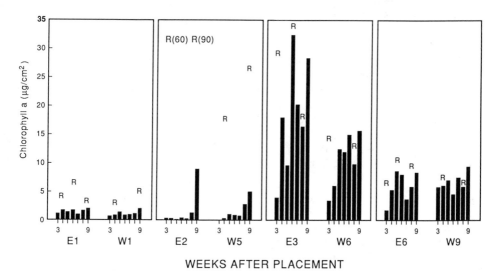

FIG. 5—*Algal (μg Chl a cm⁻²) accumulation on ceramic tiles at reference sites (E-1 and W-1), sites directly receiving effluents (E-2 and W-5), sites downstream of discharges (E-3 and W-6), and sites far downstream (E-6 and W-9). Tile were collected three to nine weeks after placement. "R" denotes Chl a cm⁻² on adjacent rocks. Values are means of four replicates.*

Algal biomass accrual on tiles was slow at some reference sites (e.g., E-1 and W-1) (Fig. 5). Larger amounts of biomass on natural substrata than on tiles may reflect refuge from grazing provided by the irregular surface of the rocks. Accrual of biomass was very slow at the highly impacted sites (Fig. 5) and never approached densities on natural substrata. Other studies [20] have shown that the green alga (SNGA) that occurred at these sites recovers quickly from biomass reductions (removal by brushing), but that recolonization following total elimination by chlorination is slow. Thus, following acutely toxic events, regrowth may be rapid on rocks where some cells have survived; however, recolonization of sterile substrata was slow. This may account for the patchy (rock-to-rock) distribution of periphyton at E-2 and W-5.

In general, algal accrual was sensitive to anthropogenic and natural factors (nutrients and grazing), but measured different aspects of the periphyton (colonization and growth) than did CAPS rates. For example, accrual was very sensitive to stress at sites near effluents, but was not as sensitive as CAPS rates or algal community composition at sites downstream of discharges (e.g., E-3 and W-6).

Chlorine Toxicity Bioassay

Results of the chlorine exposure studies showed that by day 23 the algal communities exposed to TRC at concentrations of 250 to 1000 μg L^{-1} were dominated by the same green alga (SNGA) that dominates the periphyton at E-2 and W-5, where high concentrations of TRC occur. That alga was initially rare on substrata from the two downstream sites. At lower chlorine concentrations, filamentous green algae and *Scenedesmus* sp. colonized and dominated the communities for substrata from all sites (unpublished data, C. M. Pettway, Knoxville College). Thus high TRC concentrations selected for the SNGA were found at E-2 and W-5; however, at low chlorine concentrations, this taxon was replaced by species found at downstream sites.

By day 23, algal biomass and PS had decreased dramatically at TRC concentrations >250 μg L^{-1} for communities from all sites (e.g., data for E-2 in Fig. 6). Photosynthesis was much more sensitive to TRC than was algal biomass. Even at the lowest TRC concentration tested (50 μg L^{-1}), PS/area was decreased significantly from the control. However, biomass per area did not decrease until TRC concentrations were $\geq$250 μg L^{-1}.

This bioassay strongly suggested that chlorine toxicity was responsible for the altered algal community composition at E-2 and W-5, where concentrations of TRC >250 μg L^{-1} are common. The patchy distribution of periphytic algae and the relatively low chlorophyll-adjusted PS rates appear to be due to the selection for the chlorine-tolerant SNGA found at chlorine-stressed sites and to the deleterious effects of TRC stress on that alga's PS.

Other Techniques to Evaluate Environmental Conditions

Numerous measures of algal characteristics have been used to assess biotic conditions [34]. As part of our studies, we determined the organic content of the periphyton matrix, the autotrophic index (ash-free dry wt/Chl *a*), and the trophic index (ATP/Chl *a*) at each site several times each year [details in Refs 20 and 22].

The organic content of periphyton ranged from 60 to 80% of dry weight near industrial discharges to 20 to 40% at reference sites and at sites far downstream from discharges. The periphyton organic content was strongly influenced by abiotic factors such as siltation and was difficult to relate to conditions for growth.

The autotrophic index (AFDM/Chl *a*) and trophic index (Chl *a*/ATP) have been used to

ORNL-DWG 90M-8883

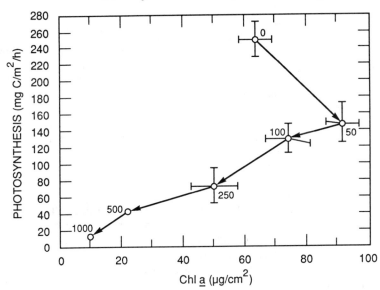

FIG. 6—*Periphyton chlorophyll* a *($\mu g\ cm^{-2}$) and photosynthesis (mg C $m^{-2}h^{-1}$) for periphyton on rocks collected from E-2 after 23 days exposure to 0, 50, 100, 250, 500, or 1000 $\mu g\ L^{-1}$ total residual chlorine in outdoor chambers. Data are means and 1 S.E. (n = 4).*

evaluate allochthonous loading to streams and to evaluate the physiological condition of the algal periphyton, respectively [35–37]. Both ratios are influenced by natural and anthropogenic factors; for example, both the algal and heterotrophic components of the periphyton contain ATP, and noncomplementary changes in either component can confound interpretation of the data. The data for these indices were in the ranges reported elsewhere; however, high temporal variability limited our ability to use these data to draw conclusions concerning microbial activity and the condition of the algal periphyton.

To further evaluate environmental conditions for the periphyton, we conducted short-term algal bioassays of water quality, following the procedure of Giddings et al. (1983) [16], measuring PS rates of the green alga *Haematococcus* sp. in water collected from selected sites in EFPC and WOC. Water was collected from nine sites in EFPC on six dates during June and July 1987. Water was also collected from the study sites in the WOC watershed drainage on three dates during June and July 1987. In the laboratory, *Haematococcus* sp. was added to water from each site, and PS was determined in a 2-h incubation with NaH[14]CO$_3$ [20].

The results showed significant differences in PS by *Haematococcus* sp. in water from the different sites across dates (ANOVA $p < 0.05$). We found occasional evidence of acute toxicity that could be corroborated with data for other communities or known discharges. For example, on one date, water from W-6 (below the ORNL-STP) completely inhibited PS by *Hematococcus* sp. While the bioassay results confirmed the presence of occasional acute toxicity, in general the assay results did not agree with data for the periphyton. The reason for these discrepancies is unknown. This suggested that stress at these sites may be intermittent, cumulative, or due to physical factors or biological interactions that are not included in this laboratory procedure.

Effects of Contaminants on Periphyton Grazers

Metals of toxicological interest in the Y-12 Plant and ORNL discharges accumulate in the periphyton in the receiving streams. Periphyton collected from E-2 and E-3 during February and May 1987 had higher concentrations of several metals than did samples collected from reference sites or from farther downstream in EFPC. Concentrations of up to 112 ppm silver (Ag) (dry wt) and 40 ppm Cd in the periphyton at E-2 were higher than those typically encountered in streams not receiving industrial inputs. Concentrations of copper (Cu) and nickel (Ni) reached 280 and 580 ppm, respectively, near discharges. Lead (Pb) was generally high throughout EFPC, likely reflecting urban runoff inputs. Zinc (Zn) ranged up to 3800 ppm in periphyton at E-3 and E-3 (values 10 to 50 times greater than at reference sites). Total Hg in the periphyton reached 607 ppm at E-2 and decreased with distance downstream. At reference sites (e.g., E-1), periphyton contained about 0.3 ppm Hg.

Periphyton in WOC near ORNL discharges had higher concentrations of some metals than periphyton farther downstream. However, the concentrations of those metals generally were not as high as for periphyton in EFPC.

Several species of periphyton grazing invertebrates were coincidentally absent from sites near industrial discharges (e.g., E-2) [22]. We measured growth and metal concentrations of two taxa of invertebrates from a reference site that fed on (1) periphyton from a contaminated site (E-2), (2) a site farther downstream (E-4), or (3) a reference site (W-1).

Periphyton from E-3 used in the invertebrate growth studies was enriched with Cd, Cu, Ni, and Zn (Table 4). Snails that consumed periphyton from E-3 had slightly higher tissue concentrations of Cd and perhaps Cu than did snails that consumed periphyton from other sites.

TABLE 4—*Periphyton metal content, ash-free dry mass (AFDM), invertebrate metal content, and relative growth rates (RGR) in a five-week feeding experiment. Values for AFDM are means ±1 S.E. (n = 3). Snail tissue metal concentrations are ppm dry wt. Replicates shown were available.*

Site	Periphyton AFDM		Metal Concentration			Relative Growth Rate	
			Periphyton		Snails	Snails	Neophylax
W-1	0.12 ± 0.02	Ag	1.4		<14	1.0	1.0
		Cd	0.67		<14		
		Cr	11		25 21		
		Cu	27		112 112		
		Ni	21		28 30		
		Zn	190		190 198		
E-4	3.08 ± 0.47	Ag	1.9	2.0	<14	3.2[a]	1.5[a]
		Cd	<4.6		<14		
		Cr	28	16	18 13		
		Cu	46	41	129 172		
		Ni	29	25	16 20		
		Zn	260	280	172 198		
E-3	1.62 ± 0.29	Ag	1.9	2.0	<14	2.9[a]	1.8[a]
		Cd	27	22	21 21		
		Cr	51	25	<16		
		Cu	110	110	190 198		
		Ni	210	82	<16 36		
		Zn	3500	2200	267 302		

[a] = RGR difference not significant, ANOVA $p < 0.01$ Fishers 5% PLSD (n = 3 or 4).

Relative growth rates of snails and *Neophylax* larvae reared on periphyton from E-3 and E-5 were greater than rates for those reared on periphyton from W-1 (Table 4). This finding likely reflected higher food availability (i.e., higher periphyton ash free dry mass) at these sites.

Although the macroinvertebrate community at E-3 is depauperate, short-term feeding studies with snails and *Neophylax* did not find significant metal accumulation or reduced growth. These results suggest that either the high metal concentrations in the periphyton at sites near industrial discharges in these systems do not affect grazing invertebrates or that chronic or life-cycle exposure studies might be needed to detect effects.

Conclusions

Periphytic algae are ubiquitous, ecologically important, and have high turnover rates. Their PS provides an easily measured and ecologically important process that responds to toxicants. Consequently, natural periphyton communities are useful for toxicity assessment.

Although evidence of toxic stress was apparent for biotic communities near industrial discharges, we knew neither the specific toxicants nor the frequency and duration of exposure. Laboratory bioassays of water quality that used algae did not relate well to information for the biotic communities at the study sites, suggesting that factors not included in the assays were important.

Because we suspected that toxic effects were intermittent or of a cumulative (chronic) and subtle nature, we employed natural communities to assess toxicity near discharges. The autotrophic index (Chl *a*/AFDM) and the trophic index (Chl *a*/ATP) for periphyton are influenced by many factors and were plagued by high temporal variability. Thus, these indices were not useful for assessing ambient toxicity. Monthly data for algal biomass on occasion suggested acutely toxic events, but, in general, the monthly sampling interval was too long (relative to the turnover time of the periphyton) and was confounded by natural factors (e.g., storms), and so could not be used to detect intermittent toxicity.

A bioassay using natural periphyton communities showed that PS and species composition were more sensitive to chlorine than was biomass and suggested that chlorine in industrial discharges could account for the limited algal taxonomic diversity directly downstream of discharges.

Periphyton PS provides an integrative measure of biotic conditions for periphyton, but is difficult to compare among communities with different amounts of biomass. We used an ANCOVA to generate chlorophyll-adjusted PS rates for the periphyton communities. The CAPS rates reflected the physiological condition of these algal communities and were relatively independent of the algal species composition, available light, and biomass. This procedure provided an ecologically meaningful metric and produced data that was correlated with information for other biotic communities at the study sites. This allowed us to track and compare the condition of periphyton communities and so was useful for ambient toxicity assessment.

Measuring algal accrual provided another integrative measure for toxicity assessment. Algal accrual was slow near discharges where other biotic communities were stressed, but was fast at sites farther from discharges, even where moderate stress on other biotic communities was apparent. Thus, algal accrual rates apparently respond differently to stress than algal taxonomic composition or physiological indices (i.e., CAPS rates).

The effects of nutrient availability were such that they confounded all comparisons of natural communities that included data for nutrient poor upstream or reference sites. Thus, for all community comparisons we found it necessary to compare data from sites near discharges (high-nutrient concentrations) with data from sites far downstream (also nutrient rich).

An invertebrate feeding study found that relative growth rates of snails and caddisfly larvae were related to food availability rather than to the metal concentration of the periphyton they consumed. As was the case for short-term algal bioassays of water quality, it is apparent that some of the factors impacting these biotic communities (i.e., resulting in low numbers, low-growth, and little diversity) were cumulative, intermittent, or otherwise of a nature not accounted for in short-term assays or laboratory studies. This again emphasizes the utility of integrative, functional measures of natural communities for ambient toxicity assessment.

Acknowledgments

The authors thank Cherise M. Pettway, Kimberly Williams, and Patrick Buckley for assistance in the field and laboratory. Our thanks to J. J. Beauchamp for statistical help. We also acknowledge the constructive criticism of this manuscript provided by A. D. Steinman, P. J. Mulholland, and three anonymous reviewers. Research sponsored by the Department of Environmental Monitoring and Compliance, Oak Ridge National Laboratory, and by the Office of Defense Waste and Transportation Management of the U.S. Department of Energy, under contract DE-AC05-840R21400 with Martin Marietta Energy Systems, Inc.

References

[1] Minshall, G. W., *Bioscience,* Vol. 28, 1978, pp. 767–771.
[2] Power, M. E., Matthews, W. J., and Stewart, A. J., *Ecology,* Vol. 66, 1985, pp. 1448–1456.
[3] Huckabee, J. W. and Blaylock, B. G. in *Metal Ions in Biological Systems,* S. K. Dhar, Ed., *Advances in Experimental Medicine and Biology,* Vol. 40, 1973, pp. 125–160.
[4] Selby, D. A., Ihnat, J. M., and Messer, J. J., *Water Resources,* Vol. 19, 1985, pp. 645–655.
[5] Donard, O. F. X., Short, F. T., and Weber, J. H., *Canadian Journal of Fisheries and Aquatic Science,* Vol. 44, 1986, pp. 167–175.
[6] Ramelow, G. J., Maples, R. S., Thompson, R. L., Mueller, C. S., Weber, C., and Beck, J. N., *Environmental Pollution,* Vol. 43, 1987, pp. 247–261.
[7] Patrick, R. in *Biological Methods for the Assessment of Water Quality, ASTM STP 528,* ASTM, Philadelphia, 1973, pp. 76–95.
[8] Tuchman, M. and Blinn, D. W., *British Phycological Journal,* Vol. 14, 1979, pp. 243–254.
[9] Eloranta, P. V., *Hydrobiologia,,* Vol. 96, 1982, p. 253–265.
[10] Hickman, M. and Klarer, D. M., *British Phycological Journal,* Vol. 10, 1975, pp. 81–91.
[11] Marcus, M. D., *Ecology,* Vol. 61, 1980, pp. 387–399.
[12] Davis, T. M., Vance, B. D., and Rodgers, Jr., J. H., *Aquatic Toxicology,* Vol. 12, 1988, pp. 83–106.
[13] Blanck, H., *Hydrobiologia,* Vol. 124, 1985, pp. 251–261.
[14] Blanck, H. and Wangberg, S.-A., *Canadian Journal of Fisheries and Aquatic Science,* Vol. 45, 1988a, pp. 1807–1815.
[15] Blanck, H. and Wangberg, S.-A., *Canadian Journal of Fisheries and Aquatic Science,* Vol. 45, 1988b, pp. 1816–1819.
[16] Giddings, J. M., Stewart, A. J., O'Neill, R. V., and Gardner, R. H. in *Aquatic Toxicology and Hazard Assessment: Sixth Symposium, ASTM STP 802,* W. E. Bishop, R. D. Cardwell, and B. B. Heidolph, Eds., American Society for Testing and Materials, Philadelphia, 1983, pp. 445–459.
[17] Milleman, R., Birge, W. J., Black, J. A., Cushman, R. M., Daniels, K. L., Franco, P. J., Giddings, J. M., McCarthy, J. F., and Stewart, A. J., *Transactions of the American Fisheries Society,* Vol. 113, 1984, pp. 74–85.

[18] Walsh, G. E., McLaughlin, L. L., Yoder, M. J., Moody, P. H., Lores, E. M., Forester, J., and Wessinger-Duvall, P. B., *Environmental Toxicology and Chemistry*, Vol. 7, 1988, pp. 925–929.
[19] Fisher, D. J., Burton, D. T., and Paulson, R. L., *Environmental Toxicology and Chemistry,*, Vol. 8, 1989, pp. 545–550.
[20] Loar, J. M., Ed., "Third Annual Report on the ORNL Biological Monitoring and Abatement Program," Draft ORNL/TM Report, Oak Ridge National Laboratory, Oak Ridge, TN, 1989a.
[21] Loar, J. M., Ed., "Second Annual Report on the ORNL Biological Monitoring and Abatement Program," Draft ORNL/TM Report, Oak Ridge National Laboratory, Oak Ridge, TN, 1988.
[22] Loar, J. M., Ed., "Second Annual Report on the Y-12 Plant Biological Monitoring and Abatement Program," Draft ORNL/TM Report, Oak Ridge National Laboratory, Oak Ridge, TN, 1989b.
[23] Boston, H. L. and Hill, W. R., *Limnology and Oceanography*, 1991a, in press.
[24] Shoaf, W. T. and Lium, B. W., *Limnology and Oceanography*, Vol. 21, 1976, pp. 926–928.
[25] Palumbo, A. V., Mulholland, P. J., and Elwood, J. W., *Limnology and Oceanography*, Vol. 32, 1987, pp. 464–471.
[26] Jeffrey, S. W. and Humphrey, G. F., *Biochemica und Physiolgia Pflanzen (BPP)*, Bd. 167, 1975, pp. 191–194.
[27] Strickland, J. D. H. and Parsons, T. R., *A Practical Handbook of Seawater Analysis,*, Fisheries Research Board of Canada, Ottawa, 1972, p. 309.
[28] Mackay, R. J., *Hydrobiologia*, Vol. 83, 1981, pp. 383–385.
[29] Stewart, A. J., Kszos, L. A., Harvey, B. C., Wicker, L. F., Haynes, G. J., and Bailey, R. D., *Environmental Toxicology and Chemistry*, Vol. 9, 1990, pp. 367–379.
[30] Davis, T. M., Vance, B. D., and Rogers, J. H. Jr., *Aquatic Toxicology*, Vol. 12, 1988, pp. 83–106.
[31] Pfeifer, R. F. and McDiffett, W. F., *Archives fur Hydrobiologia*, Vol. 75, 1975, pp. 306–317.
[32] Marker, A. F. H., *Journal of Ecology*, Vol. 64, 1976, pp. 359–373.
[33] Boston, H. L. and Hill, W. R., *Journal of the North American Benthological Society*, 1991b, in review.
[34] Stevenson, R. J. and Lowe, R. L., "Sampling and Interpretation of Algal Patterns for Water Quality Assessments," *Rationale for Sampling and Interpretation of Ecological Data in the Assessment of Freshwater Ecosystems, ASTM STP 894*, B. G. Isom, Ed., American Society for Testing and Materials, Philadelphia, 1986, pp. 118–149.
[35] Weber, C. I., "Biological Field and Laboratory Methods for Measuring the Quality of Surface Waters and Effluents," U.S. Environmental Protection Agency, Environmental Protection Agency, Cincinnati, OH, No. 670/4-73-001, 1973.
[36] Bothwell, M. L., *Limnology and Oceanography*, Vol. 30, 1985, pp. 527–542.
[37] Healey, F. P. and Hendzel, L. L., *Canadian Journal of Fisheries and Aquatic Science*, Vol. 37, 1980, pp. 442–453.
[38] Friend, D. T. C, *Ecology*, Vol. 42, 1961, pp. 577–580.

Barbara M. Judy[1], William R. Lower[2], Frank A. Ireland[3], and Garry F. Krause[4]

A Seedling Chlorophyll Fluorescence Toxicity Assay

REFERENCE: Judy, B. M., Lower, W. R., Ireland, F. A., and Krause, G. F., "**A Seedling Chlorophyll Fluorescence Toxicity Assay,**" *Plants for Toxicity Assessment: Second Volume, ASTM STP 1115*, J. W. Gorsuch, W. R. Lower, W. Wang, and M. A. Lewis, Eds., American Society for Testing and Materials, Philadelphia, 1991, pp. 146–158.

ABSTRACT: Changes in chlorophyll fluorescence yield are related to changes in the photosynthetic efficiency of plants. Under normal conditions, approximately 97% of the light energy absorbed by chlorophyll is converted to oxidation and reduction potential to drive electron transport in photosynthesis. Stress conditions such as those caused by toxic chemicals can reduce the efficiency of photosynthesis by disturbing the pigment protein apparatus or blocking the electron transport system in the chloroplast. The 6 to 10% loss of absorbed light energy due to stress conditions appears as chlorophyll fluorescence. A chlorophyll fluorescence assay can provide a simple, rapid method for measuring the effect of physical and chemical stress on the photosynthetic efficiency of plants. In this study, seedlings of soybean (*Glycine max* L. Merr.) and barley (*Hordeum vulgare* L.) were analyzed by chlorophyll fluorescence induction following treatment with three herbicides dissolved in water at three concentrations. The eluate of soil collected from a hazardous landfill was also tested. Several endpoints were measured; initial fluorescence (F_0), variable fluorescence (F_v), maximum fluorescence (F_m), F_v/F_m ratio, and electron pool (EP). This study indicated that the decline of photosynthetic efficiency of plants exposed to toxic chemicals in water or soil eluate can be measured by the chlorophyll fluorescence assay method.

KEY WORDS: chlorophyll fluorescence, fluorometer, herbicides, soil toxicity, plant bioassay

The chlorophyll fluorescence can be used to monitor a plant's response to any factor which directly or indirectly affects photosynthetic metabolism.

The present study assesses the toxicity of aqueous solutions of herbicides with different modes of action by measuring changes in chlorophyll fluorescence of barley and soybean seedlings using a new fluorometer model (SF-30, R. Brancker Research Ltd.). A substantial portion of this study involved testing the utility of the chlorophyll fluorescence method for toxicity assessment of soil samples taken from hazardous landfill with unknown compositions of chemicals.

When a dark-adapted leaf at room temperature is suddenly exposed to light, the chlorophyll fluorescence yield and the concentration of photosynthetic intermediates change. The intensity of chlorophyll fluorescence passes through a characteristic sequence of changes

[1] Research specialist, Department of Veterinary Biomedical Sciences, University of Missouri, Columbia, MO 65203.
[2] Group leader, Environmental Trace Substances Research Center, University of Missouri, Columbia, MO 65203.
[3] Research specialist, Environmental Trace Substances Research Center, University of Missouri, Columbia, MO 65203.
[4] Professor, Department of Statistics, University of Missouri, Columbia, MO 65203.

before a steady state is reached. The kinetics of this, called fluorescence induction, are complex and highly dependent on physiological conditions. This phenomenon has been known as the Kautsky effect after the investigator who first observed changes in the chlorophyll fluorescence [1].

During the past decade, interest has been growing in the practical application of chlorophyll fluorescence measurement as a rapid, sensitive, and nondestructive method for the determination of photosynthetic activity in plants under greenhouse and field conditions. Fluorescence measurements have provided a very useful tool for monitoring temperature stress, water stress, salinity, and light stress, air pollution, and herbicide toxicity in plants [2–5]. In recent years, portable instruments with microcomputers for data processing and storage have become available for field work.

Materials and Methods

Plant Material

Plants were grown in vermiculite (medium grade) in Jiffy pots under white, artificial light (400 μE/m²·s) in a growth room, maintained at 20 ± 2°C and 14/10-h light/dark cycle. Nutrients and water were applied to provide optimal growing conditions. Seedlings of two species, soybean and barley, were used in this study. Seedlings of soybean (Williams 82) and barley (Steptoe) were cut above the roots and immediately immersed in 3 cm of treatment solution (herbicide solution or soil eluate). Soybeans were exposed for 24 h in treatment solutions at the first trifoliate stage and barley at the second leaf stage.

Treatment

Three herbicides with different modes of action were dissolved in reverse osmosis double deionized (RODD) water at three concentrations: 10, 50, and 100 ppm. DCMU (Diuron) (3-[3,4-Dichlorophenyl]-1,1-dimethylurea) was obtained from Sigma, and Simazine technical (2-chloro-4,6-bis(ethylamino)-s-triazine) was obtained from Ciba-Geigy. These herbicides are well-known inhibitors of Photosystem II (PS II), which block electron transport at the secondary electron acceptors Q_B [6]. The third herbicide, Sulfometuron methyl (Oust™) [methyl-2-[[[(4,6-dimethyl-2-pyrimidinyl)amino]carbonyl]amino]-sulfonyl]benzate), was obtained from Du Pont. This herbicide suppresses and stops plant growth by arresting cell division in the growing tips of roots and shoots. All three herbicides have broad spectrum effects and are used against many broadleaf and grass weeds in certain crops.

In the second part of this study, plants were exposed to soil eluate derived from a surface layer (0 to 5 cm) of soil collected from an old landfill at Lake City Army Ammunition Plant, Lake City, Missouri. The closed landfill (1979) was used for scrap wood, paper, and also for disposal of a variety of industrial materials, including waste contaminated with explosives and solvents. Depleted uranium may also have been disposed of at this landfill. The analysis of soil done in 1983 by the United States Army Toxic and Hazardous Materials Agency (USATHAMA) determined that this site contained 23 000 μg/L volatile organics, 1300 μg/L semivolatile organics, 69.3 μg/g explosives, 110 μg/L arsenic, 170 μg/L copper, and >700 μg/L zinc. An air-dried soil sample (125 g) was mixed with 500 mL RODD water in a 1-L Teflon container and agitated at 120 rpm for 48 h at 20°C in the dark in the laboratory. After 48 h, the liquid, now referred to as soil eluate, was transferred to a 500-mL beaker. Three dilutions were formulated with RODD water, 10, 50, and 100%, and used for the chlorophyll fluorescence study.

All treatments were compared to the RODD water control treatment. The chlorophyll fluorescence measurements were made after 24 h of exposure to the treatment solutions.

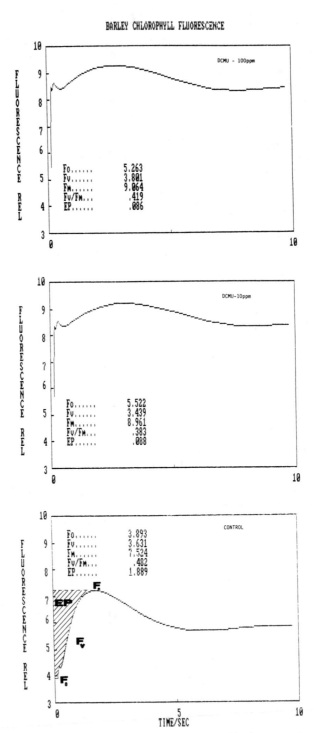

FIG. 1—*Changes in chlorophyll fluorescence induction curves of barley exposed to herbicides:* F_0 = *initial fluorescence;* F_v = *variable fluorescence;* F_m = *maximum fluorescence; EP = electron pool.*

Chlorophyll Fluorescence Measurement

When a dark-adapted leaf at room temperature is suddenly exposed to light, the chlorophyll fluorescence yield, as emitted light, shows characteristic changes (Fig. 1). The fluorescence first rises rapidly to a level 0, which is called the "constant" or "initial" fluorescence (F_0) and denotes fluorescence when the reaction center of PS II is fully oxidized. The yield of F_0 is independent of photochemical events and reflects the amount of chlorophyll. The level P characterizes the "peak" or "maximum" fluorescence yield (F_m). F_0 is followed by variable fluorescence (F_v) of PS II ($F_v = F_m - F_0$). F_v is dependent on the redox state of the Quencher Q, the first stable electron acceptor of PS II. Any action causing Q oxidation tends to result in fluorescence quenching, i.e., reduction in the emission of light. Conversely, Q reduction increases the fluorescence yield. The rate of the fluorescence rise from F_0 to P is proportional to the sum of Q reduction and oxidation rates. The increase or decrease of these fluorescence parameters, F_0, F_m, or F_v, compared to the control, can indicate an alteration in the chain of events in photosynthesis or damage to the chloroplasts. The electron pool (EP) is the area above the fluorescence curve between the $F_0 - F_m$ transients. This area characterizes the photochemical capacity of PS II, is proportional to the PS II electron acceptor pool, and determines the potential PS II photochemical activity.

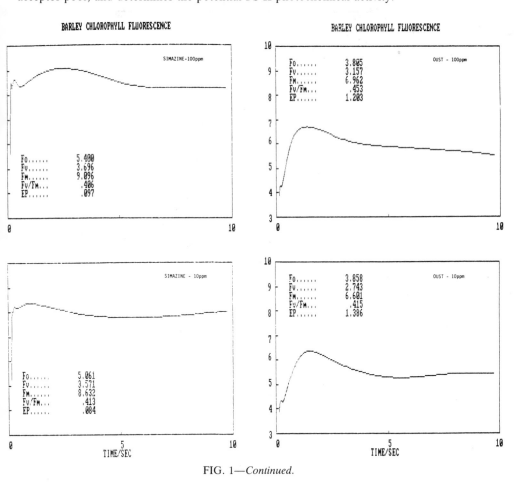

FIG. 1—*Continued.*

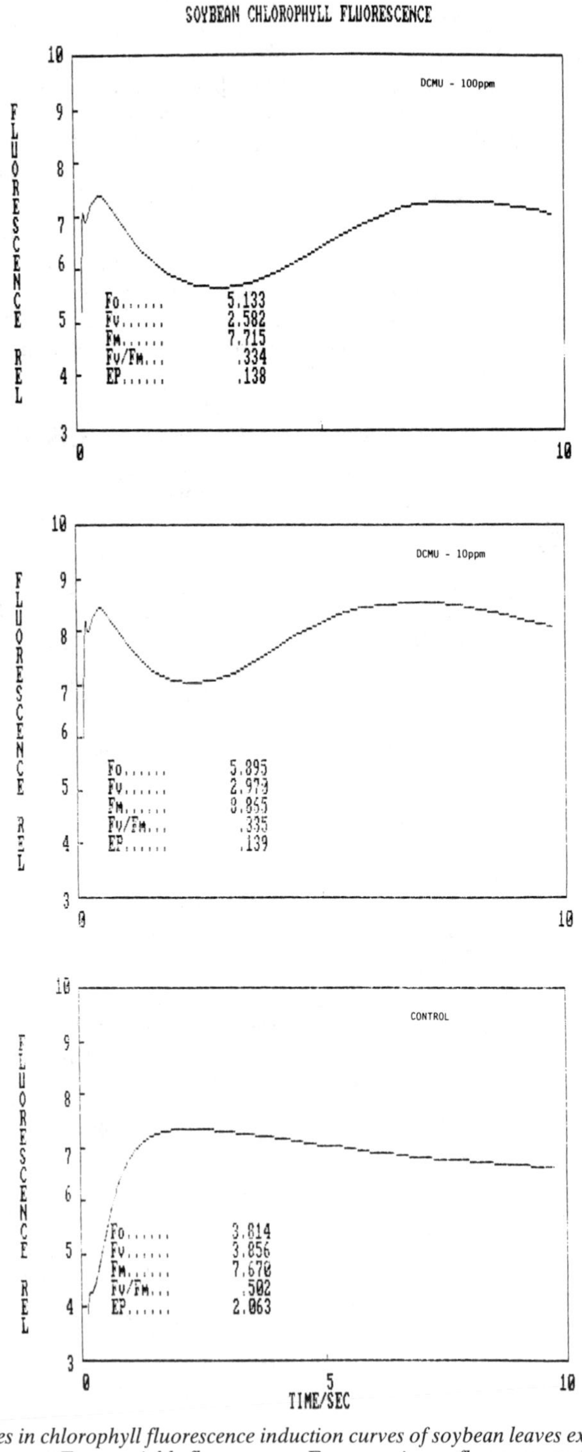

SOYBEAN CHLOROPHYLL FLUORESCENCE

FIG. 2—*Changes in chlorophyll fluorescence induction curves of soybean leaves exposed to herbicides:* F_0 = *initial fluorescence;* F_v = *variable fluorescence;* F_m = *maximum fluorescence; EP = electron pool.*

Fluorescence induction curves were recorded by a fluorometer (model SF-30, Richard Brancker Research Ltd., Ottawa, Canada) attached to a personal computer. Fluorescence was measured at room temperature from the abaxial surface of the plant leaves. The leaves were illuminated for 10 s with red light (670 nm, intensity 6.0 W/m^2) after a 15-min period of dark adaptation. The chlorophyll fluorescence was measured from the middle part of two leaves from five plants for a total of ten measurements for each treatment and control. Parameters of particular interest in this study were the F_0, F_v, F_m, ratio F_v/F_m, and EP. The shape of the curves was also considered.

Statistical Analysis

The chlorophyll fluorescence data were statistically analyzed using analysis of variance (ANOVA). An analysis of variance with factorial sources for treatment (herbicide) and concentration was computed for each variable (F_0, F_v, F_m, F_v/F_m ratio, and EP) but expressed as a percent of the peer control and for each plant species separately. An analysis of variance for barley and soybean exposed to soil eluate from the hazardous waste site was done separately. Means for both experiments were compared using a protected least significant

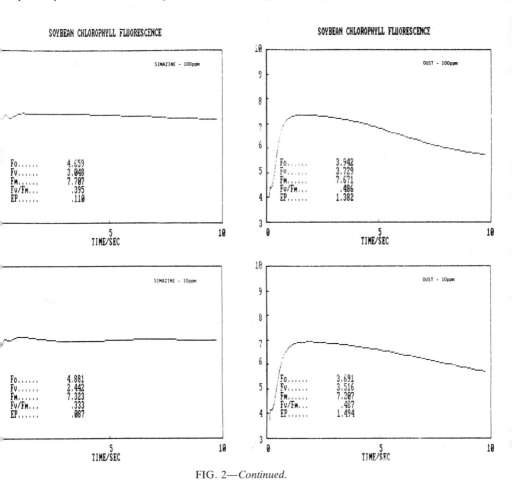

FIG. 2—*Continued.*

TABLE 1—*Percent changes in initial fluorescence F$_0$ in barley and soybean seedlings exposed to herbicides dissolved in water.*

Concentration, ppm	Barley			Soybean		
	DCMU	Simazine	Sulfometuron	DCMU	Simazine	Sulfometuron
0-CONTROL	100 a	100 a	100 a	100 a	100 a	100 a
10	138 b	132 ce	101 a	153 b	135 cd	112 e
50	138 b	126 d	101 a	150 b	139 cd	107 e
100	135 be	132 ce	99 a	144 bc	130 d	111 e

All values are the mean of 10 measurements.

	LSD ($p < 0.05$)	LSD ($p < 0.05$)
Control vs concentrations	2.7	6.4
Treatment vs treatment	3.9	9.1

NOTE: Values followed by the same letter do not differ at the 5% level according to the Protected LSD Rule.

difference (LSD) rule at $p < 0.05$ [7]. A comparison was made against the control as well as among treatments.

Results and Discussion

Herbicides

Figures 1 and 2 show the examples of fluorescence induction curves from barley and soybean seedlings exposed to 10 and 100 ppm herbicides dissolved in water. After a 24-hr exposure, barley and soybean leaves show strong changes in chlorophyll fluorescence induction curves, especially when treated with DCMU or Simazine. These herbicides are known to interfere with photochemically induced electron transport. The site of action of most herbicide electron transport inhibitors is associated closely with PS II. Consequently,

TABLE 2—*Percent changes in variable fluorescence F$_v$ in barley and soybean seedlings exposed to herbicides dissolved in water.*

Concentration, ppm	Barley			Soybean		
	DCMU	Simazine	Sulfometuron	DCMU	Simazine	Sulfometuron
0-CONTROL	100 a	100 a	100 a	100 a	100 a	100 a
10	103 a	98 ad	77 b	69 b	60 b	85 c
50	102 a	98 ad	79 bc	69 b	68 b	97 a
100	93 ac	106 a	87 bcd	67 b	77 bc	98 a

All values are the mean of 10 measurements.

	LSD ($p < 0.05$)	LSD ($p < 0.05$)
Control vs concentrations	10.5	8.0
Treatment vs treatment	14.8	11.4

NOTE: Values followed by the same letter do not differ at the 5% level according to the Protected LSD Rule.

TABLE 3—*Percent changes in maximum fluorescence* F_m *in barley and soybean seedlings exposed to herbicides dissolved in water.*

Concentration, ppm	Barley			Soybean		
	DCMU	Simazine	Sulfometuron	DCMU	Simazine	Sulfometuron
0-CONTROL	100 a	100 a	100 a	100 a	100 a	100 a
10	121 b	115 bc	89 d	109 c	96 b	98 ab
50	120 b	112 c	90 d	108 c	102 abc	102 abc
100	115 bc	119 bc	93 d	104 ac	102 abc	104 ac

All values are the mean of 10 measurements.

	LSD ($p < 0.05$)	LSD ($p < 0.05$)
Control vs concentrations	5.4	5.5
Treatment vs treatment	7.6	7.8

NOTE: Values followed by the same letter do not differ at the 5% level according to the Protected LSD Rule.

reactions coupled to PS II, such as noncyclic electron transport, where water acts as the electron donor for various electron acceptors, are inhibited [8]. The action of DCMU has been studied more intensively than any other herbicide. It blocks Q_A reoxidation and causes the maximum fluorescence level F_m to be reached very rapidly.

In this study, DCMU and Simazine caused almost identical changes in chlorophyll fluorescence induction curves in barley leaves. Both herbicides significantly increased F_0 and F_m and decreased the ratio F_v/F_m and EP (Tables 1–5). The F_0 level is known to be affected by environmental stress, which results in structural alterations at the PS II pigment level [9,10]. The EP, also called the complimentary area (CA), is thought to be a valid measure of the number of active PS II centers and characterizes PS II activity [11].

When barley leaves were exposed to Sulfometuron (Oust™), the shape of the chlorophyll induction curve was not changed, but the parameters F_v, F_m, F_v/F_m, and EP were significantly

TABLE 4—*Percent changes in ratio fluorescence* F_v/F_m *in barley and soybean seedlings exposed to herbicides dissolved in water.*

Concentration, ppm	Barley			Soybean		
	DCMU	Simazine	Sulfometuron	DCMU	Simazine	Sulfometuron
0-CONTROL	100 a	100 a	100 a	100 a	100 a	100 a
10	85 abc	85 abc	85 abc	63 b	62 b	87 d
50	84 bc	87 bcd	87 bcd	64 b	66 b	95 ae
100	81 b	89 cd	93 d	63 b	75 c	94 de

All values are the mean of 10 measurements.

	LSD ($p < 0.05$)	LSD ($p < 0.05$)
Control vs concentrations	5.6	5.0
Treatment vs treatment	7.9	7.0

NOTE: Values followed by the same letter do not differ at the 5% level according to the Protected LSD Rule.

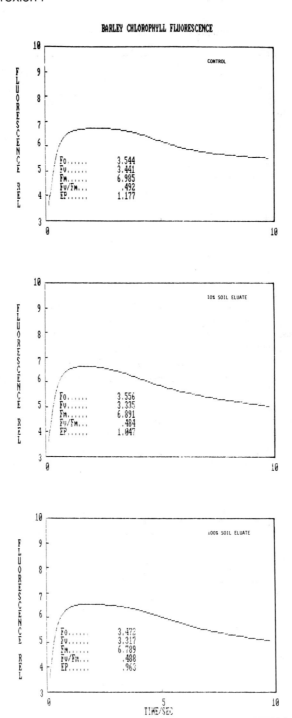

FIG. 3—*Changes in chlorophyll fluorescence induction curves of soybean and barley leaves exposed to soil eluate taken from hazardous landfill:* F_0 = *initial fluorescence;* F_v = *variable fluorescence;* F_m = *maximum fluorescence; EP = electron pool.*

SOYBEAN CHLOROPHYLL FLUORESCENCE

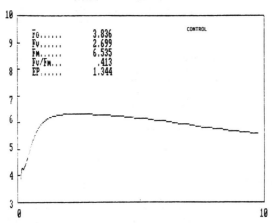

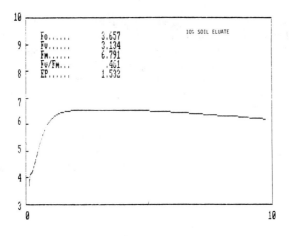

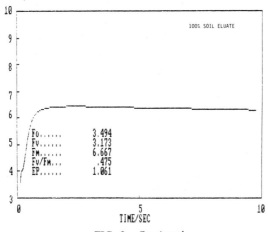

FIG. 3—*Continued.*

TABLE 5—*Percent changes in electron pool (EP) in barley and soybean seedlings exposed to herbicides dissolved in water.*

Concentration, ppm	Barley			Soybean		
	DCMU	Simazine	Sulfometuron	DCMU	Simazine	Sulfometuron
0-CONTROL	100 a	100 a	100 a	100 a	100 a	100 a
10	7 b	8 b	77 cd	10 b	4 b	74 c
50	7 b	6 b	69 d	10 b	5 b	77 c
100	6 b	6 b	80 c	8 b	5 b	79 c

All values are the mean of 10 measurements.

	LSD ($p < 0.05$)	LSD ($p < 0.05$)
Control vs concentrations	6.4	4.4
Treatment vs treatment	9.1	6.2

NOTE: Values followed by the same letter do not differ at the 5% level according to the Protected LSD Rule.

decreased compared to the control (Fig. 1, Tables 1–5). The causes of F_v decrease may be complex. Environmental stress that causes chloroplast damage can decrease the F_v yield [12]. Sulfometuron is a herbicide which does not affect photosynthesis directly. The changes in chlorophyll fluorescence obtained after exposure to a herbicide such as Sulfometuron indicate a possibility for using a chlorophyll fluorescence bioassay as a screening test for a wider range of chemicals which do not inhibit electron transport system but indirectly influence photosynthesis.

Soybean seedlings exposed to the same herbicides show slightly different changes in chlorophyll fluorescence. DCMU and Simazine strongly increased F_0 but decreased F_v. In this case, F_m as a sum of F_0 and F_v was not changed, especially in treatment with Simazine. DCMU and Simazine significantly reduced the ratio F_v/F_m about 30% and greatly reduced EP by about 90 to 95%. Soybean seedlings exposed to Sulfometuron significantly increased

TABLE 6—*Percent changes in chlorophyll fluorescence in barley seedlings exposed to soil eluate (SE) from hazardous landfill.*

Treatment	Concentration, %	Initial Fluorescence, F_0	Variable Fluorescence, F_v	Maximum Fluorescence, F_m	Ratio, F_v/F_m	Electron Pool (EP)
Control	...	100 a	100 a	100 a	100 a	100 a
Soil eluate	10	99 a	97 a	98 a	99 a	96 ab
Soil eluate	50	98 a	97 a	97 a	99 a	91 ab
Soil eluate	100	98 a	96 a	97 a	99 a	90 b

All values are the mean of 10 measurements.

LSD ($p < 0.05$) for control vs SE concentrations	2.3	6.1	3.8	2.5	9.5
LSD ($p < 0.05$) for soil eluate concentrations	3.2	8.6	5.4	3.6	13.4

NOTE: Values followed by the same letter do not differ at the 5% level according to the Protected LSD Rule.

TABLE 7—*Percent changes in chlorophyll fluorescence in soybean seedlings exposed to soil eluate (SE) from hazardous landfill.*

Treatment	Concentration, %	Initial Fluorescence, F_0	Variable Fluorescence, F_v	Maximum Fluorescence, F_m	Ratio, F_v/F_m	Electron Pool (EP)
Control	. . .	100 a	100 a	100 a	100 a	100 a
Soil eluate	10	100 ab	118 bc	107 a	109 b	117 a
Soil eluate	50	102 ab	118 bc	108 a	108 b	104 a
Soil eluate	100	95 b	116 ac	103 a	111 b	101 a
All values are the mean of 10 measurements.						
LSD ($p < 0.05$) for control vs SE concentrations		4.8	17.4	8.8	7.3	19.6
LSD ($p < 0.05$) for SE concentrations		6.9	24.6	12.4	10.4	27.7

NOTE: Values followed by the same letter do not differ at the 5% level according to the Protected LSD Rule.

F_0 by about 10%. The ratio F_v/F_m was reduced, and EP was significantly smaller relative to the control. The changes in F_v were very small, and only the lowest concentration of Sulfometuron was significantly different from the control.

The concentrations of herbicides used in this study (10, 50, and 100 ppm) did not yield results which were very different for DCMU and Simazine versus Sulfometuron. The chlorophyll fluorescence assay allows the opportunity to observe the changes in photosynthetic activity of plants almost immediately, much earlier than the occurrence of visible injury.

In the second part of this study, chlorophyll fluorescence measurements were made after exposure of barley and soybean seedlings to the soil eluate (SE) from the hazardous landfill. After 24-h exposure the shape of chlorophyll fluorescence induction curves of barley for all variables was not affected (Fig. 3, Table 6). Soybean seedlings showed some changes in F_0 at the highest concentration of SE (Fig. 3, Table 7). Also F_v was increased in all concentrations compared to the control, but the differences were significant only at 10 and 50% concentrations of SE. Due to changes in F_0 and F_v, the ratio F_v/F_m was also significantly greater than the control. The chemical analysis done in 1983 showed the presence of volatile and semivolatile organics, copper, arsenic, and zinc in the soil, but the chlorophyll fluorescence assay did not indicate consistently that the evaluated soil sample was toxic.

Conclusions

Among the variety of herbicides that interfere at different stages with plant metabolism, a large number effectively block photosynthetic electron transport between the primary and secondary plastoquinones (Q_A and Q_B) at the PS II electron acceptor side. A fluorimetric method seems to be most promising in analyzing these herbicide effects. In this study, DCMU as the prototype herbicide, as well as Simazine, caused strong changes in chlorophyll fluorescence in barley and soybean leaves. The complexity of the relationship between fluorescence emission and the physiological state of the whole photosynthetic apparatus suggest the utility of this method with a wider range of toxic chemicals. Changes in chlorophyll fluorescence were detected by fluorometry in the case of Sulfometuron, a herbicide which does not influence electron transport. The plant fluorometry technique appears to be a rapid, easy method to measure the toxicity of chemicals and thus is suited for large-scale screening analysis.

References

[1] Kautsky, M. and Hirsche, A., *Naturwissenschaften,* Vol. 48, 1931, p. 964.
[2] Björkman, O., Badger, M. R., and Armond, P., *Carnegie Inst. Washington* (monograph series), Vol. 77, Year Book 1978, pp. 262–275.
[3] Smillie, R. M., in *Low Temperature Stress in Crop Plants,* J. M. Lyons, D. Graham, and J. K. Raison, Eds., Academic Press, New York, 1979, pp. 187–202.
[4] Govindjee, Downton, W. J. S., Fork, D. C., and Armond, P. A., *Plant Science Letters,* Vol. 20, 1981, pp. 191–194.
[5] Judy, B. M., Lower, W. R., Miles, D. C., Thomas, M. W., and Krause, G. F., "Chlorophyll Fluorescence of a Higher Plant as an Assay for Toxicity Assessment of Soil and Water," in *Plants for Toxicity Assessment, ASTM STP 1091,* ASTM, 1990, pp. 308–318.
[6] Van Rensen, J. J. S., *Physiologia Plantarum,* Vol. 54, 1982, pp. 515–521.
[7] Snedecor, G. W. and Cochran, W. G., *Statistical Methods,* ISU Press, Ames, IA, 1980, 7th ed., p. 234.
[8] Moreland, D. E., *Annual Review of Plant Physiology,* Vol. 31, 1980, pp. 597–638.
[9] Schrieber, U. and Berry, J. A., *Planta,* Vol. 136, 1977, pp. 233–238.
[10] Schreiber, U. and Armond, P. A., *Biochimica and Biophysica Acta,* Vol. 502, 1978, p. 138.
[11] Lavorel, J., Breton, J., and Lutz, M., in *Light Emission by Plant and Bacteria,* Govindjee, J. Amesz, and D. C. Fork, Eds., Academic Press, San Diego, 1986, pp. 57–98.
[12] Krause, H. G. and Weis, E., *Photosynthesis Research,* Vol. 5, 1989, pp. 139–157.

Xenobiotic Uptake By Plants

Sandra M. Stewart-Pinkham[1]

Detecting Ambient Cadmium Toxicity in an Ecosystem

REFERENCE: Stewart-Pinkham, S. M., **"Detecting Ambient Cadmium Toxicity in an Eco-system,"** *Plants for Toxicity Assessment: Second Volume, ASTM STP 1115*, J. W. Gorsuch, W. R. Lower, W. Wang and M. A. Lewis, Eds., American Society for Testing and Materials, Philadelphia, 1991, pp. 161–171.

ABSTRACT: The threshold for a toxic effect from cadmium is close to its background levels. In experimental studies, chronic low-dose exposure does not necessarily increase body burden and causes effects associated with other stressors; therefore, low-dose effects from environmental sources are hard to establish using traditional approaches. The effects of ambient cadmium have increased with the precipitous drop in lead in the atmosphere. They are manifested by increased stress responses. In a community exposed to cadmium fumes from waste incineration, one can infer cadmium effects in such observed environmental stresses as increased seeding in trees, tree injuries, and learning and behavioral problems in children.

Biochemical effects of free cadmium, rather than the levels of cadmium burden or accumulation, need to be assessed in multiple biological systems, especially trees and humans. Environmentally bioavailable cadmium can be detected more accurately by assaying multiple metals in hair, leaves, lichens, and house dust. The hypothesis that cadmium is released with stress and/or mediates the stress response should be directly tested through carefully designed scientific studies.

KEY WORDS: cadmium, lead, health effects, waste incineration, plant stress

This paper provides an interdisciplinary review of recent information about low-dose cadmium toxicity in animal and plant cell systems. On the basis of this review I propose the hypothesis that cadmium is a central mediator of stress responses and/or augmentor of stress responses. Since lead does not act as a stressor and can antagonize cadmium, the fall in lead air pollution can be expected to increase cadmium toxic effects from air pollution. The role of a variety of stressors that may act as antagonists and cofactors in these effects are also discussed. Observations in a community exposed to environmental cadmium air pollution from a large municipal waste incineration plant are discussed to support the hypothesis that cadmium is exerting a stress effect in humans and trees at present levels of exposure. Finally, the need for definitive studies to answer specific testable questions raised by this hypothesis is addressed.

Background

Cadmium is a ubiquitous pollutant whose toxic level is very close to the background level [*1*]. Its toxicity in aquatic environments is well recognized, but toxic exposures from air pollution have not been established. Although the most significant source of indoor cadmium air pollution is cigarette smoke (and smoking is linked to human stress and chronic disease),

---

[1] M.D. in private practice, 1890 Northwest Blvd., Columbus, OH 43212.

cadmium, itself, is not considered a major factor in human disease. Since emissions are always mixed, injuries from air pollution are never related solely to cadmium. The most important source of outdoor cadmium air pollution is waste incineration [2], but all combustion, many industrial processes, traffic fumes, and fertilizers are also significant sources [3]. Cadmium is relatively volatile, and it is exceptionally able to cross cell membranes [4]. Anthropogenic sources are taken up by cells and can be released from plants back into the atmosphere [5]. Cadmium has long-distant transport associated with sulfur and nitrogen, the gases associated with acid rain [3]. However, the speciation of cadmium in the atmosphere has never been studied, and air filters are not being used that capture gaseous forms.

Cadmium is a transitional group IIB metal with unusual chemical characteristics [4]. It can bind to a number of compounds such that its behavior in complex systems is impossible to predict. It has a high affinity for chlorides. Cadmium chloride is unusual in that in the ocean it exists as a polar nonionic substance, which may explain its ability to cross cell membranes [3]. Its ability to complex with sulfur and nitrogen may allow it to complex with bacteria-generated gases. Since cadmium enters cells so rapidly during the log phase of growth [6], these gases may act to carry cadmium out of the cell and into the atmosphere. So far, these intriguing possibilities have not been studied.

Biochemically, cadmium damages cells, plant or animal, by changing calcium, sodium, and hydrogen transport across the cell membrane, interfering with cell energy, altering enzyme functions, altering DNA and gene expression, and inducing free radical formation by changing microsomal enzymes. It also induces protective mechanisms that increase a cell's chances of surviving adverse conditions through these effects and induction of metal-binding proteins and hormonal secretions [4–11].

Until recently, environmental air cadmium pollution was not suspected of having low-dose effects on mammalian cells. Of all the common metals, Lag found it had the lowest threshold of effect when tested in a tracheal epithelial organ culture [12]. It paralyzed all the cilia and caused necrosis of many cells at 10 μM, while lead had minimal effects at doses as high as 200 μM.

One of the most important effects discovered is that cadmium has 1000 times the affinity for membrane CaATPasae as calcium [13]. At a 1-nM level, CaATPase was inhibited. By blocking the efflux of calcium, increasing intracellular calcium, cadmium can evoke a variety of stress responses in a variety of cells. Normally there is 150 nM of free calcium in a cell. As little as 0.2 nM of free intracellular cadmium can be expected to influence free intracellular calcium, the second messenger system of cells. Stress is a chaos factor; a small change can produce a large effect [14]. Therefore, small changes in environmental cadmium levels or bioavailability are capable of producing large effects.

At 1 to 10 μM concentration, cadmium generally functions as a calcium channel blocker in nerve cells. Thus, effects found at low dose may be the opposite of those found with higher doses. In reviewing effects of cadmium determined experimentally, this discrepancy must be taken into account.

By causing somatic mutations and interfering with DNA repair [15], cadmium is associated with an important mechanism linked with aging in animals. In plant studies, cadmium also causes premature senescence and aging [16]. The generation of free radicals in cells is another mechanism of aging linked to cadmium exposure [17].

There are several key variables in cadmium toxicity. Route of exposure is important since, even at low concentrations in the air, cadmium is absorbed readily by plant leaves and translocated to other sites [18]. In animal studies, cadmium was shown to be taken up rapidly during respiration [19].

Nutritional factors are important. When cadmium enters cells rapidly during the log phase of growth, it alters the cell membrane, increasing its permeability to other divalent ions [20]. Exposure to cadmium can lead to losses of nutrient metals and uptake of toxic metals.

Age affects cadmium toxicity; the neonatal mammalian brain is extremely susceptible [4]. The effect of cadmium is different on lymphocytes of young and old animals [21].

There are large individual differences even within a species and among in-bred laboratory animals due to subtle genetic differences. When viruses are interacting with a host, cadmium can affect the outcome [22]. The susceptibility to cadmium is associated with major histocompatibility antigens in mammalian cells [23]. Cadmium is the metal with the most immune modulatory effects in mammalian cells, far more than lead or mercury.

Another important factor is the presence of other chemical agents. Cadmium reacts with a number of other toxicants, such as chloroform [24] and lindane [25], causing toxic effects in hepatocytes when they are together at doses that are not toxic when they are given singly. In mammalian cells, this is accomplished by altering the activity of liver detoxifying enzymes [26].

To understand the behavior of a metal like cadmium in the environment, it is essential to consider the presence of the total mixture of substances in which it is occurring. Calcium, zinc, lead, iron, manganese, and chromium are antagonistic to cadmium toxicity [27,28], but copper increases its toxicity synergistically in industrial wastes. Copper displaces cadmium from one of the cadmium binding proteins. The addition of a normally nontoxic concentration of copper to a sublethal concentration of cadmium causes death of plant cell cultures [29].

Phytotoxicity

Cadmium produces a wide range of effects on plants [30]. Cadmium is extremely toxic to the chloroplast, as it is to the mitochondria. Cadmium decreases photosynthesis and increases photorespiration [31]. This results in a decrease in starch storage [30], making the plant more vulnerable to stresses. This effect differs from other heavy metals like nickel, cobalt, and copper, which induce starch accumulation [32].

Cadmium exposure causes wilting [33]. Barcelo has found that this effect is associated in part with changes in calcium transport [34]. The effect on calcium transport corresponds well to the cadmium effect on calcium transport observed by Verbost et al. in animal cells [13].

If one assumes that the toxic effect in animal cells occurring at a threshold of only 1 nmol/L free cadmium is the threshold for all cells, a toxic effect of cadmium could easily occur in plant tissues at very low levels of exposure from atmospheric pollution if there were not sufficient binding and detoxifying capabilities. Barcelo found that cadmium levels in the plant leaves were much lower than in the roots, and yet there was a much higher toxic effect in the leaves. An important proportion of leaf cadmium appears to be present in low molecular weight complexes that are more mobile and toxic [35].

Trees are highly sensitive to cadmium because it affects their resistance to drought, disease, and other stresses. For this reason one might suspect that they would be particularly sensitive in detecting atmospheric conditions changing the bioavailability of cadmium and its impact on human health and the environment. Unfortunately, there have been no controlled studies of the response of trees to air cadmium exposure. By using information derived in other cell systems, it may be possible to detect changes in trees that can be linked with cadmium.

Cadmium and Stress

An agent that causes an organism to deviate from homeostasis is a stressor. There are many kinds of stress agents: physical, chemical, biological, and psychosocial. Although the inciting agents differ, the biochemical response can be the same, such as increased intracellular calcium.

Although changing intracellular calcium is one mechanism by which a stress response is evoked by cadmium, it has a remarkable capacity to affect the function of a great variety of enzymes [36]. Since all tissues of an organism contain at least small amounts of cadmium, the effect of different stress agents could be the release of very small amounts of free cadmium, triggering the biochemical changes associated with stress responses. For instance, cadmium and heat stress both induce heat shock protein in mammalian cells [37]. Could it be that heat stress releases cadmium and that this free cadmium is responsible for the induction of heat shock protein? Heat shock proteins are also produced in plant cells in response to cadmium.

Emotional stress is the most effective stimulus for increasing liver metallothionein, the major cadmium detoxifying protein in the liver [38]. Cadmium is also capable of greatly increasing gene expression [39]. If emotional stress led to a release of cadmium, that release would accelerate the effects of the stress by releasing a powerful stress agent.

Another vital effect that has recently been determined is that cadmium has 40 times the affinity as zinc for the steroid binding site on DNA [40]. In stress responses there is a fall in plasma zinc and an increase in bound zinc. This drop is associated with free radical production in granulocytes, killing bacteria, in which cadmium may play a key role [8]. When stress responses are exaggerated and poorly modulated, they become destructive. Low-level increases in environmental pollution with cadmium may contribute to just such effects.

In plants, drought is a stressor. Since cadmium blocks water translocation, cadmium and drought could be expected to have a greater effect than the effect of drought alone. This exaggerated effect would occur in genetically susceptible plants and those with nutritional deficiencies.

Exposure to a stressor can also increase hardiness and stimulate growth. In the same environment in which trees are showing destructive stress, one could expect to see trees that are thriving. The presence of cadmium could be inferred from a marked variability in the response of the ecosystem to a drought stress, with some trees dying and others growing exceptionally rapidly in response to hormonal influences triggered by the stressor but amplified by cadmium.

Another plant response to stress is increased seeding. There are no studies of effects of cadmium on seeding. However, in aquatic toxicology cadmium has profound effects on reproduction. Many times it is so lethal that there is decreased viability and decreased progeny, but in animal studies it can be linked to a blastogenic agent, cyclic cytidine monophosphate (CMP) [41], that would increase reproductive capacity. Since cadmium is such an effective mutagen by both altering DNA and interfering with DNA repair, one can appreciate the adaptive value in the liberation of free cadmium when any organism, plant or animal, is exposed to a severe stress so that mutations might result in offspring that could survive in the altered environment.

Cadmium produces a dose-dependent decrease in 5'-nucleotidase, which leads to an increase in cytidine triphosphate [4]. Cytidine triphosphate is the substrate for a cyclase enzyme that can produce cyclic cytidine monophosphate (CMP) in certain primordial cells. If one exposed primordial plant cells to low-dose cadmium, one could determine if there is an increase in cyclic CMP and the development of reproductive structures. A search for better evidence linking cadmium exposure to plant stress in the environment is needed.

Detecting Cadmium Toxicity

Currently, ambient cadmium air pollution is not being linked to human stress or chronic disease or environmental stress. The effects of experimental low-dose exposure to cadmium in animals are variable and can be caused by a variety of agents that induce stress responses. They are not associated with elevations of blood, hair, or urine cadmium levels [42].

The inability to predict cadmium behavior in complex systems was recognized by Aylet in his review of the chemistry of cadmium [4]. Using traditional approaches, it is only possible to scientifically prove low-dose cadmium effects that have been produced experimentally. Because it is particularly important to identify environmental effects that could be due to cadmium, it is necessary to consider alternative ways to implicate its actions.

Lead is currently linked to human disease [43–46]. Since cadmium induces zinc deficiency, which increases lead absorption [46], and increases bone resorption at very low concentrations of 10 nM, releasing lead from bone stores [47], the elevation of blood lead levels, which is assumed to come from increased lead exposure, can clearly come from a cadmium toxic effect in an environment previously contaminated with lead. Cytidine triphosphate is a substance that increases in the erythrocytes of children with neurotoxicity showing elevated blood lead levels [48], but it can be elevated by cadmium as well.

Since low doses of lead or cadmium can increase cadmium uptake into the brain by disrupting the blood brain barrier [49,50], where cadmium is highly toxic, one could consider that the two are synergistically toxic. In Lockett and Leary's study [49], however, the animals exposed to lead and cadmium were sedated and showed no sign of stress. Moreover, Cook found that lead, prednisone, and endotoxin-treated rats lived while cadmium, prednisone, and endotoxin-treated animals died [51]. So lead, which does not induce stress responses, could block low-dose cadmium effects associated with stress. Lead has such a protective effect against cadmium toxicity in hepatocyte culture as well [28].

Lead is not considered a significant toxic substance in plants. One way to implicate low-level ambient cadmium toxicity in an ecosystem is by finding a toxic effect attributed to lead in humans, learning and behavioral problems in children, with a stress effect in plants linked to cadmium. These effects should also be associated with an emission source of cadmium and conditions that increase its bioavailability by lowering pH and increasing chlorides, which is found with waste incineration.

In order to study effects of cadmium pollution from waste incineration, it is necessary to consider other sources of cadmium exposure and possible confounding factors. A major confounding factor could come from the dramatic drop in lead air pollution.

In this decade air lead has been reduced from 1.5 to 0.04 µg/m³ by elimination of lead from gasoline (Ohio EPA). Showman [52] found a significant but unexplained drop in zinc levels in lichens remote from sources of known air pollution from 50 µg/g in 1973 to 15 µg/g in 1987. Both lead and zinc block cadmium toxicity in cell culture [28] and block cadmium uptake into plants [53]. The change in air pollution may be increasing the toxic effect of cadmium [54].

A recent U.S. document considers that cadmium air pollution from traffic has decreased from the elimination of cadmium-contaminated lead additives [55]. Cadmium, however, is present in electroplated auto parts and vulcanized tires. Cadmium fumes are emitted easily from metal surfaces and tire wear [56]. The failure to look for gaseous forms of cadmium has resulted in decreased awareness of cadmium pollution in air.

Consideration needs to be given to other factors as well. Effects from a stress such as drought could be influenced by chromium. Chromium is a trace metal that increases cell wall elasticity and increases water stress tolerance [57]. In a region affected by drought, the exposure to cadmium air pollution, coupled with a decrease in soil chromium, would enhance the cadmium toxic effect.

Observations

Environmental Impact of a Waste Incineration Plant

In December 1983, a large municipal waste incineration plant began operating in Columbus, Ohio, a chromium-deficient area. This plant uses electrostatic precipitators that capture 70% of the cadmium emissions but none of the HCl, which is 3% of the steam [2]. Using the estimate that the cadmium concentration in garbage averages 12 μg/g, this incinerator emits 2000 kg of cadmium per year.

I found increased concentration of hair cadmium and lead but low levels of hair phosphorus and zinc in children with learning and behavior problems in a large area surrounding the plant [58]. The lead level was only 4 ppm and the cadmium level 0.4 ppm in children with neurotoxic symptoms, but this was twice the level in the unaffected children.

In the vicinity of the homes of these children, parents had noted significant changes in trees. In 1986, families noted a marked increase in the fruiting of trees. By 1990, I noticed an increase in deaths and disease in a variety of trees: apple, catalpa, crab apple, birch, beech, cedar, chinese elm, cherry, maple, mountain ash, pin oak, and white oak. Plant pathologists attributed the deaths to multiple causes: nutritional deficiencies, general pollution sensitivity, disease, or drought but never to cadmium exposure. Other trees, particularly the Bartlett pear, were thriving and growing unusually rapidly.

I have lived in this community for 30 years. I noted a marked degree of seeding in various trees. Although there was a general drought in 1987, 1988, and 1989, I noticed a marked variability in drought sensitivity. Some trees showed chlorosis. Red coloration of the veins in leaves was prominent in certain trees. The leaves showed injuries with red spots. They died prematurely. In the fall, there was an unusual degree of red coloration in the trees.

In spite of the pollution, the air filter measurements of cadmium fell from 0.014 μg/m³ in 1981 to 0.0006 μg/m³ in 1986 (personal communication from Ohio EPA). To get a better idea about the level of outside cadmium pollution, I was able to obtain analyses of the concentrations of cadmium in tree bark, lichens, and pine needles from a research laboratory using induction coupled atomic absorption spectrometry. The detection limit was 0.01 ppm. The highest level, 8-ppm cadmium, was in a lichen-laden piece of bark from a tree close to a busy street at a distance of 10 km from the waste incineration plant.

To determine levels of human exposure, cadmium levels in house dust from five houses within 100 km of the plant were analyzed. They also contained 8-ppm cadmium. This level did not vary with distance from the incinerator.

When leaves of five trees in an area 10 km from the incineration plant were measured for cadmium levels on 15 Oct. 1987, they varied from none detectable in a white oak to 0.58 ppm in a nearby red oak. With a threshold of 0.1 ppb, cadmium could be causing the toxic effect even though no measurable cadmium was detected in the leaves using atomic absorption.

The toxic effect in the white oak was manifested in the death of branches resembling multiple vascular occlusions. Barcelo found that plants treated with cadmium develop abnormal vascular bundles which become occluded with calcium oxalate crystals [34]. The tree leaves with the most cadmium were from the tree with the least evidence of toxicity, a red oak. Previous studies have shown that plants that are able to produce metal-binding peptides accumulate cadmium but do not show toxic effects [59]. In 1990, the white oak, which had only four vascular occlusions in 1987, now has three quarters of its branches dead.

Since cadmium does not settle around an emission source, one cannot look for evidence of increased effects in direct proximity to the point source, but a regional diffuse effect could be suspected. Indirect evidence for such an effect was found using U.S. Department of Justice records of the distribution of methylphenidate, a medication used for children with attention deficit disorder [*60*]. The distribution was highest in Ohio in the area within 150 km of the incineration plant. In 1990, the state of Ohio is conducting a survey of tree conditions in the state. It is possible that differential regional effects on trees will be discerned.

However, the failure to look for cadmium in a gaseous phase coupled with the failure to do biological monitoring could have led to a major underestimation of cadmium emissions from traffic. In a trip across the United States, I observed overseeding of trees across the country, frequently accentuated by busy streets. In Europe, I observed this effect in countries using unleaded gasoline and less noticeably in countries using leaded gasoline. Destruction of the chloroplast in evergreens adjacent to highways is commonly observed.

Discussion

In a study of the effects of cadmium air pollution, it would be desirable to find a control area. From reviewing the literature, however, it is clear that each area has its own uniqueness due to the variability of interacting variables. With such a diffuse pollutant there is no area that is unaffected. In a study done in Shipham, England, a village with significant cadmium ground contamination, but also large amounts of lead and zinc, virtually no health effects were found [*53*]. Their "control" village had 8 ppm cadmium in house dust, which is certainly elevated and the same as that found in this study. A careful examination of a particular area was the strategy used in this study.

In an ecosystem exposed to waste incineration, cadmium levels in lichens and house dust were elevated, 8 ppm, compared to a background crustal level of 0.2 ppm. Mean hair cadmium in neurotoxic children was also elevated 0.4 ppm, compared to the unaffected children, who had a mean hair cadmium of 0.2 ppm, but this difference did not achieve significance. A red oak had an elevated level of cadmium of 0.58 ppm in fall leaves. The available evidence suggests that the area is exposed to increased cadmium air pollution and that the air filter measurements which are lower than before the operation of the incinerator are misleading.

The effect of this exposure in children can be inferred by finding an increased utilization of a stimulant medication for children with attention deficit disorder in a diffuse area surrounding the incinerator. In another study, I found that children who respond to this stimulant, methylphenidate, have increased hair cadmium and hair lead and low hair zinc and hair phosphorus [*60*].

In order to distinguish an effect on children caused by cadmium rather than lead, trees were examined. In the same time period, trees in the area were showing a variety of stress responses, particularly excess seeding. A biochemical pathway, involving a low-dose effect of cadmium that would lead to increased seeding in trees, is linked to a finding in children with learning and behavioral problems.

The children were exposed to psychosocial stresses arising from their learning and behavioral problems. They were usually drinking chlorinated drinking water which contains chloroform, and they were exposed to pesticides and fertilizers used on grass. They were exposed to a variety of chemicals in their food and drink. They were exposed to traffic fumes. All these exposures would be expected to have a synergistic effect with cadmium. They had a decreased exposure to lead air pollution. The elimination of lead could have had an effect of increasing the effect from cadmium.

Trees were exposed to drought conditions from 1986 to 1989 and then excess rain in 1990. They were exposed to acid rain. Some were exposed to chlorinated water. Some were exposed to pesticides and fertilizers. Local and distant air pollutants of all kinds were affecting them. There was an increase in solar flares during the time of the drought. Local geological conditions contribute to chromium, selenium, cobalt, and iodine deficiency. All these exposures would be expected to have a synergistic effect with cadmium. They also had a decreased exposure to lead air pollution, which could have had an effect of increasing the effect from cadmium air pollution.

In isolation, the findings in the trees and in the children do not allow one to confidently state that observed effects can be attributed directly to the exposure to cadmium fumes from waste incineration alone or in combination with the fall in lead air pollution. Taken together, however, they are suggestive that cadmium, since it is capable of causing the observed effects at very low doses of exposure, needs to be considered a major factor in the observed effects.

The effects of cadmium on an ecosystem are highly complex and enmeshed. Since it is necessary to consider effects rather than levels, it is particularly important that accurate assessment is made of emission sources.

Hair analysis is a reasonable way to detect cadmium exposures in humans, but hair cadmium was only elevated in children with learning and behavioral problems. Stress, genetic susceptibility, and nutritional states interact with cadmium exposure in affecting hair cadmium levels and other effects. Deviation of nutrient hair mineral content away from the mean is a more reliable indicator of a low-dose cadmium effect, but it occurs in response to many stressors. When one can realiably determine the cadmium exposure, then other stressors can be considered interactive rather than confounding.

Multiple metal analyses of leaves in the spring, summer, and fall from a variety of trees, those showing stress and those that are thriving, are not available. They would be very helpful in assessing effects of cadmium in combination with other stressors.

By studying how levels of metals in humans and trees change in response to changes in air pollution exposures, it may be possible to learn how cadmium interacts with other stressors, such as endotoxins. For instance, the drop in lead air pollution appears to have fostered fungal growth in both trees and humans. Fungal endotoxin may react synergistically with cadmium in causing injury. Careful interdisciplinary studies of ecosystems in which physical, chemical, biological, and psychosocial stresss are studied simultaneously are clearly needed.

Conclusions and Need for Further Studies

To detect a subtle stress agent, it is essential to have a hypothesis to direct research. Trees showing damage in the environment should be carefully studied using the hypothesis that they are suffering from cadmium toxic effects. Ultrastructural studies may be able to show that the lesions are identical to those induced by cadmium in experimental studies.

There is a need for a worldwide biological monitoring system for cadmium using lichens for outdoor air and house dust for indoor air. Only in this way can the effects of local conditions on cadmium bioavailability be assessed. To address the actual problem of cadmium from traffic fumes, cadmium levels in lichens growing by busy streets could be monitored or special traps could be devised to detect gaseous phases that air filters miss.

Pollution comes from a combination of sources, so the role of cadmium in highly complex emissions like cigarette smoke, incineration fumes, sludges, and industrial effluent must be studied. By determining how much lime, selenium, or other materials need to be added to block its uptake, one could substantially reduce the toxic effect coming from the synergism.

In cell culture, the effects of cadmium on the uptake of other metals like lead and aluminum should be specifically studied. Using plant cell cultures or respiratory epithelium, it should be possible to see if lead blocks cadmium uptake and toxicity in conditions of stress. Using Verbost's finding that cadmium increases intracellular calcium, that effect can be looked for using flow cytometry techniques.

Experimental studies can be designed to test the hypothesis that cadmium is released in a free state during many kinds of stresses. Perhaps one can block harmful stress reactions with nutrients that block cadmium.

Acknowledgments

The author gratefully acknowledges the suggestions offered by Mohan K. Wali and the reviewers.

References

[1] Lithner, G., "Some Fundamental Relationships Between Metal Toxicity in Fresh-water, Physico-chemical Properties and Background Levels," *Science of the Total Environment*, Vol. 87/88, 1989, pp. 365–380.

[2] Lisk, D. J., "Environmental Implications of Municipal Solid Waste and Ash Disposal," *Science of the Total Environment*, Vol. 74, 1 Aug. 1988, pp. 39–66.

[3] Mislin, H. and Ravera, O., Eds., *Cadmium in the Environment. Experientia* Supplement, Vol. 50, Birhauser Verlag, Boston/Basel, 1986.

[4] Webb, M., Ed., *Chemistry, Biochemistry, and Biology of Cadmium,* Elsevier/North Holland Biomedical Press, Amsterdam, 1979.

[5] Beauford, W., Barber, J., and Barringer, A. R., "Release of Particles Containing Metals from Vegetation into the Atmosphere," *Science*, Vol. 195, No. 4278, 11 Feb. 1977, pp. 571–573.

[6] Motohashi, K. and Tsuchida, T., "Uptake of Cadmium by Pure Cultured Diatom, *Skeletonema costatum,*" *Bulletin of the Planton Society of Japan*, Vol. 21, 1974, pp. 111–117.

[7] Barcelo, J., Poschenrieder, C., Andrue, I., and Gunse, B., "Cadmium-Induced Decrease of Water Stress Resistance in Bush Bean Plants (*Phaseolus vulgaris* L. cv. Contender) I. Effects of Cd on Water Potential, Relative Water Content, and Cell Wall Elasticity," *Journal of Plant Physiology,* Vol. 125, 1986, pp. 17–25.

[8] Henderson, L. M., Chappell, J. B., and Jones, O. T., "Internal pH Changes Associated with the Activity of the NADPH Oxidase of Human Neutrophils. Further Evidence for the Presence of an H^+ Conducting Channel," *Biochemical Journal*, Vol. 251, No. 2, 15 April 1988, pp. 563–567.

[9] Morselt, A. F., Finelli, V. N., Copius-Peereboom-Stegeman, J. H., and van Veen, H. A., "Mechanisms of Damage to Liver Cells after Chronic Exposure to Low Doses of Cadmium Chloride," *Archives of Toxicology* Suppl., Vol. 11, 1987, pp. 213–215.

[10] Grady, D. L., Moyzis, R. K., and Hildebrand, C. E., "Molecular and Cellular Mechanisms of Cadmium Resistance in Cultured Cells," *Experientia* Suppl., Vol. 52, 1987, pp. 447–456.

[11] Hagen, G., Uhrhammer, N., and Guilfoyle, T. J., "Regulation of an Auxin-Induced Soybean Sequence by Cadmium," *Journal of Biological Chemistry*, Vol. 263, No. 13, 5 May 1988, pp. 6442–6446.

[12] Lag, M. and Hegeland, K., "Effects of Cadmium and Other Metals on Ciliary Activity of Mouse Trachea Organ Culture," *Pharmacology and Toxicology*, Vol. 60, 1987, pp. 318–320.

[13] Verbost, P. M., Flik, G., Pang, P. K., Lock, R. A., and Wendelaar-Bonga, S. E., "Cadmium Inhibition of the Ca2+ Pump. A molecular interpretation," *Journal of Biological Chemistry*, Vol. 264, No. 10, April 1989, pp. 5613–5615.

[14] Gleick, J., *Chaos*, Viking, New York, 1987.

[15] Nocentini, S., "Inhibition of DNA Replication and Repair in Mammalian Cells. Protective Interaction of Zinc," *Nucleic Acids Research*, Vol. 15, No. 10, 26 May 1987, pp. 4211–4225.

[16] Thimann, K. V., "The Senescence of Leaves," in *Senescence in Plants*, CRC Press Inc., Boca Raton, FL, 1980, pp. 85–115.

[17] Ochi, T., Takashi, K., and Ohsawa, M., "Indirect Evidence for the Induction of a Prooxidant State by Cadmium Chloride in Cultured Mammalian Cells and a Possible Mechanism for the Induction," *Mutation Research*, Vol. 180, No. 2, October 1987, pp. 257–266.

[18] Harrison, R. M. and Chirgawi, M. B., "The Assessment of Air and Soil as Contributors of Some Trace Metals to Vegetable Plants," Parts I, II, III, *Science of the Total Environment*, Vol. 83, 1989, pp. 13–62.

[19] Froslie, A., Norheim, G., Rambaek, J. P., and Steinnes, E., "Heavy Metals in Lamb Liver: Contribution from Atmospheric Fallout," *Bulletin of Environmental Contamination and Toxicology*, Vol. 34, 1985, pp. 175–182.

[20] Kessel, B. G., Theuvenet, A. P., Peters, P. H., Dobbelmann, J., and Borst-Pauwels, G. W., "Changes in 45Ca and 109Cd Uptake, Membrane Potential and Cell pH in Saccharomyces Cerevisiae Provoked by Cd2+," *Journal of General Microbiology*, Vol. 133, Part 4, April 1987, pp. 843–848.

[21] Fujinaki, H., "Comparison of the Effect of Cadmium on Lymphocytes of Young and Adult Mice," *Journal of Environmental Pathology, Toxicology & Oncology*, Vol. 7, 1987, pp. 39–45.

[22] Blakely, B. R., "The Effect of Cadmium on Chemical and Viral Induced Tumor Production in Mice," *Journal of Applied Toxicology*, Vol. 6, 1986, pp. 425–429.

[23] Lawrence, D., "Immunotoxicity of Heavy Metals," in *Immunotoxicology and Immunopharmacology*, J. Deans, M. I. Luster, A. E. Munson, and H. Amos, Eds, Raven Press, New York, 1985, pp. 341–349.

[24] Stacey, N. H., "Assessment of the Toxicity of Chemical Mixtures with Isolated Rat Hepatocytes: Cadmium and Chloroform," *Fundamental & Applied Toxicology*, Vol. 9, No. 4, November 1987, pp. 616–622.

[25] Khanna, R. N., Anand, M., and Gopal, K. et al., "Effect of Repeated Exposure to Lindane and Cadmium on Lindane Metabolism in Rats," *Toxicology Letters*, Vol. 42, No. 2, August 1988, pp. 177–183.

[26] Alary, J., Carrera, G., Lamboef, Y., and Escrieut, C., "Cadmium-Induced Alterations of Chlorpropham Metabolism in Isolated Rat Hepatocytes," *Toxicology*, Vol. 59, No. 2, pp. 211–223.

[27] Dive, D., Gabriel, L., Hanssens, O., and Benger-Bengome, A., "Studies of Interactions Between Components of Electroplating Industry Wastes: Influence of Nickel and Calcium on Interactions Between Cd, Cu, Cr, and Zn," *Science of the Total Environment*, Vol. 87/88, 1989, pp. 355–369.

[28] Stacey, N. H. and Klaassen, C. D., "Interaction of Metal Ions with Cadmium-Induced Cellular Toxicity," *Journal of Toxicology and Environmental Health*, Vol. 7, 1981, pp. 149–158.

[29] Robinson, N. J., Barton, K., Naranjo, C. M. et al., "Characterization of Metal Binding Peptides from Cadmium Resistant Plant Cells," in *Metallothionein II*, J. H. R. Kagi and Y. Kojima, Eds., Birkhauser Verlag, Basel, *Experientia* Suppl., Vol. 52, 1987, pp. 323–327.

[30] Barcelo, J., Vazquez, M. D., and Poschenrieder, C., "Structural and Ultrastructural Disorders in Cadmium-Treated Bush Bean Plants," *New Phytology*, Vol. 108, 1988, pp. 37–49.

[31] Lamoreaux, J. R. and Chaney, W. R., "The Effects of Cadmium on Net Photosynthesis, Transpiration, and Dark Respiration of Excised Silver Maple Leaves," *Physiologia Plantarum*, Vol. 43, 1978, pp. 231–236.

[32] Samarakoon, A. B. and Rauser, W. E., "Carbohydrate Levels and Photoassimilate Export from Leaves of *Phaseolus vulgaris* Exposed to Excess Cobalt, Nickel, and Zinc," *Plant Physiology*, Lancaster, Vol. 63, 1979, pp. 1165–1169.

[33] Lamoreaux, R. J. and Chaney, W. R., "Growth and Water Movement in Silver Maple Seedlings Affected by Cadmium," *Journal of Environmental Quality*, Vol. 6, 1977, pp. 201–205.

[34] Barcelo, J., Vazquez, M. D., and Poschenrieder, C., "Cadmium-Induced Structural and Ultrastructural Changes in the Vascular System of Bush Bean Stems," *Botanica Acta*, Vol. 101, 1988, pp. 254–261.

[35] Cataldo, D. A., Garlan, T. R., and Wildung, R. E., "Cadmium Distribution and Fate in Soybean Plants," *Plant Physiology*, Vol. 68, 1981, pp. 835–839.

[36] Vallee, B. L. and Ulmer, D. I., "Biochemical Effects of Mercury, Cadmium and Lead," *Annual Review of Biochemistry*, Vol. 41, 1972, pp. 91–123.

[37] Taketanu, S., Kohno, H., Voshinaga, T., and Tokynaga, R., "Induction of Heme Oxygenase in Rat Hepatoma Cells by Exposure to Heavy Metals and Hyperthermia," *Biochemistry International*, Vol. 17, No. 4, October 1988, pp. 665–672.

[38] Hidalgo, J., Armario, A., Flos, R., and Garvey, T. S., "Restraint Stress Induced Changes in Rat Liver and Serum MT and in Zn Metabolism," *Experientia*, Vol. 42, No. 9, September 1986, pp. 1006–1010.

[39] Dickerson, I. M., Dixon, J. E., and Mains, R. E., "Transfected Human Neuropeptide Y cDNA Expression in Mouse Pituitary Cells. Inducible High Expression, Peptide Characterization, and Secretion," *Journal of Biological Chemistry*, Vol. 262, No. 28, 5 Oct., 1987, pp. 13646–13653.

[40] Freedman, L. P., Luisi, B. F., Korszun, Z. R. et al., "Function and Structure of the Metal Coordination Sites with the Glucocorticoid DNA Binding Domain," *Nature*, Vol. 334, No. 6182, August 1988, pp. 543–546.

[41] Chan, P. T., "The Effect of Cyclic Cytidine 3',5' Monophosphate on the In Vitro Development, Hatching, and Attachment of the Mouse Blastocytes," *Experientia*, Vol. 43, No. 8, August 1987, pp. 929–930.

[42] Smith, M. J., Pihl, R. O., and Garber, B., "Postnatal Cadmium Exposure and Long Term Behavioral Changes in the Rat," *Neurobehavioral Toxicology and Teratology,* Vol. 4, 1982, pp. 283–287.

[43] Needleman, H., ed., *Low Level Lead Exposure: the Clinical Implications of Current Research,* Raven Press, New York, 1980.

[44] Harlan, W. R., Landis, J. R., and Schmouder, R. L. et al., "Relationship of Blood Lead and Blood Pressure in the Adolescent and Adult U.S. Population," *Journal of the American Medical Association JAMA,* Vol. 253, No. 4, January 1985, pp. 530–534.

[45] Silbergeld, E. K., Schwartz, J., and Mahaffey, K., "Lead and Osteoporosis. Mobility of Lead from Bone in Post-Menopausal Women," *Environmental Research,* Vol. 47, No. 1, October 1988, pp. 79–94.

[46] U.S. Environmental Protection Agency, *Air Quality Criteria for Lead,* Vol. I–IV, 1986.

[47] Bhattacharyya, M. H., Whelton, B. D., and Stein, P. H. et al., "Cadmium Accelerates Bone Loss in Ovariectomized Mice and Fetal Rat Limb Bones in Culture," *Proceedings of the National Academy of Sciences of the USA,* Vol. 85, No. 22, November 1988, pp. 8761–8765.

[48] Angle, C. R. and McIntire, M. S., "Erythrocyte Nucleotides in Children-Increased Blood Lead and Cytidine Triphosphate," *Pediatric Research,* Vol. 16, No. 4, April 1982, pp. 331–334.

[49] Lockett, C. J. and Leary, W. P., "Neurobehavioral Effects in Rats Fed Low Doses of Cadmium and Lead to Induce Hypertension," *South African Medical Journal,* Vol. 69, No. 3, 1 Feb. 1986, pp. 190–192.

[50] Burger, J. and Gochfeld, M., "Tissue Levels of Lead in Experimentally Exposed Herring Gull (*Laus argentalus*) Chicks," *Journal of Toxicology and Environmental Health,* Vol. 29, No. 2, 1990, pp. 213–233.

[51] Cook, J. A., Hoffman, E. O., and DiLuzio, N. R., "Influence of Lead and Cadmium on Susceptibility of Rats to Bacterial Challenge," *Proceedings of the Society for Experimental and Biological Medicine,* Vol. 150, 1975, pp. 741–748.

[52] Showman, R. E. and Hendricks, J. C., "Trace Element Content of *Flavoparmelia caperata* (L.) Thalli Due to Industrial Emissions," *Journal of the Air Pollution Control Association,* Vol. 39, 1989, pp. 317–320.

[53] Morgan, H., Ed., "The Shipham Report, An Investigation into Cadmium Contamination and Its Implications for Human Health," *Science of the Total Environment,* Vol. 75, No. 1, 15 Aug. 1988.

[54] Stewart-Pinkham, S. M., "The Toxicity of Cadmium Air Pollution: a Reappraisal," *Journal of Environmental Geochemistry and Health,* Suppl. Vol. 12, 1990, pp. 345–369.

[55] Carey, P. M., "Air Toxic Emissions from Motor Vehicles," technical report, U.S. Environmental Protection Agency, Washington, DC, 1987.

[56] Venugopal, B. and Luckey, T. D. *Metal Toxicity in Mammals,* Vol. 2, Plenum Press, New York, 1978, pp. 76–86.

[57] Barcelo, J., Poschenrieder, C., and Gunse, B., "Water Relations in Chromium-Treated Bush Beans (*Phaseolus vulgaris* L.) under Normal Water Supply and Water Stress Conditions," *Journal of Experimental Botany,* Vol. 37, 1986, pp. 178–187.

[58] Stewart-Pinkham, S. M., The Effect of Ambient Air Cadmium Pollution on the Hair Mineral Content of Children," *Science of the Total Environment,* Vol. 78, No. 1, January 1989, pp. 289–296.

[59] Jackson, P. J., Unkefer, C. J., Doolen, J. A., Watt, K., and Robinson, N. J., "Poly(gamma-glutamylcysteinyl)glycine: Its Role in Cadmium Resistance in Plant Cells," *Proceedings, National Academy of Sciences USA,* Vol. 84, No. 19, October 1987, pp. 6619–6623.

[60] Stewart-Pinkham, S. M., "Attention Deficit Disorder: A Toxic Effect of Cadmium," *International Journal of Biosocial and Medical Research,* Vol. 11, No. 2, 1989, pp. 134–143.

Bobby L. Folsom, Jr.[1] and Richard A. Price[1]

A Plant Bioassay for Assessing Plant Uptake of Contaminants from Freshwater Soils or Dredged Material

REFERENCE: Folsom, B. L., Jr. and Price, R. A., "**A Plant Bioassay for Assessing Plant Uptake of Contaminants from Freshwater Soils or Dredged Material,**" *Plants for Toxicity Assessment: Second Volume, ASTM STP 1115,* J. W. Gorsuch, W. R. Lower, W. Wang, and M. A. Lewis, Eds., American Society for Testing and Materials, Philadelphia, 1991, pp. 172–177.

ABSTRACT: The plant bioassay described herein is one of the modules of the decision-making framework developed by the U.S. Army Engineer Waterways Experiment Station for evaluating sediments before dredging. Collection and preparation of sediments are described as well as conduct of the plant bioassay. The bioassay is appropriate for estimating mobility of contaminants into the environment through plant uptake in wetland, marsh environments as well as drier agricultural-type upland conditions.

KEY WORDS: plant bioassay, heavy metal, contaminant, sediment, wetland, upland, freshwater, marsh

The decision-making framework (DMF) developed by Lee et al. (1990) [1] provides a method for evaluating sediments before dredging. This framework is comprised of several modules, one of which is the plant bioassay for materials proposed for upland or wetland placement. The purpose of this paper is to describe the methods and materials necessary to conduct such a plant bioassay.

Sediment Collection and Preparation

Sediments to be tested are collected from a waterway using an appropriate *sampler* that can remove the entire vertical profile of the material to be dredged. The plant bioassay requires 60.6 L (16 gal) of each sediment to be tested. Therefore, a 75.7- or 113.6-L (20- or 30-gal) drum would provide a sufficient quantity of material to conduct the required testing; however, 208-L (55-gal) steel drums are generally easier to obtain. The filled drums are sealed with airtight lids and transported to the laboratory. Temperature during shipping should be maintained at 4° ± 2°C. The drums should be steam-cleaned prior to use.

Before testing, the original, flooded sediments should be mixed thoroughly while in their respective drums. All debris, such as cans, bottles, leaves, or twigs, is removed. Generally, 1 h of mixing is required for adequate sediment homogenization. A minimum of four replicates is required to keep sample variability (coefficient of variation) low (Table 1). Subsamples from the drums represent original, flooded sediment and are used in the procedure described below.

---

[1] Soil scientist and agronomist, respectively, U.S. Army Engineer Waterways Experiment Station, Vicksburg, MI 39180-6199.

TABLE 1—*Mean heavy metal concentration, standard error, and coefficient of variation in typical freshwater sediments after the recommended 1 h of mixing.*

Heavy Metal	Number of Replicates	Mean Concentration, µg/g	Standard Error	Coefficient of Variation
Arsenic	4	22.9	0.819	7.16
Cadmium	4	28.1	1.48	10.6
Chromium	4	650	20.1	6.19
Copper	4	322	8.65	5.37
Nickel	4	137	4.39	6.42
Zinc	4	4550	52.1	2.29

Flooded Condition

A schematic diagram of the standard U.S. Army Engineer Waterways Experimental Station (WES) plant bioassay apparatus is shown in Fig. 1. The mixed original, flooded sediment is placed one 500-mL scoop at a time—to further minimize mixing variability—into each of four 7.6-L (2-gal) Bain-Marie containers (four flooded replicates). When the containers are filled with sediment, sediment diameter and sediment depth should be in a ratio of 1.5:1. Each 7.6-L Bain-Marie container is placed inside a 22.7 L Bain-Marie con-

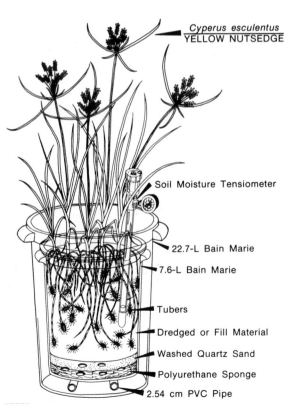

FIG. 1—*Schematic diagram of the plant bioassay apparatus. The sand layer and sponge are each 2.54 cm thick.*

tainer. Flooded conditions are maintained by keeping a 5-cm depth of deionized water or distilled water above the sediment surface of the inner container. Percent moisture on an oven-dry weight basis (ODW) (oven temperature is 105° ± 2°C for 17 h) is determined on 5- to 10-g subsamples of the flooded material in each replicate. A test sediment weight of 4500 g (ODW) per replicate generally provides sufficient plant material for maximum plant growth. Since flooded and upland biomass production (yield) is one of the DMF comparisons, each replicate must contain the same quantity of sediment (ODW). The four flooded replicates are sealed with their lids and stored at 4° ± 2°C until the upland replicates (described below) are prepared for planting.

Upland Condition

Four additional sediment replicates are prepared as described above for the flooded condition except that the sediment from each container is placed into an aluminum drying pan and allowed to air dry. Drying and preparation of the upland replicates should be completed within three weeks of collection. The sediment must be turned and mixed daily with a large plastic spatula or Teflon-coated shovel to facilitate drying. All debris, such as cans, bottles, leaves, or twigs, is removed when observed as the sediment dries. If large quantities of large rocks, gravel, and other materials are present, then a separation analysis should be conducted [2]. After air drying, most sediments form large brick-like clods that are extremely difficult to crush. Crushing and grinding of these clods is best performed using a hammermill. Personnel operating the hammermill should wear appropriate respirators and protective clothing. One pass through the hammermill is sufficient for the material to pass a 2-mm screen (U.S. Standard Sieve No. 10). Greenhouse pot experiments generally use material that has been ground to pass a 2-mm screen to approximate field macroporosity (pore space affects particle surface area, drainage, gas movement, and other soil physical properties) to estimate weathered sediment placed in an upland disposal site. The screened material is returned to a drying flat where it is remixed and subsampled for ODW analysis. Air-dried sediment (4500-g ODW) is placed (one 500-mL scoopful at a time) into each of four 7.6-L Bain-Marie containers prepared as before. Any remaining air-dried sediment in the drying flat can be placed into an appropriate container (7.6-L Bain-Marie bucket is a good choice) and stored until needed for subsequent chemical and/or physical analyses, if necessary. For air-dried replicates, soil moisture is maintained between 0.033 and 0.06 kPa (field capacity is 0.033 kPa) by checking soil moisture tensiometers in each container daily. Plants are watered when tensiometers read greater than 0.06 kPa (generally every other day). When watering is necessary, the outer container is filled up to the sediment level of the inner container with reverse osmosis (RO) or distilled water. When tensiometers read less than 0.04 MPa, the water is siphoned out of the outer container.

Material from the reference or proposed disposal site is prepared in exactly the same manner as that described above for each disposal condition (i.e., flooded and upland).

Greenhouse Operation

The replicates are randomly placed on tables in a greenhouse. Day length of 16 h is maintained by supplementing natural light with artificial light fixtures whose face is 130 cm from the top of the 22.7-L Bain-Marie container. The 130-cm height allows potential maximum plant growth to occur without contacting the light fixture or becoming so close to the light that plant tissue is damaged from excess heat. Lights should be arranged in a pattern of alternating high-pressure sodium lamp and a high-pressure multivapor metal halide lamp.

Alternating the lamps provides an even photosynthetic active radiation (PAR) distribution pattern. The PAR should be at least 1200 μEi/m^2. The temperature of the greenhouse is maintained at 32.2 ± 2°C maximum during the day and 21.1° ± 2°C minimum at night to simulate a summer environment. Relative humidity should be maintained as high as possible, but never less than 50%.

Planting and Growing Techniques

The plant used in the WES plant bioassay is *Cyperus esculentus*. Normally, *C. esculentus* (common name, yellow nutsedge) is considered a persistent major problem weed and causes yield reductions in many crops of the United States, Canada [3,4], and throughout the world [5]. It is also considered a pioneer species that readily invades disturbed areas such as dredged material disposal sites [3]. *Cyperus esculentus* reproduces primarily by tubers [6,7]. It also reproduces by bulbs and rhizomes and by seeds under certain conditions [3]. Therefore, *C. esculentus* was chosen as the plant bioassay index plant because of its natural tenacity and ability to survive in both flooded and upland conditions. In addition, *C. esculentus* also showed greatest potential for heavy metal uptake compared to other plant species [8]. *Cyperus esculentus* also has a fairly short (45-day) vegetative growth period under long days (16-h photoperiod) [9].

Each replicate of flooded and upland sediment is planted with four germinated tubers of *C. esculentus*. Suppliers of the tubers include Valley Seed Services (P.O. Box 8335, Fresno, CA 93791, phone: 209-435-2763) and Wildlife Nurseries (P.O. Box 2724, Oshkosh, WI 54903; phone 414-231-3780). Germination of *C. esculentus* is close to 50% [10,11], and twice as many tubers as needed for the experiment are set out for germination. Germination temperature is 23° ± 2°C in light (16-h length). The tubers are first rinsed in distilled water to remove substances that may inhibit sprouting of buds on the tubers [3]. The tubers are then placed between white paper towels and kept moist with distilled water until enough have sprouted to plant five tubers per container (usually seven to ten days). Sprouts should be approximately 3 cm long before planting. Plants are allowed to grow for 45 days from the time of planting.

Harvesting

Forty-five days after planting, the above-ground plant material from each replicate is cut 5 cm above the sediment surface with stainless steel scissors and placed into a labeled brown paper bag perforated by several holes to allow water vapor to escape during drying. Any flowers, stems, or seeds that may have developed are separated from the leaves, wrapped separately with white paper towels, and placed into the bag with the leaves. The bags containing the harvested plant material are dried to a constant weight in a forced-air drying oven at 70° ± 2°C (generally four or five days). All dried tissue is removed from the bags and weighed separately. Total plant yields are determined by weighing the oven-dried plant material to the nearest 0.1 g.

Digestion and Chemical Analysis of Plant Material

The dried leaves are ground in a small Wiley mill. Two grams (weighed to the nearest 0.0001 g) of the ground leaf tissue are digested using the tertiary acid digestion procedure [12] except that 2.0-g ODW tissue are used rather than 1.0 g. In some sediments, plant growth is not sufficient to provide 2.0 g of tissue. In these cases, whatever amount of tissue produced is digested. Chemical analysis of flowers, stems, and seeds is not conducted since their production is sporadic. The diluted digestates are analyzed for heavy metals by atomic absorption (AA) spectroscopy and/or heated graphite analysis (HGA) (or any method capable of providing comparable low-level analysis). Results of the digestion are calculated using the equation

tissue metal concentration ($\mu g/g$)

$$= \frac{\text{solution metal concentration } (\mu g/ML) \times \text{dilution volume (ML)}}{\text{ODW leaf tissue digested (g)}}$$

Plant uptake of organic compounds by *C. esculentus* has been limited to studies with 2,4,6-trinitrotoluene (TNT) [13–15] and polychlorinated biphenyl (PCB)[2]. In a recent literature review on potential contaminant migration pathways, Brannon et al. (1989) [16] suggested that plant uptake of organic compounds may be very important to cycling of organic compounds in confined dredged material disposal facilities and recommended further research be conducted.

Summary

Sediments are thoroughly mixed before testing. The plant bioassay procedure is generally conducted using four replicates of each disposal condition (i.e., flooded and upland) for each sediment or soil considered. Flooded plant bioassay replicates are prepared and stored until sediment has been air-dried for the upland plant bioassay replicates. Replicates are placed into a controlled greenhouse environment and allowed to grow for 45 days. Aboveground plant tissue is harvested, weighed, acid digested, and analyzed for heavy metals by atomic absorption spectroscopy. Generally, heated graphite analysis is required to obtain the heavy metal concentration data.

Acknowledgments

The test described and the resulting data presented herein, unless otherwise noted, were obtained from research conducted under the Long-Term Effects of Dredging Operations program of the United States Army Corps of Engineers by the United States Army Engineer Waterways Experiment Station. Permission was granted by the Chief of Engineers to publish this information.

The contents of this paper are not to be used for advertising, publication, or promotional purposes. Citation of trade names does not constitute an official endorsement or approval of the use of such commercial products.

References

[1] Lee, C. R., Tatem, H. E., Brandon, D. L., Kay, S. H., Peddicord, R. K., Palermo, M. R., and Francingues, N. R., Jr., "General Decisionmaking Framework for Management of Dredged Material," Miscellaneous Paper D-90, in press, U.S. Army Engineer Waterways Experiment Station, Vicksburg, MS, 1990.
[2] Headquarters, U.S. Army Corps of Engineers, "Engineering and Design: Laboratory Soils Testing. Appendix V. Grain-Size Analysis," Engineer Manual No. EM 1110-2-1906, Washington DC, 1970.

[2] Folsom, B. L., Jr. U.S. Army Engineer Waterways Experiment Station. Unpublished laboratory results.

[3] Mulligan, G. A. and Junkins, B. E., "The Biology of Canadian Weeds," *Canadian Journal of Plant Science*, No. 56, 1976, pp. 339–350.
[4] Willis, G. D., Hoagland, R. E., and Paul, R. N., "Anatomy of Yellow Nutsedge (*Cyperus esculentus*)," *Weed Science*, No. 28, 1980, pp. 432–437.
[5] Holm, L. G., Plucknett, D. L., Pancho, J. V., and Herberger, J. P., "The World's Worst Weeds, Distribution and Biology," University Press, Hawaii, 1977.
[6] Bell, R. S., Lachman, W. H., Rahn, E. M., and Sweet, R. D., "Life History Studies as Related to Weed Control in the Northeast. 1—Nutgrass," Northeast Regional Publication, Agricultural Experiment Station Bulletin 364, University of Rhode Island, Kingston, RI, 1962.
[7] Tumbleson, M. E. and Kommedahl, T., "Reproductive Potential of *Cyperus esculentus* by Tubers," *Weeds*, No. 9, 1961, pp. 646–653.
[8] Lee, C. R., Sturgis, T. C., and Landin, M. C., "A Hydroponic Study of Heavy Metal Uptake by Selected Marsh Plant Species," Technical Report D-76-5, U.S. Army Engineer Waterways Experiment Station, Vicksburg, MS, 1976.
[9] Doty, C. H. and Sweet, R. D., "Growing Nutsedge (*Cyperus esculentus*) in the Greenhouse for Research Purposes," *Proceedings of the Northeast Weed Science Society*, No. 24, 1970, pp. 302–306.
[10] Thomas, P. E. L., "Effects of Dessication and Temperature on Survival of *Cyperus esculentus* Tubers and *Cynodon Dactylon* Rhizomes," *Weed Research*, No. 9, 1969, pp. 1–8.
[11] Yip, C. P. and Sweet, R. D., "Biotypes of Yellow Nutsedge (*Cyperus esculentus* L.): I. Tuber Dormancy and Responses to Photoperiod and Herbicides," Weed Science Society of America, Champaigne, IL, Abstract No. 128, 1978, p. 59.
[12] Folsom, B. L., Jr. and Houck, M. H., "A Computerized Procedure for Predicting Plant Uptake of Heavy Metals from Contaminated Freshwater Dredged Material," Technical Note EEDP-04-12, U.S. Army Engineer Waterways Experiment Station, Vicksburg, MS, 1990.
[13] Folsom, B. L., Jr., Pennington, J. C., Teeter, C. L., Barton, M. R., and Bright, J. A., "Effects of Soil pH and Treatment Level on Persistence and Plant Uptake of 2,4,6-Trinitrotoluene," Technical Report EL-88-22, U.S. Army Engineer Waterways Experiment Station, Vicksburg, MS, 1988.
[14] Pennington, J. C., "Soil Sorption and Plant Uptake of 2,4,6-Trinitrotoluene," Technical Report EL-88-12, U.S. Army Engineer Waterways Experiment Station, Vicksburg, MS, 1988.
[15] Palazzo, A. J. and Leggett, D. C., "Effect and Disposition of TNT in a Terrestrial Plant," *Journal of Environmental Quality*, No. 15, 1986, pp. 49–52.
[16] Brannon, J. M., Pennington, J. C., Gunnison, D., and Myers, T. E., "Comprehensive Analysis of Migration Pathways (CAMP): Migration Pathways Controlling Internal Cycling and Net Mass Transport of Contaminants at Confined Dredged Material Disposal Facilities," Miscellaneous-Paper D-89, U.S. Army Engineer Waterways Experiment Station, Vicksburg, MS, 1989.

Frank A. P. C. Gobas,[1] *Lesley Lovett-Doust,*[2]
and G. Douglas Haffner[2]

A Comparative Study of the Bioconcentration and Toxicity of Chlorinated Hydrocarbons in Aquatic Macrophytes and Fish

REFERENCE: Gobas, F. A. P. C., Lovett-Doust, L., and Haffner, G. D., "**A Comparative Study of the Bioconcentration and Toxicity of Chlorinated Hydrocarbons in Aquatic Macrophytes and Fish,**" *Plants for Toxicity Assessment: Second Volume, ASTM STP 1115*, J. W. Gorsuch, W. R. Lower, W. Wang, and M. A. Lewis, Eds., American Society for Testing and Materials, Philadelphia, 1991, pp. 178–193.

ABSTRACT: This study reports the uptake and elimination kinetics, the bioconcentration, and the acute toxicity of a series of chlorinated benzenes and biphenyls in a submerged aquatic macrophyte (*Myriophyllum spicatum*) and in a fish (*Poecilia reticulata*) species. The objective of this study is to investigate the relationship between the acute lethality in fish and in aquatic plants. The study shows linear relationships between the plant-water and fish-water bioconcentration factors and the 1-octanol-water partition coefficient, indicating that plant-water and fish-water exchange are largely controlled by the chemical's tendency to partition between the lipids of the plants or fish and the water. The toxicokinetics in both the plants and the fish involve "passive" transport phenomena, which can be described by a lipid-water kinetic model. Toxicity data demonstrate that the acute lethality of chlorobenzene and chlorobiphenyl congeners in fish is associated with an internal concentration in the fish of approximately 6330 μmol/L. Based on the similarity of the lethal internal concentration among the chlorobenzene congeners and between various aquatic organisms, it is hypothesized that the acute lethal toxicity of chlorobenzenes in plants and fish are similar, which would provide a basis for the extrapolation of toxicity data between fish and aquatic plants.

KEY WORDS: aquatic macrophytes, bioconcentration, kinetics, toxicity, chlorobenzene, PCB, fish, hydrophobicity

Aquatic toxicologists are usually interested in the effects of waterborne substances. A typical aquatic toxicity test therefore involves the preparation of a series of solutions with different concentrations of the tested substance. Then, a number of individuals of a certain aquatic species are exposed to each of these solutions for a defined period of time, after which a certain toxic endpoint is determined. One of the most widely used tests is the acute lethality test, where the number of dead test organisms at the end of the test is the toxic endpoint. The concentration which causes lethality to half of the individuals in the test, i.e., the LC_{50}, expresses the "toxicity" of the tested substance. These tests exist for a variety of aquatic species such as *Daphnia magna*, brine shrimp (*Artemia*), fathead minnows (*Pime-*

---

[1] Professor, School of Natural Resources and Environmental Management, Faculty of Applied Sciences, Simon Fraser University, Burnaby, British Columbia, Canada V5A 1S6.
[2] Professor, The Great Lakes Institute, University of Windsor, Windsor, Ontario, Canada N9B 3P4.

phales promelas), and guppies (*Poecilia reticulata*). However, LC_{50} tests for aquatic plants are still under development and thus are relatively scarce at present.

In an effort to protect aquatic life, some of these acute lethality tests have now been incorporated in a legislative framework. For example, in the province of Ontario, an initiative has been launched which requires that municipal and industrial effluents undergo, on a regular basis, lethality tests with rainbow trout and a Daphnia species to ensure that the effluents are not toxic to life in the receiving water body. This approach is based on the premise that what is toxic to a rainbow trout is probably also toxic to other aquatic organisms, including macrophytes. In other words, it is assumed that the rainbow trout toxicity test is able to represent the toxic impacts to all organisms of the aquatic ecosystem. From a practical point of view, it may be necessary to adopt this approach since it is impossible to perform toxicity testing for every species exposed to the tested substance. However, it is conceivable that certain substances have little or no effect on rainbow trout but are toxic to other organisms, thus causing our safeguard for environmental protection to fail. It is likely that the chance of such an event increases when differences in the physiology and biochemistry between organisms are larger. Fundamental differences in physiology and biochemistry exist between plant and animal life. It is thus possible that the toxicity of a substance in a fish species is unrelated to that in plants.

To investigate the ability of lethality tests in fish to simulate the toxic impacts in aquatic plants, we will examine the mechanisms of uptake, elimination, and toxicity of a series of chlorobenzene (CB) and chlorobiphenyl (PCB) congeners in a plant and a fish species. The objective of this study is to determine if there is a similarity between the toxicokinetics and toxicity in aquatic plants and in fish. The scope of our study will be limited to a series of CB and PCB congeners. They represent a group of persistent industrial chemicals that are of environmental concern in many parts of the world. They were selected because they are nonreactive and considered to be very poorly metabolizable by many aquatic organisms including fish. By eliminating the potential of significant metabolic breakdown we attempt to facilitate the study of the toxicokinetics and toxicity mechanisms.

Experiments in Aquatic Macrophytes and in Fish

To investigate the mechanisms of chemical uptake, elimination, and bioconcentration in aquatic plants and in fish, we will briefly summarize the results of bioconcentration experiments in a submerged aquatic macrophyte species, *Myriophyllum spicatum,* and in the guppy *Poecilia reticulata,* which were performed in a similar fashion. A detailed description of the experiments and their results is presented elsewhere [1,2].

Bioconcentration in Myriophyllum spicatum

One hundred and twenty plants (*Myriophyllum spicatum*), with an average wet weight of 9 g and a lipid content of $0.2 \pm 0.02\%$, were exposed for 25 days in a 150-L glass tank to an aqueous solution of 1,3,5-tri-, 1,2,4,5-tetra-, penta- and hexachlorobenzene and 2,2',5,5'-tetra-, 2,2',4,4',6,6'-hexa-, 2,2',3,3',4,4',5,5'-octa- and deca-chlorobiphenyl, delivered by a continuous flow generator column. During the experiment, the plants were in a submerged state, but freely floating in the water. No soils or sediment were present. Water and plant samples were collected throughout the experiment and analyzed as described by Gobas et al. [1]. After the 25-day uptake period, the plants were transferred to a tank that contained clean water which was continuously being filtered through an activated carbon filter to remove test chemicals eliminated by the plants. Chemical elimination from the plants was followed for up to 133 days.

Typical results of the uptake experiment are illustrated in Fig. 1a for 2,2′,5,5′-tetra-chlorobiphenyl. Figure 1a shows that during the uptake period, the chemical concentration in the plants increased with time to approach a constant level toward the end of the uptake period. After the uptake period, when plants were transferred to clean, uncontaminated water, a drop of the chemical concentration in the plants was observed (Fig. 1a). During the first 37 days of the elimination period, the concentrations of all chemicals in the plants dropped exponentially with time, corresponding to a linear decrease of logarithm of the concentration in the plant with time. During the remainder of the elimination period, the drop of the chemical concentration in the plant was somewhat slower, causing a loss of the initial linear relationship between the logarithm of the concentration in the plant and time [1]. The largest drop of concentration in the plant with time was observed for 1,2,4,5-tetrachlorobenzene, the smallest for octachlorobiphenyl.

Bioconcentration in Poecilia reticulata

Following a procedure similar to that described for the submerged aquatic macrophytes, 95 to 120 guppies were exposed to aqueous solutions containing CB and PCB congeners (Table 1) for up to 20 days. During this period, water and fish samples were taken and analyzed as in Ref 2. The fish were then transferred to a depuration tank with clean water that was continuously being carbon filtered to follow the decrease of the concentration in the fish with time.

During the uptake period, the concentration of the test chemicals in the fish increased with time to approach a constant level. Figure 1b illustrates the increase of the concentration of 2,2′,5,5′-tetrachlorobiphenyl in the fish during the uptake period. For some of the PCB congeners, in particular those with log K_{ow} above 6.1, the duration of the uptake period was too short to reach a constant concentration in the fish. During the depuration experiment the chemical concentrations in the fish dropped exponentially, which is illustrated by the linear relationship between the logarithm of the concentration in the fish with time.

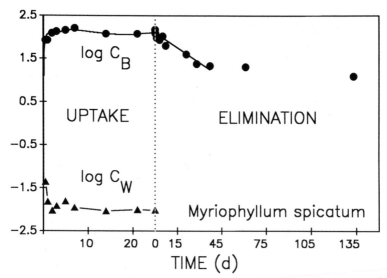

FIG. 1a—*Logarithms of the concentrations of 2,2′,5,5′-tetrachlorobiphenyl in the water* (▲), C_W (*μg/ L), and in* Myriophyllum spicatum (●), C_B (*μg/L), during the uptake and elimination experiment. The solid line illustrates the model fit.*

TABLE 1—*The logarithm of the 1-octanol-water partition coefficient log K_{OW}, the rate constant for uptake from the water k_1 (L · L^{-1} · d^{-1}), the rate constant of elimination k_2 (d^{-1}), the logarithm of the bioconcentration factor log BCF, the logarithm of the lipid-weight-based bioconcentration factor log K_L, the logarithm of the LC$_{50}$ (μmol/L), and the logarithm of the concentration at the site of action log C_T (μmol/L) of a series of chlorobenzenes and PCBs in the submerged aquatic macrophyte Myriophyllum spicatum and in the guppy Poecilia reticulata. NT means no acute lethality was observed.*

Compound	Log K_{OW}	k_1	k_2	Log BCF	Log K_L	Log LC$_{50}$	Log C_T
MYRIOPHYLLUM SPICATUM							
1,3,5-Trichlorobenzene [1]	4.02 [6]	20	0.6	1.52	4.22		
1,2,4,5-Tetrachlorobenzene [1]	4.51 [6]	93	0.54	2.24	4.94		
Pentachlorobenzene [1]	5.03 [6]	275	0.2	3.14	5.84		
Hexachlorobenzene [1]	5.47 [6]	150	0.14	3.03	5.73		
2,2',5,5'-Tetrachlorobiphenyl [1]	6.10 [7]	450	0.09	3.70	6.40		
2,2',4,4',6,6'-Hexachlorobiphenyl [1]	7.00 [7]	500	0.02	4.40	7.10		
2,2',3,3',4,4',5,5'-Octachlorobiphenyl [1]	7.80 [7]	496	0.000 8	5.79	8.49		
Decachlorobiphenyl [1]	8.26 [7]	162	0.000 3	5.73	8.43		
POECILIA RETICULATA							
Chlorobenzene	2.98 [6]					2.23 [10]	3.75
1,2-Dichlorobenzene	3.38 [6]					1.60 [10]	3.56
1,3-Dichlorobenzene	3.48 [6]					1.70 [10]	3.77
1,4-Dichlorobenzene [8]	3.38 [6]	98	1.0	1.99	3.18	1.43 [10]	3.39
1,2,3-Trichlorobenzene	4.04 [6]					1.11 [10]	3.86
1,2,4-Trichlorobenzene	3.98 [6]					1.12 [10]	3.79
1,3,5-Trichlorobenzene [8]	4.02 [6]	302	0.4	2.88	4.07	1.26 [10]	3.99
1,2,3,4-Tetrachlorobenzene	4.55 [6]					0.57 [10]	3.96
1,2,3,5-Tetrachlorobenzene [8]	4.65 [6]	1000	0.26	3.59	4.77	0.57 [10]	4.07
1,2,4,5-Tetrachlorobenzene	4.51 [6]					0.15 [10]	3.49
Pentachlorobenzene [8]	5.03 [6]	1738	0.11	4.20	5.39	-0.15 [10]	3.69
Hexachlorobenzene	5.47 [6]					NT [10]	
2,2',5,5'-Tetrachlorobiphenyl [2]	6.10 [7]	1122	0.016 2	4.84	6.03	NT [10]	
2,2',5,5'-Tetrachlorobiphenyl [9]	6.10 [7]	1202	0.015	4.90	6.09	NT [10]	
2,2',4,4',5,5'-Hexachlorobiphenyl [9]	6.90 [7]	794	0.003 98	5.30	6.49	NT [10]	
2,2',3,3',4,4',5,5'-Octachlorobiphenyl [9]	7.80 [7]	151	0.007 1	4.33	5.51	NT [10]	
Decachlorobiphenyl [2]	8.26 [7]	41.7	0.005	3.92	5.11	NT [10]	
Decachlorobiphenyl [9]	8.26 [7]	39.8	0.004	4.00	5.18		

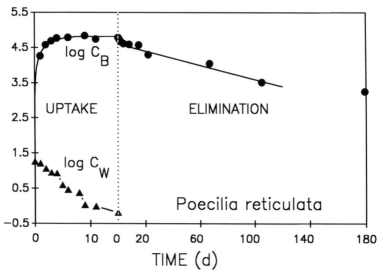

FIG. 1b—*Logarithms of the concentrations of 2,2',5,5'-tetrachlorobiphenyl in the water* (▲), C_W (*μg/ L), and in Poecilia reticulata* (●), C_B (*μg/L), during the uptake and elimination experiment. The solid line illustrates the model fit.*

Toxicokinetics in Aquatic Macrophytes and Fish

Figure 1 illustrates that the plant-water and fish-water transfer of CBs and PCBs is a reversible process. The toxicant concentration in plants or fish rises when the plants or fish are introduced to the aqueous solution and declines when the chemical is no longer present in the water. The simplest description of this process can be derived by treating the plant, the fish, and the water as single, homogeneous compartments, each containing a certain chemical concentration. If, in addition, no chemical transformation occurs and chemical transfer between the plant or fish and the water is adequately represented by first order rate constants, the following two-compartment model can be proposed to describe the chemical exchange between the water and the plants or fish

$$d(V_B \cdot C_B)/dt = k_1 \cdot V_B \cdot C_W - k_2 \cdot V_B \cdot C_B \tag{1}$$

where

C_B = the chemical concentration (μg/L) in the organism (i.e., plant or fish),
C_W = the chemical concentration (μg/L) in the water,
V_B = the volume of the plant or fish (L), and
k_1 $(L \cdot L^{-1} \cdot d^{-1})$ and k_2 (d^{-1}) = the rate constants for, respectively, chemical uptake into and chemical elimination from the plants or fish.

To fit this model to the observed time response of the chemical concentrations in the water and in the organism, Eq 1 can be integrated. This can be performed simply when the chemical concentration in the water, the volume of the plant or fish (V_B), and the rate constants of chemical uptake and elimination do not vary with time (e.g., there is no growth), resulting in

$$C_B = C_W \cdot (k_1/k_2) \cdot \{1 - \exp(-k_2 \cdot t)\} \tag{2}$$

Equation 2 illustrates that, if the organism is exposed to a constant C_W, C_B should increase logarithmically with time to approach a constant level of $C_W \cdot (k_1/k_2)$, where (k_1/k_2) is often referred to as the bioconcentration factor BCF [3–5].

When the chemical concentration in the water is zero, such as during elimination when the organisms are exposed to clean water, integration of Eq 1 gives

$$C_B = C_{B,t=0} \cdot \exp(-k_2 \cdot t) \tag{3}$$

or

$$\log C_B = \log C_{B,t=0} - (k_2/2.303) \cdot t \tag{4}$$

where $C_{B,t=0}$ is the concentration ($\mu g/L$) in the organism at the beginning of the elimination period.

If the organism is growing and the chemical concentration in the water is not constant, such as in our uptake experiments, the model (i.e., Eq 1) can be fitted to the experimental data by a numerical integration procedure, which derives the chemical mass in the organism, i.e., X_B or $V_B \cdot C_B$ (in μg), as the sum of increments in chemical mass dX_B over time intervals dt, i.e., $X_B = \Sigma dX_B$. Each dX_P is calculated from Eq 1 as

$$dX_B = (k_1 \cdot V_B \cdot C_W - k_2 \cdot V_B \cdot C_B) \cdot dt \tag{5}$$

where dt should be chosen to be sufficiently small, and C_W and V_B at every exposure time t or Σdt, correspond with the experimentally observed values. Then, values for k_1 and k_2 are selected in an iterative fashion to produce the best agreement between calculated and observed X_B. The best fit of the observed data is the one with the k_1 and k_2 values, for which the sum of the squared differences between calculated and observed X_B is the smallest. This technique ensures that the estimates of k_1 and k_2, and thus the bioconcentration factor BCF, i.e., k_1/k_2, are not affected by the duration of the exposure period or by variations of the water concentration. This method was applied to determine k_1, k_2, and the BCF in the plants and fish, which are listed in Table 1. For this purpose, the time function of the water concentration and plant or fish volume during the experiments was established by fitting the observed values to a series of linear functions, each of which connect the observed values at two consecutive exposure times. The applicability of the model is represented by the quality of the fit, which can be expressed by the deviation, E, of the model predicted from the observed values, i.e.

$$E = \frac{\sum_{i=1}^{n} \sqrt{(C_{B,i}^0 - C_{B,i}^M)^2}/C_{B,i}^M}{n} \tag{6}$$

where C_B^0 is the observed, C_B^M is the predicted concentration in the plant or fish, and n is the number of observations. The deviation between observed and fitted concentrations ranged from 12 to 43% for the plants and 10 to 60% for the guppies, which is of similar magnitude as the experimental error associated with the plant and water analysis. This demonstrates that, considering the experimental error, the reversible organism-water two-compartment model with first order rate constants (i.e., Eq 1) satisfactorily describes the chemical exchange between the plants and water and between the fish and water.

Figure 1b shows that the exponential decrease of the concentration of 2,2',5,5'-tetra-chlorobiphenyl in the guppies is in agreement with Eqs 4 and 5. The rate constants for chemical elimination, k_2, can thus be determined from the slope of the log C_B-time plot. The elimination rate constants agree with those derived from the uptake data. Figure 1a illustrates that during the first 37 days of the depuration period, the chemical concentrations in the plants also drop exponentially. However, after the first 37 days of the elimination period, the decrease of the chemical concentration in the plant tends to be somewhat slower. This does not agree with the plant-water two-compartment model. It indicates that the chemical accesses a small fraction of the plant at a slower rate than the majority of the plant. The plant may thus be more accurately represented by two compartments than by a single compartment [11]. However, for the purpose of this analysis, we will focus on the chemical kinetics during the initial 37 days, which represent the elimination of the majority of the chemical in the plant. During this time frame, the kinetics in the plant are satisfactorily described by a water-plant two-compartment model. Estimates of the elimination rate constant can thus be derived from the slope of the log C_B-time plot. A discussion of the elimination kinetics in plants is presented elsewhere [1].

This kinetic analysis demonstrates that an organism-water two-compartment model is able to give a satisfactory representation of the uptake and elimination of the investigated CBs and PCBs in both the guppies and the aquatic macrophytes. This implies that from a toxicokinetic point of view a fish and a plant can be treated as single homogeneous compartments. Studies of the anatomy and physiology of plants and fish have identified that there are many physiologically different compartments in the fish and the plant. A two- or multi-compartment model may thus be more a realistic description of the kinetics of chemicals in the plants and fish. However, the experimental detail of the uptake and elimination studies is not sufficient to distinguish between different compartments in the fish or plants. Consequently, the rate at which a chemical arrives at a target site of the fish or plant should be considered to be equal to the rate at which the chemical reaches other compartments in the organism. For the CBs and PCBs, this rate of chemical exchange is satisfactorily described by Eq 1 and the rate constants are listed in Table 1.

Bioconcentration

The plant-water and fish-water bioconcentration factors are listed in Table 1 and plotted versus the 1-octanol-water partition coefficient (K_{OW}) in Fig. 2. Figure 2 demonstrates that the plant-water bioconcentration factor and the 1-octanol-water partition coefficients follow a linear relationship, i.e.,

$$\log \text{BCF} = 0.98 \, [\pm 0.16] \cdot \log K_{OW} - 2.23 \, [\pm 0.67] \qquad n = 8, r^2 = 0.97 \qquad (7)$$

where the confidence intervals have a 95% probability. The bioconcentration factors in the guppies also follow a linear relationship with K_{OW}, but only for chemicals which have a log K_{OW} less than 6.2, i.e.,

$$\log \text{BCF} = 1.03 \, [\pm 0.24] \cdot \log K_{OW} - 1.30 \, [\pm 0.58]$$

$$n = 6, r^2 = 0.97, \log K_{OW} < 6.2$$

(8)

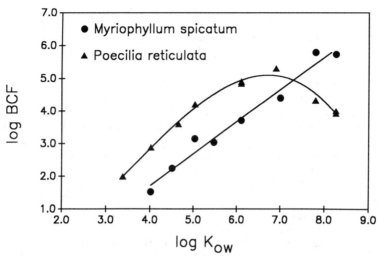

FIG. 2—*Relationship between the logarithm of 1-octanol-water partition coefficient, log* K_{OW}, *and the logarithm of the bioconcentration factor, log BCF, in* Myriophyllum spicatum *and in* Poecilia reticulata.

Equations 7 and 8 demonstrate that there is a strong relationship between the bioconcentration factors in plants and fish and the chemical's tendency to partition between water and 1-octanol. 1-Octanol is often considered to be a satisfactory surrogate phase for natural lipids. The 1-octanol-water partition coefficient therefore represents the chemical's ability to partition between lipids and water [12]. The excellent relationship between the BCF and K_{OW} suggests that chemical bioconcentration in the plant and fish is essentially a chemical partitioning process between the plant lipids and the water. This can be further illustrated by expressing the bioconcentration factor on a lipid weight basis as K_L. K_L is the ratio of the chemical concentration in extractable lipids of the plants or fish (C_L) over that in the water, i.e., C_L/C_W or $(C_B/C_W \cdot L_B)$ or BCF/L_B, where L_B (g/g) is the lipid content of the plant [i.e., 0.0020 ($\pm$0.000 23)], or fish (i.e, approximately 0.055). Table 1 illustrates that the lipid-weight-based plant-water and fish-water bioconcentration factors are approximately equal to the 1-octanol-water partition coefficient. This suggests that chemical bioconcentration occurs predominantly in the extractable lipids of the plants and fish since the solubility of the test chemicals for 1-octanol and lipids are often similar [12,13]. It can thus be concluded that, in essence, bioconcentration of the investigated chemicals in the plant and fish is a thermodynamically controlled process determined by the affinity of the chemical for the plant lipids relative to that for the water. The driving force of this process is the higher solubility of the chemical in the plant and fish lipids compared to that in the water. The lipids will absorb the chemical until the ratio of the lipid/water concentrations equals the ratio of the chemical's activities or solubilities in the plant lipids and the water. This situation is often referred to as chemical equilibrium, where the chemical potential or fugacity of the chemical in lipids and water are the same [5]. After a chemical equilibrium has been established, there is no further net uptake of the chemical into the plant or fish. It thus appears that uptake and elimination of the CBs and PCBs in the plants and fish are passive processes, controlled by the chemical's thermodynamic gradient. Active transport, i.e., transport against the thermodynamic gradient which requires energy, is insignificant.

The linear relationship between the fish-water bioconcentration factor and the 1-octanol-water partition coefficient breaks down for chemicals with a log K_{OW} exceeding 6.2. Evidence supports that this loss of linear correlation is caused primarily by (1) a reduction of the bioavailability of chemicals with very high K_{OW} during the experiment, and (2) the elimination of chemical by faecal egestion [2]. The reduction of the bioavailability is the result of the tendency of very hydrophobic (i.e., high K_{OW}) chemicals to sorb onto organic matter in the water introduced by the fish. As a result, a considerable fraction of the chemical in the water is in a sorbed or non-truly-dissolved state which cannot be absorbed by fish via the gills [2,14]. This sorption tendency and thus the fraction of the chemical concentration in a sorbed state tends to increase with increasing K_{OW}. Presently, there are no reliable techniques that can distinguish between sorbed and dissolved chemical in the water. As a result, water concentration measurements often reflect the total concentration of the chemical, not the truly dissolved chemical. The water concentration measurements thus overestimate the concentration in the water, which can actually be bioconcentrated by the fish, resulting in an underestimate of the bioconcentration factor and a loss of the linear BCF-K_{OW} relationship [2,15]. The loss of linear correlation due to a reduction in bioavailability is due to experimental difficulties regarding the measurement of the chemical concentrations in the water. It is not due to fundamental changes in the mechanism of the bioconcentration process if K_{OW} increases. The second factor that was identified to cause a loss of linear correlation is the chemical elimination in faecal matter [2,16]. In contrast to submerged aquatic macrophytes, guppies have the ability to eliminate chemicals not only to the water (i.e., via the gills), but also into faecal matter. The transfer of chemicals between the water and the fish during the bioconcentration experiment should therefore be described by

$$d(V_B \cdot C_B)/dt = k_1 \cdot V_B \cdot C_W - k_2 \cdot V_B \cdot C_B - k_E \cdot V_B \cdot C_B \qquad (9)$$

where k_E is the rate constant (d^{-1}) for chemical elimination by faecal egestion. For chemicals with log K_{OW} less than 6.2, k_2 is much larger than k_E. Chemical elimination is therefore predominantly via the gills to the water and k_E can be ignored with respect to k_2, which simplifies Eq 9 to Eq 1. However, as we will demonstrate in more detail later, k_2 tends to drop with increasing K_{OW} and becomes smaller than k_E for chemicals with a log K_{OW} exceeding 6.2 [16]. For very hydrophobic chemicals, elimination is predominantly by faecal egestion and elimination to the water, i.e., k_2 can be ignored with respect to k_E, thus simplifying Eq 9 to

$$d(V_B \cdot C_B)/dt = k_1 \cdot V_B \cdot C_W - k_E \cdot V_B \cdot C_B \qquad (10)$$

Equation 10 illustrates that for chemicals with log K_{OW} exceeding 6.2, chemical exchange is no longer between the fish and the water. Bioconcentration is therefore no longer a fish-water partitioning process, but it reflects the balance between the rates of chemical uptake from the water and chemical elimination by faecal egestion.

Factors Controlling Toxicokinetics in Plants and Fish

To explore the factors controlling the water-plant and water-fish exchange, it is interesting to plot the rate constants of chemical uptake and elimination as a function of the K_{OW} of the chemical. This has been done in Fig. 3 for the uptake rate constants in the plants and fish and in Fig. 4 for the elimination rate constants.

Figure 3 illustrates that for chemicals with a log K_{OW} below 5.5, the uptake rate constant (k_1) in the plant increases with increasing K_{OW} to approach a constant value of approximately 500 d^{-1} for chemicals with a log K_{OW} exceeding 5.5. The uptake rate constant in fish shows

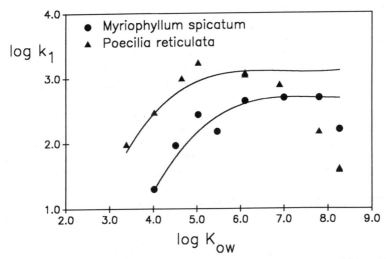

FIG. 3—*The logarithm of the uptake rate constant, log* k_1 (L · $L^{-1}d^{-1}$), *versus the logarithm of the 1-octanol-water partition coefficient, log* K_{OW}, *for* Myriophyllum spicatum *and in* Poecilia reticulata. *The solid line represents the model fit, i.e., Eq 13 for the plants and Eq 15 for the guppies.*

a similar relationship with K_{OW}. However, for very high log K_{OW} chemicals (>6.2), k_1 tends to fall instead of remaining constant at a level of approximately 1200 d^{-1}. This drop of k_1 for the very high K_{OW} chemicals is believed to be caused by the incorrect measurement of the bioavailable concentrations of these chemicals in the water and is thus an artifact of the experimental procedures used.

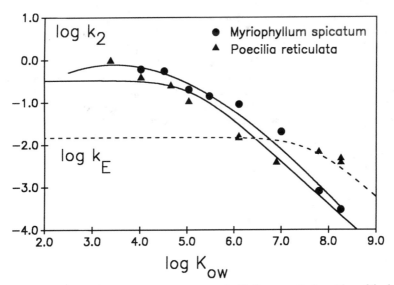

FIG. 4—*The logarithm of the elimination rate constant,* k_2 (d^{-1}), *versus the logarithm of the 1-octanol-water partition coefficient, log* K_{OW}, *for* Myriophyllum spicatum *and in* Poecilia reticulata. *The solid line illustrates the model fit, i.e., Eq 14 for the plants and Eq 16 for the guppies. The broken line represents the plot for* k_E.

Figure 4 shows that the elimination rate constant (k_2) in the plants and fish also tends to follow a "biphasic" relationship with K_{OW}. It shows that with increasing K_{OW}, the elimination rate constant drops, first slowly, but then more steeply. In particular for fish, the "biphasic" nature of the k_2-K_{OW} relationship is not as evident as that for the uptake rate constants. This may be due to the measurement of the elimination rate constants, which does not distinguish between elimination to the water, i.e., via the gills (k_2), and elimination in faecal matter (k_E).

It has been proposed that the "biphasic" nature of the relationship between the rate constants and K_{OW} in the plants and fish is the result of the fact that chemical uptake from the water and elimination to the water involves chemical transport in aqueous and lipid parts of the plants or fish [1,4,5,17,18]. Examples of lipid phases in the plants and fish are the lipid bilayers of biological membranes, the plant's waxy cuticle or the mucus layers in fish. Transport in aqueous phases may involve the cytoplasm of cells or the water flow in the gill compartment of the fish. If chemical transport in water and lipid phases occurs in series, the following equations can be derived for the uptake and elimination rate constants [1,5]

$$1/k_1 = (V_B/D_W) + (V_B/D_L)/K_{OW} \tag{11}$$

$$1/k_2 = (L_B \cdot V_B/D_W) \cdot K_{OW} + (L_B \cdot V_B/D_L) \tag{12}$$

where D_W and D_L are transport parameters representing the transport rate in the aqueous and the lipid phases of the organisms. The derivation of Eqs 11 and 12 and an explanation of the lipid-water kinetic model for plants and fish can be found elsewhere [1,5]. In essence, Eqs 11 and 12 demonstrate that the uptake and elimination tend to be controlled by transport in the lipid phases when the chemical's K_{OW} is low. However, with increasing K_{OW}, chemical transport in the aqueous phases of the plant becomes more important and ultimately dominates the kinetics. Equation 11 thus predicts that with increasing K_{OW}, k_1 increases when transfer in the lipid phases (e.g., membranes) controls the uptake kinetics and then approaches a constant level (i.e., D_W/V_B) for high K_{OW} chemicals, for which transport in water phases becomes the rate-determining step. Likewise, Eq 12 illustrates that k_2 tends to be approximately constant (i.e., $D_L/V_B \cdot L_B$) for low K_{OW} chemicals, when transfer in lipid phases of the plant is the rate-determining process, and then drops with increasing K_{OW}, when transport through water phases controls the elimination process.

To test the applicability of this lipid-water kinetic model and to quantify the lipid and water phase transport parameters, Eqs 11 and 12 can be fitted to the experimental data, resulting in

Myriophyllum spicatum:

$$1/k_1 = 0.0020 + 500/K_{OW} \tag{13}$$

$$1/k_2 = 1.58 + 0.000\,015 \cdot K_{OW} \tag{14}$$

Poecilia reticulata:

$$1/k_1 = 0.000\,78 + 30/K_{OW} \tag{15}$$

$$1/k_2 = 1.0 \pm 0.000\,095 \cdot K_{OW} \tag{16}$$

Figures 3 and 4 illustrate the excellent fit of the model to the experimental data. This indicates that the uptake and elimination kinetics in the plants and fish can be satisfactorily described by the same lipid-water kinetic model. The only lack of agreement between the model and the experimental data is for some of the very high K_{OW} chemicals. As mentioned earlier, this may be due to the fact that the measured concentration of very high K_{OW} chemicals in the water did not truly represent the chemical concentration that can be absorbed and bioconcentrated by the plants and fish.

From Eqs 13 and 14, it can be observed that the uptake and elimination kinetics of chemicals with a log K_{OW} below approximately 5.5 is predominantly controlled by transport in the lipid phases of the plant (e.g., transport across the lipid membranes). The uptake and elimination of chemicals with a log K_{OW} exceeding 5.5 are largely determined by transport in aqueous phases. The lipid and aqueous phase transfer conductivities can also be quantified. Equations 12 to 15 illustrate that in *Myriophyllum spicatum* D_W/V_B is between 133 and 500 d^{-1} and D_L/V_B is between 0.0013 and 0.0020 d^{-1} *(L_B is 0.002)*. In the guppies, D_W/V_B is between 700 and 1300 d^{-1} and D_L/V_B is between 0.03 and 0.07 d^{-1}. The differences in the water and the lipid phase conductivities demonstrate that the plant and the guppies have their own specific water and lipid phase transport parameters. The water and lipid phase transport parameters are organism specific and reflect the differences in physiology and structure between the plant and the guppies. Based on the data from this study, it is not possible to identify the water and lipid phase transport processes in terms of actual transport processes in specific parts or tissues of the plants or fish. More detailed experiment are required to establish the nature of the lipid and aqueous phase transport processes.

Toxicity in Submerged Aquatic Macrophytes and Fish

Toxic effects in organisms, such as lethality in a LC_{50} test, are the combined result of the chemical concentration in the "target site" and the toxicity of the chemical. The toxicity of the chemical reflects the chemical's activity at the site of action. Therefore, to compare the toxicity of a chemical to that of another chemical or the chemical's toxicity in a plant to that in fish, it is necessary to determine the concentration at the site of action when the effect occurs. This can be achieved by direct measurements of the chemical concentration at the site of action. However, this is often difficult and therefore rarely performed. An alternative exists when in addition to data on toxic effects, such as LC_{50} values, the kinetics of the toxicant in the organism are available. We will demonstrate this approach for the guppy.

Chemical Toxicity in the Guppy

In a typical lethality test, guppies are exposed for up to 14 days to a constant concentration of the test chemical in the water. The concentration in the water which causes mortality to 50% of the guppies, or the LC_{50}, is often reported to express the "toxicity" of the chemical. The LC_{50} values of the CBs and PCBs are listed in Table 1 and plotted versus the 1-octanol-water partition coefficient in Fig. 5. Figure 5 illustrates that a chemical of high K_{OW} tends to cause mortality at a lower concentration than a chemical of low K_{OW}. This relationship breaks down if log K_{OW} exceeds 5.5, above which no acute lethality is observed. The reciprocal relationship between the LC_{50} and K_{OW} has often been interpreted by assigning a greater toxicity to high K_{OW} chemicals. This interpretation is not entirely correct since it equates toxic effects (i.e., 50% mortality) to chemical toxicity, thus ignoring the chemical concentration in the organism that triggered the effect. To determine the relative toxicities of the CBs, we need to consider the chemical concentration in the organism when the toxic effect occurs. This can be achieved by considering the uptake and elimination kinetics of the CB and PCB congeners in the guppy. Earlier, it was shown that uptake and bioconcentration of CBs and PCBs in the guppy can be satisfactorily described by treating the guppy as a single homogeneous compartment. This implies that the chemical accesses the site of action at approximately the same rate as the rest of the fish, which is described by Eq 1. Since during the lethality test the chemical concentration in the water is constant, Eq 1 can be integrated to give Eq 2, which provides a means to determine the concentration at the site of action, C_T, as a result of exposure to a concentration of chemical in the water (C_W) for a period of time t. The target site concentration causing 50% mortality after 14 days of exposure can thus be estimated for each chemical by substituting the LC_{50} for C_W, 14 for t, and the appropriate rate constants (Table 1) in Eq 2, i.e.,

$$C_T = LC_{50} \cdot (k_1/k_2) \cdot \{1 - \exp(-k_2 \cdot 14)\} \tag{17}$$

In this fashion, the target site concentrations (C_T) for 50% acute lethality were calculated for the CBs, listed in Table 1 and plotted versus K_{ow} in Fig. 5.

Figure 5 shows that the target site concentrations of the CBs in the guppy are approximately similar at a level of 6330 (± 2770) μmol/L. They are not dependent on K_{ow}. Since each CB congener causes 50% mortality at approximately the same concentration in the guppy, it appears that the toxicity of all CB congeners is essentially the same. It is interesting that for a series of linear alcohols and ketones a similar internal concentration of 6000 μmol/L was estimated to cause 50% mortality in fathead minnows [19]. Based on the LC_{50} values of approximately 90 organic substances in fathead minnows, guppies, *Daphnia magna*, and the saltwater brine shrimp *Artemia*, Abernethy and Mackay [20] estimated that 50% mortality occurs when the chemicals reach a fairly constant volume fraction in the organisms of approximately 0.63%, which corresponds to a concentration of approximately 6000 μmol/

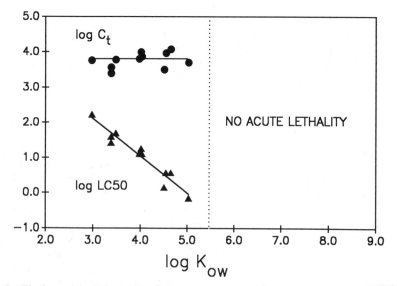

FIG. 5—*The logarithm of the LC_{50} and the internal concentration in the guppy as a function of the logarithm of the 1-octanol-water partition coefficient. For the congeners with a log K_{ow} above 5.5, no acute lethality has been observed.*

L. These results suggest that for a certain group of organic chemicals, including the CBs and PCBs, one specific chemical concentration in the organism causes 50% lethality in all of the investigated organisms.

It appears that the internal concentration causing lethality tends to be independent of the type of chemical. It has been suggested that this mode of action is related to that of narcosis or anaesthesia, which also tends to be the result of one particular concentration or chemical activity at the site of action and fairly independent on the type of chemical [21]. The chemicals that cause acute lethality at a constant internal concentration are also causing narcosis and they are therefore often referred to as "narcotics." The actual mechanisms of acute lethality or narcosis are unclear. However, it is suspected that the chemical has a "physical" effect on a lipid-like target site, possibly the membrane systems of the organism [22]. Chemicals that cause lethality at a lower internal concentration in the organisms exert their toxic action through a different mechanism. They are often reactive chemicals, which tend to interact with specific proteins or receptors in the organism.

Estimates of the internal concentrations of narcotics suggests that the acute lethality in several aquatic organisms is associated with the same internal chemical concentration of approximately of 6000 μmol/L. It is therefore tempting to speculate that this internal concentration of approximately 6000 μmol/L is universal among aquatic organisms, including plants, thus suggesting that the toxicity of the CBs is independent on the organism and similar in plant and animal life. It is possible that the chemicals interfere with fundamental molecular processes required for the proper functioning of the cell. In that case, one specific concentration in the organism may trigger acute lethality in virtually all organisms, including plants. However, it is also conceivable that the internal lethal concentration in the plant is different from that in animal life since membranes may be the site of action and plant and animal membranes are fundamentally different. Unfortunately, there are only few LC_{50} data for aquatic macrophyte species, which makes it difficult to test this hypothesis. In absence of appropriate LC_{50} data, it is interesting to examine the results of a study by Wong et al. [23] regarding the effects of CBs on the primary productivity of the freshwater green algae *Ankistrodesmus falcatus*, which are summarized in Table 2. Assuming that the 4-h exposure period was sufficiently long to reach equilibrium and that the BCF can be expressed as $L_B \cdot K_{OW}$, where L_B is the lipid content of the algae (i.e., 1%), it is possible to estimate the internal concentration C_T in the algae causing the 50% reduction in primary productivity. The estimates of C_T, which are listed in Table 2, demonstrate that all CB congeners cause a 50% reduction in primary productivity at approximately the same internal concentration in the algae of 4840 ($\pm$1430) μmol/L. Considering the error in the calculations, the internal concentration in the algae causing a 50% reduction in primary productivity is not significantly different from the internal lethal concentration in the guppies. The results further demonstrate that, similar to acute lethality in fish, congeners with a very high K_{OW}, such as hexachlorobenzene, do not demonstrate the toxic effects.

It can be concluded that in the guppy, and possibly in other aquatic organisms as well, the acute toxicity of CB congeners is essentially the same. The extent to which lethality (i.e., the effect) occurs thus reflects the concentration in the organism, which in turn is the result of the chemical concentration in the water, the rates of uptake and elimination, and the duration of chemical exposure. As demonstrated earlier, the uptake and elimination rates, and thus the relationship between the concentration in the water and that in the organism, vary between organisms such as guppies and aquatic plants and are dependent on the K_{OW} of the chemical. This explains the role of chemical properties, such as the K_{OW}, on the acute lethality and differences in "sensitivities" between organisms. For example, the LC_{50} of pentachlorobenzene is lower than that of monochlorobenzene because pentachlorobenzene has a higher bioconcentration factor, which is related to its higher K_{OW} and

TABLE 2—*The logarithm of the 1-octanol-water partition coefficient log K_{OW}, the logarithm of the EC_{50} ($\mu mol/L$), and the logarithm of the internal concentration log C_T ($\mu mol/L$) of a series of chlorobenzenes in the fresh water green algae* Ankistrodesmus falcatus. *NT means no reduction in primary productivity was observed.*

ANKISTRODESMUS FALCATUS			
Compound	Log K_{ow}	Log EC_{50}	Log C_T
Chlorobenzene	2.98 [6]	0.444	3.63
1,2-Dichlorobenzene	3.38 [6]	0.136	3.51
1,3-Dichlorobenzene	3.48 [6]	0.156	3.67
1,4-Dichlorobenzene	3.38 [6]	0.136	3.51
1,2,3-Trichlorobenzene	4.04 [6]	0.033	3.56
1,2,4-Trichlorobenzene	3.98 [6]	0.033	3.50
1,3,5-Trichlorobenzene	4.02 [6]	0.05	3.72
1,2,3,4-Tetrachlorobenzene	4.55 [6]	0.019	3.83
1,2,3,5-Tetrachlorobenzene	4.65 [6]	0.014	3.80
1,2,4,5-Tetrachlorobenzene	4.51 [6]	0.023	3.87
Pentachlorobenzene	5.03 [6]	0.005	3.73
Hexachlorobenzene	5.47 [6]	NT	

thus requires a lower concentration in the water than monochlorobenzene to reach the same lethal concentration in the organism. The PCB congeners which have very high K_{OW} are unable to trigger acute lethality in the guppies because the largest possible concentration in the water, i.e., the congeners's aqueous solubilities, is too small to achieve the internal lethal concentration in the guppies.

Conclusions

This study demonstrates that the main driving force of the uptake and bioconcentration of chlorobenzenes and PCBs in the submerged macrophyte species *Myriophyllum spicatum* and in fish, i.e., *Poecilia reticulata,* is the tendency of hydrophobic chemicals to partition between the lipids of the plant or fish and the water. The mechanism of chemical uptake and elimination in the plants and the guppies is essentially the same and appears to involve passive transfer of the chemical as a result of diffusion or convection by natural fluid flow processes in the organisms (e.g., gill flow in fish). Active uptake or elimination of the chlorobenzenes and PCBs does not appear to be a significant process. In absence of metabolic transformation, the plants and fish are therefore incapable of controlling their internal concentrations of CBs and PCBs.

When plants and fish are exposed to the same concentration of chlorobenzene or PCB congeners in the water, the lipids of the plants and fish tend to approach approximately the same concentration, reflecting the chemical's lipid-water partition coefficient. Very hydrophobic chemicals with a log K_{OW} exceeding approximately 6.2 are the only exception. Their bioconcentration potential in fish is less than that in the plants. Except for experimental problems with the measurement of the bioavailable concentration in the water, this is largely the result of chemical elimination by faecal egestion.

The rate of uptake and elimination and thus the time to reach equilibrium differs between the plants and the fish. However, the dynamics of chemical exchange in the plants and fish can be described by the same lipid-water kinetic model. This model illustrates that the uptake and elimination of chemicals with a log K_{OW} less than approximately 5.5 are largely controlled by transport in lipid phases of the plants or fish, whereas for higher K_{OW} chemicals the rate determining process is in an aqueous phase.

Acute lethality tests in guppies suggest that an internal concentration of any of the chlorobenzene congeners in the fish of approximately 6330 $\mu mol/L$ causes 50% lethality. This demonstrates that the toxicity (i.e., the activity at the site of action) of the CB and PCB congeners is the same. Although there are limited data for acute lethal effects in plants, it is conceivable that a similar internal lethal concentration applies to plants. In that case, differences in the acute lethal response in plants and fish will only reflect the uptake and elimination dynamics of the chemical in the plants and fish. Since this study suggests that the mechanism of uptake and elimination of the CBs and PCBs in the plants and the fish are similar, it is conceivable that plants and fish respond similarly to aqueous concentrations of CBs and PCBs. It should be emphasized that this similarity in toxic response may only apply to acute lethal effects, which tend to occur at relatively high concentrations in the water. There may be other, possibly nonlethal, effects that apply to fish, but not to plants or visa versa. This issue may only be resolved with continued research on the mechanisms of chemical uptake and toxicity.

Acknowledgment

The authors gratefully acknowledge the financial support of the Ontario Ministry of the Environment and the Natural Sciences and Engineering Research Council of Canada.

References

[1] Gobas, F. A. P. C., McNeil, E. J., Lovett-Doust, L., and Haffner, G. D., *Environmental Science and Technology*, in press.

[2] Gobas, F. A. P. C., Clark, K. E., Shiu, W. Y., and Mackay, D., *Environmental Toxicology and Chemistry*, Vol. 8, 1989, pp. 231–247.

[3] Branson, D. R., Blau, G. E., Alexander, H. C., and Neely, W. B., *Transactions of the American Fish Society*, Vol. 104, 1975, pp. 785–792.

[4] Gobas, F. A. P. C., Opperhuizen A., and Hutzinger, O., *Environmental Toxicology and Chemistry*, Vol. 5, 1986, pp. 637–646.

[5] Gobas, F. A. P. C. and Mackay, D., *Environmental Toxicology and Chemistry*, Vol. 6, 1987, pp. 495–504.

[6] Miller, M. M., Wasik, S. P., Huang, G. L., Shiu, W. Y., and Mackay, D., *Environmental Science and Technology*, Vol. 19, 1985, pp. 522–529.

[7] Shiu, W. Y. and Mackay, D., *Physical and Chemical Reference Data*, Vol. 15, 1986, pp. 911–929.

[8] Konemann, H. and Van Leeuwen, K., *Chemosphere*, Vol. 9, 1980, pp. 3–19.

[9] Bruggeman, W. A., Opperhuizen, A., Wijbenga, A., and Hutzinger, O., *Toxicology and Environmental Chemistry*, Vol. 7, 1984, pp. 173–189.

[10] Konemann, H., *Toxicology*, Vol. 19, 1981, pp. 209–221.

[11] Moriarty, F. in *Organochlorine Insecticides: Persistent Organic Pollutants*, Academic Press, London, 1975, Chapter 2, pp. 29–72.

[12] Gobas, F. A. P. C., Lahittete J. M., Garofalo, G., Shiu, W. Y., and Mackay, D., *Journal of Pharmaceutical Sciences*, Vol. 77, 1988, pp. 265–272.

[13] Dobbs, A. J. and Williams, N., *Chemosphere*, Vol. 12, 1983, pp. 97–104.

[14] Black, M. C. and McCarthy, J. F., *Environmental Toxicology and Chemistry*, Vol. 7, 1988, pp. 593–600.

[15] Gobas, F. A. P. C., Shiu, W. Y., and Mackay, D. in *QSAR in Environmental Toxicology II*, K. L. E. Kaiser, Ed., D. Reidel Publishing Company, Dordrecht, Germany, 1987, pp. 107–123.

[16] Gobas, F. A. P. C., Muir, D. C. G., and Mackay, D., *Chemosphere*, Vol. 17, 1988, pp. 943–962.

[17] McKim, J., Schnieder, P., and Veith, G., *Toxicology and Applied Pharmacology*, Vol. 77, 1985, pp. 1–10.

[18] Spacie, A. and Hamelink, J. L., *Environmental Toxicology and Chemistry*, Vol. 1, pp. 309–320.

[19] McCarthy, L. S. in *QSAR in Environmental Toxicology II*, K. L. E. Kaiser, Ed., D. Reidel Publishing Company, Dordrecht, Germany, 1987, pp. 207–229.

[20] Abernethy, S. G., Mackay, D. and McCarthy, L. S., *Environmental Toxicology and Chemistry*, Vol. 7, 1988, pp. 469–481.

[21] Crisp, D. J., Christie, A. O., and Ghobasy A. F. A., *Comparative Biochemistry and Physiology*, Vol. 22, 1967, pp. 629–649.

[22] Franks, N. and Lieb, W., *Nature*, Vol. 274, 1978, pp. 339–342.

[23] Wong, P. T. S., Chau, Y. K., Rhamey, J. S., and Docker, M., *Chemosphere*, Vol. 13, 1984, pp. 991–996.

General Phytotoxicology

Richard A. Brown[1] and Deborah Farmer[1]

Track-Sprayer and Glasshouse Techniques for Terrestrial Plant Bioassays with Pesticides

REFERENCE: Brown, R. A. and Farmer, D., **"Track-Sprayer and Glasshouse Techniques for Terrestrial Plant Bioassays with Pesticides,"** *Plants for Toxicity Assessment: Second Volume, ASTM STP 1115,* J. W. Gorsuch, W. R. Lower, W. Wang, and M. A. Lewis, Eds., American Society for Testing and Materials, Philadelphia, 1991, pp. 197–208.

ABSTRACT: In order to determine the potential of pesticides to cause damage to nontarget terrestrial plants, assessments of the intrinsic toxicity of the pesticide must be matched to estimates or measures of exposure. This paper describes a glasshouse bioassay which has been developed in order to assess pre- and postemergence toxicity of pesticides to a taxonomic range of crop and weed species. Different techniques used to assess pesticide effects in these tests are compared, and their contribution to the overall toxicological hazard assessment is shown. Use of the glasshouse bioassay to compare (*a*) effects of a given herbicide on different plant species, (*b*) effects of different herbicides on one plant species, and (*c*) differences in pre- and postemergence effects for a given herbicide are demonstrated. Interpretation of the ecological risks in the field based on the glasshouse toxicity data is discussed.

KEY WORDS: ecotoxicology, phytotoxicity, seedling emergence, vegetative vigor, ecological risk assessment, growth stage, pesticide, glasshouse

Over recent years, there has been an increase in concern about the hazards posed to nontarget plants by pesticides. Initially, concerns about deposition of herbicidal residues outside the intended target area were directed at damage to crops [1–2] more than to the endemic flora, though more recently concern has centered around effects on wild plants [3–4].

Currently in Europe, effects on wild plants in the target area are also of interest, leading to the development of "conservation headlands" at the edges of fields in certain areas of the United Kingdom (UK), Germany, and Denmark [5–6]. In the UK, the "Game Conservancy" has issued a series of recommendations concerning management of the crop margin for conservation purposes [6]; their proposals are for a 6-m zone at the crop edge where use of herbicides is carefully controlled. In addition to allowing the growth of some endangered plant species, this approach has been shown to promote brood size of game birds as a result of conserving the host plants of preferred chick food insects and to increase the abundance of Lepidoptera.

In the United States, the Environmental Protection Agency (USEPA) requires that an initial or Tier I test is carried out for forestry, aquatic/terrestrial nonfood uses, and other uses on a case-by-case basis to determine effects of applications at the maximum label rate or, if known, maximum environmental exposure concentration (MEEC) and three times this rate [7]. Six dicotyledonous species (including soybean and a root crop) from at least

[1] Ecology and Soil Science Section, ICI Agrochemicals, Jealott's Hill Research Station, Bracknell, Berkshire, RG12 6EY, UK.

four families and four monocotyledonous species (including maize/corn) from at least two families are treated. Detrimental effects of greater than 25% compared with control plants trigger the next stage or Tier II testing. Obviously, herbicides have an inherent potential to harm nontarget plants and would therefore be automatically tested at the Tier II level. The data produced from Tier II tests are used to determine pre- and postemergence EC_{10}, EC_{25}, and EC_{50} values and no observable effect levels (NOELs) for each species tested. Tier III field tests may then be triggered. At this point, however, herbicide effects in the glasshouse must be related to environmental exposure to determine the ecological risk; currently, there are no guidelines for this process.

The glasshouse bioassay described here has been devised by ICI Agrochemicals to conform to USEPA guidelines for Tier II terrestrial nontarget plant testing of pesticides. The methods are similar to those developed by Marshall and Birnie [3] to evaluate the effects of herbicides and plant growth regulators on a range of field margin flora.

Methods

Plant Species

In accordance with USEPA guidelines, ten species are used in any one test, four monocotyledons and six dicotyledons, representing at least six families. Maize (*Zea mays*), soybean (*Glycine max*), and a dicotyledonous root crop—sugar beet (*Beta vulgaris*) are always included. The remaining seven species are chosen to represent a diversity of potential nontarget plants.

The suitability of the species chosen is constantly being reviewed. Factors considered include: representation of important plant families, good germination, uniform growth in the glasshouse, and sensitivity to herbicides. The species under review are shown in Table 1. It is intended that ten species will be selected for use in all future tests so that the activities of different pesticides may be compared.

Chemical Application

Applications are made using a hydraulic track sprayer. The track sprayer is fitted with a single, even, stainless steel jet, and the application volume is 200 L ha^{-1}. The spray jet is prechecked for output and evenness of spray pattern on a Lurmark "Patternator." Jet travelling speeds are then calculated to give the required output for a set height. Before each usage, the track sprayer is fully calibrated, by spraying and reweighing petri dishes containing filter paper of known weight, to ensure that the target output of $\pm 10\%$ is achieved. Rates of application follow a dilution series, with the highest rate equalling the maximum field application rate and the lowest being 1% or less of the highest. In addition, application rates chosen should enable the determination of concentrations causing 10, 25, and 50% effects in comparison with untreated control plants (EC_{10}, EC_{25}, and EC_{50}) and NOEL values for seedling emergence and plant yield.

The chemical to be tested is applied in the commercial end-use formulation with the highest concentration of active ingredient (a.i.). The test material is fully characterized to indicate content of formulation adjuvants wherever possible; a.i. content of the test formulation is always checked by chemical analysis.

Seedling Emergence Test Design

Three replicates of ten seeds of each plant species for each rate to be tested are sown in seed trays in characterized potting compost (Table 2). The organic matter content is kept

TABLE 1—*Plant species used in nontarget terrestrial plant bioassay.*

Latin Name	Code	Common Name	Family	Climate[a]
	DICOTYLEDONS			
Glycine max	GLXMA	Soybean	Leguminosae	Warm
Beta vulgaris	BEAVA	Sugar beet	Chenopodiaceae	Cool
Brassica napus	BRSNN	Oilseed rape	Crucifereae	Cool
Abutilon theophrasti	ABUTH	Velvet leaf	Malvaceae	Warm
Sida spinosa	SIDSP	Spiny teaweed	Malvaceae	Warm
Xanthium strumarium	XANST	Italian cocklebur	Compositae	Cool
Xanthium spinosum	XANSP	Spiny cocklebur	Compositae	Cool
Ipomea hederacea	IPOHE	Purple morning glory	Convolvulaceae	Cool
Cassia obtusifolia	CASOB	Sicklepod	Leguminosae	Warm
Galium aparine	GALAP	Goose grass	Compositae	Cool
Sinapis alba	SINAL	White mustard	Cruciferae	Cool
Polygonum lapathifolium	POLLA	Pale persicaria	Polygonaceae	Cool
	MONOCOTYLEDONS			
Zea mays	ZEAMX	Maize	Gramineae	Warm
Triticum aeshivum	TRZAW	Winter wheat	Gramineae	Cool
Avena fatua	AVEFA	Wild oat	Gramineae	Cool
Cyperus rotundus	CYPRO	Purple nutsedge	Cyperaceae	Warm

[a] *Warm:* Temperature day/night = 24/19°C   *Cool:* Temperature day/night = 18/12°C
Humidity day/night = 70/40%   Humidity day/night = 70/40%
Photoperiod = 14 h   Photoperiod = 14 h

as low as possible, <5%, to prevent atypical levels of adsorption of the herbicide. Crop species are sown at 2-cm depth and weed species at 1-cm depth. The ten species are divided between two trays, with the five cool climate species being sown in one tray and the five warm climate species in a second. Spray applications are made to the soil surface using the

TABLE 2—*Characterization of standard potting compost illustrating the range of values experienced.*

Compost	pH	Available Nutrients, ppm			Organic Matter, %	Sand, %	Silt, %	Clay, %
		P	K	Mg				
WCA	6.4–6.9	61–153	340–580	73–104	3.9–7.5	55–67	17–28	15–19
WCB	6.2–7.1	42–227	235–529	62–119	2.0–6.4	65–79	13–30	14–19
WCC	5.9–7.0	29–78	243–492	78–123	3.2–5.0	50–69	15–31	16–19

NOTE: All preemergence plants were grown in a standard compost called Weed Compost B (WCB), but for the postemergence test some plants were grown in composts of differing fertilizer levels in which they are known to grow better. These are known as Weed Compost A (WCA) or Weed Compost C (WCC).

Analytical Methods

pH   Measured by glass electrode in 1:2.5 soil:water slurry, stirred for 30 min.
Potassium/magnesium   Shaken with ammonium nitrate for 30 min. K determined by flame photometry; Mg determined by atomic absorption spectrophotometry.
Phosphorus   Extracted at 20°C in 0.5 *M* sodium bicarbonate at pH 8.5. The inorganic P reacts with acidic ammonium molybdate to form phosphomolybdate, reduced with ascorbic acid; blue complex measured with spectrophotometer at 850 nm.
Organic matter   Oxidized with potassium dichromate/concentration H_2SO_4, followed by titration of excess dichromate with ferrous sulphate (Walkley-Black method).
Sand, silt, clay   Sand by wet sieving; silt and clay by sequential sedimentation and analysis of the supernatant (using Stoke's law of particle sedimentation). Sand = >0.05 mm, silt = 0.002 to 0.5 mm, and clay = <0.002 mm.

track sprayer. All species and all replicates for a given rate are sprayed in one pass to prevent any differences between passes and hence differences in spray deposition. Control trays are sown up in the same way and left unsprayed. The controls are not sprayed with water since the application is made at a low volume (20 mL/m^2) and any chemical effects would be much greater than any differences due to the deposition of such a small volume of water. After spraying, all seed trays are transferred to a glasshouse of the appropriate climate (Table 1) and laid out in a random block design where treatment position is randomized within each replicate. The seed trays are top watered twice a day, using a hose fitted with a rose, in order to keep the soil moist but not overwatered as this would cause excessive leaching of the pesticide. Each seed tray is placed on an individual container to prevent cross contamination between rates.

Seedling Emergence Test Assessments

Percentage emergence—The number of seedlings emerged in treated and control units are counted daily. When there has been no increase in counts for three consecutive days, the number of seedlings emerged is taken as full emergence and the first of the three days is taken as the number of days to full emergence. For each species the total number of seedlings emerged at full emergence is recorded for each replicate of each pesticide rate.

Damage—Following emergence, subsequent plant development is observed in comparison with controls, and % damage of seedlings that have emerged is assessed using the scale shown in Table 3. Damage assessments are carried out at weekly intervals for four weeks following application; the potential for a plant to recover from initial damage can therefore also be assessed. To supplement the percentage damage assessment, notes are made of the symptomology or types of damage sustained, e.g., stunt, chlorosis, necrosis, abscission, pigmentation, and senescence.

Growth stage—At four weeks after application, mean growth stage is measured for each species at each pesticide rate and for the controls. The growth stage key in Table 4 is used.

Dry weight—Following the final damage and growth stage assessments, four weeks after application, the plants are harvested by cutting the stems at soil level. For each species, all plants within each replicate of each treatment rate or control are combined. The aerial plant parts are dried to constant weight in an oven at 75°C. The weight per plant is calculated (by relating to total number of seedlings emerged), and treated groups are compared to controls.

Vegetative Vigor Test Design

Three replicates of five plants for each species for each rate to be tested are grown individually in plastic plant pots (7.5-cm diameter) in characterized potting compost (organic matter content <5%), the same as for the seedling emergence test (Table 2). The same ten species as for the seedling emergence test are used. The plants to be sprayed are selected for uniformity and vigor and applications are made to young, actively growing plants (three to four leaves). As in the seedling emergence test, all species and all replicates for a given rate are sprayed in one pass. Control plants are left unsprayed. The plants are then transferred to a glasshouse of the appropriate climate (Table 1) and laid out in a random block design where treatment position is randomized within each replicate.

The plants are watered individually with a hose, avoiding wetting the foliage and overwatering. The pots containing the treated plants are held on separate retaining trays for each rate to avoid cross contamination.

Vegetative Vigor Test Assessments

Damage—Visual assessments of damage in comparison with controls are carried out at weekly intervals for four weeks following application, using the percentage scale shown in Table 3. Symptomology is also recorded.

Growth stage—At two and four weeks after application, mean growth stage is measured for each species at each pesticide rate and for the controls. The growth stage key in Table 4 is used.

Dry weight—Following the final damage and growth stage assessments, four weeks after application, the plants are harvested and dried to determine the weight per plant as described for the seedling emergence test.

Statistical Analysis

For seedling emergence assessments, the percentage data were normalized for parametric analysis by being transformed to angles using the arcsine transformation such that

$$Y = \sin^{-1} \sqrt{\% \text{ seeds emerged}/100}$$

The final dry weight assessment data are normalized for parametric analysis by a square-root transformation according to

$$Y = \sqrt{\text{dry weight(g)}}$$

The transformed data are then subjected to a two-way ANOVA (treatment and block) and the estimate of within-plot error used to calculate least significant differences (LSD) values. For each species the treatment rate means are compared to untreated controls at the 5% significance level. The NOEL values are then taken as the highest treatment rate below which there are no significant differences from the untreated controls.

For dose-response analysis, conventionally the percentage damage scores are normalized

TABLE 3—*Damage assessment scale for glasshouse bioassays.*

Percent	Description
0	Vigorous plant, indistinguishable from control
5	Vigorous plant but slight detectable differences
10	Vigorous plant but readily distinguishable differences
15	Vigorous plant but readily distinguishable differences
20	Less vigorous with pronounced differences
25	Less vigorous with pronounced differences
30	Less vigorous with pronounced differences
40	Poor vigor with increasing severity of effects
50	Poor vigor with increasing severity of effects
60	Poor vigor with increasing severity of effects
70	Very poor vigor but still growing, recovery possible
75	Very poor vigor but still growing, recovery possible
80	Very poor vigor, still growing but recovery unlikely
85	Very poor vigor, ceased growing, recovery very unlikely
90	Not all tissue dead but further growth unlikely
95	Moribund
100	Dead

TABLE 4–*Growth stage key.*

Monocotyledons		Dicotyledons	
Definition	Code	Definition	Code
SEEDLING EMERGENCE		COTYLEDON PRODUCTION	
Coleoptile emerged	1.0	Seedling emergence	1.0
Leaf just at coleoptile		Cotyledons expanding	1.05
tip	1.1	Cotyledons expanded	1.1
LEAF PRODUCTION ON MAIN SHOOT		LEAF/WHORL PRODUCTION ON MAIN STEM	
1st leaf through		1st leaf expanding	2.05
coleoptile	2.0	1st leaf expanded	2.1
1st leaf unfolded	2.1	2nd leaf expanding	2.15
2 leaves unfolded	2.2	2nd leaf expanded	2.2
3 leaves unfolded	2.3	Etc.	
Etc.			
TILLERING		BRANCH/SHOOT PRODUCTION ON MAIN STEM	
Main shoot only	3.0	Main stem only	3.0
Main shoot and 1 tiller	3.1	One branch/shoot	
Main shoot and 2 tillers	3.2	(>0.5 cm)	3.1
Etc.		Two branches/shoots	3.2
		Etc.	
Stem elongation	4.0	Flower buds present	4.0
Booting	5.0	Flowering	5.0
Inflorescence emerged	6.0	Leaf senescence	6.0

by transformation to logits such that

$$\text{logit } \% \text{ damage} = \frac{(\% \text{ damage} + 0.5)}{(100.5 - \% \text{ damage})}$$

and plotted against $\log_e$ (treatment rate). The EC_{10}, EC_{25}, and EC_{50} values can then be interpolated from a dose-response curve fitted to the transformed data using iteratively reweighted maximum likelihood regression of the logit transformation of % damage on $\log_e$ (rate). In these tests the glasshouse assessments of damage are made by comparing the treated plants with the untreated controls of the same species. As data are only generated for the treated plants, the comparative NOEL values cannot be calculated as above, and so for this analysis they are taken as the EC_{10}. This value is considered to be representative of the between-plant variability in these tests. This is supported by glasshouse measurements of the variability of leaf numbers, plant height, and final dry weight of 30 untreated individual plants of *Sinapis alba*, *Zea mays*, and *Avena fatua* [8].

Results

The glasshouse bioassay described has been used by ICI Agrochemicals to indicate pre- and postemergence effects on terrestrial nontarget plants of a number of ICI herbicides:

glyphosate-trimesium (N-phosphonomethylglycine trimethylsulphonium salt); acetochlor (2-chloro-N-ethoxymethyl-6'-athylacet-o-toluidide); tralkoxydim {2-[1-(ethoxyimino)propyl]-3-hydroxy-5-(2,4,6-trimethylphenyl) cyclohex-2-enone}; molinate (S-ethyl hexahydro-1 H-azepine-1-carbothioate); and, napropamide [(RS)—N,N-diethyl-2-(1-naphthyloxy) propion-amide]; at and below maximum field application rates. In this section, some of the results from these tests are given and compared in order to validate the test methods.

Comparison of Assessment Techniques

The data in Table 5 show a comparison of the results obtained from the four different assessment techniques (seedling emergence, damage following emergence, growth stage, and dry weight at study termination) for a seedling emergence test with the selective herbicide molinate. The emergence of only one species (*Cyperus rotundus*) was affected up to the maximum application rate. Following emergence, however, seven species exhibited pesticide damage (*Glycine max, Brassica napus, Sida spinosa, Abutilon theophrasti, Triticum aestivum. Avena fatua, and Cyperus rotundus*) as shown by the EC values. The damage sustained resulted in decreased biomass with five of these species, as indicated by the dry weight NOEL values. By looking at the effects on numbers of leaves for the seven damaged species, it can be seen that only the five that showed decreased dry weight also showed a decrease in numbers of leaves. Assessment of symptoms may be used to clarify the estimated EC and NOEL values e.g., malformation, chlorosis, and pigmentation will not lead to a significant decrease in dry weight, whereas stunt, necrosis, and leaf abscission will. In this test, dry weight NOEL values were generally at rates between EC_{25} and EC_{50} values, suggesting that the species in this test could sustain up to 25% damage before suffering a loss of dry weight. In some cases the plant species may counteract quite severe chlorosis and necrosis by producing more side shoots/tillers or leaves; the effects observed would give EC values below the maximum field rate, but this may not be reflected in plant yield determinations. Some effects will be transient with full recovery of a field population expected, and by continuing the test for four weeks, any recovery of affected plants may be observed.

TABLE 5—*Comparison of results[a] from different techniques used to assess preemergence effects of molinate on terrestrial nontarget plants (all values are kg ha⁻¹).*

Species Code[b]	Seedling Emergence NOEL[c]	Damage[d]			Dry Weight NOEL	Growth Stage NOELS	
		EC_{10}	EC_{25}	EC_{50}		Leaves	Shoots[e]
GLXMA	>4.5	0.28	1.0	3.8	2.2	2.2	...
BEAVA	>4.5	>4.5	>4.5	>4.5	>4.5	>4.5	...
BRSNN	>4.5	0.22	3.5	>4.5	>4.5	>4.5	...
SIDSP	>4.5	0.06	0.23	0.87	0.56	2.2	...
ABUTH	>4.5	0.06	0.14	0.31	0.14	0.14	...
SINAL	Very poor germination in treated and control units						
ZEAMX	>4.5	>4.5	>4.5	>4.5	>4.5	>4.5	...
TRZAW	>4.5	0.15	>4.5	>4.5	>4.5	>4.5	>4.5
AVEFA	>4.5	0.26	0.46	0.81	0.56	0.56	0.56
CYPRO	0.56	0.18	0.47	1.3	0.56	0.56	...

[a] Seedling emergence, damage, and dry weight statistically analysed, growth stage from graphs.
[b] See Table 1 for species codes.
[c] NOEL = no observable effect level.
[d] EC = effective concentration, causing 10, 25, and 50% damage.
[e] Shoots = side-shoots/branches on main stem (dicotyledons) or tillers (monocotyledons).

The data presented in Table 6 compare damage, growth stage, and dry weight assessments following postemergence treatment with the selective preemergence herbicide napropamide. In this test the dry weight NOEL values are very similar to the damage EC_{10} values, indicating that all effects of the pesticide lead to decreased biomass as measured by dry weight. Different species were, however, affected differently. For example, affected soybean (GLXMA) plants did not show a reduction in number of leaves or side shoots, indicating that overall plant stunt was more important; whereas for wheat (TRZAW) the rate for damage EC_{10}, dry weight NOEL value, leaf reduction NOEL value, and tiller reduction NOEL value were the same 0.5 kg ha^{-1}.

Comparison of Effects on Different Plant Species

For each pesticide, "selectivity" can be demonstrated by producing a sensitivity series of the species tested. Dose-response curves may be compared as shown in Fig. 1. This graph demonstrates the "selectivity" of acetochlor, a broad-spectrum preemergence herbicide applied in a seedling emergence test. The dose-response curve for preemergence effects (% damage) on *Sida spinosa* is compared with the lack of dose-response exhibited by *Zea mays* at the same rates. EC values calculated for *S. spinosa*, as shown in Fig. 1, show the intrinsic toxicity of the herbicide to this species; maize, however, was unaffected up to the highest application rate (EC_{10} > maximum expected field rate).

Comparison of Effects of Different Pesticides

Differential sensitivity to two herbicides, glyphosate-trimesium and tralkoxydim, is shown in Fig. 2. Sugar beet exhibits dose-response effects to the broad-spectrum herbicide glyphosate-trimesium, but is unaffected by tralkoxydim, a graminicide applied up to maximum field rate in a vegetative vigor test.

TABLE 6—*Comparison of results[a] from different techniques used to assess postemergence effects of napropamide on terrestrial nontarget plants (all values are kg ha^{-1}).*

Species Code[b]	Damage[c]			Dry Weight NOEL[d]	Growth Stage NOELs	
	EC_{10}	EC_{25}	EC_{50}		Leaves	Shoots[e]
GLXMA	1.2	>4.5	>4.5	1.5	>4.5	>4.5
BEAVA	0.32	3.4	>4.5	1.5	1.5	...
BRSNN	>4.5	>4.5	>4.5	>4.5	>4.5	...
IPOHE	>4.5	>4.5	>4.5	>4.5	>4.5	1.5
ABUTH	0.31	2.0	>4.5	0.17	>4.5	...
XANST	1.7	>4.5	>4.5	1.5	1.5	1.5
ZEAMX	0.56	1.2	2.4	0.5	0.5	...
TRZAW	0.52	0.98	1.8	0.5	0.5	0.5
AVEFA	0.6	1.0	1.8	0.5	1.5	1.5
CYPRO	0.56	1.9	>4.5	0.5	>4.5	1.5

[a] Damage and dry weight statistically analyzed, growth stage from graphs.
[b] See Table 1 for species codes.
[c] EC = effective concentration, causing 10, 25, and 50% damage.
[d] NOEL = no observable effect level.
[e] Shoots = side-shoots/branches on main stem (dicotyledons) or tillers (monocotyledons).

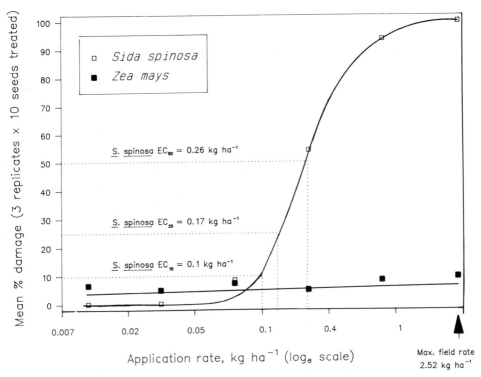

FIG. 1—*Comparing the effects (percentage damage) of a nonselective herbicide, acetochlor, on two plant species included in the seedling emergence test.*

Comparison of Pre- and Postemergence Effects of the Same Pesticide

Herbicides often vary in pre- compared with postemergence activity. In Fig. 3, soybean is unaffected (% damage) by preemergence applications of glyphosate-trimesium in the seedling emergence test, but is damaged by postemergence applications in the vegetative vigor test at rates below the maximum field rate.

Discussion

The most widely used approach to ecological risk assessment is first to consider the toxicological hazard (intrinsic toxicity) of the material in question, then to relate this to the environmental exposure [9–10]. This is consistent with Paracelsus' (1493-1541) early observations that "the dose maketh the poison" [11]. The intrinsic toxicity can be determined using the glasshouse bioassay described here; however, before relating this to environmental exposure, chemical activity under glasshouse and field conditions must be compared.

It is generally accepted that there are differences in the activity exhibited by a given herbicide under glasshouse compared to field conditions; most frequently this is a reduction in the level of effects in the field. Garrod [12] describes the concept of a "transfer factor" in herbicidal activity between glasshouse and field trials. He found that glasshouse-grown weeds were 10 to 20 times more sensitive to preemergence applications of a preemergence graminicide than weeds in the field. However, effects of a postemergence, broad-leaved

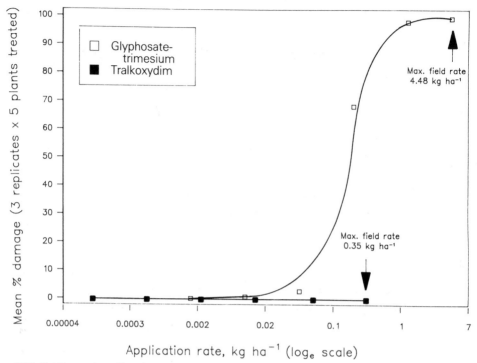

FIG. 2—*Comparing effects (percentage damage) on sugar beet of a nonselective herbicide, glyphosate-trimesium, and a graminicide, tralkoxydim, in the vegetative vigor test.*

weed herbicide were similar or four times as great on glasshouse grown weeds. Fletcher, Johnson, and McFarlane used an extensive database (PHYTOTOX) [13] to show a magnitude of variability of two between glasshouse and field grown plants. There are a number of factors affecting the differences observed; interactions occur between environmental conditions, chemical behavior, and plant physiology [12,14].

It is important to be aware of the potential differences between glasshouse and field data. However, since the "transfer factor" is known to be highly variable, it is not possible to give a single value to be used whenever interpreting glasshouse data. The use of glasshouse results will, however, give an additional "safety margin" or represent "worst case" when estimating the ecological risk in the field.

There have been a number of field experiments in which spray drift and deposition from field applications of herbicides under normal agricultural conditions were measured [2,4,15,16]. Data from these trials may be used to predict the decrease in deposition of the pesticide in question with distance from the target crop. By modeling these results, data from the glasshouse bioassay can be related to distance from the target crop at which these levels of pesticide will be found and be used to produce predictions of effects in the field.

There have also been a number of field experiments carried out relating plant herbicidal damage to distance downwind from an area sprayed using normal agricultural practice. For example, Marrs et al. [4] calculated "safe distances" for various parameters while assessing the effects of spray drift of several herbicides on species of conservation interest. Their "safe distance" calculations were for lethal effects, damage effects, and suppression of flowering. Fifteen wild plants of varying sensitivities were included in the experiment. It was concluded

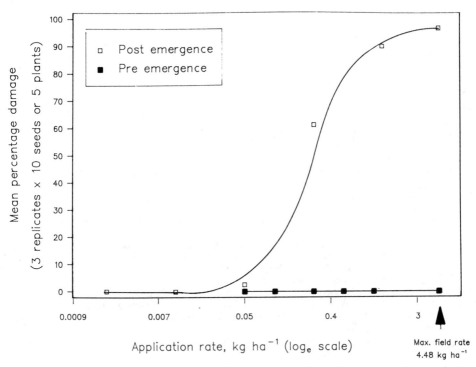

FIG. 3—*Comparing pre- and postemergence effects (percentage damage) on soybean of a nonselective, postemergence herbicide, glyphosate-trimesium.*

that a "buffer zone" of 5 to 10 m between the crop edge and the conservation area would reduce lethal effects from tractor-mounted ground sprayers to almost zero. Nordby and Skuterud [2], in a similar field experiment, compared the effects of drifting aminotriazole herbicide at varying wind speeds, boom height, and spray pressure on barley, a crop plant known to be highly sensitive to this material. Effects of a 10% reduction in green plant mass were observed up to 10 m downwind, while spraying with a boom height of 40 cm, spray pressure of 2.5 bars, and in light wind conditions of 1.5 m s^{-1}.

Such field data may be compared with predicted "safe distances" to validate the quicker and more simplified approach to determining ecological risk by relating glasshouse data directly to exposure (predicted from models or measured in the field). This process could form an important first step in determining the need for progression beyond Tier II nontarget plant testing.

Conclusions

The glasshouse and track-sprayer techniques described here provide a simple, reliable, and repeatable bioassay to aid the assessment of the ecological risk of pesticides to nontarget plants. The studies are sensitive to a wide range of effects. The data can be readily analyzed to characterize the intrinsic toxicity of a pesticide. Interpretation of the field situation by relating intrinsic toxicity data to environmental exposure may be carried out and validated by comparisons to existing field data. This technique could be readily used to carry out pesticide ecological risk assessments from glasshouse data.

Acknowledgments

The authors would like to thank Lorraine Canning, Kim Travis, Malcolm Moore, Jenni Shaw, Peter Stratton, and Pauline Cory for their assistance.

References

[*1*] "Spray Drift Damage to Crops," leaflet on agriculture, fisheries, and food, Her Majesty's Stationery Office, London, 1958.
[2] Nordby, A. and Skuterud, R., *Weed Research,* Vol. 14, 1975, pp. 385–395.
[*3*] Marshall, E. J. P. and Birnie, J. E., *Proceedings,* 1985 British Crop Protection Conference—Weeds, published by Thornton Heath, 1985, pp. 1021–1028.
[*4*] Marrs, R. H., Williams, C. T., Frost, A. J., and Plant, R. A., *Environmental Pollution,* Vol. 59, 1989, pp. 71–86.
[*5*] Schumaker, W. in *Field Margins,* BCPC Monograph No. 35, J. M. Way and P. W. Greig-Smith, Eds., Thornton Heath, UK, 1987, pp. 109–112.
[*6*] Sotherton, N. W., Boatman, N. D., and Rands, M. R. W., *The Entomologist,* Vol. 108, 1989, pp. 135–143.
[*7*] Federal Insecticide, Fungicide and Rodenticide Act (FIFRA), code of federal regulations for protection of the environment—data requirements for registration, revised as of July 1, 1986, *Federal Register,* Vol. 158, No. 40, 1986, pp. 44–92.
[*8*] ICI Agrochemicals unpublished data, 1990.
[*9*] Johnson, E. L., *The American Statistician,* Vol. 36, August 1982, pp. 232–239.
[*10*] Urban, D. J. and Cook, N. J., "Hazard Evaluation Division Standard Evaluation Procedure Ecological Risk Assessment," EPA-540/9-85-001, USEPA, Office of Pesticide Programs, Washington, DC, 1986, p. 2.
[*11*] Brown, R. A. in *Pesticides and Non-target Invertebrates,* P. C. Jepson, Ed., Intercept, Wimborne, Dorset, England, 1989, Chapter 2, pp. 19–42.
[*12*] Garrod, J. F., *Aspects of Applied Biology,* Vol. 21, 1989, pp. 51–64.
[*13*] Fletcher, J. S., Johnson, F. L., and McFarlane, J. C., *Environmental Toxicology and Chemistry,* Vol. 9, 1990, pp. 769–776.
[*14*] Davies, W. J. and Blackman P. G., *Aspects of Applied Biology,* Vol. 21, 1989, pp. 1–12.
[*15*] Yates, W. E., Akesson, N. B., and Bayer, D. E, *Weed Science,* Vol. 26, No. 6, November 1978, pp. 597–604.
[*16*] Byass, J. B. and Lake, J. R., *Pesticide Science,* Vol. 8, 1977, pp. 117–126.

Xiao-Dong Huang,[1] D. George Dixon,[1] and Bruce M. Greenberg[1]

Photoinduced Toxicity of Polycyclic Aromatic Hydrocarbons to the Higher Plant *Lemna Gibba* L. G-3

REFERENCE: Huang, X.-D., Dixon, D. G., and Greenberg, B. M., **"Photoinduced Toxicity of Polycyclic Aromatic Hydrocarbons to the Higher Plant *Lemna Gibba* L. G-3,"** *Plants for Toxicity Assessment: Second Volume, ASTM STP 1115,* J. W. Gorsuch, W. R. Lower, W. Wang, and M. A. Lewis, Eds., American Society for Testing and Materials, Philadelphia, 1991, pp. 209–216.

ABSTRACT: Polycyclic aromatic hydrocarbons (PAHs) are known to exhibit photoinduced toxicity to animals and microorganisms, especially in the presence of ultraviolet (UV) irradiation. Since higher plants readily assimilate PAHs from the environment, and since they cannot avoid UV radiation (a natural component of solar radiation), PAHs have the potential to be a potent phytotoxicant. We have begun to assess the photoinduced toxicity of two PAHs (anthracene and phenanthrene) to the higher plant *Lemna gibba*. In the presence of low-intensity visible light (60 μmol m^{-2} s^{-1}, 400 nm to 700 nm), anthracene and phenanthrene were found to be essentially nontoxic. However, both compounds were highly toxic to *Lemna* in low levels of simulated solar radiation [visible light, 30 μmol m^{-2} s^{-1}; UV$_A$ (320 nm to 400 nm), 4 μmol m^{-2} s^{-1}; UV$_B$ (280 nm to 320 nm), 0.6 μmol m^{-2} s^{-1}]. Furthermore, in the absence of plant tissue, anthracene and phenanthrene were chemically modified by simulated solar radiation. These photomodified PAHs exhibited toxicity in visible light and increased toxicity in simulated solar radiation.

KEY WORDS: anthracene, phenanthrene, PAH, organic pollutant, phototoxicity, phytotoxicity, UV radiation, solar radiation, aquatic toxicology, duckweed

Polycyclic aromatic hydrocarbons (PAHs) are a class of compounds consisting of two or more fused benzene rings [1,2]. They are known to be both mutagenic and toxic to animals and microorganisms [1,3]. PAHs are ubiquitous in nature, as indicated by their detection in sediments, soils, air, surface waters, and biota. The compounds are introduced into the atmosphere mostly by combustion and/or refining of fossil fuels. Aquatic environments typically receive PAHs from accidental releases of petroleum products, refining of fossil fuels, sewage effluents, and/or aerial deposition [1,4]. Because PAHs are very hydrophobic, they readily partition from air or water into the cuticular layers and lipoprotein membranes of plants [1,4,5]. Indeed, PAHs have been detected throughout the plant kingdom, primarily as a result of assimilation from atmospheric and aquatic environments [6,7]. Despite this, there is only limited information on PAH toxicity to higher plants [6].

PAHs are readily modified to species that are more toxic and mutagenic than the parent compounds [2,8–11]. For example, PAHs are oxidized by cytochrome P-450 to more potent toxicants [9–11]. PAHs strongly absorb UV radiation (200 to 400 nm) and are prone to light-induced chemical modification [2,12–14]. Oxidation is the most common photochemical reaction of PAHs; quinones, diols, and epoxides are the resultant products [2,12].

[1] Visiting scientist, associate professor, and assistant professor, respectively, Department of Biology, University of Waterloo, Waterloo, Ontario, Canada, N2L 3G1.

PAHs can also act as photosensitizers [12,15,16]. After absorbance of a photon they are elevated to an excited singlet state, which can degenerate to an excited triplet state. PAH triplet states have enough energy and sufficient half-lives to react with ground-state triplet oxygen [16], forming activated singlet oxygen and ground-state PAH. Active oxygen is recognized as a major source of damage to living systems [17]. Since intact ground-state PAH is recovered from the reaction, this cycle will continue until the PAH is degraded.

Not surprisingly, light (particularly UV radiation) has been found to significantly enhance the toxicity of PAHs to animals, algae, and lower plants [8,15,16,18–20]. However, photoinduced toxicity of PAHs to higher plants has not been systematically addressed. Because plants cannot avoid the UV component of solar radiation, and since they cannot escape contact with PAHs, they may prove to be very susceptible to the chemicals. This research was undertaken as an initial examination of the photoinduced toxicity of two representative PAHs (anthracene and phenanthrene) to the aquatic higher plant *Lemna gibba*.

Experimental Procedure

Light Conditions for Plant Growth

Prior to chemical treatment, *Lemna gibba* L. G-3 plants were maintained aseptically on half-strength Hutner's medium [21] and grown phototrophically under 50 μmol m^{-2} s^{-1} of continuous cool-white fluorescent light. Note that micromoles in light quantity measurements always refers to micromoles of photons. A mole of photons is an einstein, but the latter is no longer a supported SI unit. The irradiation sources for plant growth during chemical treatments were either visible light [cool-white fluorescent, (400 nm to 700 nm) 60 μmol m^{-2} s^{-1}] or simulated solar radiation (SSR) [cool-white fluorescent light, 30 μmol m^{-2} s^{-1}; UV$_A$, (320 nm to 400 nm) 4 μmol m^{-2} s^{-1}; UV$_B$, (280 nm to 320 nm) 0.6 μmol m^{-2} s^{-1}]. The UV radiation was provided by 350 nm (UV$_A$) and 300 nm (UV$_B$) photoreactor bulbs (Rayonet). The spectral output of the SSR source (Fig. 1) was measured with a radiometer (Photodyne) using interference filters to isolate specific wavelengths of radiation.

Plant Growth Rates

For lemnacea plants, the number of leaves is proportional to the fresh weight [23]. Therefore, the doubling frequency, as determined by counting leaves, was used as a quantitative measure of growth rates. The doubling rate of the plants was approximately 48 h in visible light and 96 h in SSR.

Exposure of Plants to PAHs

Plants were placed on 10 mL of fresh half-strength Hutner's medium in a 5-cm petri dish. Phenanthrene or anthracene (Sigma) were dissolved to 2 mg/mL in dimethyl sulfoxide (DMSO) and delivered to the growth medium. Except where noted, the final PAH concentration was 2 μg/mL. The DMSO carrier allowed supersaturation of the medium with the chemicals. DMSO (final concentration, 0.1% [v/v]) had no effect on *Lemna* growth and was present in the medium for all control experiments. The PAH-containing medium was replaced every 48 h to keep nutrient and chemical levels approximately constant. For some experiments, the chemicals were pretreated in SSR or darkness for 24 h in growth medium prior to incubation with plant tissue. Growth of the plants in the PAH-containing medium was monitored by counting the number of leaves every second day over a 14-day interval. A doubling frequency was derived from the raw data to produce a growth rate [23].

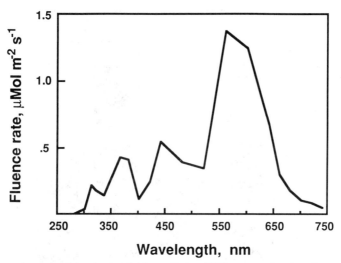

Wavelength, nm

FIG. 1—*Spectral output of the simulated solar radiation (SSR) source. Combined emission spectrum of one cool-white fluorescent lamp, one 350-nm photoreactor bulb, and one 300-nm photoreactor bulb filtered through a standard plastic petri dish top. The plants were grown in covered plastic petri dishes. The radiation from the 300-nm lamp was filtered through two layers of cheese cloth. This spectrum approximates sunlight at the earth's surface with respect to the amounts of UV_B and UV_A radiation relative to visible light [see Ref 29].*

Absorption Spectra

After a 24 h incubation in darkness or SSR, anthracene and phenanthrene (2 µg/mL) were extracted from 10 mL of growth medium into 1 mL of chloroform. Absorption spectra were measured with a Perkin-Elmer Lambda-3 spectrophotometer.

Results

Photoinduced Toxicity of Anthracene and Phenanthrene to Plants

Two PAHs were chosen for toxicity testing with *Lemna*, one known to show strong photoinduced toxicity to fish (anthracene) and the other reported to have mild photoinduced toxicity towards fish (phenanthrene) [15]. In low-intensity visible light, the growth rate of *Lemna* in the presence of either compound was found to be on par with the control plants (Fig. 2). However, in low levels of SSR, anthracene and phenanthrene both inhibited growth of *Lemna* (Fig. 2). Interestingly, inhibition of growth of *Lemna* by anthracene in SSR was not significantly different from that of phenanthrene, which contrasts with animal systems where phenanthrene has been found to show little photoinduced toxicity [cf. 15,16].

Concentration Dependence of Anthracene Toxicity

The concentration (2 µg/mL) of anthracene or phenanthrene used above was relatively high in relation to PAH loads in contaminated aquatic environments [1,4]. To determine the threshold response of *Lemna* to anthracene in SSR, a dose-response curve was constructed (Fig. 3). A typical log-linear response was detected for chemical concentration versus toxicity. Anthracene significantly inhibited growth of *Lemna* at a level of 0.2 µg/mL. Concentrations in this range have been found in aquatic environments [1,4]. As well,

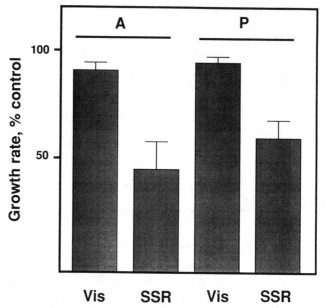

FIG. 2—*Growth of* Lemna gibba *plants in the presence of PAHs. Plants were cultured in anthracene* (2 μg/mL) (A) *or phenanthrene* (2 μg/mL) (P) *under visible light* (Vis) *for 7 days or under SSR for 14 days. Growth rates were monitored by counting the number of fronds. Histograms represent relative growth rates of treated plants compared to control plants grown under identical conditions in the absence of PAHs. Growth rates were determined as a doubling frequency. Error bars are standard error of the mean* (n = 3 *to 5 repeats with 3 to 6 plants per repeat*).

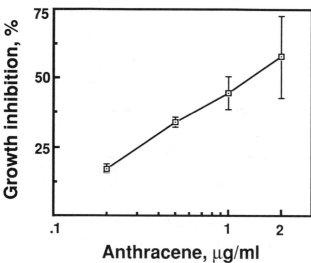

FIG. 3—*Dose response curve for anthracene toxicity to* Lemna gibba. *Plants were cultured in the presence of different concentrations of anthracene. Plants were grown in SSR as described in Fig. 2. Percent inhibition of growth due to anthracene is relative to the growth rate of control plants cultured in the absence of anthracene. Growth rates were determined as a doubling frequency. Error bars are standard error of the mean* (n = 3 *to 5 repeats with 3 to 6 plants per repeat*).

one can estimate the maximal PAH content in the tissue by assuming 100% absorption of the chemical from the medium by the plants. Thus, 10 mL of a 0.2 μg/mL PAH solution combined with 50 mg of plants would yield a maximal tissue concentration of 0.04 μg/mg, which is comparable to amounts found in plants grown in contaminated environments [6].

Toxicity of Photomodified PAHs to Lemna

Some modified PAHs are known to be more toxic to animals than the parent compounds [9–11]. Since anthracene and phenanthrene are most likely photomodified in SSR, it is important to determine whether the intact compounds, the photomodified compounds, or both are the active phytotoxicants. Both anthracene and phenanthrene were found to undergo photochemical modification under SSR. This was readily observed as changes in the absorbance spectra of the compounds and occurred in growth medium without plant material present. For example, after a 24-h exposure to SSR, there was a dramatic alteration in the absorbance properties of anthracene (Fig. 4). The half-life for this conversion was 4 h (data not shown). Photoconversion of phenanthrene was slower than photomodification of anthracene (data not shown). We have not yet determined the structure of the photomodified PAHs, but it is thought that the intact compounds were most likely oxidized [see 2,12].

Photomodified anthracene and phenanthrene were tested for their toxicity to *Lemna*. The chemicals were pretreated for 24 h in SSR, and the plants were subsequently grown in the presence of these modified chemicals under visible light and SSR (Fig. 5). Phototransformation of anthracene and phenanthrene rendered the chemicals toxic in visible light. In fact, the toxicity of phototransformed anthracene in visible light was comparable to the nontreated chemical in SSR. Moreover, the photomodified compounds had dramatically

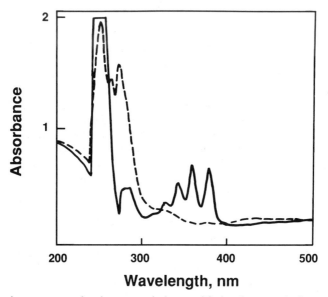

Wavelength, nm

FIG. 4—*Absorbance spectra of anthracene and photomodified anthracene. Anthracene (2 μg/mL in growth medium) was incubated for 24 h in darkness (———) or in SSR (- - - - -). The chemical was extracted from 10 mL growth medium into 1 mL of chloroform and scanned with a UV/vis spectrophotometer.*

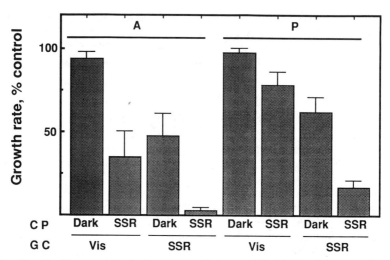

FIG. 5—*Growth of* Lemna gibba *in the presence of photomodified PAHs. Prior to application to the plants, anthracene (2 μg/mL in growth medium) (A) and phenanthrene (2 μg/mL in growth medium) (P) were pretreated (CP) for 24 h either in SSR or darkness (dark). The plants were then cultured in the pretreated chemicals. Light conditions for plant growth (GC) were visible light (Vis) or SSR. Histograms represent growth rates relative to control plants (see Fig. 2). Error bars are standard error of the mean (n = 3 to 5 repeats with 3 to 6 plants per repeat).*

higher levels of phytotoxicity in SSR than the unaltered chemicals. Therefore, transformed PAHs are actively toxic to plants.

Discussion

In this study, initial progress was made on evaluating the effects of PAHs on higher plants. We have shown that PAHs are toxic to plants in the presence of UV irradiation and that *Lemna* will be an excellent system to examine the phytotoxicity of PAHs. A major advantage of the *Lemna* system is the speed (one to two weeks) with which the toxicity assays can be performed. There are other examples where *Lemna* has been used successfully for toxicity assessment. The toxicity of herbicides has been measured with *Lemna gibba* [24]. Lockhart et al. [25] examined the bioconcentration of organic molecules with *Lemna minor*. The toxicity of oils and sewage effluents has been evaluated with *Lemna gibba, Lemna minor,* and/or *Lemna perpusilla* [26,27]. Thus, *Lemna* presents itself as a very versatile plant source for the evaluation of both organic and inorganic phytotoxicants.

The concentration of anthracene required for measurable inhibition of plant growth was approximately 200 ng/mL (200 ppb). This level is relevant to terrestrial and aquatic plants. Precipitation can bring large amounts of PAHs out of the atmosphere, both from the gaseous state and bound to particulate matter. Upon contact with plants, the chemicals will readily partition into the cuticular and membrane phases of the plants [1,5,28], which could result in local lesions and cell death. While this alone may not kill an entire plant, it could cause genetic mutations, slow growth, lower crop yields, and/or weaken the plant, thus increasing sensitivity to other forms of stress.

A requirement for anthracene and phenanthrene toxicity to plants is UV radiation. Since the net content of UV radiation present in the SSR source employed (4.6 μmol m^{-2} s^{-1}) was much lower than the total UV fluence in solar radiation currently reaching the surface of the earth (>150 μmol m^{-2}s^{-1} [29]), plants are routinely exposed to sufficient UV radiation to induce PAH toxicity. Furthermore, UV$_B$ radiation, a spectral region strongly absorbed by most PAHs, is increasing in the environment due to depletion of the ozone layer. UV radiation can activate PAHs either before or after contact with plant tissue. Since UV radiation is known to penetrate to the interior of leaves [30], it is possible that PAHs assimilated by plants in an unmodified state can be activated by solar radiation inside the plant. Additionally, PAHs which have been photooxidized exterior to the plant can be assimilated and cause toxicity without further UV irradiation (c.f. Fig. 5). As such, photomodified compounds have the potential to damage plant tissue that is not normally exposed to light (e.g., roots). It is clear from this study that PAHs pose a real and present danger to plant life.

Although photomodified anthracene and phenanthrene both showed enhanced toxicity compared to the unaltered compounds, they each behaved differently relative to their intact form (cf. Fig. 5). Phototransformed anthracene was equally or more toxic in visible light than the nonphotomodified compound in SSR, while intact phenanthrene was more toxic in SSR than the phototransformed chemical in visible light. In SSR, both intact PAHs showed similar levels of toxicity. However, photomodified anthracene lowered the growth rate of *Lemna* in SSR to almost zero, while phototransformed phenanthrene only depressed the growth rate to 17% of the control. Anthracene toxicity, therefore, can be primarily attributed to the photooxidized compound. Conversely, the pattern of toxicity of phenanthrene cannot be explained by a single mechanism. Its toxicity may also derive from its ability to act as a photosensitizer. Both the quantum yield of triplet-state formation and the triplet-state lifetime are greater for phenanthrene than they are for anthracene [16]. This would give excited-state phenanthrene sufficient time to react with triplet oxygen, forming activated singlet oxygen and ground-state phenanthrene. Thus, it appears that PAHs pose a danger to plants both as photooxidized chemicals and as photosensitizers. The interaction and relative contribution of these two modes of toxicity for PAHs are an area of research which we are currently pursuing.

Acknowledgments

We wish to thank Mike Wilson for many helpful discussions. X.-D. Huang was supported by the Chinese Government as a visiting scholar. This work was supported in part by operating grants from the Natural Sciences and Engineering Research Council of Canada to B. M. Greenberg (OGP 43179) and D. G. Dixon (OGP 8511).

References

[1] Cook, R. H., Pierce, R. C., Eaton, P. B., Lao, R. C., Onuska, F. I., Payne, J. F., and Vavasour, E., *Polycyclic Aromatic Hydrocarbons in the Aquatic Environment: Formation, Sources, Fate and Effects on Aquatic Biota*, NRCC 18981, National Research Council of Canada, Ottawa, 1983.

[2] Nikolaou, K., Masclet, P., and Mouvier, G., *Science of the Total Environment*, Vol. 32, 1984, pp. 103–132.

[3] Cooke, M. and Dennis, A. J., *PAHs: Formation, Metabolism and Measurement*, Battelle Press, Columbus, OH, 1983.

[4] Suess, M. J., *Science of the Total Environment*, Vol. 6, 1976, pp. 239–250.

[5] Southworth, G. R., Beauchamp, J. J., and Schmieder, P. K., *Water Research*, Vol. 12, 1978, pp. 973–977.

[6] Edwards, N. T., *Journal of Environmental Quality*, Vol. 12, 1983, pp. 427–441.

[7] Grimmer, G. and Duvel, D., *Zeitschrift fur Naturforschung*, Vol. 25b, 1970, pp. 1171–1175.

[8] Kochevar, I. E., Armstrong, R. B., Einbinder, J., Walther, R. R., and Harber, L. C., *Photochemistry and Photobiology*, Vol. 36, 1982, pp. 65–69.

[9] Yang, S. K., McCourt, D. W., Roller, P. P., and Gelboin, H. V., *Proceedings of the National Academy of Science,* Vol. 73, 1976, pp. 2594–2598.

[10] Harvey, R. G., Cortez, C., Sugiyama, T., Ito, Y., Sawyer, T. W., and DiGiovanni, J., *Journal of Medicinal Chemistry,* Vol. 31, 1988, pp. 154–159.

[11] Shimada, T. and Nakamura, S.-I., *Biochemical Pharmacology,* Vol. 36, 1987, pp. 1979–1987.

[12] Zepp, R. G. and Schlotzhauer, P. in *Polynuclear Aromatic Hydrocarbons,* P. W. Jones and P. Leber, Eds., Ann Arbor Science Publishers, Ann Arbor, MI, 1979, pp. 141–158.

[13] Johnsen, S., Gribbestad, I. S., and Johansen, S., *Science of the Total Environment,* Vol. 81/82, 1989, pp. 231–238.

[14] Sinha, B. K. and Chignell, C. F., *Photochemistry and Photobiology,* Vol. 37, 1983, pp. 33–37.

[15] Newsted, J. L. and Geisy, J. P., *Environmental Toxicology and Chemistry,* Vol. 6, 1987, pp. 445–461.

[16] Morgan, D. D., Warshawsky, D., and Atkinson, T., *Photochemistry and Photobiology,* Vol. 25, 1977, pp. 31–38.

[17] Halliwell, B. and Gutteridge, J. M. C., *Free Radicals in Biology and Medicine,* Clarendon, Oxford, England, 1985.

[18] Blackburn, G. M. and Taussig, P. E., *Biochemical Journal,* Vol. 149, 1975, pp. 289–291.

[19] Schoeny, R., Cody, T., Warshawsky, D., and Radike, M., *Mutation Research,* Vol. 197, 1988, pp. 289–302.

[20] Forrest, V., Cody, T., Caruso, J., and Warshawsky, D., *Chemico-Biological Interactions,* Vol. 72, 1989, pp. 295–307.

[21] Posner, H. B. in *Methods in Developmental Biology,* F. A. Witt and N. K. Wessels, Eds., Crowell, New York, 1967, pp. 301–317.

[22] Landolt, E. and Kandeler, R., *The Family of Lemnaceae,* Vol. 2, Veroffentlichungen Des Geobotanischen Institutes Der Eidg. Tech. Hochschule, Zurich, Switzerland, 1987.

[23] Greenberg, B. M., Huang, X.-D., and Dixon, D. G. in *Proceedings,* International Symposium on Aquatic Ecosystem Health, submitted November 1990.

[24] Hughes, J. S., Alexander, M. M., and Balu, K. in *Aquatic Toxicology and Hazard Assessment (10th volume), ASTM STP 971,* W. J. Adams, G. A. Chapman, and W. G. Landis, Eds., ASTM, Philadelphia, 1988, pp. 531–547.

[25] Lockhart, W. L., Billeck, B. N., de March, B. G. E., and Muir, D. C. G. in *Aquatic Toxicology and Hazard Assessment (6th Symposium), ASTM STP 802,* W. E. Bishop, R. D. Cardwell, and B. B. Heidolph, Eds., ASTM, Philadelphia, 1983, pp. 460–468.

[26] King, J. M. and Coley, K. S. in *Aquatic Toxicology and Hazard Assessment (8th Symposium), ASTM STP 891,* R. C. Bahner and D. J. Hansen, Eds., ASTM, Philadelphia, 1985, pp. 302–309.

[27] Wang, W. and Williams, J. M., *Environmental Toxicology and Chemistry,* Vol. 7, 1988, pp. 645–652.

[28] Edwards, N. T., *Environmental Toxicology and Chemistry,* Vol. 5, 1986, pp. 659–665.

[29] Henderson, S. T. in *Daylight and Its Spectrum,* Adam Hilger Ltd., Bristol, England, 1977.

[30] Caldwell, M. M., Robberecht, R., and Flint, S. D., *Physiologia Plantarum,* Vol. 58, 1983, pp. 445–450.

Frank A. Ireland,[1] *Barbara M. Judy,*[2] *William R. Lower,*[3]
Mark W. Thomas,[4] *Gary F. Krause,*[5] *Amha Asfaw,*[5]
and William W. Sutton[6]

Characterization of Eight Soil Types Using the *Selenastrum Capricornutum* Bioassay

REFERENCE: Ireland, F. A., Judy, B. M., Lower, W. R., Thomas, M. W., Krause, G. F.,
Asfaw, A., and Sutton, W. W., **"Characterization of Eight Soil Types Using the *Selenastrum
Capricornutum* Bioassay,"** *Plants for Toxicity Assessment: Second Volume, ASTM STP 1115,*
J. W. Gorsuch, W. R. Lower, W. Wang, and M. A. Lewis, Eds., American Society for Testing
and Materials, Philadelphia, 1991, pp. 217–229.

ABSTRACT: A recently developed bioassay protocol was used to characterize seven different
soil types and Promix BX. The green alga *Selenastrum capricornutum* was used as the test
organism. A ruggedness test which altered seven variables was conducted to identify procedural
stipulations that should be carefully controlled in the *Selenastrum capricornutum* bioassay.
Stimulatory or inhibitory effects as measured by population growth were compared to a control
using two different nutritional methods termed proportional amendment (PA) and complete
amendment (CA). Population growth, calculated effect (CE) values, and EC_{50} values are given
for each soil. A wide range of effects were documented indicating a substantial difference
among soils with respect to their ability to stimulate or inhibit algal population growth.

KEY WORDS: *Selenastrum capricornutum*, bioassay, proportional amendment, complete
amendment, calculated effect, EC_{50}, ruggedness test, soils

The single cell alga *Selenastrum capricornutum* is currently used extensively for the bioassay of water contamination [1–7]. An extension of this bioassay technology is the application
to problems of contaminated soil and sediment and the testing of their aqueous eluates. A
number of steps are of value before a test or an extension of a test is ready for use. One
important step is the systematic evaluation by an appropriate, well-defined and specific
methodology which determines the operational characteristics of a specific protocol of the
test. One appropriate methodology is found in "Guidelines for Conducting Single Laboratory
Evaluations of Biological Methods," developed for the USEPA [8]. In addition, the single
laboratory evaluation using the guidelines to determine the operational characteristics of a
protocol developed for *S. capricornutum* starting with soil as the original sample material
has been completed by our laboratory [9]. An aspect of this first step is the determination

[1] Research animal scientist, Department of Animal Sciences, Dixon Springs Agricultural Center,
University of Illinois, Simpson, IL 62985.
[2] Research specialist, Department of Veterinary Biomedical Sciences, University of Missouri, Columbia, MO 65201.
[3] Group leader, Bioenvironmental Section, Environmental Trace Substances Research Center, University of Missouri, Columbia, MO 65203.
[4] Vegetation specialist, American Cyanamid Company, Birmingham, AL 35216.
[5] Professor and instructor, respectively, Department of Statistics, University of Missouri, Columbia,
MO 65203.
[6] Senior biologist, Environmental Research Laboratory, U.S. Environmental Protection Agency,
Athens, GA 30613.

of those procedural variables that must be carefully controlled. If the test procedure is "rugged" it will not be susceptible to the inevitable, modest departures in routine that occur. This determination is called a ruggedness test. Another step is to evaluate methods of nutritional amendment when more than one method is available. Another step is to establish a body of information on the effects on algal growth of eluates from soils not contaminated with xenobiotics. Considerable variation of algal growth might be expected with eluates of different soils due to different levels of macro and micronutrients.

Materials and Methods

The protocol for testing the toxicity of a soil or sediment aqueous eluate was in part derived from the literature, in part from work done by the Environmental Trace Substances Research Center of the University of Missouri, and in part from a procedure developed at EPA-Corvallis.

The basic protocol is as follows: a quantity of soil (125 g), air dried for 24 h at room temperature, is mixed with 500 mL reverse osmosis-deionized water and agitated in a capped Teflon container at 120 rpm for 48 h at 20°C in the dark. The aqueous soil eluate (SE) is filtered through several filters of decreasing porosity size with a final sterile filtration through a 0.45-μm filter. Two 125-mL flasks are each aliquoted with 50 mL of various amounts of the filtered soil eluate, e.g., 80, 50, 10, and 1% for a total of eight flasks of soil eluate amended with growth medium. The constituents of the growth medium are found in Table 1. The flasks are inoculated with 10^4 cells/mL $\pm$ 10^3 cells/mL and assayed for growth as changes in cell number per mL of eluate with a Coulter counter after incubation for 96 h at 24°C with continuous agitation at 100 rpm and illumination at 400 μE. There are two methods of amending the soil eluate with growth medium which are discussed in this paper. In the proportional nutrient amendment (PA), the growth medium at 100% concentration is added to obtain the four different concentrations of 1 mL SE and 99 mL growth medium, 10 mL SE and 90 mL growth medium, 50 mL SE and 50 mL growth medium, and 80 mL SE and 20 mL growth medium for 1, 10, 50, and 80% SE, respectively. In the complete nutrient amendment (CA), all dilutions of SE, 1, 10, 50, and 80%, are in 100% growth medium.

A variety of soils and Promix BX were tested. The soils were collected in Missouri and Alabama. Soils 1 and 8 came from farms owned by the University of Missouri in eastern Boone County, Missouri. Soil 2 was obtained from a private farm in northwestern Boone County, Missouri. The soils, designated as 3, 4, 6, and 7, were collected in Alabama from Crenshaw, Covington, Escambia, and Coffee counties, respectively.

Promix BX, designated as soil 5, is an artificial potting mixture manufactured by Premier Brands, Inc., and consists of shredded sphagnum peat moss, horticultural vermiculite, perlite, dolomitic limestone, calcium nitrate, phosphorus, potash, phosphate, trace elements, and a wetting agent.

Prior to testing, two different determinations were made of the soils and Promix. Table 2 describes the agricultural soil composition and texture in terms of the sand, silt, and clay content and designates the soils as clay loam, loam, etc. Table 3 describes the physical and chemical composition of the soils.

Soil 8, a clay loam, was used for the ruggedness test of the single laboratory evaluation.

The effect of SE and level of growth medium on growth of *S. capricornutum* is determined as the calculated effect (CE). There are two similar equations, one for the proportional nutrient amendment and one for the complete nutrient amendment. The equation of the calculated effect (CE) for proportional amendment (PA) at each concentration of soil eluate is:

$$CE_{(PA)} = \frac{(SM - IN) - P(NC - IN)}{P(NC - IN)} \times 100 \tag{1}$$

TABLE 1—*Composition of algal growth medium.*

Stock Solution[a]		Nutrient Composition Prepared Medium	
Compound	Concentration, g/L	Element	Concentration, mg/L
MACRONUTRIENTS			
Stock Solution I			
NaNO$_3$	25.500	N	4.200
NaHCO$_3$	15.000	Na	11.001
		C	2.143
Stock Solution II			
K$_2$HPO$_4$	1.044	K	0.469
		P	0.186
Stock Solution III			
MgSO$_4$ · 7H$_2$O	14.700	S	1.911
Stock Solution IV			
MgCl$_2$ · 6H$_2$O	12.164	Mg[b]	2.904
CaCl$_2$ · 2H$_2$O	4.410	Ca	1.202
MICRONUTRIENTS			
Stock Solution V			
H$_3$BO$_3$	185.520	B	32.460
MnCl$_2$ · 4H$_2$O	415.610	Mn	115.374
ZnCl$_2$	3.271	Zn	1.570
CoCl$_2$ · 6H$_2$O[c]	1.428	Co	0.354
CuCl$_2$ · 2H$_2$O[c]	0.012	Cu	0.004
Na$_2$MoO$_4$ · 2H$_2$O[c]	7.260	Mo	2.878
FeCl$_3$ · 6H$_2$O	160.000	Fe	33.051
Na$_2$EDTA · 2H$_2$O	150.000	···	···

NOTE: The growth medium is prepared by adding 1.0 mL of each of the five stock solutions to reverse osmosis-deionized sterile water and brought to 1 L volume.
[a] Includes Na from NaNO$_3$.
[b] Includes Mg from MgSO$_4$ · 7H$_2$O.
[c] Make up in 100-fold or more concentrations and add each in appropriate amount to stock solution V.

where

SM = cells/mL at each dilution of soil eluate,
IN = cells/mL of the inoculum,
NC = cells/mL of the negative control (growth medium), and
P = the proportion of growth medium. P takes the values of 0.20 with 80% soil eluate, 0.50 with 50% soil eluate, 0.90 with 10% soil eluate and 0.99 with 1% soil eluate.

The equation of calculated effect for complete amendment (CA) is:

$$CE_{(CA)} = \frac{(SM - IN) - (NC - IN)}{(NC - IN)} \times 100 \qquad (2)$$

TABLE 2—*Composition and texture of seven soils.*

Soil	Sand, %	Silt, %	Clay, %	Texture
1	51.6	28.8	16.6	Loam
2	27.6	56.4	16.0	Silt loam
3	76.4	12.8	10.8	Sandy loam
4	24.4	32.8	42.8	Clay
6	52.4	33.6	18.8	Loam
7	36.4	14.8	48.8	Clay
8	21.4	50.7	27.9	Clay loam

where

SM = cells/mL at each concentration of soil eluate,
IN = cells/mL of inoculum, and
NC = cells/mL of negative control (growth medium).

The value P is not used because the growth medium is at 100% at all concentrations of soil eluate. A positive calculated effect value indicates a stimulation of cell growth, and a negative calculated effect value indicates an inhibition of cell growth.

The calculation of the EC_{50} (EC_{50} as percent) values by regression uses the CE value as Y, the dependent variable, and the concentration of soil eluate, X, as the independent variable. The value EC_{50} means the experimentally derived chemical concentration that is calculated to affect 50% of the test criterion. In this presentation, the EC_{50} is calculated by regression and the test criterion is cell number. An EC_{50} value between 0 and 100% indicates

TABLE 3—*Soil test results for eight soils.*

	Soil							
	1	2	3	4	5	6	7	8
pH	7.5	6.4	4.7	7.4	5.8	4.8	4.3	5.5
NA[a]	0.0	0.5	3.0	0.0	1.5	13.0	5.0	3.5
OM, %[b]	2.5	3.1	1.4	0.6	4.9	1.7	2.9	4.6
P, ppm	5.7	65.5	4.0	1.5	6.8	1.5	14.0	1.1
Na, ppm	117.6	117.5	120.0	144.0	9.5	134.5	124.0	9.4
Ca, ppm	3303.0	1005.0	325.0	4195.0	119.0	875.0	155.0	185.3
Mg, ppm	81.3	123.5	48.0	151.0	11.8	570.5	21.5	27.4
N, ppm	25.0	31.0	4.0	6.0	49.0	17.0	29.0	45.5
K, ppm	137.1	195.5	66.5	25.5	14.1	134.0	36.5	20.9
Zn, ppm	4.2	8.1	0.1	0.3	2.7	0.0	0.8	7.1
Fe, ppm	8.4	36.6	14.4	3.6	25.0	7.8	22.3	70.5
Mn, ppm	1.1	8.0	4.3	0.9	8.8	1.2	6.3	65.6
Cu, ppm	0.6	0.7	0.1	0.1	0.6	0.03	0.2	0.9
SO₄-S, ppm	46.0	8.3	3.5	68.3	50.5	9.0	20.0	7.2
CEC[c]	⋯	⋯	⋯	⋯	8.8	⋯	⋯	⋯
EC[d]	⋯	0.3	0.1	0.3	1.3	0.1	0.1	15.6

NOTE: Nitrogen concentrations are estimates based on % organic matter.
[a] Neutralizable acidity, meq/100 g.
[b] Organic matter.
[c] Cation Exchange Capacity, meq/100 g.
[d] Electrical conductivity, mmho/cm.

some degree of inhibition of growth, with a lower value indicating greater toxicity of the SE and a value close to 100% indicating essentially no toxicity. If the CE values are positive, the EC_{50} values should not be considered or calculated. For the purpose of this study (comparison of different soils without any toxicant) a positive CE value, especially with higher concentrations of SE, indicated a stimulatory effect, and the EC_{50} was arbitrarily set at $+100.0\%$. Also, if the CE value shows slight inhibition (low negative value), the EC_{50} is greater than 100% and indicates a benign soil eluate, but not one which stimulates cell growth. In this case the EC_{50} value was also set to $+100.0\%$. This allows use of a range of EC_{50} values from 0 to 100%, and direct comparison and averaging of results could then be made.

Statistical Analysis

Three separate tests, each in duplicate flasks for a total of six flasks, were run for each soil. *S. capricornutum* population growth expressed in cells $\times 10^6/mL$ was normalized with a log transformation. These log transformed means of the three cell counts per flask were statistically analyzed using analysis of variance (ANOVA). A protected least significant difference (LSD) was computed at $p < 0.05$. The variation was documented based on an ANOVA of log transformed cell count means for the interactions of soils, nutritional amendment methods, and dilutions. A comparison was made against the control as well as among treatments.

Results and Discussion

There are three aspects of this study: first, a ruggedness test to determine the susceptibility to the inevitable, modest departures in routine that occur; second, the effect of the two methods of nutritional amendment on alga population growth and EC_{50} values as percent; and third, the effect of the eluates of different soils on cell growth.

Ruggedness

An important step in the development of a protocol is to identify those procedural variables that must be carefully controlled. A procedure called the "ruggedness test" was developed for this purpose. The ruggedness test is part of "Guidelines for Conducting Single Laboratory Evaluations of Biological Methods," developed by W. D. McKenzie and T. A. Olsson for the USEPA [8]. If the test procedure is "rugged" it will not be susceptible to the inevitable, modest departures in routine that occur, and the final test result will not be altered by these slight variations. If the results are altered by small procedural variations, it is important to emphasize in the protocol that a specific step must be strictly followed or, in some cases, to indicate the limits of allowable variability. The experimental design used during the ruggedness test, based on the design by W. J. Youden, does not seek to study each separate variable in an individual sequential fashion, but rather to provide for the introduction of a number of changes (protocol variations) simultaneously [10]. A single ruggedness test was run using the proportional nutritional amendment and three concentrations of 80, 10, and 1% soil eluate (Tables 4 and 5).

Eight complete assays were conducted as part of the single ruggedness evaluation. In this study each test sample contained 100 mg of the organic compound sodium dodecyl sulfate that had been added to the 125-g aliquot soil samples immediately prior to the test. The "protocol directed" conditions are designated as *A* through *G*, and the "varied conditions"

TABLE 4—*Experimental design for a seven-variable ruggedness test.*

Assay Determination Number	Variables	Assay Result
1	A B C D E F G	s
2	A B c D e f g	t
3	A b C d E f g	u
4	A b c d e F G	v
5	a B C d e F g	w
6	a B c d E f G	x
7	a b C D e f G	y
8	a b c D E F g	z

were designated as *a* through *g*. The evaluation is concerned with identifying respective variations in the final test result due to the specific procedural differences, i.e., *A-a*, *B-b*, *C-c*, *D-d*, *E-e*, *F-f*, and *G-g*. Each of the eight determinations consisted of a single assay conducted using eight respective aliquots of a single test material (in this study 100 mg of sodium dodecyl sulfate). The final test results are indicated as *s*, *t*, *u*, *v*, *w*, *x*, *y*, and *z*.

The average of $A = (s + t + u + v)/4$ compared with the average of $a = (w + x + y + z)/4$ serves as a rapid means of assessing the effect of changing variable *A* to *a*. Because each of the two groups of four determinations contains the other six variables (twice at the upper case level and twice at the lower case level), the effect of these variables (if present) tends to cancel out, leaving only the effect of changing variable *A* to *a*. The relative effect of the other variables was estimated by examining the following averages.

Only a single ruggedness test was run here, although a laboratory may wish to conduct multiple complete ruggedness tests, for example ten successive independent complete analyses for each of the ruggedness variables chosen. Under these conditions, a standard deviation from the ten analyses may be used to evaluate the differences between any two variables. For example, if the difference between *A-a* is greater than two times the standard deviation, the testing laboratory will have an indication that the particular variable is affecting the test result. A set of ten ruggedness tests can be very expensive and was outside of the scope of the research reported on here. With only one ruggedness test, the values with the greatest difference between the protocol-directed procedure and its variation (e.g., *A-a*) was used as an indication of an important protocol procedure that must be adhered to.

Most of the modest procedural alterations should typically have little or no effect on the test result. However, a comparison of the respective differences (*A-a*; *B-b*; etc.) provides information on which variables, if any, have the greater effects.

Moderate procedural variations definitely have altered the final algal result when the *y* test result occurs at certain critical steps. Steps that apparently must be followed strictly as written (variables *D*, *E*, and *F*) concern the amount of water added to the dry soil, the pH of the resulting soil eluate that is actually used during the assay, and the specific eluate concentrations (dilutions) prepared for the assay dilution series. The protocol consequently was revised to emphasize strict adherence to these critical instructions.

Nutritional Amendment Methods Usability for Soil Toxicity Evaluation

S. capricornutum population growth at the end of the 96-h test, expressed in cells $\times 10^6$ mL^{-1}, was normalized with a log transformation (Table 6). The population growth in seven

TABLE 5—*Ruggedness variables selected for* S. capricornutum *bioassay.*

Protocol-Directed Instructions	Altered Instructions, Variable
A. 125 g of soil as sample material	a. 110 g of soil as sample material
B. 1 × 10⁴ cells/mL inoculum	b. 0.75 × 10⁴ cells/mL inoculum
C. wet soil sample air dried for 24 h at room temperature	c. wet soil sample air dried for 20 h at room temperature
D. 500 mL of water added to air-dried soil sample	d. 550 mL of water added to air soil sample
E. soil eluate pH 6–10	e. soil eluate pH 10
F. growth medium 20, 90, and 99%	f. growth medium 20, 70, and 99%
G. test of soil eluate for 96 h	g. test of soil eluate 92 h

$$B = \frac{(s + t + w + x)}{4} \qquad b = \frac{(u + v + y + z)}{4}$$

$$C = \frac{(s + u + w + y)}{4} \qquad c = \frac{(t + v + x + z)}{4}$$

$$D = \frac{(s + t + y + z)}{4} \qquad d = \frac{(u + v + w + x)}{4}$$

$$E = \frac{(s + u + x + z)}{4} \qquad e = \frac{(t + v + w + y)}{4}$$

$$F = \frac{(s + v + w + z)}{4} \qquad f = \frac{(t + u + x + y)}{4}$$

$$G = \frac{(s + v + x + y)}{4} \qquad g = \frac{t + u + w + z)}{4}$$

TEST RESULTS USING DIFFERENT COMBINATIONS OF PROTOCOL VARIABLES

Determination	Test Result
s	30.5
t	79.1
u	29.9
v	43.4
w	21.8
x	45.1
y	90.8
z	47.9

RELATIVE EFFECT OF INDIVIDUAL VARIATIONS

$A - a =$	-5.7	$E - e =$	-20.4
$B - b =$	-8.9	$F - f =$	-25.3
$C - c =$	10.6	$G - g =$	7.8
$D - d =$	27.0		

soils, all except soil 1, was less than in the control (growth medium), especially in higher concentration of SE (Tables 6 and 7). The results show an interesting wide range of algal growth response to different soils. The population growth was also expressed as a CE value using the two methods of nutritional amendment (Tables 6 and 7). The calculated effect can be negative or positive. A negative value indicates a decrease in cell number, i.e., an inhibitory effect compared to the control (growth medium). A positive value indicates an

TABLE 6—*Statistical significance of means for log transformed cell counts.*

Soil	Proportional Amendment Method					Complete Amendment Method				
	Control	1% SE	10% SE	50% SE	80% SE	Control	1% SE	10% SE	50% SE	80% SE
1	14.25	14.47	14.73[b]	14.92[b]	14.81[b]	14.25	14.42	14.77[b]	14.97[b]	14.96[b]
2	14.48	14.38	14.21	13.30[a]	13.34[a]	14.48	14.44	14.09[a]	13.80[a]	13.64[a]
3	14.49	14.39	14.29	13.74[a]	12.48[a]	14.49	14.44	14.23	13.98[a]	13.74[a]
4	14.59	14.56	14.42	13.90[a]	12.98[a]	14.59	14.50	14.54	14.50	14.58
5	14.42	14.55	14.25	13.98[a]	13.73[a]	14.42	14.48	14.18	13.97[a]	12.89[a]
6	14.83	14.73	14.53	13.41[a]	11.46[a]	14.83	14.76	14.64	14.14[a]	14.11[a]
7	14.61	14.63	14.25[a]	12.11[a]	10.66[a]	14.61	14.54	14.45[a]	12.52[a]	12.13[a]
8	14.63	14.61	14.52	...	13.03[a]	14.72	14.59	14.70	...	14.61

NOTE: LSD for control *vs* treatments, 0.318 ($p < 0.05$). LSD for treatment *vs* treatment, 0.45 ($p < 0.05$).
[a] Significant inhibition for control *vs* SE concentrations.
[b] Significant enhancement for control *vs* SE concentrations.

increase in cell number compared to the control and is regarded as an indication of a stimulatory effect. The EC_{50} values were calculated using the regression of the four CE values against the four SE dilutions. Only negative calculated effect values should be used to calculate EC_{50} (Table 8). In the case of positive or low negative CE values, the EC_{50} was set to 100.0%. This correction allowed the use of a range of EC_{50} value from 0 to 100%, and resulting comparisons and averaging of results could then be made.

According to the results by the CA method, the soils and Promix (soil 5) can be divided into three groups. The first group contains soils with an EC_{50} value of <50.0%. Only soil 7, collected in Coffee County, Alabama, caused such a toxic effect. The EC_{50} values were similar, 41.8 and 47.3% for the CA and PA method, respectively. It is difficult to explain such a strong inhibitory reaction of soil on algal growth, but there is always the possibility of the presence of some unknown toxic material in soil.

The second group, with EC_{50} values between 51 to 85%, includes soils 2, 3, 5, and 6 and can be described as soils with slightly toxic effects. Among these soils, the EC_{50} value of Promix (soil 5) was low (56.2%). At 1% SE, Promix did not inhibit cell growth, but higher concentrations of SE caused inhibition in algal growth. The CE values for 10, 50, and 80% SE were -21.2, -35.4, and -78.2%, respectively. A low EC_{50} value, coupled with the high negative CE value, are indications that Promix aqueous eluate inhibits cell growth of *S. capricornutum*. One explanation for this might be the high concentrations of tannins, associated with Sphagnum moss and bark, which are major constituents of Promix, and sulfur (SO_4-S). Similar results were observed for soil 2, collected in Missouri ($EC_{50} = 61.9\%$). The inhibitory effect expressed as CE value was not as pronounced as Promix at 1, 10, and 50% but for 80% SE had a value of -55.9%. This soil contains a higher amount of Zn (8.1 ppm). The response of algal growth for these two soils, 5 and 2, was different using the PA nutritional amendment method. In the case of both soils, a significant decrease ($p < 0.05$) in cell numbers was observed in 50 and 80% SE, but the calculations used for the PA method did not detect this. The CE formula, presented in the Materials and Methods section for the PA method, acts as a neutralizing mechanism since the amount of growth medium is corrected by the appropriate multiplications ($P = 0.99, 0.9, 0.5$, and 0.2). In order to perform this calculation, one must assume that there is a direct 1:1 correlation between population growth and the amount of growth medium [11]. This assumption is not precisely correct, and the P factor did not accurately correct the response of the algae to the amount of growth medium present, especially if slightly toxic material was evaluated.

The last group of soils contained innocuous soils with $EC_{50} > 85\%$. Soils 1, 4, and 8 indicated essentially no toxicity. Soil 4, collected in Covington County, Alabama, and soil 8, collected from a farm of the University of Missouri, Columbia, are innocuous and did not stimulate cell growth (low CE values). The cell numbers on higher concentrations (50 and 80%) of SE for the PA method were significantly lower ($p < 0.05$) compared to the control or to this same concentration for the CA method. A reduction in cell growth at 50 or 80% SE (at 50 or 20% growth medium present) is probably more of a function of the lack of sufficient nutrients than a reduction in cell number due to a toxic effect of the soils themselves.

The eluate from soil 1 substantially increased growth of *S. capricornutum* above that of the negative control (growth medium). The SE of soil 1, collected near Columbia, Missouri, from a university farm, acts as a growth medium itself, due probably to the high concentration of Ca, K, and neutral pH.

Conclusions

The ruggedness test phase demonstrated that certain moderate procedural variations definitely altered the *S. capricornutum* bioassay results when the variations were introduced

TABLE 7—Growth of Selenastrum capricornutum expressed in number of cells × 10⁶/mL in four dilutions of soil eluate (SE) for both proportional amendment (PA) and complete amendment (CA) experimental methods for eight different soils.

Soil	Number of Experiments	Proportional Amendment Method (PA)					Complete Amendment Method (CA)				
		Control	1% SE	10% SE	50% SE	80% SE	Control	1% SE	10% SE	50% SE	80% SE
1	1	1.45	1.97	2.31	2.44	2.19	1.45	1.78	2.30	2.77	2.65
	2	1.52	1.96	2.63	3.69	3.67	1.52	1.73	2.99	3.61	3.89
	3	1.64	1.87	2.58	3.03	2.46	1.64	1.96	2.58	3.20	3.01
	x̄	1.54	1.93	2.51	3.05	2.77	1.54	1.82	2.62	3.19	3.18
2	1	2.42	2.24	1.80	0.64	0.95	2.42	2.26	1.84	1.16	1.42
	2	1.96	2.57	2.62	0.81	0.91	1.95	2.00	1.89	1.25	0.83
	3	1.56	0.95	0.69	0.41	0.28	1.56	1.44	0.65	0.65	0.50
	x̄	1.98	1.92	1.70	0.62	0.71	1.98	1.90	1.46	1.02	0.92
3	1	1.55	1.41	1.12	0.65	0.11	1.55	1.41	0.98	0.71	0.62
	2	2.00	2.03	1.88	1.22	0.89	2.00	2.12	1.97	1.70	1.34
	3	2.43	1.95	1.98	0.99	1.82	2.43	2.16	1.82	1.37	0.95
	x̄	1.99	1.80	1.66	0.95	0.98	1.99	1.90	1.59	1.26	0.97
4	1	2.05	1.97	1.61	0.86	0.47	2.05	1.87	1.71	1.66	1.76
	2	1.82	1.82	1.44	1.05	0.38	1.82	1.69	1.89	1.80	2.06

3	2.74	2.62	2.72	2.49	2.72	0.45	1.44	2.62	2.62	2.72
x̄	2.19	2.03	2.11	2.02	2.20	0.43	1.13	1.89	2.14	2.20
5 1	0.45	1.65	1.94	3.55	2.87	1.00	1.77	1.91	3.77	2.87
2	0.35	0.81	1.21	1.67	1.48	0.80	0.84	1.32	1.74	1.48
3	0.40	1.21	1.28	1.24	1.47	0.96	1.10	1.44	1.37	1.47
x̄	0.40	1.22	1.48	2.15	1.94	0.92	1.24	1.56	2.29	1.94
6 1	0.71	0.77	2.31	2.80	3.00	0.03	0.27	2.31	2.67	3.00
2	1.66	1.79	2.47	2.87	2.97	0.13	0.95	2.09	2.66	2.97
3	2.05	1.94	2.09	2.10	2.37	0.22	1.16	1.76	2.19	2.37
x̄	1.47	1.50	2.29	2.59	2.78	0.13	0.79	2.05	2.51	2.78
7 1	0.02	0.04	1.54	2.23	2.74	0.01	0.02	1.32	2.46	2.74
2	0.59	0.63	2.16	1.92	2.03	0.20	0.68	1.77	2.02	2.03
3	0.55	0.92	2.02	2.06	1.94	0.04	0.37	1.51	2.29	1.94
x̄	0.39	0.53	1.91	2.07	2.24	0.08	0.36	1.53	2.26	2.24
8 1	2.93	⋯	2.62	2.06	2.22	1.32	⋯	2.16	2.08	2.22
2	1.68	⋯	2.25	2.28	2.75	0.17	⋯	2.04	2.27	1.73
3	1.03	⋯	2.25	2.78	2.91	0.41	⋯	1.88	2.31	3.01
x̄	1.88	⋯	2.37	2.37	2.63	0.63	⋯	2.03	2.22	2.32

TABLE 8—Results of Selenastrum capricornutum bioassay expressed as calculated effect (CE) value for eight different soils and EC_{50}.

Soil	Proportional Amendment Method (PA)					Complete Amendment Method (CA)				
	1% SE	10% SE	50% SE	80% SE	EC_{50}	1% SE	10% SE	50% SE	80% SE	EC_{50}
1	25.2	81.6	298.0	805.2	100.0[a]	18.7	79.9	108.5	107.9	100.0[a]
2	-4.2	-6.7	-37.8	70.9	100.0[a]	-4.0	-28.5	-49.0	-55.9	61.9[a]
3	-8.1	-8.4	-4.5	-3.3	82.2	-4.7	-21.2	-37.8	-51.5	70.9
4	-1.6	-6.0	1.3	-1.1	100.0[a]	-8.1	-4.2	-7.8	0.5	100.0[a]
5	15.3	-5.9	28.6	155.7	100.0[a]	7.2	-21.2	-35.4	-78.2	56.2
6	-8.7	-18.0	-40.4	-76.9	62.2	-7.2	-17.3	-44.2	-44.9	74.4
7	3.6	-21.1	-65.1	-87.1	47.3	-6.0	-11.0	-73.7	-81.0	41.8
8	1.6	2.9	...	36.4	100.0[a]	-9.6	-7.6	...	-23.8	86.0

NOTE: The results are averaged from three experiments.
[a] EC_{50} value corrected to +100.0.

at certain critical steps. In the second part of the experiment, the seven soils and Promix BX were tested by the *S. capricornutum* assay using two nutritional amendment methods. A wide range of responses were documented indicating a substantial difference among soils with respect to their ability to stimulate or inhibit algal population growth. Soil 1 acted as a growth medium and substantially increased algae cell numbers. Soils 4 and 8 were examples of soils which do not stimulate algal growth but also were not toxic. The other soils showed some degree of toxicity, especially soil 7.

The result for Promix BX and soil 2 were contradictory using two proportional amendment methods PA and CA. The *P* factor used in the PA method for correction of the amount of growth medium present did not always accurately amend the response of the algae to the amount of nutrients.

S. capricornutum proved to be a very sensitive bioassay organism, and care should be taken if the comparison of soil eluates derived from different soils has to be made. If working with contaminated soil, the control test with similar but not contaminated soil is recommended to avoid data misinterpretation.

References

[1] Green, J. C., Miller, W. E., Debacon, M., Long, M. A., and Bartels, C. L., *Environmental Toxicology and Chemistry*, Vol. 7, 1988, pp. 35–39.
[2] Turbak, S. C., Allson, S. B., and McFeters, G. A., *Water Resources*, Vol. 20, 1986, pp. 91–96.
[3] ASTM E 1218-90, "Guide for Conducting Static 96-Hour Toxicity Tests with Microalgae," 1990.
[4] Porcella, D. B., "Protocol for Bioassessment of Hazardous Waste Sites," EPA/600/2-83-054, U.S. Environmental Protection Agency, Corvallis, OR, 1983.
[5] Organization for Economic Cooperation and Development (DECD), Algae Growth Inhibition Test No. 201, in *OECD Guidelines for Testing of Chemicals, Section 2, Effects of Both Systems*, Paris, France, 1981.
[6] Payne, A. G. and Hall, R. H. in *Aquatic Toxicology (Second Symposium), ASTM 667*, L. L. Marking and R. A. Kimerle, Eds., American Society for Testing and Materials, Philadelphia, 1979.
[7] Miller, W. E., Greene, J. C., and Shiroyama, T., "Experimental Design, Application and Data Interpretation Protocol," EPA/600/9-78-018, U.S. Environmental Protection Agency, Corvallis, OR, 1979.
[8] McKenzie, W. D. and Olsson, T. A., "Guidelines for Conducting Single Laboratory Evaluations of Biological Methods," EPA/600/SA-83-056, U.S. Environmental Protection Agency, Las Vegas, NV, 1984.
[9] Lower, W. R., Thomas, M. W., Judy, B. M., and Sutton, W. W., "Freshwater Assay Using Soil Eluates as Sample Material (Single Laboratory Evaluation)," EPA/600/53-90/038, EPA Environmental Research Laboratory (ERL), Athens, GA, 1990.
[10] Youden, W. J. in "Precision Measurement and Calibration," H. H. Ku, Ed., U.S. Department of Commerce, National Bureau of Standards, Washington, DC, 1969, pp. 151–158.
[11] Green, T. C., Bartles, C. L., Warren-Hicks, W. T., Parkhurst, B. R., Linder, G. L., Peterson, S. A., and Miller, W. E., "Protocols for Short-Term Toxicity Screening of Hazardous Waste Sites," U.S. Environmental Protection Agency, Corvallis, OR, 1989.

Loren J. Larson[1]

The Influence of Test Length and Bacteria on the Results of Algal Bioassays with Monophenolic Acids

REFERENCE: Larson, L. J., "The Influence of Test Length and Bacteria on the Results of Algal Bioassays with Monophenolic Acids," *Plants for Toxicity Assessment: Second Volume, ASTM STP 1115,* J. W. Gorsuch, W. R. Lower, W. Wang, and M. A. Lewis, Eds., American Society for Testing and Materials, Philadelphia, 1991, pp. 230–239.

ABSTRACT: Static bioassays lasting 11 to 16 days were conducted on *Chlorella pyrenoidosa* using vanillic acid, syringic acid, and 4-hydroxybenzoic acid. Vanillic and 4-hydroxybenzoic acid were initially inhibitory, then after 3 to 5 days became stimulatory compared to control. Bioassays with syringic acid were all stimulatory except higher concentrations, which resulted in 100% mortality. Bacteria-free cultures of *C. pyrenoidosa* were stimulated by vanillic acid at low concentration and inhibited at higher levels with no shift in response observed. It was concluded that degradation of test material was responsible for the shift from inhibition to stimulation. This work concludes that the length of tests, as well as whether or not careful monitoring for bacterial contamination was conducted, can have a significant impact on bioassay results.

KEY WORDS: Chlorella, syringic, vanillic, 4-hydroxybenzoic, humics

The three phenolic acids used in this study were selected based on an earlier study [1] which identified their accumulation in laboratory cultures of an aquatic macrophyte, *Ceratophyllum demersum* L. Subsequent qualitative bioassays with the green alga *Chlorella pyrenoidosa* demonstrated an algicidal potential for these compounds.

Phenolic acids have been shown to affect uptake of nutrients [2–4], mitochondrial phosphorylation [5], and membrane polarization [6,7]. Their activity is correlated with the octanol-water partitioning coefficient (log *P*), suggesting that these compounds may be concentrating in the lipid layer of membranes, thereby either reducing their integrity for the partitioning of ions [8] or affecting enzymes occurring within the lipid layer.

Some phenolics, including vanillic, syringic, and 4-hydroxybenzoic acids, appear to act both as synergists and inhibitors for the plant auxin catabolic enzyme, auxin-oxidase, depending on concentration [9–11]. Para-hydroxylated acids tend to be cofactors in the degradation of indole-3-acetic acid (IAA). Meta- or ortho-hydroxylate acids have the opposite effect, acting as IAA synergists. At lowered concentrations these effects reverse, making *p*-hydroxylated acids IAA promoters. Algal growth is positively correlated with IAA concentration [12–15].

Methods and Materials

Bioassays were conducted under static conditions in 250-mL Erlenmeyer flasks containing 200-mL Bold's Basal Medium (pH = 8.0 to 8.3) [16]. Analytical grade test material was

[1] Senior environmental scientist, James M. Montgomery, Consulting Engineers, Inc., 545 Indian Mound, Wayzata, MN 55391.

added to sterilized distilled water to provide stock solutions, which was then added to test chambers in appropriate volumes. Equal volumes of distilled water were added to controls (<0.5 mL) to discount dilution effect. All concentrations presented are nominal. An assumption was made that the effect of phenolics was at the molecular level, thus equal molar concentrations were tested to allow for direct comparisons between compounds per equivalent toxic unit.

Flasks were inoculated with *C. pyrenoidosa* (Emerson; UTEX No. 252) from stock cultures in log-growth phase. Inoculation was accomplished by transferring an aliquot (25 μL) of stock culture to test chambers using Micro-titer™ equipment (Cooke Engineering Company, Alexandria, VA). One treatment involved inoculation with bacteria-free *C. pyrenoidosa*, and care was taken to protect these flasks from contamination. This included using aseptic techniques for the transfer of samples from the flasks. Three times weekly a sample was taken from each of these chambers and examined by microscope. If bacterial contamination was apparent, flasks were discarded.

Treatments were cultured in triplicate at 23°C with constant illumination of 400 μc. Growth of each culture was estimated daily by fluorescence (Turner Model 111), over 10 to 16 days. Samples of 1 to 2 mL were taken for fluorescence analysis and were afterwards discarded. Growth data were analyzed by two-way mixed model ANOVA.

Results

Test Length Bioassays

Results of the bioassays are presented in Figs. 1A, 1B, and 1C and Table 1. EC_{50}s, calculated using a trimmed Spearman-Karber analysis [17], remained relatively stable for all three compounds in 24 to 96 h. Because of the lack of sufficient intermediate responses in tests conducted with syringic acid, EC_{50}s for this compound are only presented as a range.

Bioassays with vanillic and 4-hydroxybenzoic acid appear to be very similar (Figs. 1A and 1B, respectively). Lowest concentration (0.1 mM) of both compounds stimulated the growth of *C. pyrenoidosa*. At higher concentrations (0.3 and 0.4 mM), inhibition was initially observed, but at 3 to 5 days culture growth was stimulated, surpassing both control and 0.1-mM treatments. These cultures showed not only an accelerated rate of log-phase growth, but also a more sustained log-growth phase.

Bioassays with syringic acid (Fig. 1C) at 0.1, 0.3, and 0.4 mM were stimulatory after 7 days. Initial culture inhibition was not observed as with vanillic and 4-hydroxybenzoic acid. Additional treatments of 0.5 and 2.5 mM (not included in Fig. 1C) resulted in 100% mortality of algae. At 16 days, the 0.4-mM treatment still appeared to be in log-growth phase. This phase appeared to have ended after 13 days in the 0.3-mM treatment.

On all days (except Day 0), the algal growth of all test concentrations was statistically different ($P < 0.01$) from that of the control and all other test concentrations. ANOVA of the growth data for all bioassays indicated the effects of both nominal concentration and day of the experiment were significantly correlated ($P < 0.001$) to culture growth. Also highly correlated ($P < 0.001$) were the interaction of nominal concentration and day of the experiment.

Bacteria-Free Culture Bioassays

The comparison between bacteria-free cultures and cultures in which no effort was taken to exclude bacteria was made only using vanillic acid. The results of these bioassays are summarized in Figs. 2A and B. EC_{50}s, again calculated using a trimmed Spearman-Karber analysis, are provided in Table 2.

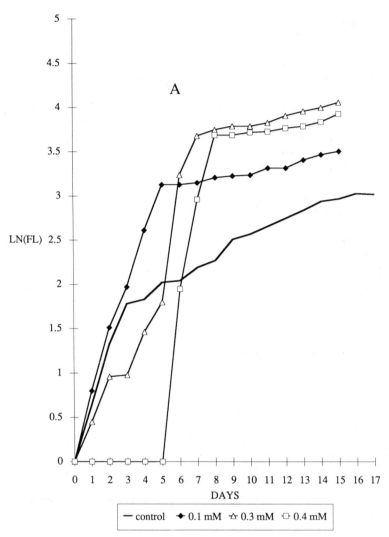

FIG. 1—*Growth of* C. pyrenoidosa, *measured as* in vivo *fluorescence, in vanillic acid* (A), *4-hydroxybenzoic acid* (B), *and syringic acid* (C). *At all data points, 95% confidence interval is smaller than point size.*

The cultures of algae which were not maintained as bacteria-free (Fig. 2A), as discussed in the previous section, showed a strong inhibition to growth through Day 5 at 0.4 mM (see Fig. 1A). Some inhibition was observed through Day 5 at 0.3 mM. However, after Day 5 both treatments resulted in a stimulation in growth compared to the control. At a concentration of 0.1 mM, growth of the alga was stimulated on all days.

In contrast, the bacteria-free cultures (Fig. 2B) produced a fairly consistent response of the treatments compared to the control through the entire test (11 days). On every day (except Day 0) culture growth was inhibited by 0.3-mM vanillic acid and stimulated by 0.1 mM. The resulting EC$_{50}$s (Table 2) indicate a lesser impact of vanillic acid under the bacteria-free conditions.

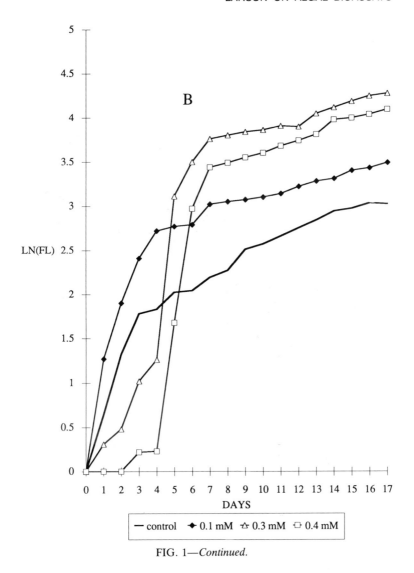

FIG. 1—*Continued.*

Discussion

The results of this study indicate that, at least for the chemicals tested, both the test length and the occurrence of bacteria in the cultures can influence the bioassay results.

Test Length Effects

In this series of tests, the bioassay results, as EC_{50}, were statistically correlated to day of experiment. The results were therefore influenced not only by the nominal concentration of the treatments, but also by the length of the experiment.

As seen in Table 1, the EC_{50} results calculated from the short-term results (24 to 96 h)

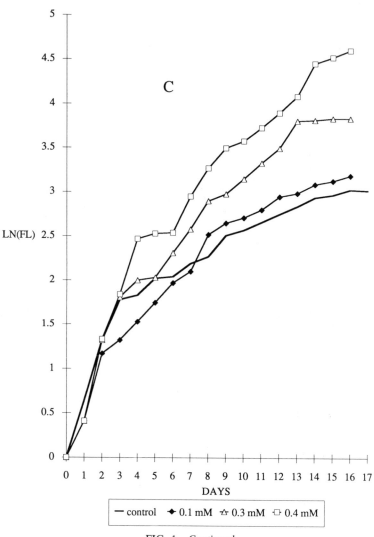

FIG. 1—*Continued.*

were fairly consistent over time. However, after 96 h the results changed significantly such that calculation of EC_{50} was not possible.

This trend can be explained in part by the degradation of the test chemicals during the experiments. This issue is addressed in detail in the subsequent section of this paper.

The correlation between test results and length of experiment can be compound specific. The test materials used for this work are known algal hormones, as discussed at the beginning of the paper. Therefore, their interaction with the growth of experimental cultures can change as actual concentration changes through biodegradation or any other possible fate mechanism, such as volatilization. Other test materials, such as herbicides or heavy metals, may not produce similar results. However, the understanding of algal physiology is far from complete, therefore this analogous interaction with other compounds, particularly organics, cannot be discounted.

TABLE 1—$EC_{50}s$ and 95% C.I. (in parentheses) at various time intervals, based on reduction in growth measured as in vivo fluorescence of Chlorella pyrenoidosa.

Compound	Unit	Duration, h				
		24	48	96	120	>120
Vanillic acid	mM	0.28 (0.26–0.3)	0.28 (0.27–0.31)	0.30 (0.29–0.32)	0.3–0.4	>0.4
	mg/L	47.0 (43.7–50.4)	48.7 (45.4–52.1)	50.4 (48.7–53.8)	50.4–67.2	>67.2
4-Hydroxybenzoic acid	mM	0.25 (0.23–0.27)	0.22 (0.21–0.24)	0.31 (0.28–0.34)	>0.4	>0.4
	mg/L	34.5 (31.7–37.3)	30.4 (29.0–33.1)	42.8 (38.6–46.9)	>55.2	>55.2
Syringic acid[a]	mM	0.4–0.5	0.4–0.5	0.4–0.5	0.4–0.5	0.4–0.5
	mg/L	79.3–99.1	79.3–99.1	79.3–99.1	79.3–99.1	79.3–99.1

NOTE: C.I. = confidence interval.
[a] Because of the lack of intermediate inhibitory response in tests with this chemical, the $EC_{50}s$ are presented as a range.

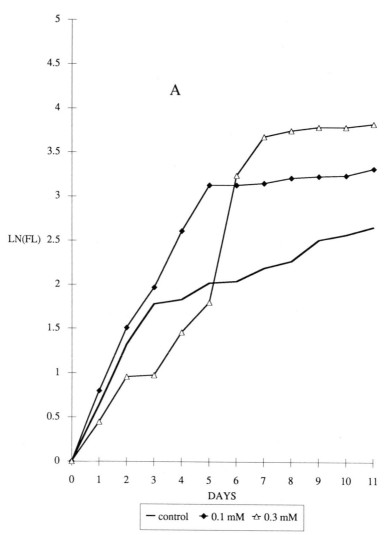

FIG. 2—*Growth of* C. pyrenoidosa, *measured as* in vivo *fluorescence, in vanillic acid under normal laboratory culture conditions* (A) *and under bacteria-free conditions* (B). *At all data points, 95% confidence interval is smaller than point size.*

Bacterial Effects

It is certainly reasonable to conclude that the bacterial degradation of the test compounds in the current study could explain the shift in response of the algal cultures. At higher concentrations these compounds are reported to synergize enzymatic degradation of IAA, which inhibits culture growth [9–11]. As concentration decreases, their activity reportedly shifts to inhibition of IAA degradation, thus stimulating cultural growth. This hypothesis is supported by the results of bacteria-free bioassays in which there was no shift from inhibition to stimulation, presumably due to more stable toxicant levels. Further studies in which the concentration of toxicant is regularly monitored are certainly warranted.

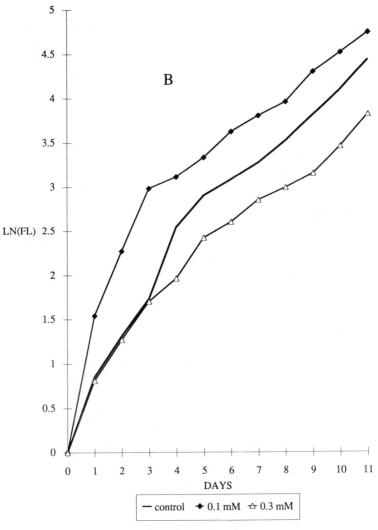

FIG. 2—*Continued.*

TABLE 2—*EC_{50}s and 95% C.I. (in parentheses) from normal and bacteria-free cultures, based on reduction in growth measured as* in vivo *fluorescence of* Chlorella pyrenoidosa.

Compound	Units	Duration, h		
		24	48	96
Vanillic acid (normal culture)	mM	0.28 (0.26–0.3)	0.28 (0.27–0.31)	0.30 (0.29–0.32)
	mg/L	47.0	48.7	50.4
Vanillic acid (bacteria-free culture)	mM	>0.4	>0.4	>0.4
	mg/L	>67.3	>67.3	>67.3

NOTE: C.I. = confidence interval.

A strong shift in response was not observed in bioassays with syringic acid due to a lower rate of degradation for this compound. Earlier studies [1] indicated that while vanillic acid and 4-hydroxybenzoic acid could be used as a sole carbon source for aerobic microbes, syringic acid could not. This is the result of stearic hindrance in the dimethoxylated ring, or because the dimethoxy-ring does not allow for rapid orthodihydroxylation, a condition necessary for degradation of the aromatic ring [18,19].

Syringic acid can be biodegraded under certain conditions. The degradation of all three of these compounds by anaerobic sewage-degrading bacteria under strict anaerobic conditions was reported by Healy and Young [20]. Under those conditions, syringic acid was degraded at the highest rate (97% degraded in 19 days), followed by vanillic acid (85% in 40 days) and 4-hydroxybenzoic acid (65% in 32 days). Acclimation lag periods ranged from 5 days for syringic acid to 15 days of 4-hydroxybenzoic acid. These tests were conducted at 35°C. The biochemical pathway utilized in the anaerobic degradation process could avoid the orthodihydroxylated condition necessary for aerobic degradation.

In the current study, the absence of bacteria in the cultures had two apparent impacts. First, the effect of the treatments on culture growth remained constant over the 11-day experiment. Secondly, the apparent effect of vanillic acid on the growth of the algal cultures (as EC_{50}) was less than in the normal experimental cultures.

Conclusions

In summary, using three phenolic acids, the results of static bioassay experiments were impacted by both the length of the experiment, particularly after 96 h, and by the occurrence of bacteria in the experimental cultures. This indicates that frequent culture growth measurements are appropriate to identify similar trends, and that monitoring for bacterial contamination, particularly using organic test materials, and measuring toxicant concentrations are appropriate test protocols.

Acknowledgments

This research was supported by the Department of Biological Sciences. St. Cloud State University, St. Cloud, MN. Statistical analyses were performed by the University of Notre Dame Computer Center. The assistance of David R. Felstul of James M. Montgomery, Consulting Engineers, Inc. in the preparation of this manuscript is gratefully acknowledged.

References

[1] Larson, L. J., "Identification and Verification of Algal Inhibitors Produced by *Ceratophyllum demersum* L.," Masters thesis, St. Cloud State University, St. Cloud, MN, 1983.
[2] Glass, A. D., *Plant Physiology*, Vol. 54, 1974, pp. 858–868.
[3] Glass, A. D., *Phytochemistry*, Vol. 14, 1975, pp. 2127–2130.
[4] McClure, P. R., Gross, H. D., and Jackson, W. A., *Canadian Journal of Botany*, Vol. 56, 1978, pp. 764–767.
[5] Stenlid, G., *Phytochemistry*, Vol. 9, 1970, p. 225.
[6] Glass, A. D., *Journal of Experimental Botany*, Vol. 25, 1974, pp. 1104–1113.
[7] Levitan, H. and Baker, J. L., *Science*, Vol. 176, 1972, pp. 1423–1425.
[8] Mitchell, P., *Nature*, Vol. 191, 1961, pp. 144–148.
[9] Hare, R. C., *Botany Review*, Vol. 30, 1964, pp. 129–165.
[10] Imbert, M. P. and Wilson, L. A., *Phytochemistry*, Vol. 9, 1970, pp. 1787–1794.
[11] Zenk, M. H. Muller, G., *Nature*, Vol. 200, 1963, pp. 761–763.
[12] Kefeil, V. I., *Natural Plant Growth Inhibitors and Phytohormones*, The Hauge, Boston, 1978.
[13] Ahmad, M. R. and Winter, A., *Planta*, Vol. 88, 1968, pp. 61–66.

[*14*] Bunt, J. S., *Nature,* Vol. 192, 1961, pp. 1274–1275.

[*15*] Iswaki, H., *Journal of the Oceanography Society of Japan,* Vol. 27, 1971, pp. 152–157.

[*16*] Baker, A. F. and Bold, H. C., "Phycological Studies, X.Taxonomic Studies in the Oscillatoria," University of Texas, Austin, TX, 1970.

[*17*] Hamilton, M. A., Russo, R. C., and Thurston, R. V., *Environmental Science Technology,* Vol. 11, 1977, pp. 714–719.

[*18*] Barz, W. and Hoesel, W., in *Biochemistry of Plant Phenolics,* T. Swaine, J. B. Harborne, C. F. Van Sumere, Eds., Plenum Press, NY, 1979, pp. 234–252.

[*19*] Butt, V. S., in *Biochemistry of Plant Phenolics,* T. Swaine, J. B. Harborne, C. F. Van Sumere, Eds., Plenum Press, NY, 1979, pp. 351–369.

[*20*] Healy, J. B. Jr. and Young, L. V., in *Microbial Degradation of Pollutants in Marine Environments,* EP-600/0-79-012, A. W. Bourquin, P. H. Pritchard, Eds., U.S. Environmental Protection Agency, Washington, DC, 1979, pp. 348–359.

Minocher Reporter,[1] *Merline Robideaux,*[1] *Paul Wickster,*[1]
John Wagner,[2] *and Lawrence Kapustka*[3]

Ecotoxicological Assessment of Toluene and Cadmium Using Plant Cell Cultures

REFERENCE: Reporter, M., Robideaux, M., Wickster, P., Wagner, J., and Kapustka, L.,
"Ecotoxicological Assessment of Toluene and Cadmium Using Plant Cell Cultures," *Plants
for Toxicity Assessment: Second Volume, ASTM STP 1115*, J. W. Gorsuch, W. R. Lower, W.
Wang, and M. A. Lewis, Eds., American Society for Testing and Materials, Philadelphia,
1991, pp. 240–249.

ABSTRACT: Cell suspension cultures from a variety of plant species can be grown from callus
for use in ecotoxicological testing. The effects of toluene and cadmium on plant cell growth
were investigated. The accumulation of cadmium in the soluble cell "cytoplasmic" and in the
cell residue compartment in two cell types was also verified by elemental analysis using inductively coupled atomic plasma emission spectroscopy (ICPAES). Plant cell cultures in suspension can be most useful in tracing the fate of labeled toxic chemicals and in an isolated
environment. Both tracer fate tests and toxicity tests with plant cell cultures can then aid in
the optimal design of more costly experiments with whole plants. Callus and suspension cultures
can be maintained with their individual characteristics through many cell transfers. Alternately,
these characteristics can be preserved by placing appropriately treated cells at liquid nitrogen
temperatures. A majority of these cultures can be grown in standard media that can be made
in individual labs or purchased from regular suppliers. The use of standard cultures make it
possible to address comparability for routine ecotoxicological testing of chemicals between
different labs.

KEY WORDS: plant cell suspension cultures, toxicity testing, metabolism, toluene, cadmium

The increasing amounts of anthropogenic chemicals in the air, water, and ground are of
concern to us all. A data base has been established concerning the effects of the increase
of many of these chemicals in animals and animal cells; a parallel data base for plant cells
and plants has been recently developed [1]. The effect on plant growth and metabolism for
some agricultural chemicals has been investigated [1,2]. However, our knowledge of the
fate of most xenobiotic chemicals on plants and vegetation after inadvertent release or even
planned accumulation at waste sites is inadequate. The concern is increased because of the
possibility of toxic effects of the unknown metabolites from these chemicals on animals and
man through the ecological food chain.

The need for the preliminary testing of the response of plant cells in culture to xenobiotic
compounds is desirable for a number of reasons. The tests can be carried out under controlled
and reproducible conditions. Because the cultures are axenic, the response and the metabolism of the chemicals of interest by plant cells alone can be studied. Should the occasion
arise, the combined effects of plant cells and the microbes commonly found in association
with selected plants can be studied in a pattern of increasingly complex experiments where

[1] NSI Technological Services, USEPA-ERL, Corvallis, OR 97333.
[2] Ames Laboratory IAG, USEPA-ERL.
[3] U.S. EPA, Environmental Research Laboratory, Corvallis, OR 97333.

a single variable is changed at a time [3]. The chemicals, if they are known to be toxic to humans, can be studied in a compact, isolated facility devoted to plant cell culture. The fate of the chemicals traced quantitatively within the culture flasks of interest thus minimizes risk to workers. The selection of plant cells and the chemicals can be made to cover a broad range of conditions. In many instances, the conditions of culture make it possible to trace the metabolites of these chemicals without using radioactive labels. The changes in the physiology of the cells used for these tests can also be monitored. At the simplest level, cultures exposed to toxic or xenobiotic chemicals can be monitored for growth alone.

Data are presented from two distinct experiments. In the first, the growth of two types of plant cells exposed to toluene was documented. In the second, the fate of cadmium in a photosynthetic cell line was determined and compared with a cell line from the black locust.

Materials and Methods

Chemicals

High-purity chemicals were used to make media for cell culture studies. Media B5 and MS [4-6] and plant hormones were also obtained from Sigma[4] for comparison with media made at ERL-Corvallis. Callus cultures were maintained on "Phytagel"-agar substitute (Sigma) plates.

Plant Cells

The plant cells utilized were: alfalfa [*Medicago sativa* cv Vernal (L.)]; soybean [*Glycine max* cv Williams (L.)]; rose (*Rosa* cv *Paul's Scarlet*); black locust [*Robinia pseudo-acacia* (L.)]; and the apocynaceous plant cells (*Mandevilla pohliana*). The reasons for selection of a variety of plant cells are given in the discussion.

Alfalfa and soybean cells were started in callus culture from the hypocotyl hook of the germinating seeds. The seeds were obtained from Oregon State University, either from the Department of Horticulture or from the Department of Soil Science. Callus cultures for black locust were started from a leaf node of an adult plant obtained from the ozone study project at ERL-Corvallis. Callus cells of Paul's Scarlet Rose were obtained at the Corvallis ERL [7,8]. Mandevilla calli were started from a leaf internode of an adult plant. Callus cells were maintained on B5 or MS medium [4,5] and transferred to fresh media every three weeks. Suspension cultures were started from callus cells and the liquid media changed every ten days. The callus cells were grown in the dark at 27°C. Suspension cultures of the cells were also grown at 27°C and on gyratory shakers operated at 120 rpm. The cell suspensions were grown in 80-mL medium using 250-mL flasks. Replicate plates from Mandevilla cells were split and grown in the dark as well as in the light at 27°C.

Cells grown as calli in the presence of light received a flux of 166 $\mu M \cdot M^{-2} \cdot s^{-1}$, while the cells grown in suspension in the presence of light received 16 $\mu M \cdot M^{-2} \cdot s^{-1}$. The light measurements were monitored utilizing a LI 85B, Li Cor, Inc radiometer-photometer. The cultured cells were examined with a Leitz Orthoplan microscope using phase optics.

All suspension cultures were monitored every two days for microbial growth by a microscopic examination and by plating of suspension medium from the cells on standard agar plates containing yeast-mannitol or Bacto-Peptone. Plant cell growth was measured as increase in fresh weight using 20-mL samples from four replicate flasks per group. Weights

---

[4] Mention of trade names or commercial products does not constitute endorsement or recommendation for use.

were also taken after drying the cells at 80°C to constant weight. Replicates for the wet weight measurements were more consistent, and therefore wet weight results were also used to determine the elemental analysis of cell residue, which was dried, weighed, and then acid digested (see below). Toluene-treated cells were grown in the presence of 500-ppm toluene. The experiment was terminated after twelve days due to the concern over the continued presence of toluene at the end of this period. Toluene was therefore sampled from the growth medium of all the treated flasks of rose and alfalfa cells at the end of the experiment.

Chemical Analysis

Toluene Analysis—Toluene analysis on the filtered suspension medium was performed using a Varian 2700 Aerograph GC equipped with a 6-m OV-210 column. Conditions were as follows: Carrier gas He (10 mL/min); H_2 flow 3 mL/min (20 psi); air flow 30 mL/min (30 psi); Detector F1D; Attenuation 2 Range 2. Between 5 to 50 μL were injected. The peaks were resolved and the retention time was ~3 min. In this experiment neither the metabolites of toluene nor the rate of toluene decrease were followed.

Digestion Procedure—Between 0.3 and 2.0 g of wet weight cell tissue were sampled at each time point by withdrawing 20 mL of cell suspension. The culture medium was aspirated and used for analysis after suitable dilution and filtration through 0.4-μ filters. The cells were triple washed and then frozen and thawed three times in redistilled RO (reverse osmosis or reagent grade) water and the "cytoplasmic" contents removed for metal analyses as above. The residue was digested in all glass Pyrex No. 9826 screw-capped test tubes using 2 mL of double distilled concentrated nitric acid (G. F. Smith Chemical Co., No. 621) followed by 1 mL of 50% assay-hydrogen peroxide (Hach Chemical Co.; No. 144-11). The samples were first digested overnight with nitric acid at 130°C, cooled to room temperature, and then digested overnight with the peroxide. The samples were cooled and diluted (ca. 51.8 mL) with reagent grade water. The final volume of the sample was determined by the net weight of the liquid after digestion in each test tube.

Cadmium—Cadmium was analyzed by an Applied Research Model 3580 inductively coupled plasma atomic emission spectrophotometer (ICP-AES) system, simultaneously along with other elements. The system consisted of an ICP argon excitation source operated at 27 MHz and 1200 W of power, a 1-m vacuum, direct reading spectrometer with 1080 lines per mm ruled grating, and a Digital Equipment Corporation Model PDP 11/23 dedicated minicomputer for data acquisition and readout. A fixed pneumatic nebulizer (J. E. Meinhard Associates Inc. Model TR 30-C3) connected to a peristaltic pump (Gilson Model Minipuls III) was used to aspirate liquid samples through the torch assembly.

Results

The two cell cultures used for studying effects of Cd are shown for *Mandevilla pohliana* cell line 408 (Figs. 1a and 1b) and for the black locust, *Robinia pseudoacacia* (Figs. 1c and 1d). The *Mandevilla* cells formed tight spheres which enlarged in suspension cultures. When the cell masses were cut smaller, the new clusters did not separate but continued to grow as spheres. The cells were redistributed in additional flasks after aseptically reducing them into smaller fragments and expressing them through a screen or subjecting them to a 30-s homogenizer treatment at a low setting. This treatment is not destructive for this particular cell line. The green callus cells, with photosynthetic reaction centers, were maintained green in suspension by growing at light flux a tenth the intensity of the callus growing on agar plates (see Materials and Methods). The green cells assumed spheroidal-shaped masses within two days of being placed in shake cultures. (This spheroid-forming ability was also

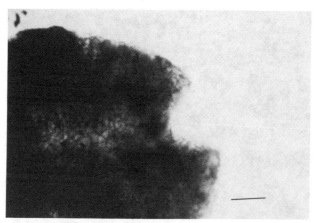

FIG. 1a—*A tight ball of* Mandevilla *cells is shown. The cells were green with dark areas from the accumulated cadmium and dosed at 200 μM for three days (bar = 1 mm).*

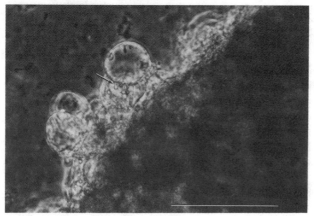

FIG. 1b—*Closeup of the edge of the cell mass shown in 1*a. *The arrows indicate the dark precipitate from the cadmium (bar = 200 μm).*

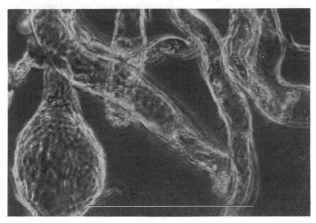

FIG. 1c—*Shows control cells of the black locust. Note the length of these cells and the manner in which they attach (Bar = 200 μm).*

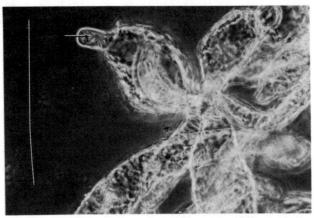

FIG. 1d—*Black locust cells after exposure to 200-μM cadmium with arrows indicating the presumed deposit of the metal (bar = 200 μm).*

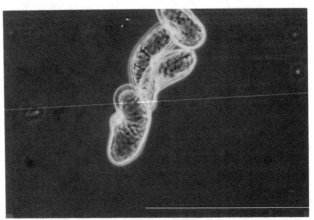

FIG. 1e—*Normal cells of alfalfa in culture. These cells grow rapidly, do not adhere tightly, and have a characteristic shape with prominent cytoplasmic strands (bar = 200 μm).*

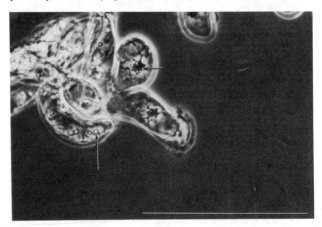

FIG. 1f—*Alfalfa cells after overnight exposure to cadmium (200 μM). The arrows clearly indicate that there are deposits on the cytoplasm and in the vacuolar membrane after the short interval. Magnification as in 1e.*

seen in nongreen cells of the *Mandevilla* from the start of such cultures two years ago; it is a characteristic of this cell line.) The amyloplast and the light greenish chloroplasts can barely be seen in the photograph of the tiny ball of small associated cells (Figs. 1*a* and 1*b*). Mass doubling of this cell line, in regular passage, was nearly two days (Fig. 3). The cell growth of this line was slowed in both control and experimental group *only once* in 15 months as in the experiment shown (Fig. 2).

The long narrow cells of the suspensions of the black locust cell line (Figs. 1*c* and 1*d*) were associated in small numbers and spread out over the distance of the microscopic field on the wet-mounted slides used for their observations. The black locust cells were colored

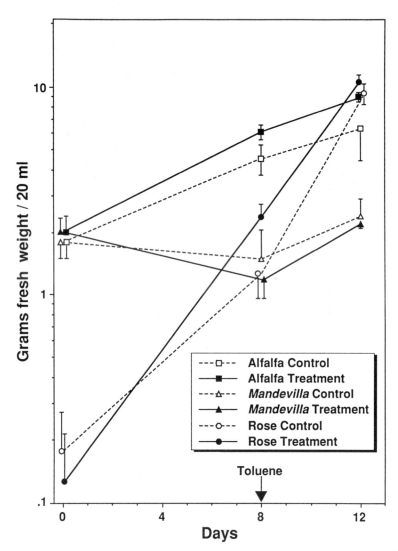

FIG. 2—*Patterns of weight increase in control cells and cells exposed to a dose of 500 ppm of toluene. The three types of cells were exposed in the same experiment and in the same incubator. The slow growth of the green* Mandevilla *cells in the first half of this experiment was an exception. Neither these cells, the alfalfa cells, or the rose cells showed differences in weight gain after dosing with toluene* (arrow). *The bars indicate standard deviation.*

red in calli; they gradually lost some color when grown as suspensions. The suspension cultures of these locust cells had mass doubling of 4.5 days in regular passage.

Alfalfa cells (Figs. 1c and 1d) grew rapidly and do not associate easily. These cells were characterized by their kidney shape and prominent cytoplasmic strands. Alfalfa cells were used in early tests with Cd, and Figs. 1e and 1f show the manner in which this metal was deposited in the first 12 h within the alfalfa cells. The dark deposit of the metal was found on vacuolar membranes and on endoplasmic reticulum. Mass doubling of the alfalfa lines took approximately two days.

Cells of Paul's Scarlet Rose were also used for determining the effects of toluene. These cells have been in culture for many years and have been well characterized [7,8]. Mass doubling of these cells took approximately 40 h (Note: Cell doubling times, e.g., with tritiated thymidine, for our cell lines have not yet been determined).

Cells in suspension were exposed to 500-ppm toluene after a seven-day pretest culture to assure growth. Both wet and dry weights were taken at the start and the end of the experiment. Cell weights between the controlled and treated groups were not found to be significantly different over the duration of the assay (Fig. 2). The sampled media from the cell suspension cultures indicated that toluene was present at the end of the experiment in all the test flasks. Examination of the control cells and toluene-treated cells with the aid of the light microscope did not indicate obvious differences in their appearance.

Four types of plant cells were exposed to similar concentrations of Cd. The Cd $(NO_3)_2$ concentrations ranged from 50 to 500 μM (representing 5.6 to 56 ppm). We selected green, growing *Mandevilla* with small tightly adhered cells and large and long, slow growing black locust cells for this report (see Figs. 1a, 1b, 1c, and 1d). The cells were sampled at three and seven days when the experiments were terminated. Cell growth, measured as increase in wet weight, was inhibited at Cd ranges greater than 150 μM. The Cd concentrations are given in μM to compare the results with those reported in the literature [9]. As shown in Fig. 3 with *Mandevilla*, cell weight was not significantly increased at 100-μM Cd by the end

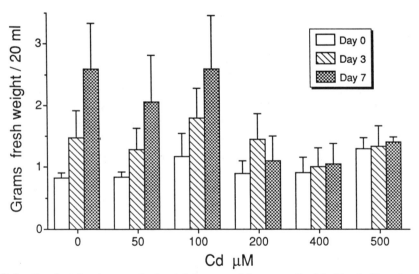

FIG. 3—*Bar chart showing gram fresh weight increase of the green cells of the* Mandevilla *at increasing dosage with cadmium. The weights were taken at the start of the experiment, after three and seven days. Bars show standard deviation. The increase in weight at three days with the low doses of Cd was not significant at the 0.95 level (see text).*

of our test, although a weight gain had been previously reported with tobacco [9] with a Cd level of 100 μM at three days. In our *Mandevilla* study, a weight gain at three days was noted but was not significant at a 0.95 confidence level (Fig. 3). The cell weight did not selectively increase for black locust cells at 100-μM Cd and three days (data not shown).

The Cd analyzed in the medium showed a linear increase proportionate to the dose in the flasks containing *Mandevilla* cells. The Cd increases for the cytoplasmic compartment and the cell residue were not linear (Tables 1a and 1b). This is in contrast to the data from experiments with locust cells (Tables 2a and 2b). The locust cells showed a monotonic increase in accumulation of Cd in both the cytoplasmic compartment and the cell residue. These locust cells also accumulated less ppm Cd than the *Mandevilla*. The cell residue analyses were expressed in terms of both dry and wet cell weights in the automatic computer program of the ICPAES unit but are shown as ppm wet weight in the tables for easier comparisons between different types of experiments.

Discussion

The selected cell lines of Figs. 1a to 1f for the experiments were chosen to cover a broad range of types and then to identify specific types for use with the chemicals to be tested. The rose cells were from a line which was well characterized and standardized. The alfalfa cells were selected because it is used extensively in temperate pastures and is being studied in a number of laboratories for the genetic basis of its association with symbiotic, nitrogen-fixing *Rhizobium meliloti*. Alfalfa cells can also be regenerated and the plants studied for normal development between controls and experimental groups. Soybean cells used in our

TABLES 1a and 1b—*These tables show the increase of cadmium at increasing dosage in the nutrient medium, the "cytoplasmic" fraction, and the residue from the green* Mandevilla *cell cultures after three and seven days. The cytoplasmic and cell residue accumulation were not linear. The standard deviations are given in parentheses.*

TABLE 1a—CADMIUM CONCENTRATION OF *MANDEVILLA* CELLS—3 DAYS

Cd Dose, μM	Medium, ppm	Cytoplasm, ppm	Cell Residue, ppm
0	0.02 (0.003)	0.05 (0.004)	0.10 (0.04)
50	3.9 (0.10)	5.1 (0.7)	12.8 (1.2)
100	6.3 (0.4)	17.4 (4.3)	27.5 (1.2)
200	9.6 (0.8)	60.2 (16.8)	157.9 (23.6)
400	24.2 (1.7)	30.7 (16.9)	298.0 (65.0)
500	26.8 (1.6)	28.3 (2.7)	280.5 (49.2)

TABLE 1b—CADMIUM CONCENTRATION OF *MANDEVILLA* CELLS—7 DAYS

Cd Dose, μM	Medium, ppm	Cytoplasm, ppm	Cell Residue, ppm
0	0.03 (0.009)	0.18 (0.05)	0.05 (0.01)
50	3.7 (0.28)	23.3 (5.6)	13.9 (5.4)
100	6.1 (0.10)	63.7 (24.1)	24.6 (5.0)
200	11.1 (0.15)	34.2 (11.1)	213.4 (7.3)
400	23.8 (1.97)	46.1 (8.9)	399.5 (102.0)
500	27.2 (2.20)	65.9 (6.7)	379.9 (46.9)

TABLES 2a and 2b—These tables show the increase of cadmium in black locust cells at increasing dosage in the nutrient medium. In this case there was a linear increase in the "cytoplasmic" and the residue fractions. Less metal accumulated per gram fresh weight than in the green cells of Tables 1a and 1b.

TABLE 2a—CADMIUM CONCENTRATION OF ROBINIA CELLS—
3 DAYS

Cd Dose, μM	Medium, ppm	Cytoplasm, ppm	Cell Residue, ppm
0	0.02 (0.001)	0.02 (---)	0.21 (0.002)
50	4.98 (0.144)	0.12 (0.01)	16.76 (2.29)
100	10.17 (0.16)	0.19 (0.04)	31.86 (10.05)
200	21.22 (1.19)	0.36 (0.04)	87.51 (6.4)

TABLE 2b—CADMIUM CONCENTRATION OF ROBINIA CELLS—
7 DAYS

Cd Dose, μM	Medium, ppm	Cytoplasm, ppm	Cell Residue, ppm
0	0.02 (---)	0.002 (0.002)	0.14 (0.04)
50	4.91 (0.07)	0.16 (0.06)	18.27 (6.36)
100	9.79 (0.42)	0.33 (0.06)	40.65 (8.6)
200	20.30 (0.76)	0.61 (0.05)	75.05 (9.66)

culture were started from the seeds used for plant toxicity tests at ERL-Corvallis. The Mandevilla cells were chosen because the same callus can be split and grown in darkness or light rapidly and easily for comparisons between green and nongreen tissue. These cells can also be regenerated conveniently in the laboratory at ERL-Corvallis (private communication K. G. Wilson, Miami University, Oxford, OH). The black locust cells are an example of cells from an indigenous woody plant.

On the basis of increases in cell mass, different cells in suspension culture were found to tolerate up to 500-ppm toluene for a ten-day test period or, interpreting from past experience, through at least three cell doublings. As shown in Fig. 2, there were no significant differences in growth rates from the sampling of rose, Mandevilla, or alfalfa cells at different times.

The wet weights of Mandevilla cells at different levels of Cd were examined for normal weight distribution and homogeneity of variance before performing Dunnett's test (also confirmed by William's isotonic regression model). At the Cd dosages of 50 and 100 μM, the wet weights of the cells were not different from the controls ($p = 0.95$). There was a significant difference, compared to control weights, at Cd dosage of 200 μM and above. The cell weights between 200, 400, and 500-μM Cd dosage groups were not statistically different, i.e., there was no proportional decrease. The distribution of Cd was indicated in cells shown in Figs. 1b, 1d, and 1f. The crude separation of this Cd into "cytoplasmic" fraction and cell residue is given in Tables 1a, 1b and Tables 2a, 2b, for the Mandevilla cells and the black locust, respectively. The Cd in the media increased in proportion to the dosage during the experiment, less the amount taken up by the cells. The "cytoplasmic" Cd changed with dosage at three and seven days in a pattern that is interpreted as forming Cd intermediates, presumably compounds like phytochelatin and related peptides [10], before being deposited in the insoluble cell matrix or forming less soluble salts within the cell matrix.

The cell wet weights of the slow-growing black locust did not change. The black locust cells were started as sparse cultures with 0.15 to 0.3 gfw (grams fresh weight) per 20-mL medium. At the end of the experiment, the cells showed 0.3 to 0.5 gfw per 20 mL. The black locust cells accumulated less Cd in the "cytoplasmic" extract as well as the cell residue (Tables 2*a* and 2*b*). (It can be argued that the cells of the locust did not grow rapidly and hence did not accumulate Cd to the extent of the *Mandevilla* cells. There is a physiological reason for this difference. Unpublished results from our experiments show cation accumulation and distribution between the two cell types to be different.)

In conclusion, the present study was limited in not addressing the methodology for observing the distribution of the test compounds within soluble polar, nonpolar, and nonextractable residue [*11*]. The study has not further explored the increased protein and RNA synthesis at low Cd levels [*9*]. The intracellular distribution of Cd in the vacuolar compartment [*12*] has to be confirmed because this mechanism provides an alternate mechanism to alleviate Cd stress in addition to binding Cd to organic acid or a cytosol peptide such as "phytochelatin" [*10,12*]. The study confirms the drastic effect of Cd on plant cells at levels above 100 μM. This study addresses the usefulness of cell culture techniques to assess toxicity levels of xenobiotics in plants and shows that more than a single type of cell from an established cell line should be used in assessments concerning xenobiotics. The cell culture approach is especially safe in the cases where hazardous chemicals have to be assessed for their toxicity on plants in an isolated setting or when preliminary data are necessary on their mode of action on plant cells.

Acknowledgment

We wish to thank C. Wickliff and J. Fletcher for the Rose cell culture and K. G. Wilson for the *Mandevilla* plant. J. Fletcher also suggested the usefulness of the black locust for toxicity studies.

References

[*1*] Royce, C. L., Fletcher, J. S., and Riser, P. G., *Journal of Chemical Information and Computer Science*, Vol. 24, 1984, pp. 7–10.

[*2*] Mumma, R. O. and Davidonis, G. H. in *Progress in Pesticide Chemistry*, Vol. 3, D. H. Huston and T. R. Roberts, Eds., Wiley, Chichester, England, 1983, pp. 255–262.

[*3*] Reporter, M. in *Cell Culture and Somatic Cell Genetics of Plants*, Vol. 1. I. K. Vasil, Ed., Academic Press, Orlando, FL, 1984, pp. 586–597.

[*4*] Murashige, T. and Skoog, F., *Physiologia Plant.*, Vol. 15, 1962, pp. 473–479.

[*5*] Gamborg, O. L., Miller, R. A., and Ojima, K., *Experimental Cell Research*, Vol. 50, 1968, pp. 151–158.

[*6*] Gamborg, O. L. in *Cell Culture and Somatic Cell Genetics*, Vol. 1, I. K. Vasil, Ed., Academic Press, Orlando, FL, 1984, pp. 19–26.

[*7*] Nesuis, K., Uchytil, L. F., and Fletcher, J. S., *Planta*, Vol. 106, 1972, pp. 173–176.

[*8*] Wickliff, C. and Fletcher, J. S., this volume.

[*9*] Hirt, H., Casari, G., and Barta, A., *Planta*, Vol. 179, 1989, pp. 414–420.

[*10*] Reporter, M. in *Models in Plant Physiology and Biochemistry*, Vol. 3, D. W. Newman and K. G. Wilson, Eds., CRC Press, Boca Raton, FL, 1987, pp. 95–102.

[*11*] Harms, H. and Langebartels, C., *Plant Science*, Vol. 45, 1986, pp. 157–165.

[*12*] Krotz, R. M., Evangelou, B. P., and Wagner, G. J. *Plant Physiology*, 1989, pp. 780–787.

Carlos Wickliff and John S. Fletcher

Tissue Culture as a Method for Evaluating the Biotransformation of Xenobiotics by Plants

REFERENCE: Wickliff, C. and Fletcher, J. S., **"Tissue Culture as a Method for Evaluating the Biotransformation of Xenobiotics by Plants,"** *Plants for Toxicity Assessment: Second Volume, ASTM STP 1115,* J. W. Gorsuch, W. R. Lower, W. Wang, and M. A. Lewis, Eds., American Society for Testing and Materials, Philadelphia, 1991, pp. 250–257

ABSTRACT: Suspension cultures of *Rosa* cultivar (cv.) Paul's Scarlet were used as a model system to examine the metabolism of 1,3-dinitrobenzene (DNB), an industrial waste compound. In a 3-day period, 90% of the DNB supplied (96 nmol) was metabolized by approximately 12 g (fresh weight) of cells. The primary end product of DNB metabolism partitioned into the insoluble residue fraction of the extracted cells. This study demonstrates the usefulness of plant tissue cultures in evaluating the chemical influence of plants on xenobiotics.

KEY WORDS: suspension cultures, dinitrobenzene, metabolism, plant tissue culture

In assessing the environmental influence and fate of xenobiotics, consideration must be given both to the phytotoxicity of chemicals and the capacity for plants to chemically alter organic pollutants. Although there are many deficiencies in our knowledge of how xenobiotics influence the growth and development of plants, there is an even greater deficiency in understanding how plants influence the fate of xenobiotics released into the environment. Due to this deficiency, it is difficult for regulatory agencies to accurately assess the impact and fate of the broad spectrum of chemically diverse xenobiotics that exist in contaminated areas or that will be released in the future as new products, including insecticides and herbicides.

Plants account for the largest portion of biomass in terrestrial ecosystems and have been shown to modify cycling patterns and chemical fate of many organic and inorganic hazardous materials [1–4]. Thus, there is a need to collect more data on plant-chemical interactions. Such data are needed to develop models and assessment schemes for accurately predicting the impact and fate of xenobiotics released into terrestrial ecosystems. Plant tissue cultures have many features which make them useful in collecting phytotoxicity and biotransformation data. They provide a rapid test system whereby small amounts of hazardous chemicals can be evaluated in a controlled area under safe conditions. Tissue cultures are also well suited for extraction and chemical analysis due to the lack of fibrous tissue, and there is minimal interference from microorganisms, which sometimes confuse bioassay interpretation. These features permit a large assortment of plants and chemicals to be examined within a short period of time.

[1] Biologist, U.S. Environmental Protection Agency, Environmental Research Laboratory, Corvallis, OR 97333.

[2] Professor of botany, Department of Botany and Microbiology, University of Oklahoma, Norman, OK 73019-0245.

Algal cultures have been used in xenobiotic uptake, fate-and-effect, and metabolism studies [5–7]. However, far less progress has been made in culture systems of terrestrial plant cells. Plant tissue cultures allow the laboratory testing of the same species that will be subjected to the same chemicals found in hazardous waste sites. Therefore, cell cultures can better be used to predict the fate of such chemicals than algae, which will not be grown at these waste sites. There are data that indicate that algae and different types of plants have different metabolic potentials relative to some substrates [5]. However, one purpose of this research is to further refine the tissue culture method as a tool for xenobiotic uptake, fate-and-effect, and metabolism studies. Therefore, this laboratory study was undertaken.

In the present study, suspension cultures of *Rosa* cv. Paul's Scarlet were used as a model system to examine the influence of 1,3-dinitrobenzene (DNB) on plant growth, and the influence of plant cells on the biotransformation of the DNB. The results of this study show how a single test system (a tissue culture test) can be used to simultaneously obtain data on the actions of both the chemical and the plant on each other.

Materials and Methods

This investigation used nonphotosynthetic suspension cultures of *Rosa* cv. Paul's Scarlet grown in the dark at 25°C on a rotary shaker set at 150 rotations/min (rpm). Individual cultures were provided with 80 mL of minimal organic medium [8] designed for maximum growth of the cultures. The cultures were transferred biweekly and harvested for growth and chemical analysis as previously described [9].

The toxicity of 1,3-dinitrobenzene (DNB) to rose cultures was estimated by testing the cultures with 0.1, 1.0, 10, and 100 µg/mL concentrations of the chemical in a range-finding test. All additions were made aseptically with a µL-syringe by adding an ethanol solution of DNB to the medium containing 11-day-old cultures since the cells are in logarithmic growth phase at this time and doubling at maximum rate; a carrier control containing only ethanol (0.9%) was included with the tests. Three days after adding DNB to the cultures, the absence of contaminating organisms was verified by standard microbial plating techniques. The rose cells were then separated from their medium by filtration through Miracloth and then glass-fiber filter paper (1.6-µm effective retention), rinsed with fresh culture media and then with deionized water, dried (24 h at 70°C), and weighed (to the nearest 0.1 mg). The dry weight data were used to determine DNB effect upon cell growth through a one-way analysis of variance (ANOVA) followed by Dunnett's test to compare treatments to the control.

The metabolism of DNB by plant cells was studied by adding 96 nmol of ^{14}C-1,3-DNB (10.2 µCi/µmol) in an ethanol solution with a µL syringe to 11-day-old cultures to yield a concentration of 0.2 µg of DNB per mL; this concentration was shown previously to be nontoxic. The ^{14}C-DNB was purchased from Sigma Chemical Company, St. Louis, Missouri[3] and was shown to be 98% pure by thin layer chromatography (TLC) [10] and high-pressure liquid chromatography (HPLC) [11]. The cells and media were harvested and separated by filtration through Miracloth and glass fiber filter paper (as above) and the cells dried and weighed as described above. To provide a "killed control," a parallel group of 11-day-old cultures was autoclaved at 121°C for 40 min and allowed to cool to room temperature before being treated with ^{14}C-DNB and harvested at the same time and in the same manner as the living cells (described above).

The cells and the medium collected from each culture were both subjected to extraction

---

[3] Mention of trade names or commercial products does not constitute endorsement or recommendation for use.

and solvent fractionation as previously described [12]. This procedure partitioned substances into three fractions: chloroform soluble (nonpolar compounds), methanol/H_2O soluble (polar compounds), and an insoluble residue. Each of these fractions was sampled and analyzed for radioactivity by liquid scintillation spectroscopy. The insoluble residues were oxidized in a Packard model 306 combustion oxidizer[3] prior to being radioassayed. Standard samples containing known ^{14}C were ignited in the oxidizer preceding the samples and at the end of the analysis period to check the overall efficiency of the oxidation and counting [13]. The residue samples were oxidized in 20 to 50-mg portions, and the $^{14}CO_2$ resulting from the oxidation was collected in 9-mL absorbing solution and 13-mL scintillation cocktail added. The absorbing solution was Carbosorb, while Permafluor V was the scintillation cocktail[3]. Both materials were obtained from Packard Instrument Company[3] and mixed in proportions recommended by the manufacturer for plant samples after combustion and absorption of $^{14}CO_2$.

Liquid fractions that contained large quantities of ^{14}C were analyzed further by thin layer chromatography (TLC) on silica gel plates [10]. The plates spotted with nonpolar compounds (chloroform fraction) were developed with toluene-acetonitrile (3:1), whereas those receiving polar compounds (methanol/H_2O fraction) were developed in toluene-ethanol-methanol (2:1:1). The positions of radioactive compounds on the developed plates were determined by autoradiography using Kodak[3] diagnostic X-ray film. Regions of the silica gel possessing radiation were scraped from the plates, oxidized and assayed for ^{14}C content by liquid scintillation spectroscopy. Aliquants of the ^{14}C-DNB dosing solution were included on each chromatogram to serve as reference points for the culture extracts and to establish the purity of the parent substance.

Results and Discussion

Growth

A comparison of the dry weights of cells in the cultures in response to increasing concentrations of 1,3-dinitrobenzene (DNB) showed that culture growth was not significantly ($P < 0.05$) influenced at DNB concentrations <1 ppm (Fig. 1). The R^2 for the one-way ANOVA was 0.95 and P-value for the test was 0.0001. This finding is consistent with the effect of DNB on soybean growth in previous experiments with whole plants[4]. Soybean growth was not affected by DNB concentrations from 0.1 to 1.0 ppm in the nutrient solution. However, the EC_{50} value (concentration reducing growth 50%) for soybean growth was 2.17 ppm, while that for rose culture growth was 25.2. Thus, rose cultures are much less sensitive than soybean plants to DNB. Replicating this experiment with soybean cell cultures will be necessary to determine whether the relatively low sensitivity is due to the bioassay itself or to the plant species. This method may be made more useful as an impact assessment tool if a more sensitive species is used in parallel with the rose cells, which may be decreasing the toxicity with a more extensive biotransformation of the chemical (see "Metabolism" below).

At the higher concentrations of 1,3-DNB (10 and 100 ppm), dry weights of the rose cultures were declined (Fig. 1), indicating substantial cell destruction. Cell dry weight increased 47% in the carrier control (0.9% ethanol) over the 3-day growing period (Fig. 1), which was similar to increases in dry weights of the negative controls (i.e., lacking either ethanol or DNB; not shown). Thus, the presence of the low concentrations of ethanol used as a carrier did not complicate interpretation of the experimental results. There appeared to be a slight stimulation of cell growth at the lower DNB concentrations. Subsequent metabolic studies were conducted using DNB at a concentration of 0.2 ppm, which was nontoxic.

---

[4] Personal communication.

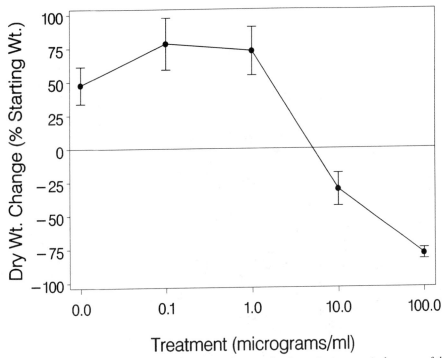

Treatment (micrograms/ml)

FIG. 1—*Toxicity of 1,3-dinitrobenzene to rose cells. The value for each treatment is the mean of three replicates. The error bars indicate standard deviation of the means. The ethanol carrier control is given as the 0.0 treatment in the figure. Culture growth with ethanol (0.9%) did not differ from growth without ethanol (see text for explanation).*

Metabolism

The distribution patterns of ^{14}C among various fractions recovered from living versus autoclaved rose-cell cultures showed that living plant cells maintained under axenic conditions were capable of metabolizing DNB (Table 1). In the living cultures, the percent distribution among the chloroform, methanol-H_2O, and insoluble residue fractions was 9, 36, and 55%, respectively. In contrast, the amount of ^{14}C recovered from the autoclaved cultures was almost all (94%) in the nonpolar chloroform fraction. The parent compound (DNB) is nonpolar and highly soluble in chloroform. Thus, the small amount of chloroform-soluble ^{14}C recovered from the living cultures versus that recovered from the autoclaved cultures indicated that the majority of the DNB had been biotransformed by the living cells. All (100%) of the added ^{14}C was recovered from the autoclaved cells, while only 74% was recovered from the living cultures. This suggests the formation by living rose cells of a gaseous metabolite(s) of DNB, which was not recoverable using the extraction procedures.

The ability of the plant cultures to metabolize DNB was further substantiated by chromatographic analysis of the soluble fractions that contained the greatest amounts of [14]C (Table 2). When the methanol-H_2O extract recovered from the living cells was subjected to TLC, radioactivity was present at five locations on the chromatogram. None of these locations, though, had the *Rf* of DNB. In an earlier study, several unidentified DNB metabolites were recovered from polar extracts of soybean roots [10]. Two of those metabolites had *Rf* values identical to the *Rf* values found in extracts of rose cells when comparable extraction solvent (ethanol-water) and similar TLC eluent (benzene/ethanol, 5:1)

TABLE 1—Distribution of ^{14}C in extracts prepared from cells and medium following a 72-h exposure of 11-day-old living and dead cultures to 0.2 ppm ^{14}C-1,3-dinitrobenzene.

Culture	Component	Fraction											
		Chloroform			MeOH-H$_2$O			Residue			Total		
		dpm × 10^{-3}											
Living	Cells	67^a	13^b	(4)^c	441	24	(28)	777	288	(49)	1285	305	(81)
	Medium	76	23	(5)	128	37	(8)	90	48	(6)	294	69	(19)
	Total	143	12	(9)	569	55	(36)	867	276	(55)	1579^d	236	(100)
Autoclaved^e	Cells	46	2	(2)	7	2	(<1)	42	8	(2)	95	8	(4)
	Medium	2067	75	(92)	93	5	(4)	5	4	(<1)	2165	3	(96)
	Total	2113	77	(94)	100	4	(4)	47	12	(2)	2260^f	69	(100)

NOTE: DPM = distintegrations per minute.

^{a} Values are means based on three replicates.

^{b} Standard deviations of the means.

^{c} Percent of the total counts recovered.

^{d} Value represents 74% of the provided 2.2 × 10^6 dpm.

^{e} Autoclaved for 40 min at 121°C (See "Materials and Methods" for explanation).

^{f} Value represents 103% of the provided 2.2 × 10^6 dpm.

TABLE 2—Thin-layer chromatography of selected fractions recovered from 1,3-dinitrobenzene treated cells.

Culture	Component	Fraction	Rf 0.0	0.1	0.2	0.3	0.4	0.5	0.6	0.7	0.8	0.9	10.0
Living	Cells	Chloroform[a]	(42)(44)[c]							(8)		(6)	
		Methanol-H2O[b]	(39)(15)			(11)			(12)		(23)		
	Medium	Chloroform	(5)(13)							(23)		(46)	
		Methanol-H2O	(54) (7)				(9)	(9)		(9)(2)	(9)		(4)
Dead	Cells	Chloroform	(24)		(6)	(13)						(43)	
		Methanol-H2O				(3)	(4)	(9)(6)	(5)	(6) (36)(36)(28)			
	Medium	Chloroform	(2)										
		Methanol-H2O	(100)									(98)	
1,3-Dinitrobenzene Standards													
Toluene-acetonitrile (3:1)			(5)									(95)	
Toluene-ethanol-methanol (2:1:1)			(1)								(2)		(97)

[a] All chromatograms prepared from chloroform fractions were eluted with toluene-acetonitrile (3:1).
[b] All chromatograms prepared from methanol-H2O fractions were eluted with toluene ethanol-methanol (2:1:1).
[c] Values in parentheses are the percent of the total ^{14}C recovered from developed chromatograms at the corresponding Rf value above.

were used. These results suggest that diverse species of plants may metabolize DNB in a similar manner.

Analysis of the methanol-H_2O fraction recovered from the medium of living cultures of rose cells showed that the extract contained at least seven radiolabeled constituents, three of which had Rf values corresponding to those of compounds recovered from the cells, but none matching the Rf of DNB. When the most prominent ^{14}C-labeled constituent recovered from the autoclaved rose cell cultures (chloroform extract of the medium) was chromatographed, we found that 98% of the radioactivity had an Rf value corresponding to that of pure DNB.

A plant tissue culture system may demonstrate the potential for a given species to metabolize different classes of compounds. It is unknown whether the same species expresses all or only a portion of this potential under a given set of environment conditions. The metabolites produced may depend upon combinations of environmental factors (e.g., temperature, type of media, etc.).

While diverse species of plants may metabolize DNB similarly, other compounds may be metabolized to divergent derivatives depending on the plant species. For example, benzo(a)pyrene (BaP) was metabolized differently by a suspension culture of soybean cells and the green alga *Selenastrum Capricornutum* [2,7]. Products of the algal metabolism of BaP were the *cis*-11,12- and *cis*-7,8-diols, while soybean converted BaP to the 1,6- and 3,6-quinones [2,7].

The metabolism of other compounds, such as the insecticide DDT, the ascaricide Kelthane, and the herbicides bentazon and 2,4-D, was studied with soybean tissue culture systems [2,14,15], and wheat was studied with 2,4-D [2]. Soybean metabolized 14% of applied 2,4-D, whereas wheat metabolized 82% of the same applied amount of this herbicide, largely to (70%) polar compounds. The soybean suspension cultures metabolized both DDT and bentazon to polar metabolites, but the metabolism of bentazon (79% in 6 h) was more extensive [14,15]. DDT was more resistant to metabolism by a suspension culture of parsley than by soybean [4]. Velvetleaf cells, used in parallel with soybean [15] did not metabolize bentazon in 6 h; all the radiolabel was recovered as the parent bentazon compound.

Independent investigators found that different wheat tissue cultures metabolized pentachlorophenol (PCP) similarly [2,16]. The percentages of polar metabolites were nearly identical (50 and 48%) in the separate investigations; the nonpolar metabolites were so low as to be nearly undetectable in both studies. The percentages of nonextractable residues were similar in the two investigations (50 and 40%). These results demonstrate the replicability of tissue culture test systems.

Most products of biotransformations are less toxic than the parent compound and/or have been metabolized to CO_2 since living organisms attempt to detoxify exogenously applied poisonous compounds [17], although in some instances a product of biotransformation or an intermediate may be more toxic than the parent compound [17]. Thus, biotransformation data from tissue culture experiments can be used to assess the impact of synthetic organic chemicals on the environment based on the ability of plants to metabolize them into substances less toxic (detoxification). For example, a compound that is rapidly transformed (i.e., detoxified) would be expected to have a less detrimental effect on living organisms than one that is resistant to biotransformation, which would allow the unchanged parent toxic compound to remain in the environment.

Conclusions

Suspension cultures of rose cells biotransformed approximately 90% of the provided DNB within three days. These results, along with similar reports in the literature for other plant cultures and xenobiotics [2,18,19], show that plant culture systems can be simple, safe, and economical for studying a broad spectrum of toxic waste compounds for which we have little plant-biotransformation data [20]. Plant cells can be cultured in suspension in small (250-mL) flasks containing a small volume (80 mL) of nutrient media which occupy a small amount of laboratory space. Experiments with toxic and/or radioactively labeled test chemicals can thus be conducted in a hood meeting laboratory safety requirements. Data on biotransformation of the material can be obtained more rapidly in a manner insuring greater safety to the operator than would be the case with larger quantities of material or with whole-plant test systems.

Other advantages of the system are: (1) only 11 days are required to culture the plant material to experimental size (logarithmic growth phase); (2) several treatment replicates can be maintained simultaneously; (3) more tests can be completed in a given amount of time than would be possible with whole plants; and (4) the information gained using this system will suggest compounds warranting further or more extensive testing with whole plants. A plant tissue culture system may also be a useful addition to laboratory tests evaluating the chemical fate of new synthetic compounds.

References

[1] McFarlane, C. and Pfleeger, T., Environmental Toxicology and Chemistry, Vol. 9, 1990, pp. 513–520.

[2] Harms, H. and Langebartels, C., Plant Science, Vol. 45, 1986, pp. 157–165.

[3] Scheel, D., Schäfer, W., and Sandermann, H., Jr., Journal of Agricultural and Food Chemistry, Vol. 32, 1984, pp. 1237–1241.

[4] Callahan, M. A. and Slimak, W. M., "Water-Related Environmental Fate of 129 Priority Pollutants," EPA 440/4-79-0296, U.S. Environmental Protection Agency, Washington, DC, 1979.

[5] Gentile, J. M., Johnson, P., Lippert, M., and Shafer, T., "Comparative Metabolism of Promutagens by Algae and Tobacco, Cotton, and Carrot Cells in Culture," this symposium.

[6] Lee, R. F., Valkins, A. O., and Seligman, P. F., Environmental Science and Technology, Vol. 23, 1989, pp. 1515–1518.

[7] Warshawsky, D., Radike, M., Jayasimhulu, K., and Cody, T., Biochemical and Biophysical Research Communications, Vol. 152, No. 2, 1988, pp. 540—544.

[8] Nesius, K. K., Uchytil, K. K., and Fletcher, J. S., Planta, Vol. 106, 1972, pp. 173–176.

[9] Mohanty, B. and Fletcher, J. S., Physiologia Plantarum, Vol. 42, 1978, pp. 221–225.

[10] McFarlane, C., Nolt, C., Wickliff, C., Pfleeger, T., Shimabuku, R., and McDowell, M., Environmental Toxicology and Chemistry, Vol. 6, 1987, pp. 847–856.

[11] McCrady, J. K., McFarlane, C., and Lindstrom, F. T., Journal of Experimental Botany, Vol. 38, Nov. 1987, pp. 1875–1890.

[12] Fletcher, J. S., Groeger, A. W., and McFarlane, J. C., Bulletin of Environmental Contamination and Toxicology, Vol. 39, 1987, pp. 960–965.

[13] Wickliff, C., McFarlane, J. C., and Ratsch, H., Environmental Monitoring and Assessment, Vol. 4, 1984, pp. 43–51.

[14] Scheel, D. and Sandermann, H., Jr., Planta, Vol. 33, 1977, pp. 315–320.

[15] Sterling, T. M. and Balke, N. E., Weed Science, Vol. 36, 1988, pp. 558–565.

[16] Schifer, W. and Sandermann, H., Jr., Journal of Agricultural and Food Chemistry, Vol. 36, 1988, pp. 370–377.

[17] Hodgson, E. and Dauterman, W. C. in Introduction to Biochemical Toxicology, E. Hodgson and F. E. Guthrie, Eds., Elsevier, New York, 1980, Chap. 4, pp. 67–91.

[18] Sandermann, H., Jr., Scheel, D., and Trenck. T. H. V. D., Ecotoxicology and Environmental Safety, Vol. 8, 1984, pp. 167–182.

[19] Harms, H., Landbauforschung Völkenrode, Vol. 31, 1981, pp. 1–6.

[20] Fletcher, J. S., Johnson, F. L., and McFarlane, J. C., Environmental Toxicology and Chemistry, Vol. 7, 1988, pp. 615–622.

Peter Schröder[1] and Reinhard Debus[2]

Responses of Spruce Trees (*Picea Abies. L. KARST*) to Fumigation with Halone 1211—First Results of a Pilot Study

REFERENCE: Schröder, P. and Debus, R., **"Responses of Spruce Trees (*Picea Abies. L. KARST*) to Fumigation with Halone 1211—First Results of a Pilot Study,"** *Plants for Toxicity Assessment: Second Volume, ASTM STP 1115*, J. W. Gorsuch, W. R. Lower, W. Wang, and M. A. Lewis, Eds., American Society for Testing and Materials, Philadelphia, 1991, pp. 258–266.

ABSTRACT: Fumigation experiments with young cloned spruce trees were performed in a pilot study to test the trees' ability to react to pollutant impact. After 41 days of exposure to 10-ppb Halone 1211 (difluoro-chloro-bromo-methane), changes in needle fresh weight, protein content, and pigment pattern were observed compared to a control fumigation with purified air. These changes, however, seemed nonspecific reactions to a pollutant climate because they were also obtained after fumigation with nitrogen dioxide (NO_2). Only in the halone-fumigated trees was up to fourfold increases of the activity of glutathione-S-transferase, a constitutive detoxification enzyme in spruce observed. The significance of this reaction for biomonitoring of halogenated compounds is discussed.

KEY WORDS: difluoro-chloro-bromo-methane, *Picea abies*, fumigation, glutathione-S-transferase, Halone 1211, spruce

During recent years, the influence of halo-organic compounds on animals and plants has gained considerable interest. Most halo-organic compounds in the atmosphere are of anthropogenic origin. Their emission started at the beginning of the industrial revolution with the growing need for pesticides, solvents, heat exchangers, propellants, and fire repellants. Whereas pesticide and herbicide use were rapidly put under government control, the manufacturing, processing, and use of the other above-mentioned substances were not restricted for several decades. As a consequence, these substances, with lifetimes in the range of years or even decades, are emitted into the atmosphere in steadily increasing amounts (1.8 million tons per year in the FRG [*1*]), and their background mixing ratios in Europe are presently found to be in the range of 50 to 500 parts per trillion (ppt), with a trend towards increasing values [*2*]. Recently, evidence has been presented that these gases play an important role in the ozone depletion of the stratosphere, in the greenhouse effect, and that they might also influence the viability of animals and plants.

Halone 1211 is the trade name for difluoro-bromo-chloro-methane, used mainly as a fire repellant in commercially available fire extinguishers because of its inertness and inflammability. Its present global atmospheric concentration is estimated to be in the range of 1 ppt, with higher concentrations in the Western Hemisphere, and its lifetime is estimated to be in the range of decades [*3*].

---

[1] Plant physiologist, Fraunhofer Institute for Atmospheric Environmental Research, Kreuzeckbahnstrasse 19, D-8100 Garmisch-Partenkirchen, FRG.

[2] Plant physiologist, Fraunhofer Institute for Ecotoxicology and Environmental Chemistry, Grafschaft, D-5948 Schmallenberg, FRG.

During fire fighting, large amounts of Halone 1211 are emitted, and, because of its high molecular weight, it is rapidly deposited on surfaces, including plants. Halone 1211 is highly lipophilic and will quickly accumulate in the cuticles of plant leaves and in cell membranes. This path of uptake is common for various lipophilic hydrocarbons [4].

Once having entered the plant, lipophilic compounds are able to diffuse into the cytoplasm and enter cell compartments like mitochondria and chloroplasts [4,5]. There they can act as uncouplers or inhibitors of photosynthesis and respiration, similar to the action of various pesticides. In addition, activation by ultraviolet (UV) radiation might lead to reactive intermediates also influencing essential metabolic functions of plants [5,6,7].

Many animals have been reported to possess detoxification mechanisms for pesticides and other xenobiotics. Several crop plant species have also been found to detoxify certain herbicides, for example of the triazine or chloroacetanilide family [7]. In both animals and plants, one of the most important detoxification pathways starts with the action of glutathione-S-transferases (GST, Enzyme Code (EC) 2.5.1.18), a group of enzymes catalyzing the nucleophilic attack of reduced glutathione (GSH) to electrophilic xenobiotics. Glutathione-S-transferases were first studied in trees by Balabaskaran and Muniandy in *Hevea brasiliensis* [8]. The first investigations on GST in boreal forest trees (*Picea spec.*) were performed by Schröder et al. [9,10].

The present study was conducted to elucidate reactions of spruce trees fumigated with low concentrations of chemically inert halocarbons, taking Halone 1211 as a model pollutant, and to show a possible use of a simple enzyme assay for biomonitoring the impact of organic compounds.

To exclude effects generated by the fumigation chamber itself and the fumigation with a pollutant gas as such, spruces were exposed to purified pressurized air and to 30-ppb nitrogen dioxide (NO_2), which is known to affect plant physiology [11,12] in control experiments.

Materials and Methods

Plant material—For the experiments, six-year-old cloned Norway spruce trees (*Picea abies* KARST; Harz-Vorland, Hessische Forstliche Versuchsanstalt, Hannoversch-Gmünden, FRG) were used. The plants were grown in flower pots outdoors and transferred into the fumigation cuvettes for the experiments.

Chemicals—Halone 1211 was obtained from Messer-Griesheim (Hagen, FRG); CDNB (1-chloro-2, 4-dinitrobenzene), PVPP (polyvinylpolypyrrolid), Nonidet®, and reduced glutathione (GSH) were from Sigma Chemie (Steinheim, FRG). All other chemicals used were research grade commercial materials.

Fumigation and microclimate—Eight six-year-old clone trees of similar outward appearance, height, development stage, and branching pattern were chosen for the experiments. Two trees were exposed to 10 ppb of Halone 1211 in a cuvette under controlled environmental conditions in a plant growth chamber (Haereus, Hanau, FRG) for 41 days. The experimental setup for the fumigation cuvettes is shown in Fig. 1.

As Fig. 1 illustrates, fumigation was done with calibrated gaseous pollutant standards (1) mixed with purified air to result in the appropriate concentration. The gas mixture was adjusted (2) by a flow controller (Aadco, Rockville, MD). Before entering the fumigation cuvette, the gas was led through a tube with a sampling port (3) to allow the collection of gas samples for exact determination of the mixing ratio by gas chromatography (GC). The cuvettes (4) were constructed from glass plates (4 mm thick) and glued from the outside with silicone rubber. The cuvette volume was 0.15 m³. A sensor (5) for microclimate measurements was introduced into the cuvettes to monitor relative humidity and temperature.

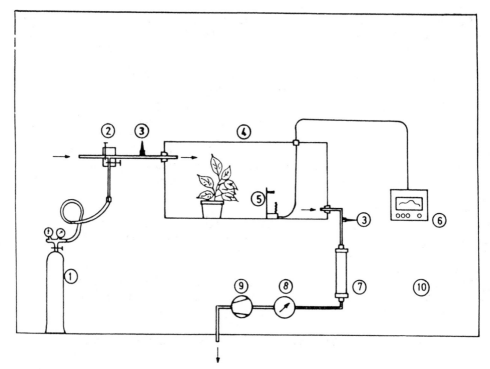

FIG. 1—*Display of the fumigation system.*

The values were recorded (6) by a strip-chart-recorder (Philips, Eindhoven, The Netherlands).

In order to remove the pollutants from the air, the air stream was passed through an activated carbon filter (7) before release of the gas into the environment. The flow rate through the cuvettes was recorded (8) by a gas meter (Elster, Mainz, FRG) to provide exact control of the pollutant doses applied to the plants. Total gas flow through each of the cuvettes was 30 L/min. A pump (9) (Reciprotor/Danmark) was connected to the system to withdraw the gas mixture and rinse the cuvettes. The setup as a whole was situated (10) in a plant growth chamber (Haereus, Hanau, FRG).

The plants were kept in a 12-h light period (max light intensity 3100 μmol m^{-2}s^{-1}) including 3-h of dawn (130 to 500 μmol m^{-2} s^{-1} within 3 h) and 2 h of dusk (500 to 130 μmol m^{-2} s^{-1} within 2 h) at temperatures of 9°C during the night and 23°C during the day. In the cuvettes, the trees were supplied with water by a wet wick system (Fackert & Helfert, Moers, FRG). In addition to this experiment, the following controls were conducted simultaneously in the same chamber in separate cuvettes: (*a*) fumigation with pressurized, purified air, halocarbon concentration $\ll$ 1 ppt (*n* = four trees, 41 days); (*b*) fumigation with pressurized, purified air, containing 30 ppb of NO$_2$ (*n* = two trees, 35 days). After 35 and 41 days, respectively, the trees were removed from the cuvettes. Needles and branches were checked for visual damage, needle year classes were separated, needle fresh and dry weights (after 48 h at 85°C) were determined, and the samples were stored in liquid nitrogen for later analysis.

Pigments—For the investigation of pigment contents, 0.5 g of the needles were inserted into 5 mL of dimethyl formamide and stored for 48 h at 4°C. Then the solvent was extracted, centrifuged, diluted 1:10 (v:v) with 80% aqueous methanol, and separated by high pressure liquid chromatography (HPLC) (Gynkotec, FRG) on a 250 by 4 mm Spherisorb C 6 column 5 μm (Latek, FRG). Separation was via a linear binary gradient of 80 to 100% methanol within 10 min. Detection was by Gynkotec UV/VIS monitor at 455 nm. The amounts of the respective pigment were calculated from standards of chlorophylls a and b and of β-carotene (Sigma Chemie, Steinheim, FRG) and from spectrophotometric determinations of single xanthophylls[3]. Chlorophyll/carotenoid and chlorophyll/β-carotene ratios were taken as an indication for possible damage [13].

Enzyme extracts—The frozen needles were homogenized as previously described [10] under liquid nitrogen with mortar and pestle and extracted at 4°C in an Ultra-Turrax (IKA, Stauffen, FRG) with 1% of soluble PVPP (K30) in ten volumes (wt/v) of 0.1 M potassium phosphate buffer, pH 7.8, containing 1% Nonidet®, and 5-mM ethylene-diamine-triacetic acid (EDTA). The crude extract was centrifuged at 20 000 × g. Proteins in the supernatant were precipitated by stepwise addition of solid ammonium sulfate to 45, 55, and 80% saturation. During this procedure, the pH was maintained at 7.8 by dropwise adding 10 N sodium hydroxide (NaOH). After each step the extracts were centrifuged at 20 000 × g and the pellets were resuspended in 1 mL 20-mM potassium phosphate buffer, pH 7.8. The extracts were desalted and further purified by passing them through gel-filtration columns (PD 10, Pharmacia, Freiburg, FRG).

Spectrophotometric enzyme assay and protein determination—GST activity was determined at 25°C in a standard spectrophotometric test at 340 nm [14]. Aliquots of the enzyme extract were incubated with 0.1 M potassium phosphate buffer, pH 6.4, 1-mM GSH and 1-mM CDNB (e_{340nm}mM^{-1} cm^{-1} = 9.6), resulting in a total assay volume of 1.1 mL. Controls lacking enzyme extract or GSH were measured. Enzyme activity is expressed in units of kat (Katal). One katal represents the enzymatic formation of 1 mol of endproducts per second. Protein in the enzyme extracts was determined by the standard method described by Bradford [15] using bovine serum albumin as a reference protein.

Statistical treatment of results—Because of the low number of replicates in this pilot study, all results were subjected to statistical treatment using the STATGRAPHICS®-PC-software (STSC Inc., Rockville, MD). For each set of data analysis of variance, testing was done before comparison of the data by the t-test provided in the program package. Duncan's multiple range test was used to test results from less than eight measurements [16].

Results

Weight determination—Needles of spruce trees exposed to Halone 1211 did not show any significant differences in their fresh and dry weights when compared to NO_2 and pure air fumigated control plants subjected to the same environmental conditions. The fresh-to-dry weight ratio in two-year-old needles was 1.83, 1.79, and 1.84; in one-year-old needles it was 1.16, 1.66, and 1.32 for the halone, pure air, and NO_2 treatments, respectively. The extremely low fresh-to-dry weight quotient in the halone-treated, one-year-old needles might be an indication for a water loss under the influence of the pollutant. It is, however, not significant and requires further investigation.

Pigment pattern—After exposure to 30-ppbv NO_2 for 35 days, the pigment pattern for two-year-old needles did not differ significantly from the pure air treatment (Fig. 2a). This is in accordance with previously achieved results from trees of the same clone[3]. In the younger needles, however, an overall decrease of about 50% in both chlorophylls ($p = 0.05$) and β-carotene ($p = 0.1$) was observed (Fig. 2b). A similar, but not significant decrease was also observed in Lutein and Neoxanthin contents of these needles.

---

[3] Debus, unpublished results.

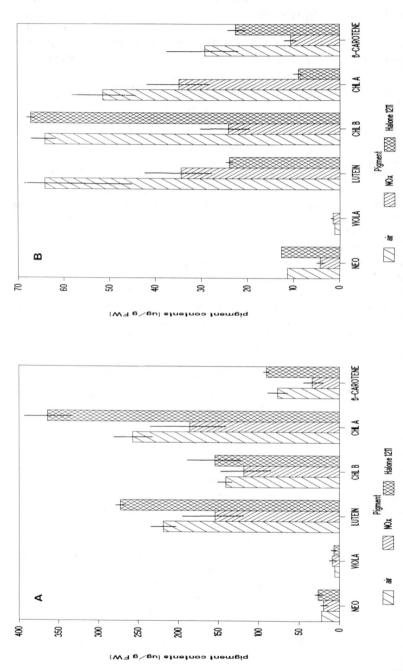

FIG. 2—Pigment patterns of halone or NO_x fumigated and control trees after 35 and 41 days of treatment. Bars are means of three to five determinations ± standard deviation. Treatments were: purified air (hatched bar); 30-ppb NO_2 (dense hatched bar); 10-ppb Halone 1211 (cross-hatched bar). 2A: pigment contents of two-year-old needles on a g fresh weight (FW) basis. 2B: pigment contents of one-year-old needles on a g FW basis.

Fumigation with Halone 1211 changed the pigment pattern strongly: in the two-year-old needles, chlorophyll a was 30 to 50% (p = 0.05) higher than in both controls (Fig. 2a). Lutein and β-carotene were 77 and 170% higher than in the NO_2 treatment (both at p = 0.01). This also influenced the chlorophyll/carotenoid ratio. It was found to be 1/1.12 in the control plants, whereas a ratio of 1/1.3 was obtained after halone fumigation.

Pigment contents in one-year-old needles were found to be generally lower than in two-year-old needles (Fig. 2b), and the halone-fumigated plants showed significantly lower lutein (air: p = 0.05; NO_2: p = 0.1) and chlorophyll a values (air: p = 0.01; NO_2: p = 0.05). The other pigments remained essentially at the control levels.

Protein contents—Both halone and NO_2 exposure led to a decrease in protein contents in the crude extracts of older needles (Table 1). In the younger needles, however, only halone fumigation decreased the protein values significantly. This pattern is not in accordance with the protein values obtained after ammonium sulfate precipitation; in the 40 to 80% saturation step, NO_2-fumigated as well as halone-treated plants showed 2.3 to 2.5-fold higher protein values than the pure air control when calculated on a fresh weight basis (Table 1).

Glutathione-S-transferase activity—The decrease of protein content in two-year-old needles, which was found after exposure to 30-ppb NO_2 for 35 days, was accompanied by a slightly increased GST activity in the same fraction (Fig. 2a). In the crude extracts of one-year-old needles, a similar increase of GST activity compared to the purified air control was observed (Fig. 3b), corresponding to an increased protein value in these samples. Both tendencies, however, are not statistically significant. As nitrate is no substrate for the enzyme, this result could be expected. The result excludes, however, possible nonspecific alterations of the GST activity as response to fumigation as such.

In crude needle extracts of halone-fumigated trees, the activity of glutathione-S-transferases, when assayed with CDNB, was significantly higher as compared to both the purified air and the NO_2 treatments (Fig. 3a). In two-year-old needles, GST was found to possess 2.3 times the activity of the purified air treated and 1.4 times the activity of the NO_2-fumigated needles. In the younger halone-fumigated needles, activity was 1.4 and 2.0 times higher than in the air and the NO_2-treated ones, respectively.

After partial purification of the enzyme by ammonium sulfate precipitation and gel filtration chromatography, similar values were obtained (Fig. 3b) for both needle year classes. When compared to the GST activity in needles exposed to pure air and in NO_2-enriched atmosphere, the halone-fumigated needles showed significantly increased values in all sam-

TABLE 1—*Protein contents of fumigated and control trees. Crude extracts and 40 to 80% ammonium-sulfate fractionations were prepared as described in the "Materials and Methods" section. Values are means of 4 to 10 measurements ± standard error.*

		Protein content, mg/gFW	
Sample	Fumigation	1988	1989
Crude extract	Synthetic air	$17.80 \pm 1.04^{++}$	8.42 ± 2.58^{ns}
	NO_x	$13.96 \pm 0.22^{**}$	$11.96 \pm 0.04^{**}$
	Halone	16.32 ± 0.65	5.90 ± 0.31
40 to 80%	Synthetic air	$0.77 \pm 0.23^{+}$	$0.32 \pm 0.08^{+}$
Ammonium-sulfate	NO_x	1.80 ± 0.42^{ns}	0.80 ± 0.22^{ns}
Precipitation	Halone	1.35 ± 0.04	0.60 ± 0.12

NOTE: $^{+}$ = significant difference between purified air control and halone treatment; * = significant difference between NO_2 fumigated control and halone treatment; $^{+,*}$ = P = 0.1; $^{++,**}$ = P = 0.05; ^{ns} = difference not significant. FW = fresh weight.

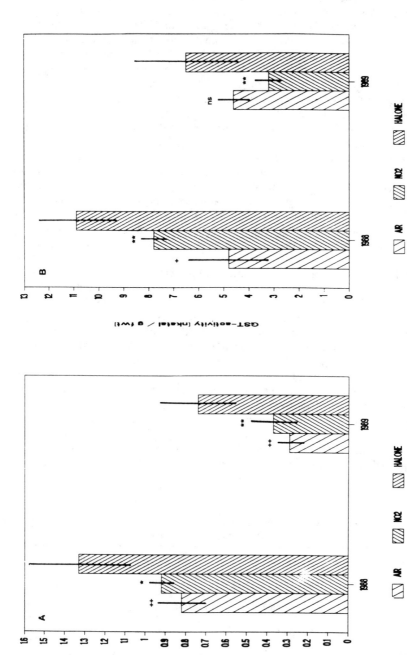

FIG. 3—Glutathione-S-transferase activity in crude enzyme extracts of Picea abies trees subjected to different fumigation treatments on a fresh weight basis. Values are means of four to six determinations ± standard deviation. 3A: GST activities in crude enzyme extracts of needles; 3B: GST activities in spruce needles after ammonium sulfate precipitation to 40 to 80% saturation. $^+$ = significant difference between purified air control and halone treatment; * = significant difference between NO_2-fumigated control and halone treatment; $^{+,*}$ = P = 0.1; $^{++,**}$ = P = 0.05; ^{ns} = difference not significant.

ples, at least towards one of the controls. The effect was extremely prominent in the youngest needles, where increases of 255 and 200% (at $p = 0.01$) were obtained.

Discussion

The present study summarizes preliminary results of a pilot experiment conducted with cloned spruce trees in order to qualify conifer reactions towards halo-organic pollutants. The results indicate that 10 ppb of Halone 1211 impact the fresh-to-dry weight ratio as well as the pigment pattern and the protein contents of young spruce trees' needles. These parameters, however, are rather nonspecific and changes might be observed whenever conifers are exposed to pollutants either of natural or anthropogenic origin. This could be shown in measurements obtained in plants fumigated with NO_2 under the same environmental conditions.

NO_2 also caused a decrease of pigments in both investigated needle year classes. Similar changes were only obtained in young needles of halone-fumigated trees. In these needles, the pigment pattern showed heavy decreases in chlorophyll a and increases in chlorophyll b, an accessory pigment protecting chlorophyll a from radiative damage. In older needles, chlorophyll a and lutein contents were higher than in the controls, with the other pigments remaining at control levels. Obviously, Halone 1211 is able to disturb the light-harvesting complexes in the chloroplasts in some way or other. Nevertheless, there is at least some doubt whether pigment analyses alone or in combination with protein values will be specific measures for halo-organic pollutant impact.

More clear-cut evidence was obtained when GST activity was measured in young spruce trees. Whereas fumigation with NO_2 yielded slight increases of the GST activity in the older needles, while values of young needles were at or even below the control level, a significant increase in the GST activity under the influence of Halone 1211 could be observed. These increases could be expected since glutathione-S-transferases are constitutive detoxification enzymes [7], the action of which can be induced when the plant is under attack by electrophilic compounds such as halogenated hydrocarbons and other xenobiotics. Increasing GST activity after exposure to xenobiotics has already been shown in different animals [17] and in plants [7], especially in corn [18] and sorghum [19]. A fourfold increase of GST under the influence of 10 ppb of a gaseous, inert halocarbon, however, has not yet been reported in any plant.

The results of the present investigation seem to offer a simple method to check impacts of xenobiotic compounds on spruce. Furthermore, they show that two-year-old spruce needles are much more effective in terms of their detoxification capacity. Higher pigment and protein contents in these needles have also been found. This is in agreement with previous results indicating a higher metabolic stability and viability of two-year-old spruce needles.

The findings presented need, however, replications at the same or a shorter time scale to provide more information for test development. Future investigations must include a screening of related substrates for the same specific reaction and the elucidation of the activation pathway, as well as the quantification of the reaction. The activity of special isoenzymes, induced after exposure to certain xenobiotics, can then be used as biomarker to classify the pollutant situation of a stand.

Acknowledgments

The authors thank H. Rennenberg for critical discussions and G. Host and F. J. Nathaus for expert technical assistance.

References

[1] Debus, R., Dittrich, B., Schröder, P., and Volmer, J., *Biomonitoring organischer Luftschadstoffe—Literaturstudie,* Ecomed Verlag, Landsberg, FRG, 1989, p. 7.

[2] Enquete Commission, *Schutz der Erdatmosphäre: Eine internationale Herausfroderung,* Dt. Bundestag, Bonn, 1988, pp. 158–159.

[3] Fiedler, H., *Z. Umweltch. Ökotoxikol.,* Vol. 1, 1989, p. 2.

[4] Figge, K. in *Direct Effects of Dry and Wet Deposition on Forest Ecosystems—in Particular Canopy Interactions,* Commission of the European Communities, Brussels, Belgium, 1986, pp. 245–291.

[5] Schlicht, G. and Schulte-Frohlinde, D., *Photochemistry and Photobiology,* Vol. 16, 1972, pp. 183–188.

[6] Frank, H. and Frank, W., *Experientia,* Vol. 42, 1986, pp. 1267–1269.

[7] Lamoureux, G. L. and Rusness, D. G. in *Glutathione: Chemical, Biochemical, and Medical Aspects,* Part B, D. Dolphin, R. Poulson, and O. Avramovic, Eds., Wiley, New York, 1989, pp. 153–195.

[8] Balabaskaran, S. and Muniandy, N., *Biochemistry,* Vol. 23, 1984, pp. 251–256.

[9] Schröder, P., Lamoureux, G. L., Rusness, D. G., and Rennenberg, H., *Pesticide Biochemistry and Physiology,* 1990, in press.

[10] Schröder, P., Rusness, D. G., and Lamoureux, G. L. in *Sulfur Nutrition and Sulfur Assimilation in Higher Plants,* H. Rennenberg, C. Brunold, L. J. deKok, and I. Stuelen, Eds., SPB Academic Publishers, The Hague, Netherlands, 1990.

[11] Freer-Smith, P. H., *New Phytologist,* Vol. 88, 1981, pp. 223–237.

[12] Lichtenthaler, H. K., *Naturwissenschaftliche Rundschau,* Vol. 37, 1984, pp. 271–277.

[13] Lichtenthaler, H. K., Schmuck, G., Döll, M., and Buschmann, C. in *PEF-Bericht,* Vol. 2, PEF-Projektleitung, KFK Karlsruhe GmbH, Karlsruhe, 1986, pp. 81–105.

[14] Habig, W. H., Pabst, M. J., and Jacoby, W. B. *Journal of Biological Chemistry,* Vol. 249, 1974, pp. 7130–7139.

[15] Bradford, M. M., *Analytical Biochemistry,* Vol. 72, 1976, pp. 248–256.

[16] Zöfel, P., *Statistik in der Praxis,* Universitäts-Taschinbücher, G. Fischer, Stuttgart, 1988, pp. 136–141.

[17] Pickett, C. B. and Lu, A. Y. H., *Annual Review of Biochemistry,* Vol. 58, 1989, pp. 743–764.

[18] Edwards, R. and Owen, W. J., *Planta,* Vol. 169, 1986, pp. 208–215.

[19] Rubin, B., Dirino, O., and Casida, J. E., *Journal of Agricultural Food Chemistry,* Vol. 33, 1985, pp. 489–494.

W. James Fleming,[1,2] *M. Stephen Ailstock,*[3] *Jeffrey J. Momot,*[1] *and C. Michael Norman*[3]

Response of Sago Pondweed, a Submerged Aquatic Macrophyte, to Herbicides in Three Laboratory Culture Systems

REFERENCE: Fleming, W. J., Ailstock, M. S., Momot, J. J., and Norman, C. M., **"Response of Sago Pondweed, a Submerged Aquatic Macrophyte, to Herbicides in Three Laboratory Culture Systems,"** *Plants for Toxicity Assessment: Second Volume, ASTM STP 1115,* J. W. Gorsuch, W. R. Lower, W. Wang, and M. A. Lewis, Eds., American Society for Testing and Materials, Philadelphia, 1991, pp. 267–275.

ABSTRACT: The phytotoxicity of atrazine, paraquat, glyphosate, and alachlor to sago pondweed (*Potamogeton pectinatus*), a submerged aquatic macrophyte, was tested under three types of laboratory culture conditions. In each case, tests were conducted in static systems, the test period was four weeks, and herbicide exposure was chronic, resulting from a single addition of herbicide to the test vessels at the beginning of the test period. The three sets of test conditions employed were: (1) axenic cultures in 125-mL flasks containing a nutrient media and sucrose; (2) a microcosm system employing 18.9-L buckets containing a sand, shell, and peat substrate; and (3) an algae-free system employing 0.95-L jars containing reconstituted freshwater and a nutrient agar substrate. The primary variable measured was biomass production.

Plants grew well in all three test systems, with biomass of untreated plants increasing by a factor of about 5 to 6.5 during the four-week test period. Biomass production in response to herbicide exposure differed significantly among culture systems, which demonstrates the need for a standardized testing protocol for evaluating the effects of toxics on submerged aquatic plants.

KEY WORDS: aquatic plants, sago pondweed, *Potamogeton pectinatus,* herbicide toxicity, aquatic test systems, alachlor, atrazine, glyphosate, paraquat

Regional declines of submerged aquatic macrophyte populations in North America [1–3] and elsewhere [4,5] have generated much concern and research interest. Aquatic macrophytes are primary producers and system stabilizers for complex aquatic ecosystems. Such systems are often among the most productive in the world and contribute substantially to regional economics by supporting rich shell and fin fisheries [6]. They also contribute intangible benefits as improved water quality, shoreline stabilization, oxygen production, and recreational opportunities [7].

Protection of wetland systems, of which submerged aquatic macrophytes are often

[1] U.S. Fish and Wildlife Service, Patuxent Wildlife Research Center, Laurel, MD 20708.

[2] Present address: North Carolina Cooperative Fish and Wildlife Research Unit, Box 7617, N.C.S.U., Raleigh, NC 27695.

[3] Anne Arundel Community College, Environmental Center, 101 College Parkway, Arnold, MD 21012.

integral components, is now a national priority. Although there have been numerous bioassays developed to detect and predict deleterious effects of toxic substances on fish, invertebrates, and algae, standardized test protocols have not been developed for the protection of submerged aquatic macrophytes [8,9]. This is not to say that the effects of toxic chemicals on aquatic macrophytes have not been explored, but rather to indicate that there have been considerable differences in assessment techniques used. Differences among test conditions undoubtedly have influenced interstudy variability in dose-response relationships [10]. Inability to maintain healthy, growing plants in laboratory microcosm systems for prolonged periods of time could be a primary factor contributing to differences in plant responses to toxic substances [10,11].

Herein we report on the responses of plants to herbicides introduced into three test systems, each system based on a different culture procedure for aquatic plants. This report represents a progression of our activities in pursuit of an optimal test system for submerged aquatic plants. Important criteria that we identified for optimal test systems included the ability of.culture systems to maintain healthy, growing plants, the capability to control extrinsic factors, and high test precision. We have attempted to compare the responses of plants in each of the three systems even though test conditions were slightly different and experiments were not conducted concurrently. The herbicides we selected were chosen because they differed from each other in mode of action.

Procedures

Plant Material

Sago pondweed (*Potamogeton pectinatus*) turions were collected at three locations in Chesapeake Bay in 1986. Turions were sterilized by rinsing in a 9 to 1 distilled water and sodium hypochlorite solution for 5 min, followed by careful removal of epidermal tissue and sequential treatment with antibiotics until axenic material was obtained. Turions were then sprouted in a nutrient rich media (Murashige Shoot Multiplication Medium B, Carolina Biological Supply, and 10-g/L sucrose in deionized water). Vegetative material was propagated from rhizome tips, and one clonal line from each of the three collection sites was selected for experimentation. Plants for study were propagated from terminal sections of rhizomes comprised of a rhizome tip and two vegetative nodes. They were grown individually in 50-mL culture tubes containing 25 mL of the propagation media. After two to three weeks, plants were screened for uniformity of growth within clonal lines. Within each experiment (toxicity trial), plants from the three clonal lines were distributed equally among experimental groups. Three clonal lines, instead of only one, were employed in order to include a limited amount of genetic variability in the tests.

Test Conditions and Herbicide Preparation

All toxicity trials were conducted indoors at 20 to 23°C under full spectrum florescent lighting providing about 70 μmol/m^2/s PAR (photosynthetically active radiation). Continuous light was provided in the microcosm system, whereas the heterotrophic and autotrophic systems were provided with 12-h light: 12-h dark. The experimental period was four weeks. Toxicity trials were not run concurrently but rather over the course of several months.

All trials were conducted under static conditions. Herbicides were introduced into each system at the beginning of an experimental period; there was no additional supplementation with herbicides nor was there any removal of the herbicide-laced water or culture media. In the microcosm system, water was added to replace losses from evaporation; evaporation was minimal in the other test systems, which were essentially closed systems.

Concentrated herbicide stock solutions were made by dissolving technical grade alachlor (2-chloro-2',6'-diethyl-N-[methoxymethyl] acetanilide), glyphosate (N-[phosphonomethyl] glycine), or paraquat (1,1'-dimethyl-4,4'-bipyridinium ion) in water, and atrazine (2-chloro-4-[ethylamino]-6,9-isopropylamino]-s-triazine) in a small quantity of acetone, which was then added to water and heated to volatilize the acetone. Logarithmic dilutions of these concentrated stock solutions were added to the culture containers in a test system. Toxicity trials consisted of four or five herbicide concentrations and a zero dose level as a control. All stock solutions were made fresh for each specific test. We did not attempt to confirm herbicide concentrations by analytical methods; determinations in previous studies were within 15% of the predicted concentrations.

Biomass production was the primary variable of interest. The biomass production of individual plants (P) was calculated as $P = [(W_F - W_I) \div W_I] \times 100$, where W_F = the fresh weight of the plant at the end of the study period and W_I = the initial weight of the plant. Growth was analyzed by two-way analysis of variance and Tukey's multiple means separation test with $\alpha = 0.05$. Because the initial size and number of plants varied among studies, biomass production was standardized among studies by expressing the percent biomass production of individual plants exposed to herbicides as a percent of the mean percent biomass production of their respective controls.

Dry weights are not presented because they so closely mirrored fresh weights $(r = 0.71;$ $P < 0.05)$ and were impractical for initial plant weights.

Test Systems

We initially examined a culture system that we were successfully employing for the production of plant material. Preliminary phytotoxicity work with this test system suggested a lack of sensitivity to photosynthetic-inhibiting herbicides. Further examination revealed that plants grown under these conditions were not photosynthetically active. They were heterotrophic, depending on sucrose in the growth media. Concern about the appropriateness of heterotrophic plant responses for predicting responses of normally autotrophic plants prompted us to test plant responses in a more traditional microcosm system. Finally, to accomplish our original goal of a standardized test system not subjected to algal fouling and other shortcomings of a microcosm-type system, we developed a system capable of supporting autotrophic growth with minimal influence by uncontrolled biotic factors.

Heterotrophic System—The plant propagation media described previously was used as the culture media. Herbicide stock solutions were filter sterilized and added to sterilized media to achieve the desired herbicide concentration. Herbicide-laced media (50 mL) was then pipetted into sterilized 125-mL Erlenmeyer flasks, and a single plant was added per flask. There were nine replicates for each dose level with each of the three clonal lines represented three times. This was an axenic system.

Microcosm System—Microcosms consisted of a 18.9-L (5-gal) plastic buckets containing 1 L of peat, covered by 1 L of crushed oyster shell and 2 L of washed white sand. Synthetic freshwater (96-mg $NaHCO_3$, 60-mg $CaSO_4 \cdot 2H_2O$, 60-mg $MgSO_4$, 4-mg KCl per L of deionized water) was added to each bucket. With the addition of the herbicide stock solution, a total liquid volume of 14 L was obtained. One explant from each of the three clonal lines was planted in each bucket. There were five replicates of each dose level for each herbicide.

Autotrophic System—The sample unit for this system consisted of a 0.94-L (1-qt) canning jar into which a small (60-mL) jar for rooting plants was placed. The "rooting jar" contained 30 mL of nutrient agar (4.4-g/L Murashige Minimal Organic Media in distilled water, adjusted to pH = 5.7, and 8-g/L agar) capped with about 30 mL of sand. Synthetic freshwater (700 mL, prepared as above) was added to each canning jar. All components of the system were heat sterilized. An explant was inserted into the nutrient agar layer of each rooting jar and the plants allowed to root for one week prior to herbicide exposure. Each experimental unit was connected to an aeration system delivering CO_2 enriched air (about 2% CO_2) through a system of microtubing. Experimental design and replicate numbers were the same as for the heterotrophic system. Filter-sterilized herbicide stock solutions were added to each canning jar (50-mL/jar) to bring the total volume of solution/jar to 750 mL. This system was considered to be algal free but not axenic because we could not completely sterilize the gas delivery system.

Statistical Interpretation—Data were analyzed by analysis of variance with mean separation by Tukey's test for means. Homogeneity of variance was tested with Bartlett's test. The significance level for all tests was 0.05.

Results

Test System Characteristics

Untreated plants in all three test systems appeared healthy, displaying a uniform green color. Microbial contamination of the heterotrophic system was about 10 to 15%; unless microbial growth was severe, plants were not eliminated from the study. Algal fouling was obvious in all units of the microcosms system, becoming more prominent as experiments progressed. Fouling was not uniform within or among microcosm experimental groups and was not so severe that sampling units had to be eliminated. Microbial contamination of the autotrophic system was not obvious.

There was a notable morphologic difference among plants in the different culture systems. Although quantitative data were not collected, plants in the microcosm and autotrophic systems produced an abundance of roots, many of which were greater than 4 cm in length. Plants in the heterotrophic system grew no or few roots; if present, roots were often short and nubby.

Average biomass of untreated plants (controls) in all three test systems increased substantially during the four to five-week growth period (Table 1). Differences in biomass production (percent increase) among test systems were not significant. However, the growth interval for plants in the autotrophic system was one week longer, and the biomass increase tended to be less compared to the other two systems. From previous work (Ailstock et al., in preparation), we estimate that the biomass of plants in the autotrophic system should be adjusted downward by about 22% to account for the extra week of growth.

TABLE 1—*Growth of sago pondweed* (Potamogeton pectinatus) *in three types of laboratory culture systems. The growth period was five weeks (including a one-week preexperimental rooting period) for the autotrophic system and four weeks for the other systems. The mean values presented represent the mean values of four studies, for which there were a total of 20 to 36 sampling units. The percent increase in biomass was not different among culture systems (P > 0.05, one-way ANOVA).*

| Culture System | N | $\overline{X}$ Weight, g | | % Increase | |
		Initial	Final	$\overline{X} \pm SD$	CV
Autotrophic	4	1.03	5.73	497 ± 68	14
Heterotrophic	4	0.51	3.42	655 ± 316	48
Microcosm	4	4.49[a]	30.36[a]	634 ± 267	42

[a] Three plants constituted the sampling unit compared to one plant for the autotrophic and heterotrophic systems.

The autotrophic system produced the most uniform growth among untreated plants. The coefficient of variation of the four trial means in the autotrophic system was 14%, about one third of that for the other two systems (Table 1). Biomass production and homogeneity of variance did not differ among trials in the autotrophic system. However, they were significantly different in the microcosm and heterotrophic systems; within-trial variability accounted for about two thirds of the total variability and between-trial variability accounted for the other third for these two systems.

Response to Herbicides

Plant responses to alachlor, atrazine, and paraquat were significant for herbicide concentration, test system type, and the interaction of these two factors (Table 2). For glyphosate, plant responses were significant for herbicide concentration and the interaction of herbicide concentration and test system type.

The microcosm system consistently yielded depressed plant growth at herbicide concentrations equal to or lower than those depressing growth in the other two systems (Table 2, Fig. 1). Low concentrations of herbicides produced an apparent stimulatory effect on growth, especially notable in the heterotrophic system.

Discussion

Bioassays to assess the responses of aquatic macrophytes to toxics have been conducted under a wide variety of test conditions. Our findings indicate that variability among culture systems used in aquatic plant toxicity tests can lead to significant differences in the responses of sago pondweed to herbicides. Not only were there significant differences in plant responses among test systems, but the significant interaction term between dose and test system for all chemicals indicates that plants in different systems were not responding in the same fashion to changes in dose. For example, low concentrations of alachlor resulted in a stimulation of biomass production in the heterotrophic system and a reduction in biomass production in the microcosm system. The reasons for these differences were not investigated but could include differences in light schedules, system energetics (capability to support autotrophic growth versus heterotrophic growth), and the amount of herbicide available, which is a function of herbicide concentration and volume.

Agricultural herbicide toxicity to aquatic macrophytes has been studied most for atrazine; however, the results are equivocal. Atrazine concentrations of 30 to 907 µg/L have been estimated to inhibit the growth of *Potamogeton perfoliatus* by 50% [12–15]. Atrazine's toxicity to other species of aquatic macrophytes, including *P. pectinatus*, generally fall within this broad range of concentrations [11,13,15]. These studies were conducted in different types of microcosm test systems and under a variety of environmental conditions (light, salinity, etc.). Extrinsic factors, as phytoplankton and algae, were not controlled, although they were believed to have influenced the results in some studies [8,13].

Short-term bioassays for submerged aquatic macrophytes have received much interest in recent years [8]. Much of the interest is generated by regulatory agencies looking for quick estimates of effluent phytotoxicity or for monitoring water bodies receiving effluents. Some of the problems of rapid bioassays are reviewed by Bowmer [8]. Aquatic plants are very resilient and, unlike animals, it is difficult to distinguish when plants have died; therefore, chronic test protocols seem appropriate. Bowmer [8] doubted the value of rapid bioassays for predicting the impact on aquatic plants of very low concentrations of chemicals over a long period of time. Correll and Wu [11] and Cunningham et al. [12] emphasized the importance of conducting tests over a period of several weeks or months. Phytotoxicity test

TABLE 2—*Biomass production^a ($\overline{X} \pm SD$) of sago pondweed (Potamogeton pectinatus) exposed to herbicides for four weeks in three culture systems. Within rows, means with similar capital letters are not significantly different (one-way ANOVA, Tukey's test for means, P > 0.05).*

| Chemical | Culture System | Biomass of Controls^b, g | | Herbicide Concentration, ppm | | | | | |
		$\overline{X} \pm SD$	% Inc^c	0	0.001	0.01	0.1	1.0	10.0
Alachlor	Autotrophic	4.80 ± 1.33	493	100^A ± 28	112^A ± 28	104^A ± 26	76^{A,B} ± 30	79^{A,B} ± 34	54^B ± 31
	Heterotrophic	2.49 ± 1.01	763	100^A ± 41	117^A ± 48	128^A ± 51	121^A ± 44	107^A ± 39	38^B ± 41
	Microcosm	28.62 ± 2.33	1022	100^A ± 8	NT^d	80^B ± 17	54^C ± 4	26^D ± 4	11^D ± 1
Atrazine	Autotrophic	6.58 ± 1.60	465	100^A ± 24	104^A ± 16	103^A ± 35	50^B ± 12	23^B ± 7	NT
	Heterotrophic	2.47 ± 0.76	331	100^{A,B} ± 31	116^{A,B} ± 30	128^A ± 43	96^{A,B} ± 35	80^B ± 34	33^B ± 9
	Microcosm	16.95 ± 3.60	438	100^A ± 21	112^A ± 13	89^A ± 22	52^B ± 20	12^C ± 2	NT
Glyphosate	Autotrophic	5.70 ± 1.40	437	100^A ± 25	95^A ± 21	105^A ± 21	96^A ± 29	97^A ± 13	99^A ± 34
	Heterotrophic	4.50 ± 1.17	1046	100^A ± 26	110^A ± 7	112^A ± 10	111^A ± 5	112^A ± 12	35^B ± 7
	Microcosm	43.39 ± 8.82	479	100^A ± 20	NT	NT	85^A ± 13	113^A ± 22	54^B ± 6
Paraquat	Autotrophic	5.83 ± 2.03	593	100^B ± 35	154^A ± 31	121^{A,B} ± 33	31^C ± 6	23^C ± 8	22^C ± 5
	Heterotrophic	4.21 ± 1.30	481	100^A ± 31	107^A ± 18	113^A ± 22	76^B ± 10	33^C ± 10	20^C ± 11
	Microcosm	32.47 ± 3.78	595	100^A ± 12	NT	21^B ± 4	0^e	0^e	0^e

^{a} Percent increase in biomass (= biomass production) standardized to control values; for the 0-ppm group, the percent increase is by definition 100% of the increase in biomass of controls.

^{b} Plant weights after four weeks of growth (heterotrophic, microcosm) and five weeks of growth (autotrophic). Plants in the autotrophic system were allowed to grow for one week prior to the introduction of herbicides.

^{c} Percent increase in biomass of controls during the test period.

^{d} Not tested.

^{e} Plants died.

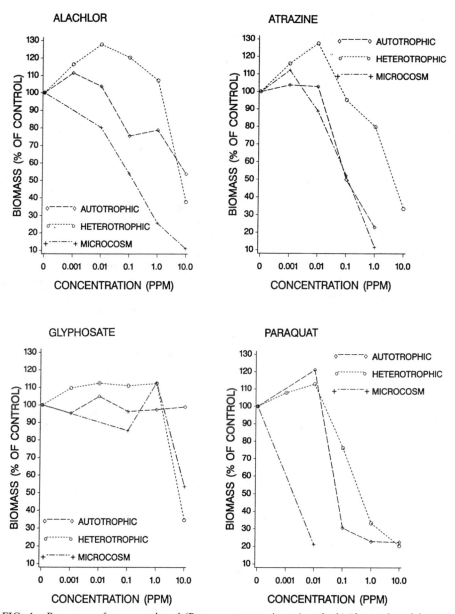

FIG. 1—*Responses of sago pondweed* (Potamogeton pectinatus) *to herbicides in three laboratory culture systems.*

systems therefore should be capable of supporting plant growth for weeks to months rather than days.

At this time it would be premature to recommend specific test systems or protocols for phytoxicity tests with submerged aquatic macrophytes. However, of the three culture systems that we examined, the autotrophic system appears to offer the most promise for a phytotoxicity testing. This system supports vigorous growth of undosed plants for up to eight weeks (unpublished data) and is not confounded by other biotic factors. In the current study it produced the least intertrial variability (CV = 14%) among untreated plants. The potential usefulness of this system is further supported by the response of plants to introduced toxics. The growth of sago pondweed was reduced in the presence of atrazine concentrations that inhibited biomass production of aquatic plants in other test systems [11–15]. Because atrazine is a photosynthetic inhibitor, one would expect biomass production to decrease with a decrease in photosynthetic activity. In the autotrophic and the microcosm systems, biomass production was reduced by about 50% at 0.1-ppm atrazine, a concentration yielding a 90% reduction in photosynthesis (unpublished data). For comparison, atrazine concentrations of at least ten times this amount were required to produce a similar reduction in biomass production in the heterotrophic system in which plants relied on sucrose rather than photosynthesis for energy. For experiments in which sensitivity is the most important criteria, the microcosm culture system should be considered.

One negative of the autotrophic system is the high intratrial variability in individual plant responses in both the control and herbicide treatment groups. This high variability was observed both in the autotrophic and heterotrophic systems and was unexpected considering the control of extrinsic factors and the use of cloned plants. We can provide no explanation for this. The lower variability within the microcosm system was probably an artifact of the numbers of plants per experimental sampling unit. The sampling unit for the microcosm studies contained three plants versus individual plants in the other two systems.

A second negative of the autotrophic system, and also the heterotrophic system, is the extra care and equipment necessary to work under axenic conditions. The microcosm system required only buckets, water, sand, peat, and shell—a minimal investment even for the most modest operation. However, our experiences have convinced us that cloning of plant materials for experimental use is a valuable technique for assuring the year-round abundance of plant material for testing. If the equipment and supplies are acquired to clone plants using tissue culture methods, the additional equipment costs are minimal for conducting a test in the autotrophic system. The labor required to initiate a test in the autotrophic system was about twice that of the microcosm system.

The purpose of a laboratory bioassay is to maximize precision by minimizing or controlling variability within the test system. Data produced under such controlled conditions can be used to rank the phytotoxicity of chemicals and effluents. It should not be expected that laboratory aquatic plant bioassays should yield results that would be directly comparable to those produced in complex aquatic systems with uncontrolled and unidentified environmental variables.

In conclusion, the responses of plants to herbicides differed depending on type of culture system in which plants were maintained. This finding, coupled with the wide range of response levels reported in the literature to individual chemicals, suggests that standard protocols need to be developed and adopted for routine phytotoxicity testing with submerged aquatic macrophytes.

Acknowledgments

We wish to thank Art Lake, Paula Henry, and Pete McGowan for their assistance and contributions to the project.

References

[1] Orth, R., *Aquatic Botany,* Vol. 2, 1976, pp. 141–159.
[2] Orth, R. J. and Moore, K. A., *Science,* Vol. 222, 1983, pp. 51–53.

[3] Orth, R. J. and Moore, K. A., *Estuaries*, Vol. 7, 1984, pp. 531–540.
[4] Phillips, G. L., Eminson, D., and Moss, B., *Aquatic Botany*, Vol. 4, 1978, pp. 103–126.
[5] Den Hartog, C. and Polderman, P. J. G., *Aquatic Botany*, Vol. 1, 1975, pp. 141–147.
[6] Kahn, J. R. and Kemp, W. M., *Journal of Environmental Economic Management*, Vol. 12, 1985, pp. 246–263.
[7] Ward, L. G., Kemp, W. M., and Boynton, W. R., *Marine Geology*, Vol. 59, 1984, pp. 85–103.
[8] Bowmer, K. H., *Australian Journal Marine and Freshwater Research*, Vol. 37, 1986, pp. 297–308.
[9] Sortkjaer, O., *Ecological Bulletin*, Vol. 36, 1984, pp. 75–80.
[10] Fleming, W. J., Momot, J. J., and Ailstock, M. S., Chesapeake Research Consortium, Publication No. 129, 1988, pp. 431–438.
[11] Correll, D. L., and Wu, T. L., *Aquatic Botany*, Vol. 14, 1982, pp. 151–158.
[12] Cunningham, J. J., Kemp, W. M., Lewis, M. R., and Stevenson, J. C., *Estuaries*, Vol. 7, 1984, pp. 519–530.
[13] Kemp, W. M., Means, J. C., Jones, T. W., and Stevenson, J. C., U.S. Environmental Protection Agency, Washington, DC, 1985, pp. 503–567.
[14] Kemp, W. M., Boynton, W. R., Cunningham, J. J., Stevenson, J. C., Jones, T. W., and Means, J. C., *Marine Environmental Research*, Vol. 16, 1985, pp. 255–280.
[15] Forney, D. R. and Davis, D. E., *Weed Science*, Vol. 29, 1981, pp. 677–685.

Maria A. Favali,[1] Maria G. Corradi,[1] and Fabrizia Fossati[1]

X-Ray Microanalysis and Ultrastructure of Lichens from Polluted and Unpolluted Areas

REFERENCE: Favali, M. A., Corradi, M. G., and Fossati, F., **"X-Ray Microanalysis and Ultrastructure of Lichens from Polluted and Unpolluted Areas,"** *Plants for Toxicity Assessment: Second Volume, ASTM STP 1115,* J. W. Gorsuch, W. R. Lower, W. Wang, and M. A. Lewis, Eds., American Society for Testing and Materials, Philadelphia, 1991, pp. 276–284.

ABSTRACT: The thalli of 24 lichen species, collected from unpolluted and polluted suburban areas, were studied by light and electron microscopy in order to detect their morphological and physiological alterations due to air pollution. The effects on photosynthesis were determined by measuring the chlorophyll content of the algae using the autofluorescence test and by the ultrastructural observations of the thylakoid system. The element distribution in the thalli was analyzed by semiquantitative X-ray microanalysis applied to a scanning electron microscope. From our results, it seems that air pollution causes a remarkable drop in chlorophyll content as recorded by measuring the algal autofluorescence and appreciable alterations of the photosynthetic apparatus such as the swelling of the thylakoids observed by electron microscopy. Moreover, X-ray microanalysis shows that the elements always detected in significant amounts are: aluminium (Al), silicon (Si), phosphorus (P), sulfur (S), chlorine (Cl), and calcium (Ca). The element concentration varies in the different portions of the thalli: magnesium (Mg) is located mainly in the algal layers and Al in the reproductive structures (apothecia) and in the cephalodia, where cyanobacteria are present. The amount of Mg and iron (Fe) is lower in the thalli of *Cladonia* from unpolluted areas than in those from polluted areas; *Cladonia furcata* and *Dermatocarpon miniatum* have a high Si content; *Cladonia* has also the highest amount of heavy metals and S. Among the techniques used in our research, electron microscopy is the most sensitive one for the documentation of cellular damages; X-ray microanalysis is indispensable to detect heavy metals, but the autofluorescence test seems to be the most useful for a rapid control of the effect of the air pollution on lichen morphology and physiology. *Lecanora dispersa, Lecanora hagenii, Physcia adscendens,* and *Xanthoria parietina* are the more tolerant lichens studied.

KEY WORDS: air pollution, lichens, light and electron microscopy

It is well known that lichens are capable of accumulating large quantities of pollutants in their thalli; therefore, they appear to be one of the most appropriate bioindicators of air pollutants [sulfur dioxide (SO_2), nitropen oxides (NO_x), ozone (O_3)] and monitors of air deposition [1]. Lichens are particularly sensitive to SO_2 both in the field and when subjected to experimental exposure [2,3]. While laboratory studies have been usually carried out with single air pollutants at high concentrations [4], the situation is much more complex in the field where several pollutants can interact simultaneously. Therefore, a clear explanation of the effect of the dominant atmospheric wetfall elements (Mg, Ca), the dustfall elements [P, Si, Al, Fe, Zn (zinc), K (potassium), Pb (lead)], and sulfur dioxide (SO_2) on the morphology and physiology of the thalli has not been reported. The aims of our experiments were to study some lichen species grown in places with different air pollution conditions and to compare their morphology and physiology in order to assess their usefulness as

[1] University of Parma, Institute of Botany, Viale delle Scienze, 43100 Parma, Italy.

practical bioindicators of air quality. The test for direct monitoring of the elements present in the thalli was semiquantitative X-ray microanalysis; the tests for indirect bioindication were ultrastructural and physiological changes of the algal cells observed by fluorescence and electron microscopy [1].

Materials and Methods

The lichens were collected in four unpolluted regional parks located in the Apennines Mountains (Northern Italy) and in the polluted suburbs of Parma (Northern Italy), where the same species of lichens could be found [5,6,7]. The samples were collected in summer and winter during 1987–89. The rain water acidity measured in the polluted areas reached the highest values (pH = 3.84) during the winter seasons [8]. The lichens were classified according to Ozenda and Clauzade [9]. The morphology and ultrastructure of the thalli of 24 species were studied: *Alectoria capillaris, Anaptychia ciliaris, Cladonia convoluta, Cladonia coniocracea, Cladonia rangiformis, Cladonia furcata, Cladonia fimbriata, Cladonia ochrochlora, Collema cristatum, Dermatocarpon miniatum, Lecanora carpinea, Lecanora dispersa, Lecanora hagenii, Lobaria pulmonaria, Peltigera aphthosa, Peltigera canina, Peltigera horizontalis, Peltigera degenii, Physcia adscendens, Solorina saccata, Squamarina crassa, Toninia coeruleonigricans, Usnea soredifera, Xanthoria parietina*. Only a few species inhabited the polluted areas: *C. fimbriata, C. furcata, L. dispersa, L. hagenii, P. adscendens,* and *X. parietina*.

The techniques used for indirect bioindication were light (LM), fluorescence (FM), and electron (EM) microscopy; the method used for direct bioindication was elemental X-ray microanalysis.

Light (LM) and fluorescence (FM) microscopy. Whole thalli and free-hand sections were observed with a Zeiss Axioscope light microscope equipped with a fluorescent attachment in order to determine subjectively the chlorophyll content of the algae by their autofluorescence and to compare their morphology. No staining was required as the algae exhibit primary fluorescence when irradiated with blue excitation light.

Transmission electron microscopy (TEM). Small pieces of the thalli adjacent to the sections used for LM and FM determinations were fixed in 0.1 M phosphate-buffered 3% glutaraldehyde, pH 6.9, for 3 h and postfixed in 1% osmium tetroxide at 4°C for 2 h, dehydrated and embedded in Epon-Araldite. Ultrathin sections, stained with lead citrate, were examined with a Hitachi 300 electron microscope at 80 KV.

Scanning electron microscopy (SEM). The samples were fixed in 0.1 M phosphate-buffered 3% glutaraldehyde for 3 h at 4°C. After fixation, some samples were dehydrated in ethanol, critical-point dried according to the technique described in a previous paper [10], then observed in a Jeol scanning electron microscope, type JSM/U3.

SEM X-ray microanalysis. Some of the specimens prepared by the critical-point drying technique were coated only with carbon, placed on graphite mounts, and observed in a Jeol scanning electron microscope equipped with an energy-dispersive spectrometer (EDS) system. The technical conditions were the same as previously described [10]. Counts were accumulated for 100 s for all spectra, the element contents were analyzed in four cross sections of each thallus, and the data were elaborated by a computer [10].

Results and Discussion

Our survey of lichen distribution showed that a few species inhabited suburban polluted areas in Italy, among them *C. fimbriata, C. furcata, L. dispersa, L. hagenii, P. adscendens,* and *X. parietina* were the most frequent [5,7]. The reason for the retreat of lichens from the town is probably due to the inhibitory effect of air pollution for spore germination; that was demonstrated for *P. distorta*, while *X. parietina* and *L. hagenii* were less sensitive [11]. Therefore, our data were obtained by comparing some of these less sensitive species still growing in the polluted suburbs with the same species collected in unpolluted areas. Free-hand sections of *P. adscendens* (Fig. 1*a* and 1*b*) and of *X. parietina* (Fig. 1*c* and 1*d*) from unpolluted areas were observed directly with LM (Fig. 1*a* and 1*c*) and with FM (Fig. 1*b* and 1*d*): in these thalli the algae areas were numerous and their autofluorescence (Fig. 1*b* and 1*d*) was much brighter than that of the few algae present in lichens from polluted areas (Fig. 1*f* and 1*h*). In fact, healthy phycobionts have a characteristic primary red fluorescence under blue excitation light, and this color changes to orange and finally to white with increasing degrees of chlorophyll destruction. It has been reported that urban population of *L. muralis* had thicker medullary and thinner algal layers than did rural populations; sulphur dioxide may have been a major stress factor [12].

The advantage of fluorescence microscopy is the ease of sample preparation, which makes it a useful method in practical bioindicator studies of air quality.

In ultrathin sections of the thallus of *C. furcata* from unpolluted areas, it is possible to observe that the algal cells exhibit a normal ultrastructure of their thylakoids (Fig. 2*a*), while the algae of lichens from polluted areas show disarranged and swollen thylakoids (Fig. 2*b*). This alteration supports the results obtained by LM and FM, suggesting that the main effect of air pollution on lichens is on photosynthetic processes as other authors have pointed out [13,14].

In the SEM preparations of *S. saccata*, several crystals were observed attached to the fungal hyphae (Fig. 3*b* and 3*c*); they are calcium oxalate crystals as demonstrated by X-ray microanalysis (Fig. 3*d*). It is thought that these crystals may arise as an end-product of metabolism and have been secreted through the plasma membrane and, therefore, represent an example of Ca extrusion phenomena in fungi and lichens. Ca represents material from substratum and/or local dust [15]; therefore, it is difficult to relate the origin of the crystals only to air pollution. In our research, Ca oxalate crystals were observed in lichens from both polluted and unpolluted places, and also in the algae. This is the first time, to our knowledge, that Ca oxalate crystals have been reported in phycobiont cells.

There are very few algae known that deposit intracellular Ca oxalate [16], but they are not phycobionts. Only a better understanding of the origin and role of these crystals can explain the ecological observation that lichens growing on basic substrata such as limestone and mortar tolerate higher levels of air pollution than lichens growing on acid substrata [13].

In Fig. 4 are reported the original spectra obtained by X-ray microanalysis performed on cross sections of the thalli of *C. fimbriata* (4*a*: spectrum of lichen from unpolluted area; 4*b*: spectrum of lichen from polluted area). In Table 1 is shown the intensity % of the elements always detected in significant amounts in the thalli of the lichens studied by X-ray microanalysis. The amount of Mg and Fe is lower in the lichens from unpolluted areas than in those from polluted areas. Mg is located mainly in the algal layers, and Al is found mainly in the apothecia (reproductive structures of the fungi) and cephalodia (particular structure where cyanobacteria are present). Among the species of *Cladonia* studied, *C. furcata* from polluted areas has the highest content of Si and S; *C. fimbriata* has the highest content of Fe, which is the dominant element of the dust particles [15]. Also in the phycobiont of *C. nitei* and *C. rangiferina*, Fe was present; therefore, these species are considered "ideal" indicators [1].

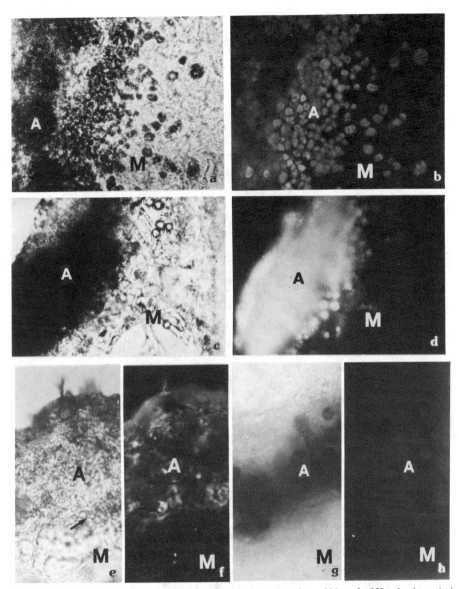

FIG. 1—*Free-hand sections of the thalli of* Physcia adscendens (a *and* b) *and of* Xanthoria parietina (c *and* d) *from unpolluted areas. Free-hand sections of the thalli of* P. adscendens (e *and* f) *and* X. parietina (g *and* h) *from polluted areas. The sections are observed directly with LM* (a, c, e, *and* g) *and FM* (b, d, f, *and* h). *Note the wide algal layers* (A) *and the bright autofluorescence of the algae in lichens from unpolluted places, especially of* X. parietina (c *and* d). *On the contrary, few algae* (A) *are present in the thalli from polluted areas and their autofluorescence is very poor* (f *and* h). *The bright spots are crystals and not algae* (e *and* f, *arrow*). NOTE: (a, b, c, d, e, *and* f ×140; g *and* h; ×230; A = *algal layer; B = medullar layer, where only the hyphae are present*).

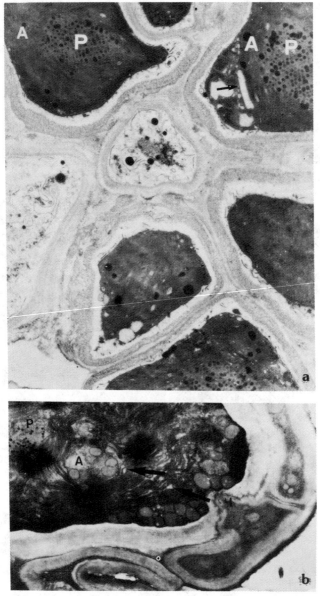

FIG. 2—*Ultrathin sections of* Cladonia furcata *observed by TEM. In the thalli from unpolluted area* (a) *the algae* (A) *show well-preserved thylakoids and pyrenoids* (P). *Note the crystal* (arrow) *in the algal cell. In the thalli from polluted areas* (b) *the algae* (A) *have swollen and disarranged thylakoids* (arrows). *Note the intraparietal haustoria* (H) *of the fungal hypha in the algal cell* (a and b ×14 000).

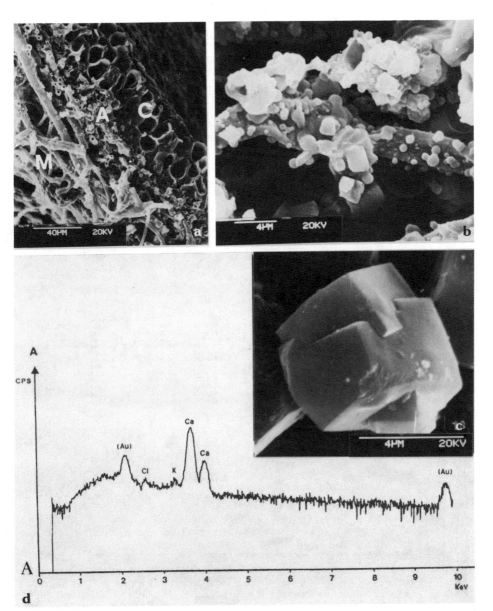

FIG. 3—*SEM preparations of transversal sections of* Solorina saccata. *Recognizable are the cortical* (C), *algal* (A), *and medullar* (M) *layers of the thallus. Several crystals are attached to the hyphae* (b). *At higher magnification it is possible to see their structure; these crystals are of calcium oxalate as demonstrated by X-ray microanalysis. A crystal of calcium oxalate* (c) *and its related spectrum obtained by X-ray microanalysis* (d) *are reported.*

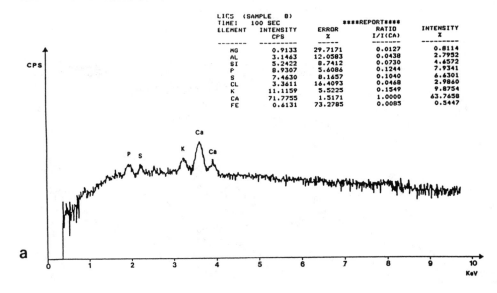

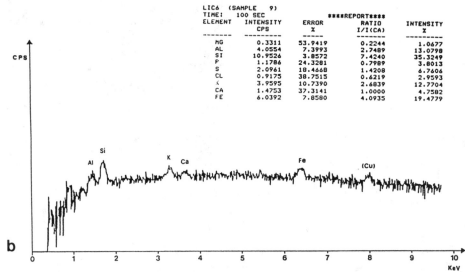

FIG. 4—*X-ray microanalysis spectra performed on gross sections of the thalli of* C. fimbriata *(a = spectrum of unpolluted sample;* b = *spectrum of polluted sample). Fe and Si are present only in* b, *Ca in both samples, but the intensity % is higher in* a *than in* b.

Conclusions

Our data demonstrate that some lichen species, such as *C. furcata, L. hagenii, P. ascendens,* and *X. parietina,* are more tolerant than others to air pollution; therefore, they can be used as bioindicators for air quality. Moreover, it seems that comparisons of light, fluorescence, and electron microscopic observations of the thalli are useful in assessing air pollution injury

TABLE 1—*Investigated species of lichens by EDAX.*

Lichens	Samples	Mg	Al	Si	P	S	Cl	K	Ca	Fe
C. fimbriata										
Unpolluted areas	thallus	...	2.8	4.6	7.9	6.6	2.9	9.8	63.7	...
Polluted areas	thallus	...	13.0	35.3	3.8	6.7	2.9	12.7	4.7	19.5
Polluted areas	thallus	...	6.9	16.3	4.2	8.2	2.5	23.0	28.4	9.5
C. furcata										
Unpolluted areas	thallus	3.7	13.5	20.4	9.0	5.1	5.1	1.5	31.4	...
Polluted areas	thallus	3.1	19.7	48.7	3.9	8.4	...	7.7	2.8	4.7
Polluted areas	thallus	4.0	18.5	31.3	7.8	11.6	3.2	5.0	16.4	1.9
C. ochrochlora										
Unpolluted areas	thallus	...	6.2	9.4	17.4	18.2	4.2	32.7	10.1	...
unpolluted areas	thallus	0.6	1.8	2.5	2.3	2.8	...	3.2	85.7	...
D. miniatum										
Unpolluted areas	thallus	4.0	14.2	25.6	7.0	19.6	...	14.3	12.2	1.3
P. aphthosa										
Unpolluted areas	thallus	3.4	7.6	12.0	6.5	7.5	6.3	43.6	12.6	...
Unpolluted areas	cephalodia	1.6	7.9	10.8	6.1	8.7	12.2	22.7	29.5	...
Unpolluted areas	thallus	...	...	1.2	1.4	0.8	...	2.8	92.0	...
Unpolluted areas	cephalodia	2.5	11.3	15.5	11.9	11.6	7.3	14.2	25.4	...
P. canina										
Unpolluted areas	thallus	2.2	7.8	11.2	10.9	9.4	7.6	33.6	17.0	...
P. degenii										
Unpolluted areas	thallus	4.0	9.1	11.7	8.9	16.3	3.1	37.4	9.3	...
Unpolluted areas	thallus	4.3	13.7	20.6	8.5	15.8	4.0	23.5	7.8	1.4
P. horizontalis										
Unpolluted areas	thallus	2.2	10.6	17.9	11.3	11.6	6.0	20.6	18.8	...
S. saccata										
Unpolluted areas	thallus	1.2	4.0	11.1	1.7	5.9	4.1	5.5	62.8	3.5
Unpolluted areas	apothecia	3.9	12.4	14.1	11.2	14.6	2.8	1.6	38.0	1.3
Unpolluted areas	thallus	2.2	8.6	13.7	6.7	11.3	4.8	22.8	28.4	...
Unpolluted areas	apothecia	2.8	11.9	21.6	6.0	8.4	4.7	6.5	37.2	...

of lichens. The combination of LM and FM methods is very useful for rapid and practical bioindicator studies of air quality. Electron microscopy is the most sensitive technique used for the identification of the causal agents of injuries and very early symptoms such as the alterations of the photosynthetic membranes in the algae.

Acknowledgments

The authors thank A. De Marchi and R. Musetti for kindly providing lichen species.

References

[*1*] Kovács, M. and Podani, J., "Bioindication: A Short Review on the Use of Plants as Indicators of Heavy Metals," *Acta Biologica Hungarica,* Vol. 37, No. 1, 1986, pp. 19–29.

[2] Holopainen, T. and Kauppi, M., "A Comparison of Light, Fluorescence and Electron Microscopic Observations in Assessing the SO₂ Injury of Lichens under Different Conditions," *Lichenologist,* Vol. 21, No. 2, 1989, pp. 119–134.

[*3*] Sharma, P., Bergman, B., Hallbom, L., and Hofsten, A. von, "Ultrastructural Changes of *Nostoc* of *Peltigera canina* in Presence of SO₂," *New Phytologist,* Vol. 92, 1983, pp. 573–579.

[4] Sigal, L. L., "Lichen Research and Regulatory Decisions," *Bryologist,* Vol. 87, 1984, pp. 185–192.

[5] Caniglia, G., Silvan, L., and Baradello, L., "Contributo alla conoscenza dei licheni del Cansiglio," Suppl. n.1., *Giornale Botanico Italiano,* Vol. 117, 1983, pp. 93–94.

[6] Favali, M. A., Caniglia, G., Bassi, R., and Benetti, C., "Contribution to Cansiglio Lichen Flora. IV. Transmission Electron Microscopy and Chlorophyll Content," Suppl. n.2., *Giornale Botanico Italiano,* Vol. 118, 1984, pp. 146–147.

[7] Favali, M. A., Fossati, F., and Realini, M., "Studio sulla natura delle pellicole osservate sul Duomo e sul Battistero di Parma," *Proceedings,* International Symposium "The Oxalate Films: Origin and Significance in the Conservation of Works of Art," Centro Nazionale Ricerche "Gino Bozza," Politecnico di Milano, 1989, pp. 261–370.

[8] Zanella, G., "Le pioggie acide," *Parmambiente,* n.5/6, 1989, pp. 4–6.

[9] Ozenda, P. and Clauzade, G. *Les Lichens. Etude biologique et flore illustree,* Masson, Paris, 1970.

[*10*] Favali, M. A., Barbieri, N., Bianchi, A., Bonecchi, R., and Conti, M. "X-ray Microanalysis of Leaf Tumours from Maize Plants Experimentally Infected with Maize Rough Dwarf Virus: Scanning and Electron Microscopy," *Virology,* Vol. 103, 1980, pp. 357–368.

[*11*] Brown, D. H. and Beckett, R. P. "Uptake and Effect of Cations on Lichen Metabolism," *Lichenologist,* Vol. 16, No. 2, 1984, pp. 173–188.

[*12*] Seaward, M. R. D., "Performance of *Lecanora muralis* in an Urban Environment," in *Systematics Association Lichenology: Progress and Problems,* Vol. 8, 1976, D. H. Brown, D. L. Hawkskowth, and R. H. Bailey, Eds., London, New York: Academic Press, pp. 323–354.

[*13*] Hill, D. J., "Experimental Study of the Effect of Sulphite on Lichens with Reference to Atmospheric Pollution," *New Phytologist,* Vol. 73, 1974, pp. 1193–1205.

[*14*] Eversman, S. and Sigal, L. L., "Effects of SO₂, O₃, and SO₂ and O₃ Combination on Photosynthesis and Ultrastructure of Two Lichen Species," *Canadian Journal of Botany,* Vol. 65, 1987, pp. 1806–1818.

[*15*] Garty, J., Galun, M., and Kessel, M., "Localization of Heavy Metals and Other Elements Accumulated in the Lichen Thallus," *New Phytologist,* Vol. 82, 1979, pp. 159–168.

[*16*] Mann, H., Mann, S., and Fyfe, W. S., "Aragonite Crystals in *Spirogyra sp. (Chlorophyta),*" *Journal of Phycology,* Vol. 23, 1987, pp. 506–509.

Biochemical and Genetic Applications

Michael J. Plewa[1]

The Biochemical Basis of the Activation of Promutagens by Plant Cell Systems

REFERENCE: Plewa, M. J., **"The Biochemical Basis of the Activation of Promutagens by Plant Cell Systems,"** *Plants for Toxicity Assessment: Second Volume, ASTM STP 1115,* J. W. Gorsuch, W. R. Lower, W. Wang, and M. A. Lewis, Eds., American Society for Testing and Materials, Philadelphia, 1991, pp. 287–296.

ABSTRACT: Plant activation is the process by which promutagenic agents are activated into mutagens by plant systems. Many promutagens are activated by plants as well as by the familiar mammalian microsomal monooxygenase systems. However, several environmentally important agents are preferentially activated by plant cells. Plants have become a reservoir for the deposition and accumulation of environmental xenobiotics. With the widespread use of agricultural chemicals on crop plants and with the global exposure of plants to pollutants, the possibility that plant-activated agents may be introduced into the human food chain is a cause of concern. Environmentally relevant agents should be evaluated with plant assays. The plant cell/microbe coincubation assay uses cultured plant cell suspensions as the activating system and bacteria or yeast cells as the genetic indicator organism. After a treatment time, the microbes are plated on selective medium. In this way the activation system and the genetic system can be independently studied. In addition, the viability of the plant cells and the microbial cells can be independently determined so that the toxicity of a test agent can be evaluated. We have employed cultured tobacco, cotton, carrot, maize, and *Tradescantia* cells to study the activation of test agents and complex environmental mixtures. In addition to screening, this assay is being used in basic research to elucidate the biochemical mechanisms of plant activation. The results of experiments using the peroxidase inhibitors acetaminophen and diethyldithiocarbamate showing repression of TX1-cell activation of *m*-phenylenediamine and 2-aminofluorene indicate that a TX1-cell peroxidase pathway is involved in the plant activation of aromatic amines.

KEY WORDS: plant activation, promutagen, antimutagen, *Salmonella* assay, cultured plant cells, peroxidation, 2-aminofluorene, *m*-phenylenediamine, acetaminophen, diethyldithiocarbamate

Environmental toxicology is a field of science that is appreciated by the informed general public. Unfortunately there is an alarming unawareness of the roles that plants play in the accumulation, metabolism, and environmental distribution of xenobiotics. The purpose of this paper is to discuss the use of cultured plant cells to identify promutagens and the biochemical pathways involved in their metabolism. A promutagen is a chemical that is not mutagenic in itself but can be biologically transformed into a mutagen. Plant activation is the process by which a promutagen is activated into a mutagen by a plant system [1]. The ability of plants to activate environmental promutagens into genotoxic metabolites that may penetrate the human food chain becomes even more meaningful with the realization of the diversity of chemicals to which plants are intentionally and unintentionally exposed. Of even greater importance is the possible global effects due to the impact of environmental toxins upon the plant kingdom.

---

[1] Professor of genetics, Institute for Environmental Studies, University of Illinois at Urbana-Champaign, Urbana, IL 61801.

Oxidative Metabolism in Higher Plant Systems

In mammalian systems the majority of enzymes participating in oxidative desulfuration, dealkylation, epoxidation, or ring hydroxylation are microsomal in nature. Currently it is unknown if microsomal cytochrome P-450 in plants have enzymatic characteristics similar to those of mammalian liver. The optical and magnetic properties of plant cytochrome P-450 are similar to those of hepatic microsomes [2]. Although limited data exist about the inducibility of plant cytochrome P-450, it is unknown if there is an equivalent inducible system to hepatic monooxygenases.

Plant peroxidases catalyze the oxidation of a diverse class of xenobiotics. Higashi [2] described two types of oxidative reactions in which plant peroxidases are participants. The first type is the peroxidative reaction that requires H_2O_2 and the second type is the oxidative reaction that uses O_2. Although peroxidases are ubiquitous in plants, their participation in the *in vivo* metabolism of foreign compounds are only now being investigated [3,4].

Aromatic Amines as Model Plant Promutagens

Aromatic amines, including 2-aminofluorene (2AF) and 2-acetylaminofluorene, are well-characterized mammalian promutagens and procarcinogens [5]. Their carcinogenic activity and *in vivo* reactivity are dependent upon metabolic activation. The first stage in the mammalian activation of these agents is *N*-hydroxylation [5]. In the presence of mammalian hepatic microsomes, the *N*-hydroxylation of aromatic amines is dependent upon the cytochrome P-450 enzyme system that functions as a terminal monooxygenase [6]. Plant cells contain exceedingly little cytochrome P-450. An excellent review on plant cytochrome P-450 was recently published by Higashi [7].

m-Phenylenediamine (*m*-PDA) is an aniline derivative and was chosen because of its structural similarity to the metabolites of many pesticides such as nitrophenols and aniline herbicides. *m*-PDA is commonly used in dyes for the textile industry, in hair dye formulations, in the manufacture of rubber and plastic products, and as a curing agent for epoxy resins. *m*-PDA was identified as a promutagen in the *Salmonella*/mammalian microsome assay after activation by hepatic microsomes obtained from rats pretreated with Aroclor 1254 or phenobarbital [8]. When assayed for mutagenicity at the thymidine kinase locus of L5178Y mouse lymphoma cells, *m*-PDA induced a positive concentration-related response [9]. *m*-PDA was activated by plant cell suspension cultures in the plant cell/microbe coincubation assay [10].

Carcinogenicity studies conducted on *m*-PDA produced conflicting results. The incidence of lung tumors in rats treated with a dermal application of a hair dye formulation containing 0.17% *m*-PDA was not significantly different from the controls, and no evidence of teratogenicity was observed [11]. Also Weisburger et al. [12] demonstrated that *m*-PDA was not carcinogenic in mice and rats.

Procedure

Chemicals

Highly purified (>99%) 2AF (CAS No. 153-78-6) was purchased from Chemservices Chemical Co., West Chester, PA. *m*-PDA (CAS No. 108-45-2), acetaminophen (CAS No. 103-90-2), and diethyldithiocarbamic acid, sodium salt (CAS No. 148-18-5) were purchased from Sigma Chemical Co., St. Louis, MO. Hydrogen peroxide (CAS No. 7722-84-1) was purchased as a 30% solution from Fisher Chemical Co, Fair Lawn, NY. All chemicals were stored under darkened conditions at 4°C or as suggested by the manufacturer.

Cell Systems

Long-term plant cell suspension cultures of tobacco (*Nicotiana tabacum*) cell line TX1 were maintained in MX medium, a modified liquid culture medium of Murashige and Skoog [13]. *Salmonella typhimurium* strain TA98 was the genetic indicator organism used [14]. A TX1 cell culture (the activating system in the coincubation assay) was grown at 28°C to early stationary phase, and the cells were washed and suspended in MX⁻ medium. MX⁻ medium is a modified Murashige and Skoog [13] liquid culture medium that lacks plant growth hormone. The fresh weight of the plant cells was adjusted to 100 mg/mL and stored on ice (<30 min) until used.

An overnight culture of *S. typhimurium* was grown from a single colony isolate in 100 mL of Luria broth (LB) at 37°C with shaking. The bacterial suspension was centrifuged and washed in 100 mM potassium phosphate buffer, pH 7.4. The titer of the suspension was determined spectrophotometrically at 660 nm, was adjusted to 1×10^{10} cells/mL, and the culture placed on ice.

Plant Cell/Microbe Coincubation Assay

This assay is based on employing living plant cells in suspension culture as the activating system and microbes as the genetic indicator organism [15]. The plant and microbial cells are coincubated together in a suitable medium with a promutagen. The activation of the promutagen is detected by plating the microbe on selective media; the viability of the plant and microbial cells may be monitored as well as other components of the assay [16].

In the coincubation assay, each reaction mixture consisted of 4.5 mL of the plant cell suspension in MX⁻ medium, 0.5 mL of the bacterial suspension (5×10^9 cells), and a known amount of the promutagen in ≤25 μL dimethylsulfoxide or sterile distilled water. Concurrent negative controls consisted of plant and bacterial cells alone, heat-killed plant cells plus bacteria and the promutagen, and both buffer and solvent controls.

These components were incubated at 28°C for 1 h with shaking at 150 rpm. After the treatment time, the reaction tubes were placed on ice. Triplicate 0.5 mL aliquots (approximately 5×10^8 bacteria) were removed and added to molten top agar supplemented with 550 μM histidine and biotin. The top agar was poured onto Vogel Bonner minimal medium plates, incubated for 48 h at 37°C, and revertant *his*⁺ colonies were scored. The remainder of the reaction mixture was used to determine the viability of the plant and bacterial cells. One volume of cold 250 mM sodium citrate buffer, pH 7, was added to each reaction tube which was then placed on ice. 0.5 mL of this suspension was removed and mixed with 2 mL of MX⁻ medium. The viability of the TX1 cells was immediately determined using the phenosafranine dye exclusion method [17]. The viability of the bacterial cells was determined by adding 1 mL of the cold reaction mixture to 1 mL of cold 100 mM phosphate buffer, pH 7.4. A dilution series using phosphate buffer was conducted so that approximately 300 to 500 cells were added to each of three molten LB top agar tubes and poured upon LB plates. After incubation at 37°C for 24 to 36 h, the bacterial colonies were counted.

Results and Discussion

Activation of Aromatic Amines

The concentration-response curves for the TX1 cell activation of 2AF and *m*-PDA are presented in Fig. 1. The concentrations of 2AF in the experimental reaction tubes (●) ranged from 0 to 0.5 μmol/reaction tube (0 to 100 μM). The concentrations of *m*-phenylenediamine in the experimental reaction tubes (■) ranged from 0 to 5 μmol/reaction tube

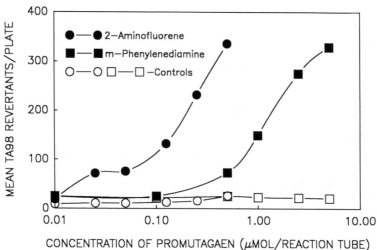

CONCENTRATION OF PROMUTAGAEN (μMOL/REACTION TUBE)

FIG. 1—*Activation of 2AF and* m-*PDA by TX1 cells.*

(0 to 1 mM). The negative control reaction tubes consisted of an identical concentration range of 2AF (O) or m-PDA (□) without TX1 cells or coincubated with heat-killed TX1 cells. As illustrated in Fig. 1, 2AF is a more potent promutagen than m-PDA. As little as 25 nmol 2AF/reaction tube caused a significant increase in mutant TA98 cells. Thus, the plant/cell microbe coincubation assay is a sensitive method to screen for plant-dependent promutagens.

The Effect of Toxicity on the Plant Cell/Microbe Coincubation Assay

In the plant cell/microbe coincubation assay, it is essential to monitor the viability of the plant and microbial cells. Our hypothesis was that any agent that was toxic to the cells would cause a reduction of TA98 revertants by killing the activating system or the genetic indicator organism. We tested this hypothesis by preparing a series of reaction tubes in which the TX1-cell populations were composed of a series of different ratios of live to heat-killed TX1 cells. These cells were exposed to 500 μM m-PDA and coincubated with 5 × 10^9 TA98 cells in a total volume of 5 mL. After 1 h at 28°C while shaking, 500 μL of the mixture was added to supplemented top agar and poured onto VB minimal plates. The data presented in Fig. 2 indicate a direct linear relationship ($r = 0.99$) of the amount of TX1-cell activation of m-PDA and the percent viable TX1 cells in the reaction tube. Thus the coincubation assay is highly sensitive to toxicity in the cultured plant cells. Any test agent, enzyme inhibitor, or presumptive antimutagen that reduced the viability of the plant cells will cause a reduced yield in the number of microbial mutants per plate.

A similar experiment was conducted to evaluate the effect of varying the number of TA98 cells in reaction tubes that contained the normal titer of TX1 cells and 500 μM m-PDA. The number of TA98 cells varied from 5 × 10^9 to 1 × 10^6 per reaction tube. This reduction of TA98 cells was an attempt to mimic the effect of bacterial cell killing on the number of revertants per plate after 1 h of coincubation. The data illustrated in Fig. 3 demonstrate that only after a reduction from 5 × 10^8 to 5 × 10^7 cells plated did a reduction in the resulting number of revertants per plate occur. This phenomenon is due to the number of induced and spontaneous plate mutants per round of cell division that arise after plating

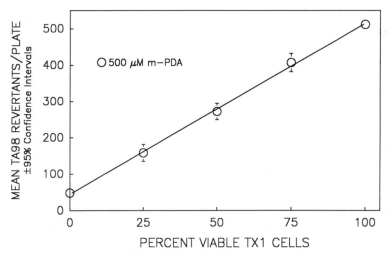

FIG. 2—*The relationship of the number of induced TA98 revertants per VB plate by TX1-cell activated m-PDA as a function of the percent viable TX1 cells.*

in supplemented top agar [18]. Thus the assay is much less sensitive to agents which exert toxic effects only on the bacteria.

Use of Enzyme Inhibitors to Identify Candidate Plant Cell Metabolic Pathways

We used specific monooxygenase and oxidase inhibitors to identify biochemical pathways that are involved in the activation of specific promutagens [3,4]. The basic hypothesis of this approach is that nontoxic concentrations of an inhibitor of an enzyme involved in the activation pathway of a specific promutagen could suppress the activation of that promu-

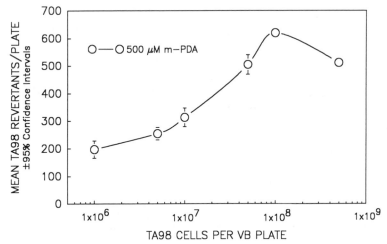

FIG. 3—*The relationship of the number of induced TA98 revertants per VB plate by TX1-cell activated m-PDA as a function of the number of TA98 cells plated.*

tagen. By using a variety of specific inhibitors, individual pathways involved in the activation of a promutagen could be defined.

Inhibition of TX1-Cell Activation by Acetaminophen

Acetaminophen is a competitive substrate of peroxidase and cytochrome P450. Recent studies by Potter and Hinson [19] established that horseradish peroxidase catalyzed the 1-electron oxidation of acetaminophen to N-acetyl-p-benzosemiquinone imine. Since plant peroxidases could use acetaminophen as a hydrogen donor, we decided to test if acetaminophen could compete for the TX1-cell peroxidase and affect the rate of the TX1-cell activation of promutagens. The inhibition of the plant activation of m-PDA and 2AF by acetaminophen is presented in Fig. 4. The positive controls consisted of TX1 and TA98 cells coincubated with 500 µM m-PDA (Fig. 4a) or 50 µM 2AF (Fig. 4b), resulting in 450 and 220 TA98 revertants/plate, respectively. The experimental tubes contained the plant and bacterial cells, 500 µM m-PDA or 50 µM 2AF and a known amount of acetaminophen. The negative controls consisted of the bacterial cells, a promutagen, and varying amounts of acetaminophen without TX1 cells. Concentrations of acetaminophen above 2.5 mM significantly inhibited the activation of m-PDA by TX1 cells ($p \geq 0.01$; Fig. 4a ●), while concentrations of 10 to 50 mM acetaminophen significantly inhibited the TX1 cell activation

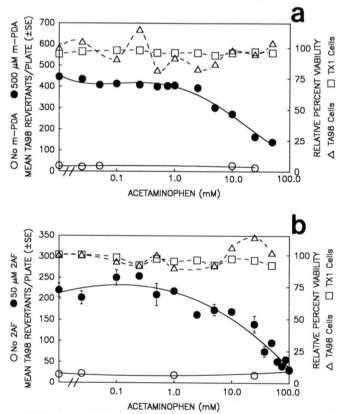

FIG. 4—*The inhibition of the TX1-cell activation of* m-PDA *and* 2AF *by acetaminophen.*

of 2AF ($p \geq 0.05$; Fig. 4*b* ●). The inhibition of the TX1-cell activation of *m*-PDA and 2AF occurred at nontoxic concentrations of acetaminophen. These data, presented in Fig. 4, indicate that acetaminophen is an effective inhibitor of the TX1 activation of *m*-PDA and 2AF.

Inhibition of TX1-Cell Activation by Diethyldithiocarbamate

Diethyldithiocarbamate, a metal chelator, was listed as an antimutagen that functioned as a general blocking agent [20]. Gichner and Veleminsky [21] demonstrated the elimination of the mutagenicity of dimethylnitrosamine in *Arabidopsis* by 1 to 20 m*M* diethyldithiocarbamate. They observed no direct interaction between dimethylnitrosamine and diethyldithiocarbamate when measured spectrophotometrically.

The ability of diethyldithiocarbamate to repress the plant activation of *m*-PDA and 2AF was determined using the same experimental design as was used with acetaminophen. Diethyldithiocarbamate (25 μ*M* to 50 m*M*) was introduced into reaction tubes with plant cells, bacterial cells, and 500 μ*M* *m*-PDA. The inhibition of revertant TA98 colonies was a function of increased diethyldithiocarbamate concentration with 50% inhibition between 750 μ*M* and 1 m*M* (Fig. 5*a* ●). At a concentration of 250 μ*M* diethyldithiocarbamate, a significant inhibition of the plant activation of *m*-PDA was noted. No decrease in the relative viability

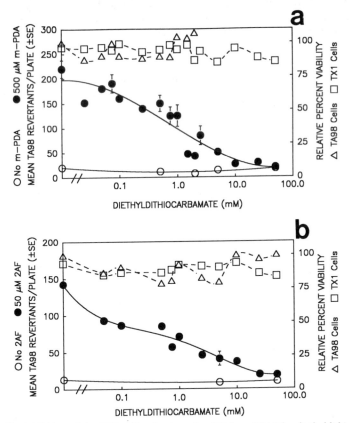

FIG. 5—*The inhibition of the TX1-cell activation of* m-*PDA and 2AF by diethyldithiocarbamate.*

of the TX1 (□) or TA98 (△) cells attended the inhibition curve (Fig. 5a). Diethyldithio-carbamate was titrated (50 μM to 50 mM) in coincubation reaction tubes with plant cells, bacterial cells, and 50 μM 2AF (Fig. 5b). The inhibition of revertant TA98 colonies was a function of increased diethyldithiocarbamate concentration with 50% inhibition between 750 μM and 1 mM. At the lowest concentration of diethyldithiocarbamate (50 μM), a significant inhibition of 2AF activation was noted (●). No consistent decrease in the relative viability of the TX1 or TA98 cells was associated with the inhibition curve. Diethyldithio-carbamate alone was not mutagenic (○). These data suggest that diethyldithiocarbamate specifically inhibits tobacco cell enzymes that are involved in the metabolic activation of both promutagens.

Inhibition of TX1-Cell Peroxidase by Acetaminophen and Diethyldithiocarbamate

The plant cell/microbe coincubation assay permits the direct biochemical analysis of metabolites of promutagens as well as monitoring specific enzymes within reaction tubes [22]. An aliquot can be removed from each reaction tube for biochemical analysis while concurrently conducting the mutagenic analysis. An example of monitoring enzyme function is presented in Fig. 6. A suspension of stationary-phase TX1 cells was washed in MX⁻ and adjusted to a cell titer of 100 mg/mL fresh weight. The suspension was divided into two samples. One sample was homogenized with a PolyTron homogenizer, and the cellular debris was removed by centrifugation at 10 000 × g for 2 min at 4°C. The supernatant fluid was kept on ice after an aliquot was taken for protein analysis. The other sample was treated with 50 mM acetaminophen at 28°C for 1 h without any promutagen or bacteria. This concentration of acetaminophen inhibited the activation of both 2AF and m-PDA (Figs. 4a and 4b). The cells were removed by centrifugation and washed with MX⁻ medium. These treated and washed cells were homogenized. The peroxidase activity of the two supernatant fluid was measured using H₂O₂ as the substrate (Fig. 6); the formation of tetraguaiacol was quantified by absorbance at 470 nm [23]. A significant decrease in peroxidase activity was

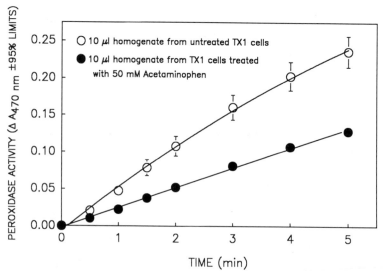

FIG. 6—*The peroxidase activity in cell homogenates from untreated TX1 cells and from TX1 cells treated with 50 mM acetaminophen.*

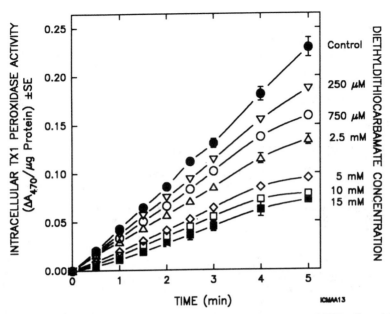

FIG. 7—*The peroxidase activity in cell homogenates from untreated control TX1 cells and from TX1 cells treated with varying amounts of diethyldithiocarbamate.*

observed with the homogenate isolated from the TX1 cells that were treated with acetaminophen (Fig. 6). Acetaminophen at concentrations that significantly inhibited the activation of 2AF and *m*-PDA by TX1 cells also inhibited the activity of cellular peroxidase. These data indicate that peroxidases function as a major pathway for the plant activation of aromatic amines.

Intact TX1 cells were exposed *in vivo* to diethyldithiocarbamate concentrations from 250 μ*M* to 25 m*M*. TX1 cell homogenates were prepared, and both peroxidase activity and protein content were measured. TX1 cells exposed to diethyldithiocarbamate expressed reduced peroxidase activities when normalized on a protein basis (Fig. 7). The diethyldithiocarbamate concentrations which caused a 50% reduction in TX1 cell peroxidase activity (750 μ*M* to 2.5 m*M*) also caused a 50% inhibition of the TX1 cell activation of *m*-PDA and 2AF.

Conclusions

There is a direct linear relationship between the level of plant cell viability and the level of plant activation. Acetaminophen and diethyldithiocarbamate suppressed the tobacco cell activation of *m*-PDA and 2AF. Using the same concentrations which suppressed activation, acetaminophen and diethyldithiocarbamate were potent inhibitors of tobacco cell peroxidase activity under *in vivo* conditions. One mechanism for the antimutagenic effects of acetaminophen and diethyldithiocarbamate was through their inhibition of cellular peroxidases.

Acknowledgments

This research was supported by the U.S. Air Force Office of Scientific Research grant No. AFOSR-88-0336 and by U.S. EPA grant No. R-815008.

References

[1] Plewa, M. J. and Gentile, J. M. in *Chemical Mutagens: Principles and Methods for Their Detection*, Vol. VIII, F. J. de Serres and A. Hollaender, Eds., Plenum, New York, 1982, pp. 401–420.

[2] Higashi, K., *Mutation Research*, Vol. 197, 1988, pp. 273–288.

[3] Wagner, E. D., Gentile, J. M., and Plewa, M. J., *Mutation Research*, Vol. 216, 1989, pp. 163–178.

[4] Wagner, E. D., Verdier, M. M., and Plewa, M. J., *Environmental and Molecular Mutagenesis*, Vol. 15, 1990, pp. 236–244.

[5] Miller, E. C. and Miller, J. A., *Annals of the New York Academy of Science*, Vol. 163, 1969, pp. 731–750.

[6] Nagata, C., Kodama, M., Kimura, T., and Nakayama, T., *GANN Monograph on Cancer Research*, Vol. 30, 1985, pp. 93–110.

[7] Higashi, K., *GANN Monograph on Cancer Research*, Vol. 30, 1985, pp. 49–66.

[8] Ames, B. N., Kammen, H. O., and Yamasaki, E., *Proceedings of the National Academy of Science, USA*, Vol. 72, 1975, pp. 2423–2427.

[9] Palmer, K. A., Denunzio, A., and Green, S., *Journal of Environmental Pathology and Toxicology*, Vol. 1, 1977, pp. 87–91.

[10] Lhotka, M. A., Plewa, M. J., and Gentile, J. M., *Environmental and Molecular Mutagenesis*, Vol. 10, 1987, pp. 79–88.

[11] Burnett, C., Lanman, B., Giovacchini, R., Wilcott, G., Scala, R., and Keplinger, M., *Food and Cosmetic Toxicology*, Vol. 13, 1975, pp. 353–357.

[12] Weisburger, E. K., Russfield, A. B., Homburger, F., Weisburger, J. H., Boger, E., Van Dongen, C. G., and Chu, K. C., *Journal of Environmental Pathology and Toxicology*, Vol. 2, 1978, pp. 325–356.

[13] Murashige, T. and Skoog, F., *Physiologium Plantarum*, Vol. 15, 1962, pp. 473–497.

[14] Maron, D. M. and Ames, B. N., *Mutation Research*, Vol. 113, 1983, pp. 173–215.

[15] Plewa, M. J., Weaver, D. L., Blair, L. C., and Gentile, J. M., *Science*, Vol. 219, 1983, pp. 1427–1429.

[16] Plewa, M. J., Wagner, E. D., and Gentile, J. M., *Mutation Research*, Vol. 197, 1988, pp. 207–219.

[17] Widholm, J. M., *Stain Technology*, Vol. 47, 1972, pp. 189–194.

[18] Seiler, J. P. in *Progress in Mutation Research*, Vol. 2, A. Kappas, Ed., Elsevier, Amsterdam, 1981, pp. 149–193.

[19] Potter, D. W. and Hinson, J. A., *Drug Metabolism Reviews*, Vol. 20, 1989, pp. 341–358.

[20] Ramel, C., Alekperov, U. K., Ames, B. N., Kada, T., and Wattenberg, L. W., *Mutation Research*, Vol. 168, 1986, pp. 47–65.

[21] Gichner, T. and Veleminsky, J., *Mutation Research*, Vol. 139, 1984, pp. 29–33.

[22] Blair, L. C., Slife, F. W., Felsot, A., and Plewa, M. J., *Pesticide Biochemistry and Physiology*, Vol. 21, 1984, pp. 291–300.

[23] Maehly, A. C. and Chance, B. in *Methods of Biochemical Analysis*, Vol. I, D. Glick, Ed., Interscience Publishers, New York, 1954, pp. 357–424.

William R. Lower,[1] Frank A. Ireland,[2] and Barbara M. Judy[3]

32P-Postlabeling for DNA Adduct Determination in Plants

REFERENCE: Lower, W. R., Ireland, F. A., and Judy, B. M., "**32P-Postlabeling for DNA Adduct Determination in Plants,**" *Plants for Toxicity Assessment: Second Volume, ASTM STP 1115,* J. W. Gorsuch, W. R. Lower, W. Wang and M. A. Lewis, Eds., American Society for Testing and Materials, Philadelphia, 1991, pp. 297–308.

ABSTRACT: Barley (*Hordeum vulgare* L.) and Australian saltbush (*Atriplex semibaccata*) plants and deoxyribonucleic acid (DNA) isolated from barley were exposed to the known carcinogen/mutagen, N-methyl-N-nitrosourea (MNU). DNA was isolated, digested to a mixture of deoxyribonucleoside 5'-monophosphates, postlabeled with [γ-32P] ATP, and separated on polyethyleneimine (PEI)-cellulose thin layer chromatography plates. Autoradiograms of the chromatograms showed the presence of normal DNA constituents as well as additional spots not present on autoradiograms of untreated DNA. These data suggest that DNA adducts are formed in growing plants and in plant DNA exposed to a carcinogen. In addition, a discussion of the 32P-postlabeling procedure and similarities and differences encountered using plants and plant DNA are addressed.

KEY WORDS: plants, DNA, adducts, carcinogen, mutagen, 32P-postlabeling, N-methyl-N-nitrosourea, barley, saltbush

Chemicals which form covalent bonds *in vivo* with DNA are potential animal carcinogens and mutagens [1], and the formation of certain DNA adducts has been highly correlated with carcinogenesis or mutagenesis [2]. Until recently, all of the research involved either exposing animals or DNA of animal origin to test compounds. DNA adduct detection by 32P-postlabeling followed by thin-layer chromatography separation is currently the most sensitive method available and has been used extensively with animal material [1–6]. Experiments in our laboratory are the first to report DNA adducts in plants using 32P-postlabeling [7] and are directed at applying these techniques toward the identification of DNA adducts in plants as a method in itself and a potential alternative to the use of animals in monitoring. The method can also be used to monitor the effects of chemicals on plant cells. Techniques developed for the detection of DNA adducts without the use of radiolabeled carcinogens have broadened the number of compounds which can be monitored for their ability to bind to DNA. A sensitive method, applicable to a large number of chemicals present in the human environment, involves postlabeling DNA constituents using [32P] ATP and polynucleotide kinase followed by detection using thin-layer chromatography (TLC). The 32P-approach has been successfully applied to approximately 100 test chemicals in animals, with DNA adducts detected with each carcinogen tested [1].

Information is available on the 32P-postlabeling procedures, although much of it is scattered

[1] Group leader, Environmental Trace Substances Research Center, University of Missouri, Columbia, MO 65203.
[2] Research specialist, Dixon Springs Agricultural Center, University of Illinois, Simpson, IL 62985.
[3] Research specialist, Department of Veterinary Biomedical Science, University of Missouri, Columbia, MO 65201.

throughout the literature and, in our experience, is difficult to assemble and interpret. This paper brings together this information and discusses the identification of DNA adducts of plants by the ^{32}P-postlabeling procedure and similarities and differences encountered in our laboratory compared to results reported in the literature.

Methods of Exposure

The binding of a compound to DNA can either be by direct action or may require transformation of the compound to an active form. In the case of direct action carcinogens, *in vitro* exposure of plant and animal DNA should result in the formation of DNA adducts. However, *in vivo* exposure may result in a different adduct picture due to alterations of the compound by the target species. The activation of chemicals into carcinogens or mutagens may be different between animals and plants. While the characterization of metabolizing enzymes in animals is much more advanced than in plants, both are capable of activating certain promutagens/procarcinogens into active compounds [3,8–11]. Plants can, therefore, be utilized for *in vitro*, *in vivo*, and *in situ* testing of mutagens and of those carcinogens which are mutagens. *In vitro* testing of compounds consists of incubating DNA with the test compound. Care should be taken to ensure that the conditions of exposure are favorable for the formation and detection of any adducts which may occur. The half-life of the compound, under specific conditions of pH, light, and temperature, will play an important role [11,12]. This information can be used in determining the appropriate length of time to expose the DNA to the compound.

In *in vivo* and *in situ* exposure, in addition to these factors, the possibility of uptake must be considered when exposing plants to mutagenic compounds. Uptake can occur by various routes, e.g., through roots, stems, or leaves, but may vary with the compound and the species of plant tested. Uptake will also depend on the physical-chemical characteristics of the chemicals tested, i.e., solubility, hydrolysis rate, photolysis rate, log K_{ow}, etc. In our laboratory, we have routinely exposed plants by cutting them off at soil level and placing them, stem down, into a beaker containing the compound. The length of exposure has been determined by considering the half-life of the compound at the conditions of exposure. This method of exposure should be suitable where there exists the possibility of a root barrier to the compound. If the mutagen is known to be light sensitive, the beaker is enclosed in foil to prevent photodegradation. Another important consideration is whether the compound requires activation. While it is generally accepted that plants can activate certain compounds, there may exist differences in various parts of plants as well as species of plants in their ability to activate specific promutagens or carcinogens. This issue requires further research in the characterization of plant enzymes.

Extraction of DNA

The method for isolation of eukaryotic DNA routinely employed in our laboratory is a modification of previously reported methods [13–17]. Plant tissue (3 g) is pulverized by mortar in liquid nitrogen. The powder is transferred to four centrifuge tubes and the liquid nitrogen allowed to evaporate. The powder is then suspended in 16 mL of extraction buffer (100 mM Tris, 50 mM EDTA, 500 mM NaCl, and 10 mM β-mercaptoethanol, final ph 8.0) containing 1.25% sodium dodecyl sulfate and incubated at 65°C for 10 min. It is then incubated at 38°C for 2 h with proteinase K (0.2 mg/mL). The protein is further removed by adding 5 mL of 5 M potassium acetate, incubating at −20°C for 20 min followed by centrifugation at 25 000 g for 20 min. The supernatant is then poured through cheesecloth into clean centrifuge tubes containing 10 mL of ice-cold isopropanol per tube. The contents of the tubes are mixed, incubated at $-40°C$ for 30 min, and the DNA pelleted by centrifuging at 20 000 g for 15 min. The supernatant is poured off and the pellets lightly dried by inverting the tubes. The DNA pellets are redissolved in 1 mL of 50 mM Tris, 10 mM EDTA, pH 8.0, and transferred to an Eppendorf microcentrifuge tube. Ribonuclease A (0.1 mg/mL) and ribonuclease T_1 (1000 units per mL) are added and the mixture incubated at 38°C for 1 h. Ribonuclease A is prepared in 10 mM Tris-HCl, pH 7.5, and heated to 80°C for 10 min prior to use. One milliliter of phenol:chloroform (1:1; v/v) saturated with 100 mM Tris-HCl, pH 8, is added and the mixture centrifuged at 12 000 g for 10 min at 4°C. (It is essential that the pH be 8.0 to prevent the DNA from being lost in the interphase [13]). Antioxidants are not added to the phenol because of the possibility of quinoline oxidation products forming DNA adducts. Phenol oxidation increases as the pH is raised; therefore, phenol equilibrated to a pH of 8.0 is stored frozen at $-20°C$ [4]. Following centrifugation, the lower phenol layer is removed with a plastic pipet. The tubes are recentrifuged for 1 min and the aqueous phase removed. The aqueous layer is then extracted with an equivalent volume of chloroform:isoamyl alcohol (24:1) and centrifuged at 12 000 g for 10 min at 4°C. The aqueous phase is removed and the DNA precipitated from it by adding an equivalent volume of ice cold isopropanol and incubating at $-40°C$ for 30 min or until strands of DNA are visible. The DNA is pelleted by centrifuging at 20 000 g for 15 min. The supernatant is poured off and the pellets washed three times with ice-cold 80% ethanol. The pellets are allowed to lightly dry and are redissolved in distilled water. Water is used to allow for the DNA to later be concentrated; however, because no EDTA is in the solution to inhibit DNase, the DNA should be stored at less than $-20°C$. We routinely store our DNA preparations at $-70°C$. The DNA concentration is estimated spectrophotometrically, and the homogenity, size, intactness, and purity are checked by electrophoresis. Because RNA is more readily labeled with polynucleotide kinase than DNA [16], a determination of RNA content in the samples should be conducted. It has been suggested that RNA contamination may be cleaned up by precipitation of the DNA with ethoxyethanol [4].

Digestion and Radiolabeling of DNA

The digestion and radiolabeling of DNA for the detection of methylated adducts is described in Fig. 1. Deoxyribonucleic acid (2 μg) is digested to deoxynucleoside 3'-mono-

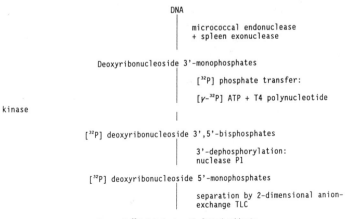

FIG. 1—*Procedure for detection of* ^{32}P-*methylated adducts.*

phosphates by incubating at 37°C for 2 to 3 h with 4 μg each of micrococcal endonuclease (EC 3.1.31.1) and spleen phosphodiesterase (EC 3.1.16.1). Five microliters of 200 mM succinate, pH 6.0, and 5 μL of 100 mM CaCl$_2$ are added and the volume made up to a total of 50 μL with sterile, distilled water. Micrococcal endonuclease is made up to 0.24 units/μL in distilled water and dialyzed against sterile distilled water at 4°C for 16 h prior to use. Dialysis removes the ammonium sulfate suspension, which can inhibit the kinase reaction. Spleen phosphodiesterase is centrifuged for 10 min at 13 000 g, the ammonium sulfate pipetted off, and the enzyme pellet redissolved in sterile distilled water. After dialysis against distilled water for 16 h, the volume is brought up to a final concentration of 2 μg/μL with sterile, distilled water.

Postlabeling of DNA digests is accomplished by the direct phosphorylation reaction whereby polynucleotide kinase catalyzes the transfer of the gamma phosphate of ATP onto the 5′-hydroxyl ends of the 3′-monophosphates. For the labeling of methylated adducts, we dilute aliquots of DNA digests to 0.01 μg/μL. We add 2 μL of this preparation to a solution (18 μL) containing [^{32}P]-ATP (100 Ci/mmol; total radioactivity of 22.3 μCi), 60 units of T4 polynucleotide kinase (EC 2.7.1.78), 2 μL of 400 nM bicine, 100 mM MgCl$_2$, 100 mM dithiothreitol, 1 mM spermidine, pH 9.0, and 5.7 μL of sterile water. The solution is mixed briefly and incubated at 37°C for 1 h. We routinely digest the 3′,5′-bisphosphates resulting from the ^{32}P-labeling to the 5′-monophosphate level by taking a 10-μL aliquot of 3′,5′-bisphosphate solution and adding 4 μL of 0.2 M sodium acetate, pH 5.0, 1 μL of 0.5 mM ZnCl$_2$ [dissolved in 0.01 mM hydrochloric acid (HCl)] and 1 μL of nuclease P1 (2 μg/μL) followed by incubation at 38°C for 1 h. This has been reported to reduce background interference on the chromatographic plates [10].

The digestion and labeling of adducts induced by aromatic compounds, e.g., benzo[a]pyrene [5], as described in Fig. 2, consists of digesting 2.5 μg of DNA to 3′-monophosphates using 5 μg each of micrococcal endonuclease (0.6 U) and spleen phospho-diesterase (0.01 U) in 12 μL of 20 mM succinate, 10 mM CaCl$_2$, final pH 6.0. We then incubate the mixture at 38°C for 3 h. Adduct enrichment is then accomplished by taking a 12 μL aliquot of the 3′-monophosphate digest and adding 1.2 μL of nuclease P1 (5 μg/μL), 3 μL of 0.25 M sodium acetate, final pH 5.0, and 1.8 μL of 0.3 mM ZnCl$_2$ and incubating at 37°C for 1 h. After incubation, 2.4 μL of 0.5 M Tris base is added. The ^{32}P-labeling of the aromatic nucleotide adducts is accomplished by taking 20.4 μL of the nuclease P1-treated digest and adding 2.4 μL of a solution containing 200 mM Bicine-NaOH, 100 mM MgCl$_2$, 100 mM dithiothreitol, 10 mM spermidine, final pH 9.6, 2.4 μL of [γ-^{32}P]-ATP (110 μCi/μL), 1.2 μL (4 U) of T4 polynucleotide kinase and incubating for 45 min at 37°C. This procedure enhances the sensitivity of the assay, down to a reported level of 1 adduct in 10^{10} nucleotides, by labeling adducts derived from a larger quantity of DNA and by the high specific activity of the [θ-^{32}P]-ATP (6000 Ci/mmol). This method is one of the most sensitive methods for adduct detection presently available; however, other methods may be more applicable to the identification of specific adducts. In addition, nuclease P1 is reported to remove normal nucleotides prior to ^{32}P-labeling by dephosphorylating normal, but not ad-ducted, nucleotides to nucleosides. With this method, only adducted nucleotides are ^{32}P-labeled; however, some adducts may be susceptible to and not detected by this method [3].

Chromatography

The method currently used in our laboratory for identification of methylated adducts is by two-dimensional chromatography on polyethyleneimine (PEI)-cellulose thin layers (20 by 20 cm). After the addition of 2 μL of a solution of 5′-monophosphates (dpA, dpG, dpC, and dpT; 4 μg/μL each) to the labeled digest, a 6 μL aliquot is applied to a PEI sheet at

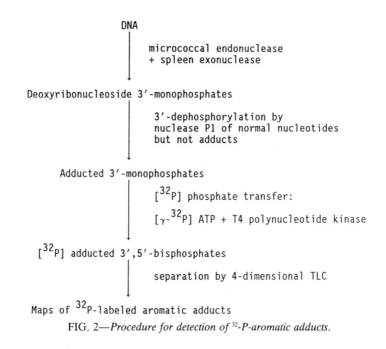

FIG. 2—*Procedure for detection of* [32]*-P-aromatic adducts.*

1.5 cm from the bottom and 2.5 cm from the left hand edge. The chromatograms are developed with water to the origin followed by 0.15 M ammonium formate, pH 3.5, to 6 cm on a Whatman No. 1 wick attached at 19 cm from the bottom edge by staples, the total wick length being 7 cm. We normally place a row of staples across the entire width of the wick in order to assure even contact with the PEI plate. The sheets are then dried, the wicks removed, and the plates washed three times for 5 min each in distilled water. The sheets are again dried in a current of cool air, developed to 1.5 cm with water followed by 0.15 M Tris-HCl, 7 M urea, final pH 8.0, to 10 cm on a wick. The sheets are dried, washed twice in distilled water for 5 min each, and dried again. Our experience has been that a longer than previously reported migration [10,18] in the second dimension provides better separation without loss due to migration off the plate.

The chromatography method for detection of aromatic adducts is a modification of previously reported systems [4,5]. After the removal of 2 µL of the nuclease P1 neutralized digest, the remainder is spotted onto a PEI-cellulose sheet at the origin, located 1.5 cm from line 1 and the right edge (see Fig. 3). By designing a 20 by 20 cm TLC plate as described in Fig. 3, two samples can be spotted on the same plate. A Whatman No. 1 wick (20 by 13 cm) is attached at the top edge of the plate with staples. Migration in the first dimension (D1) is from the bottom to the top as the chromatogram is positioned in Fig. 3. The bottom edge of the TLC plate is dipped for a few millimeters in distilled water. (The water runs in front of the strong solvent, which is reported to aid chromatography and prevents salt crystallization at the solvent front [4]). This is repeated before the chromatogram is migrated in each dimension with any solvent. The chromatogram is then placed in 1.0 M NaH$_2$PO$_4$, pH 6.5 in a TLC tank and developed overnight. The chromatograms are placed in the tank with the wicks hanging over the sides of the tank and the lid placed on top of the wick. The lid and sides of the tank are protected from the radioactive wick by covering them in plastic film. The next day the wicks and upper part of the TLC plate are removed and discarded

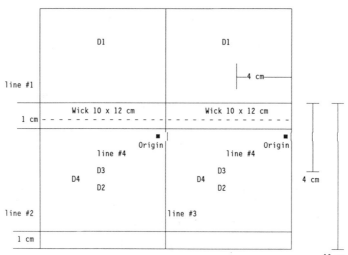

FIG. 3—*Thin-layer chromatography plate for aromatic detection.*

by cutting at the drawn line (Fig. 3, Line 1). The cut edge is irrigated with distilled water to remove any radioactive flecks generated by cutting. The chromatograms are then washed, on a shaker, two times for 5 min each in distilled water, and then blow-dried. Dimensions two, three, and four, as described in this paper, are equivalent to dimensions three, four, and five, as previously reported [4,5]. Development in the second dimension (D2) (see Fig. 3) is with 5.3 M lithium formate, 8.5 M urea, pH 3.5. The chromatogram is then cut at the top (Fig. 3, Line 2) and the cut edge irrigated with distilled water. The chromatograms are washed twice on a shaker in distilled water for 5 min each followed by 5 min in 10 mM Tris-HCl, pH 7.5, to condition the pH of the chromatogram for the next solvent. The chromatograms are then blow-dried and cut at Line 3 (Fig. 3). Development in the third dimension (D3), as indicated in Fig. 3, is in 1.2 M LiCl, 0.5 M Tris-HCl, 8.5 M urea, final pH 8.0. The chromatograms are migrated to the top followed by washing twice in distilled water, and then blow-dried. A Whatman No. 1 wick (10 by 12 cm) is attached at the dotted line indicated in Fig. 3. Migration in the fourth dimension (D4) is in 1.7 M NaH$_2$PO$_4$, final pH 6.0, for 14 h, overnight, or until the solvent reaches the top of the wick. The wick is removed by cutting immediately below the staples. The origin is removed by cutting along Line 4 (Fig. 3). The chromatogram is washed twice in distilled water, and then blow-dried.

Chromatography Plates

We have experimented with polyethyleneimine (PEI)-cellulose plates from different manufacturers, Macherey Nagel in Germany, supplied by Brinkmann Instruments, Westbury, New York, and those manufactured by E. Merck, Darmstadt, Germany. We have seen marked differences in the separation of nucleotides produced by these plates. Plates produced by Macherey Nagel, when washed by agitation on a shaker in methanol (2 min) followed by two changes of water (2 min each) prior to use, produced reliable separations. It has been reported that due to commercial plates being more retentive than laboratory-produced plates [5], a 50% increase in solvent strengths may be needed to produce similar separations [4]. In our laboratory, we have found that plates produced by E. Merck did not produce reliable separations when prepared in this manner. However, when they were prepared by washing in methanol followed by propanol and distilled water (5 min each), reliable and reproducible separation, though slightly different from Macherey Nagel produced plates, could be achieved. Washing with methanol results in easier sample spotting and faster running of the plates due to the removal of hydrophobic material, while the water removes degradation products from the plate which can cause poor migration [4]. The propanol is believed to remove slightly more hydrophobic material from the plate than the methanol.

Autoradiography

Using radioactive ink, we mark the chromatograms with alignment dots at all of the corners except the origin. The plates are then wrapped in plastic film and placed in a cassette containing intensifying screens. The length of exposure will vary with the temperature and type of intensifying screens. Exposure at $-70°C$ is reported to produce a two-fold increase in sensitivity above exposure at $20°C$ and an enhancement factor of 10 to 20 can be achieved by the use of two intensifying screens when using ^{32}P [19]. We normally use two types of screens in our laboratory, Kodak X-OMATIC Regular and DuPont Cronex Lightning Plus. While the DuPont screens are reported to be more sensitive [19], we have obtained satisfactory results from the Kodak screens by increasing the length of film exposure. For the identification of methylated adducts, we routinely expose plates for a 2-h period at $-70°C$ followed by removal of areas containing radioactivity produced by normal nucleotides. The remainder of the plate is then exposed for a period ranging from 35 to 90 h at $-70°C$ in order to locate areas containing radiolabeled adducts. Where more than one exposure time is desired, we place two sheets of film in the cassette and transfer one to another cassette after the shorter exposure period. This method is very applicable to the method of aromatic adduct identification because only the adducts are present on the TLC plate during film exposure.

X-ray Film and Development

We use Kodak XAR-5 film and a commercial film processor (Kodak RP X-OMAT), but development in tanks with Kodak D-19 developer may be used [4]. The sensitivity of commercial X-ray processors, if set up for these specific purposes, is reported to potentially be equal to tank development.

Identification of DNA Adducts

The identification of specific DNA adducts may be accomplished by various techniques including the use of thin-layer chromatography, high-performance liquid chromatography (HPLC), gas chromatography (GC), gas chromatography/mass spectrometry (GC/MS), immunological detection using antibodies, and synchronous fluorescence spectrophotometry [1,2,20–24]. Each method has its own advantages and disadvantages. With thin layer chromatography (TLC), the advantages include the use of small quantities of DNA and the ability to postlabel the adducts with ^{32}P. A major advantage with the post-labeling technique is that it is not adduct specific and can detect multiple adducts from exposure to complex mixtures [24]. A disadvantage exists in the identification of specific adducts. Currently, comigration of identified adducts with DNA adduct standards is being utilized when the adduct standards are available. To our knowledge, there is only one commercially available DNA adduct (7R, 8S, 9S-trihydroxy-10R,-(N²-deoxyguanosyl-3'-phosphate)-7,8,9,10-tetrahydrobenzo[a]pyrene; NCI Chemical Carcinogen Repository at Midwest Research Institute, Kansas City, Missouri). Other adduct standards are either produced by the laboratory or are provided through cooperation with other investigators. Cancer research could benefit by an increase in the availability of more DNA adduct standards. The calculation of Rf values, the distance traveled by the solute divided by the distance traveled by the solvent, can be useful in identifying specific DNA adducts by TLC; however, comigration as a method of specific adduct identification has limitations and drawbacks. Two compounds of different structure may migrate together in one solvent system and migrate separately in another solvent system [25]. Also, with unknown samples, absolute quantitation of adduct levels may be impossible due to the undetermined relative labeling efficiency [24].

The use of HPLC, GC, and GC/MS techniques have been applied for the identification of specific DNA adducts. These methods, while obtaining high resolution, have historically been less sensitive than other methods; however, highly sensitive techniques are being developed [24].

Antibodies generated to DNA adducts have been used to detect adducts formed by various compounds [23,24]. A major limitation of the antibody approach is the assay specificity. Antibodies must be generated for each adduct of interest in order for this technique to be widely used as a monitor. Additionally, the potential of cross reactivity with other adducts or the surrounding DNA structure is another drawback.

Synchronous fluorescence spectrophotometry has been used for aflatoxins and their metabolites [20], but the sensitivity is lower than other methods of identification [24]. A major disadvantage is that this technique is only applicable to compounds which fluoresce.

Plant DNA Adducts

Barley (*Hordeum vulgare* L.) and Australian saltbush (*Atriplex semibaccata*) plants, grown indoors in a plant growth room, were exposed to the known mutagen/carcinogen, N-methyl-N-nitrosourea, (MNU), by cutting the plants at soil level and placing the stems in a beaker

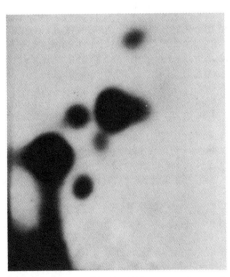

FIG. 4—*Polyethyleneimine-cellulose map of barley DNA exposed to methyl-nitroso-urea for 3.5 h at 37°C.*

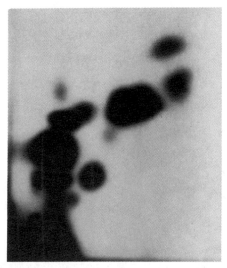

FIG. 5—*Polyethyleneimine-cellulose map of calf thymus DNA exposed to methyl-nitroso-urea for 3.5 h at 37°C.*

containing 25-mM MNU, pH 4.5, for 40.5 h at 22°C. The plants were exposed under a 16-h light: 8-h dark photoperiod and the MNU protected from photodegradation by wrapping the container in foil. After exposure, the plants were removed from the solution, washed with distilled water, placed in a sealed plastic bag, and stored at −70°C until extraction of DNA.

Deoxyribonucleic acid from untreated barley plants and calf thymus were exposed to MNU by incubating 4 µg of DNA with 80 mM MNU in 200 mM Tris-HCl, pH 8.0, for 3.5

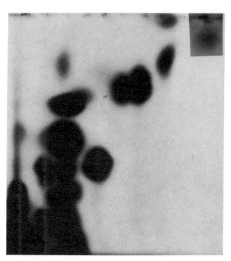

FIG. 6—*Polyethyleneimine-cellulose map of DNA digests of barley plant treated with methyl-nitroso-urea for 40 h. Spots corresponding to normal nucleosides are present, as well as additional spots not seen on autoradiograms of control DNA. Autoradiography was at −70°C for 17 h. An additional spot (dark area in upper right) was identified after 85.5-h exposure at −70°C.*

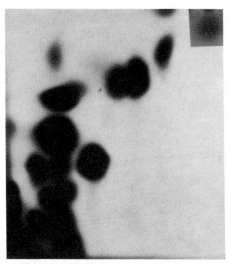

FIG. 7—*Polyethyleneimine-cellulose map of DNA digests of Australian saltbush plant treated with methyl-nitroso-urea for 40 h. Spots corresponding to normal nucleosides are present, as well as additional spots not seen on autoradiograms of control DNA. Autoradiography was at − 70°C for 17 h. An additional spot (dark area in upper right) was identified after 85.5-h exposure at − 70°C.*

h at 37°C. The barley DNA was obtained from barley grown in our laboratory. The calf thymus DNA was purchased from Sigma Chemical Company. The DNA was precipitated from the solution by adding an equivalent volume of ice-cold isopropanol. As described for methylating compounds, DNA was digested, labeled with ^{32}P, and the resulting nucleosides separated by thin layer chromatography on PEI-cellulose plates. Autoradiograms of digests from exposed barley and calf thymus DNA (Figs. 4 and 5) and DNA isolated from exposed barley and saltbush plants (Figs. 6 and 7) show spots produced by [^{32}P] deoxyribonucleoside

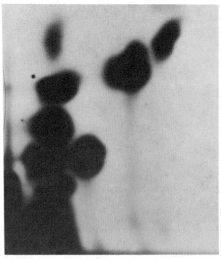

FIG. 8—*Polyethyleneimine-cellulose map of control barley DNA digests indicating spots produced by normal nucleosides.*

5'-monophosphates of normal DNA constituents as well as additional spots not present on autoradiograms of untreated DNA (Fig. 8). While further characterization of these products will be accomplished by cochromatography with known compounds using thin-layer chromatography and/or high performance liquid chromatography methods, the spots are DNA adducts based on their exhibiting a similar migration pattern to previously identified DNA adducts formed by exposing DNA of animal origin to methylating compounds [10,18].

Discussion

Postlabeling of DNA adducts has been used to test the ability of compounds to bind to DNA in animal systems [1,2,5,18,26–29]. Experiments in our laboratory are the first to report evidence of DNA adduct formation with plant DNA [3] and in plants exposed to mutagenic/carcinogenic compounds using the ^{32}P-postlabeling procedure.

The use of DNA adduct identification in plants, as opposed to animals, has many readily apparent advantages. As a monitor of the environment, the first advantage is the availability and ease of collection. Plants are readily available, being found in the home and workplace as well as in most environments. Secondly, the organism does not have to be sacrificed as is often the case with animals. Plants are stationary, allowing for repeated sampling of a location over time. This method is quick and relatively inexpensive as compared to the conduct of mutation experiments. Plants also allow for the monitoring of airborne, water or soil routes of exposure, although the uptake by plants is not the same as in animals.

Because the pattern of DNA adducts is reported to be specific for a compound [5,10], this method may be useful not only as an indication of exposure but also as an aid in identification of the source.

The development of a bioassay for mutagens/carcinogens, in the environment as well as the home and workplace, utilizing DNA adducts of plant origin would be an advantageous, complimentary, alternative method to the oftentimes lengthy and expensive mutagenicity and carcinogenicity tests currently employed. The ability of plants and animals to form identical adduct patterns may not be necessary for the successful use of a plant system. In fact, a recent article showed evidence that rat mammary cells form different DNA adducts when exposed in vitro and in vivo to benzo[a]pyrene and that these adducts were different from those formed by human mammary cells in situ [28]. These differences and similarities between the ability of plants and animals to form carcinogenic DNA adducts needs further investigation. Another factor pointing to the potential of utilizing plant DNA adducts is that the incidence of mutation in some higher eukaroytic and microbial systems is not correlated with molecular dose of the compound or with the level of all adducts, but has been shown to be highly correlated with the level of only certain DNA adducts [29]. Therefore, if DNA adducts with a high correlation for carcinogenesis in animals are also identified with plant DNA, there exists the potential for the development of a plant bioassay system.

Acknowledgments

We wish to thank Eastman Kodak Co. for generous donations of autoradiography cassettes and intensifying screens. We also thank Om Sehgal for providing equipment and Bruce Dunn for helpful discussions.

References

[1] Randerath, K., Randerath, E., Agrawal, H. P., Gupta, R. C., Schurdak, M. E., and Reddy, M. V., *Environmental Health Perspectives,* Vol. 62, 1985, pp. 57–65.

[2] Beranek, D. T., Weis, C. C., and Swenson, D. H., *Carcinogenesis,* Vol. 1, 1980, pp. 595–606.

[3] Reddy, M. V. and Randerath, K., *Environmental Health Perspectives,* Vol. 76, 1987, pp. 41–47.

[4] Dunn, B. P. in *Manual of ^{32}P-postlabeling Procedures,* June 1988, p. 7.

[5] Reddy, M. V. and Randerath, K., *Carcinogenesis,* Vol. 7, 1986, pp. 1543–1551.

[6] Randerath, E., Agrawal, H. P., Reddy, M. V., and Randerath, K., *Cancer Letters,* Vol. 20, 1983, pp. 109–114.

[7] Ireland, F., Lower, W., Sehgal, O., and Judy, B., *Proceedings, 43rd Southern Weed Science Society,* 1990, pp. 366–372.

[8] Anderson, V. A., Plewa, M. J., and Gentile, J. M., *Mutation Research,* Vol. 197, 1988, pp. 303–312.

[9] Plewa, M. J., Wagner, E. S., and Gentile, J. M., *Mutation Research,* Vol. 197, 1988, pp. 207–219.

[10] Reddy, M. V., Gupta, R. C., Randerath, E., and Randerath, K., *Carcinogenesis,* Vol. 5, 1984, pp. 231–243.

[11] Schoeny, R., Cody, T., Warshawsky, D., and Radike, M., *Mutation Research,* Vol. 197, 1988, pp. 289–302.

[12] Veleminsky, J. and Gichner, T., *Mutation Research,* Vol. 10, 1970, pp. 43–52.

[13] Blin, N. and Stafford, D. W., *Nucleic Acids Research,* Vol. 3, 1976, pp. 2303–2308.

[14] Dellaporta, S. L., Wood, J., and Hicks, J. B., *Plant Molecular Biology Reporter,* Vol. 1, 1983, pp. 19–21

[15] Junghans, H. and Metzlaff, M., *Biotechniques,* Vol. 8, 1990, p. 176.

[16] Mathew, C. G. P. in *Methods in Molecular Biology, Vol. 2, Nucleic Acids,* Humana Press, Clifton, NJ, 1984, Chapter 5, pp. 31–34.

[17] Zimmer, A. and Newton, J. in *Maize for Biological Research,* University Press, University of North Dakota, Grand Forks, NC, 1983, pp. 165–168.

[18] Randerath, K., Reddy, M. V., and Gupta, R. C., *Proceedings, National Academy of Science,* Vol. 78, 1981, pp. 6126–6129.

[19] Swanstrom, R. and Shank, P. R., *Analytical Biochemistry,* Vol. 86, 1978, pp. 184–192.

[20] Harris, C. C., LaVeck, G., Groopman, J., Wilson, V. L., and Mann D., *Cancer Research,* Vol. 46, 1986, pp. 3249–3253.

[21] Hemminki, K., *Arch. Toxicology,* Vol. 52, 1983, pp. 249–285.

[22] Mazzullo, M., Bartoli, S., Bonora, B., Colacci, A., Grilli, S., Lattanzi, G., Niero, A., Turina, M. P., and Parodi, S., *Environmental Health Perspectives,* Vol. 82, 1989, pp. 259–266.

[23] Muller, R. and Rajewsky, M. F., *Cancer Research Clinical Oncology,,* Vol. 102, 1981, pp. 99–113.

[24] Santella, R. M., *Mutation Research,* Vol. 205, 1988, pp. 271–282.

[25] Jahnke, G. D., Thompson, C. L., Walker, M. P., Gallagher, J. E., Lucier, G. W., and DiAugustine, R. P., *Carcinogenesis,* Vol. 11, 1990, pp. 205–211.

[26] Asan, E., Fasshauer, I., Wild, D., and Henschler, D., *Carcinogenesis,* Vol. 8, 1987, pp. 1589–1593.

[27] Randerath, E., Agrawal, H. P., Weaver, J. A., Bordelon, C. B., and Randerath, K., *Carcinogenesis,* Vol. 6, 1985, pp. 1117–1126.

[28] Seidman, L. A., Moore, C. J., and Gould, M. N., *Carcinogenesis,* Vol. 9, 1988, pp. 1071–1077.

[29] Schy, W. E. and Plewa, M. J., *Mutation Research,* Vol. 211, 1989, pp. 231–241.

Baljit S. Gill,[1] Shahbeg S. Sandhu,[2] Lorraine C. Backer,[1] and Bruce C. Casto[3]

Application of a Plant Test System in the Identification of Potential Genetic Hazards at Chemical Waste Sites

REFERENCE: Gill, B. S., Sandhu, S. S., Backer, L. C., and Casto, B. C., "**Application of a Plant Test System in the Identification of Potential Genetic Hazards at Chemical Waste Sites,**" *Plants for Toxicity Assessment: Second Volume, ASTM STP 1115,* J. W. Gorsuch, W. R. Lower, W. Wang and M. A. Lewis, Eds., American Society for Testing and Materials, Philadelphia, 1991, pp. 309–317.

ABSTRACT: In situ evaluations of biological effects of chemicals from industrial waste sites provide an assessment of potential health hazards to humans under the complexities of natural environment. For such studies, plants, being cost effective and suitable for multimedia exposure, are ideal for preliminary investigations for hazard identification. Several native plant species have been used as biomonitors for providing an integrated response to the bioavailability of chemical toxicants in the real-world environment.

We have utilized the *Tradescantia micronucleus* (Trad-MCN) assay for evaluating genetic hazards at a chemical waste site contaminated with agricultural insecticides scheduled for cleanup under the Superfund program. The chemical analysis of soil samples from the site indicates presence of lindane, beta BHC, and heptachlor in the subsurface samples. *Tradescantia* plants were planted at five locations to evaluate the mutagenic effects of the total environment, i.e., soil, water, and air. In addition, stem cuttings were also placed at these locations to sample the genetic impact of vapor phase organics in the atmosphere. Surface and subsurface samples were obtained from these locations for chemical and biological analysis in the laboratory. Concurrent laboratory and field controls were maintained in Durham, NC.

Results of *Tradescantia* exposures at the site showed significantly higher frequencies of micronuclei from contaminated plots before remediation, but no genetic activity was detected after the remedial action. Plants exposed to soil samples in the laboratory yielded nonsignificant results except for one subsurface sample before remediation and two surface samples after remediation. From comparison of plants exposed in the laboratory and under natural conditions, we concluded that genetically effective levels of volatile organics were present at the site before remediation. This study demonstrates the utility of simple bioassays for detecting environmental hazards at chemical waste sites.

KEY WORDS: *Tradescantia micronucleus* assay, chemical waste site, genetic hazard, superfund sites

The toxicological effects associated with exposure to chemicals commonly found at industrial waste disposal sites have been determined in laboratory studies. However, hazardous

[1] Project leader, Environmental Health Research and Testing, Inc., P.O. Box 12199, Research Triangle Park, NC 27709.
[2] Research biologist, Genetic Toxicology Division, U.S. Environmental Protection Agency, Research Triangle Park, NC 27711.
[3] Director of research, Environmental Health Research and Testing, Inc., P.O. Box 12199, Research Triangle Park, NC 27709.

waste sites typically contain combinations of highly concentrated heavy metals, pesticides, organic solvents, and other chemicals. The complex interactions among the particular chemicals, methods of disposal, the organisms present, and environmental factors such as soil composition and water runoff as well as fluctuations in the temperature and radiation make the conditions associated with a particular waste site difficult to reproduce in the laboratory. Epidemiological studies have demonstrated specific health problems associated with human and animal exposure to hazardous agents produced by industrial processes [1,2]. An increased risk of cancer associated with drinking water contaminated with chemicals that leached from disposal sites into groundwater has been recently reported [3]. However, while this type of study is potentially able to relate specific exposures to subsequent adverse health effects, it does not provide a way to estimate health risks before a disease is actually manifested. Thus, a more direct method to determine the potential harmful effects from exposure to the complex environment of a hazardous waste site is required.

The potential health risks associated with exposure to hazardous waste can be determined by on-site monitoring of environments in question. The biological effects of exposure to hazardous wastes can be measured in situ utilizing indigenous populations or specific test organisms brought onto the site. Thompson et al. [4] demonstrated that cotton rats living on hazardous waste sites had a significantly higher frequency of both structural and numerical chromosomal aberrations when compared to the control population living away from the site. Further examples using animal sentinels include the observation of a high incidence of neoplasia in adult brown bullhead catfish exposed to high concentrations of polycyclic aromatic hydrocarbons in the Black River, a tributary of Lake Erie in north-central Ohio [5] and the detection of an inverse relationship between the frequency of sister chromatid exchanges in mice caught in corn cribs and the distance between the capture site and various industrial sites near Ontario, Canada [6].

The use of plants as biomonitors represents an additional potential source of in situ biomonitoring assays. Plant bioassays that have been utilized in environmental studies include: the *Tradescantia* stamen hair assay [7] used to monitor ambient air for mutagenic activity; the forward and reverse mutation of the gene for amylose synthesis in corn (*Zea mays*) used to determine the mutagenic effects of herbicides applied to fields [8]; and cytogenetic analysis used to detect an abnormally high frequency of chromosomal aberrations in a wild population of royal fern (*Osmunda regalis*) growing in a river contaminated with paper recycling waste [9]. In addition, plants can be grown directly on a hazardous waste site, and a variety of assays, including the induction of mutations and chromosomal aberrations, can be used to determine mutagenic activity associated with exposure to contaminated soil, air, and water [10].

Many chemicals likely to be present at hazardous waste sites are clastogens. Thus, an in situ biomonitoring assay that detects chromosome breakage would be particularly useful as a bioindicator of genetic hazards of the ambient environment. The *Tradescantia micronucleus* assay (Trad-MCN) detects chromosome breakage as well as aneuploidy induction in the form of micronuclei observed at the tetrad stage in pollen mother cells [11]. This assay has been used to test 150 chemicals for their clastogenic/aneuploidy-inducing activity. In addition, the assay has been used in situ to monitor areas containing sludge, pesticides, diesel exhaust, and obscurant smokes [12–15]. The Trad-MCN assay detects chemical pollutants in the liquid and gaseous forms through roots (in soil) and foliage (in ambient environment). The present study was designed to demonstrate the applicability of the Trad-MCN assay for in situ evaluation of potential genetic risks associated with exposure to hazardous waste sites.

Procedure

Study Site

For the present study, a site in North Carolina, listed as a national priority for remediation through the Environmental Protection Agency's Superfund program, was chosen. The area (NC-1) had previously served as a trash dump. In 1984, the site was found to be contaminated with pesticides, including DDT, toxaphene, and BHC, that had been buried in trenches. The sampling locations and the concentrations of pesticides found in soil samples during the initial survey in 1986 are shown in Fig. 1 and Table 1, respectively. The chemical analysis of soil samples was performed using EPA method 8080 for pesticides/PCBs.

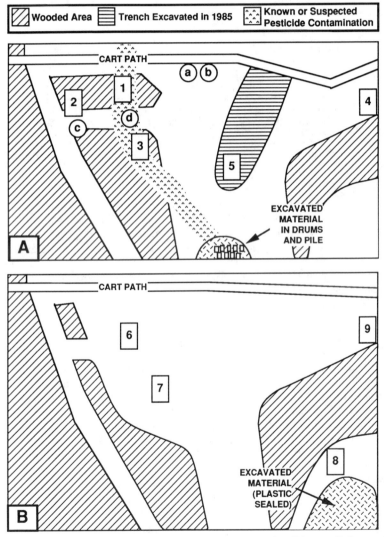

FIG. 1—*Locations of the test plots (1-5) before (A) and (6-9) after (B) remediation at the North Carolina site (NC-1) and sampling locations (a to d) before remediation in 1986.*

TABLE 1—*Chemical analysis of soil samples collected by EPA Region IV from the North Carolina pesticides waste site (NC-1).*

Chemical	Sample, mg/kg			
	FC-03[a] (a)	FC-04[a] (b)	FC-13[b] (c)	FC-14[a] (d)
Aldrin	66	23	0	0
BHC				
Alpha	65	46	0	150
Beta	49	37	0	15
Delta	31	49	0	25
Gamma	52	54	0	18
DDD	150	75	0	69
DDE	...	...	0	4
DDT	330	37	0.06	1200
Heptachlor	18	37	0	3
Toxaphene	670	500	0	...
Total	1431	858	0.06	1484

NOTE: Source is report of geophysical survey and sampling activities during the week of 12 May through 16 May 1986 (EPA, Region IV).
[a] 1.2 to 1.8 m deep.
[b] 9.1 m deep.

Five test plots were located at NC-1: two on the suspected contaminated trench, and three others 9, 45, and 137 m away. Each plot measured 1.8 by 1.8 m and was surrounded by a 1.5-m-high wire fence. Surface and subsurface (1.2-m deep) samples were taken for bioanalysis in the laboratory. After remediation, two plots were located on the excavated site, and two others were located in the surrounding area. Surface samples were collected for biological and chemical analyses from each plot. The locations of the plots before and after remediation are shown in Fig. 1.

Plant Bioassay

The procedures for maintaining the *Tradescantia* plants and for analyzing micronucleus frequency in the tetrads have been described [11]. Briefly, for in situ exposure, five clones of *Tradescantia* clone (#4430) were planted in each test plot for evaluating the effects induced by the multimedia exposure. Inflorescences were harvested at the 8-10-buds stage and fixed for 24 h in ethanol:acetic acid (3:1). The fixed material was stored in 70% ethanol. *Tradescantia* stem cuttings bearing young inflorescences were placed in a cup of water and exposed at the site for 30 h before fixing for evaluating the genetic effects induced by the vapor phase organics. In the laboratory, stem cuttings bearing young inflorescences were placed in slurries of soil samples collected from test plots for a 30-h exposure. Slides were prepared from the fixed material using the aceto-carmine squash technique. Micronuclei were scored in the early tetrad stages of pollen mother cells.

In the present study, 15 to 20 inflorescences comprised a sample. Three hundred tetrads were scored from each of five slides prepared from a treatment sample for a total of 1500 tetrads per plot. Data were recorded as the number of micronuclei (MCN) per 100 tetrads.

TABLE 2—*Chemical analysis of surface soil samples of test plots at the North Carolina site (NC-1) after the remedial action.*

| | | Pesticide, mg/kg | | | | | |
| | | BHC | | | | | |
Plot	Aldrin	Alpha	Beta	DDD	DDT	Heptachlor	Toxaphene	Total
6	0.36	0.31	0.58	0.22	13.36	0.75	3.40	18.98
7	0.18	0.88	0.16	0.41	2.92	0.39	1.99	6.93
8	0.72	7.64	3.49	9.75	43.93	2.60	10.90	79.03
9	0	0.01	0.02	0.12	0	0	0	0.15

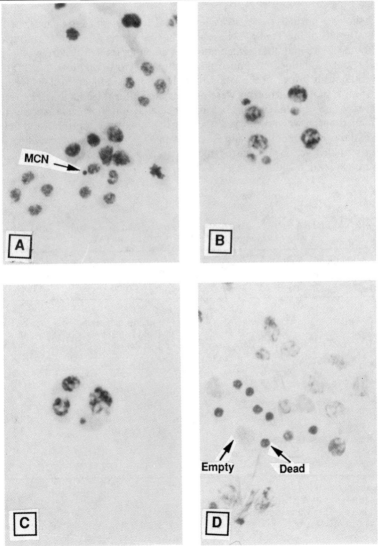

FIG. 2—*Photographs showing:* (A) *Normal tetrads and a tetrad with MCN,* (B) *Tetrad with multiple MCN;* (C) *Tetrad with unequal division, and* (D) *Tetrads with dead and empty cells.*

A change in frequency of MCN/100 tetrads was considered statistically significant (at $P <$ 0.05) if the difference between the mean of the control population and the mean of the treated population was at least twice as large as the standard error of the difference between the two means [11].

Results and Discussion

The chemical analysis of soil samples collected after remedial action (see Fig. 1 for sample location) is shown in Table 2. The chemical analysis of soil samples from Plots 6, 7, and 9 shows total chemicals below the target cleanup level of 50 ppm (total chemicals) set by EPA for remediation. Plot 8 outside the excavated area shows somewhat higher concentration of total chemicals.

Figure 2 shows a normal *Tradescantia* tetrad and tetrads with micronuclei or other abnormalities. Results of the Trad-MCN assay for in situ plantings and stem cuttings exposed at the site both before and after remediation are given in Table 3. There was a significant increase ($P < 0.05$) in induction of micronuclei in plants grown on Plots 1, 2, 3, and 5 when compared to controls grown in the laboratory. Differences in micronucleus frequency between the sample plots and controls were not detected after remediation. The results from the on-site exposure of stem cuttings before remedial action show a statistically significant increase in micronuclei when compared to either the laboratory control or the field control kept outside the Environmental Health Research and Testing laboratory building (Durham, NC). After remedial action, no clastogenic activity was detected when *Tradescantia* stem cuttings were exposed at the site.

TABLE 3—*Results of TRAD-MCN assay for on-site planting and stem cuttings. Exposure before and after remediation at the North Carolina site (NC-1).*

Planting in Soil		Stem Cuttings	
Plot	MCN/100 Tetrads	Location	MCN/100 Tetrads
BEFORE REMEDIATION			
1[a]	9.3 ± 0.53*	On site	19.3 ± 1.47*
2	17.9 ± 4.65*	Field control	7.3 ± 0.50
3	10.8 ± 1.10*	Lab control	7.1 ± 0.52
4	9.8 ± 2.56		
5	9.6 ± 1.06*		
LAB control[b]	7.1 ± 0.52		
AFTER REMEDIATION			
6	5.6 ± 0.23	On site	9.2 ± 1.12
7	5.2 ± 0.72	Field control	10.0 ± 2.04
8	7.7 ± 0.66	Lab control	8.1 ± 1.41
9	6.6 ± 1.00		
Field control[c]	6.3 ± 0.50		
LAB control	6.8 ± 0.45		

* Significance ($P \leq 0.05$).
[a] Values are the mean ± SE of 1500 tetrads per plot.
[b] LAB Control = tap water control.
[c] Field Control = Control plot outside Environmental Health Research and Testing Laboratory.

TABLE 4—*Results of TRAD-MCN assay for the soil slurry samples collected before and after remediation at the North Carolina site (NC-1).*

Surface Samples		Subsurface Samples	
Sample	MCN/100 Tetrads	Sample	MCN/100 Tetrads
BEFORE REMEDIATION			
1A[a]	4.7 ± 0.79	1B	8.3 ± 0.80
2A	7.9 ± 0.46	2B	6.7 ± 0.89
3A	14.3 ± 2.97	3B[b]	23.8 ± 2.98*
4A	11.4 ± 2.42	4B	3.5 ± 0.68
5A	5.7 ± 0.53	5B	5.1 ± 0.54
Control	8.7 ± 0.76	Control 1	7.7 ± 0.78
		Control 2	4.4 ± 0.47
AFTER REMEDIATION (Only Surface Samples)			
6	6.0 ± 1.56		
7	4.9 ± 0.25		
8	6.5 ± 0.91*		
9	6.2 ± 0.95*		
Control	4.0 ± 0.52		

* Significance ($P \leq 0.05$).
[a] Values are the mean ± SE.
[b] Samples 3B, 4B, and 5B were tested separately with concurrent Control 2.

Soil samples collected at the site were made into a slurry for exposure of the stem cuttings in the laboratory. The results from soil samples collected before and after remediation and used for exposure in the laboratory (soil slurries) are reported in Table 4. Before remediation, a significant increase in the number of MCN/100 tetrads was observed only in the subsurface sample taken from Plot 3. This sample also produced unequal size nuclei in tetrads as well as dead and empty cells; an indication of toxicity (Fig. 2). After remediation, samples from Plots 8 and 9 showed statistically significant increases in micronucleus frequency when compared with the laboratory control.

As demonstrated by this study, the biological results following exposure to hazardous wastes may not directly relate to physical measurements such as the concentration of a pesticide or group of pesticides in the surface soil. However, after remediation, the Trad-MCN assay did not detect any chromosomal damage. These results suggest that the original increases in micronucleus frequency may have been induced by volatile organics present on the site before, but not after, remediation. The *Tradescantia* clone (#4430) used in this research is particularly sensitive to gaseous pollutants [11]. This sensitivity, and the absence of increased micronucleus frequency after exposure of the plants to soil slurries of samples from original contaminated sites in the laboratory, appear to corroborate the suggestion that the original increase in micronucleus frequency was due to volatile pollutants arising from pesticides located below the surface (Table 1).

In summary, the Trad-MCN assay detected a genotoxic effect from contamination at the North Carolina hazardous waste site and also confirmed that remediation of the subsurface soil substantially decreased the genotoxic hazard associated with pesticides buried on the site. However, results from the Trad-MCN assay, i.e., positive for induction of MCN before remediation, but not after, suggested that volatile organics, possibly undocumented solvents, as well as pesticides, contributed to the induction of micronuclei.

In conclusion, the results of this study demonstrate the utility of a simple bioassay to complement the chemical analysis for a more comprehensive assessment of biological hazards from chemical waste sites.

Acknowledgments

Although the research described in this article has been supported by the United States Environmental Protection Agency (through Contract 68-02-4456 to Environmental Health Research and Testing, Inc.), it has not been subjected to Agency review and therefore does not necessarily reflect the views of the Agency and no official endorsement should be inferred. Mention of trade names or commercial products does not constitute endorsement or recommendation for use.

We thank Flint Worrell of the North Carolina Department of Human Resources, Solid and Hazardous Waste Management Branch, and Ned Jessup of the Waste Management Division, U.S. EPA, Region IV, for their help in selecting the site and assistance during our field work. We also appreciate the assistance of L. Tom Heiderscheit of the Technology Transfer Division and Shirley Milton of the Genetic Toxicology Division, U.S. EPA, in the preparation of this manuscript.

References

[1] Wang, H.-W, You, X.-J, Qu, Y.-H., Wang, W. F., Long, Y.-M., and Ni, J.-A., *Cancer Research,* Vol. 44, 1984, pp. 3101–3105.
[2] Glickman, L. T., Domanski, L. M., Maguire, T. G., Dubielzig, R. R., and Churg, A., *Environmental Research,* Vol. 32, 1983, pp. 305–313.
[3] Griffith, J., Duncan, R. C., Riggan, W. B., and Pellom, A. C., *Archives of Environmental Health,* Vol. 44, No. 2, 1989, pp. 69–74.
[4] Thompson, R. A., Schroder, G. D., and Connor, T. H., *Environmental and Molecular Mutagenesis,* Vol. 11, No. 3, 1988, pp. 359–368.
[5] Harshbarger, J. C., Cullen, L. J., Calabrese, M. J., and Spero, P. M., *Marine Environmental Research,* Vol. 14, 1984, pp. 535–536.
[6] Nayak, B. N. and Petras, M. L., *Canadian Journal of Genetics and Cytology,* Vol. 27, 1985, pp. 351–356.
[7] Schairer, L. A., Van't Hof, J., Hayes, C. G., Burton, R. M., and de Serres, F. J., *Environmental Health Perspectives,* Vol. 27, 1987, pp. 51–60.
[8] Plewa, M. J., Wagner, E. D., Gentile, G. J., and Gentile, J. M., *Mutation Research,* Vol. 136, 1984, pp. 233–245.
[9] Klekowski, E. and Levin, D. E., *Environmental Mutagenesis,* Vol. 1, 1979, pp. 209–219.
[10] Sandhu, S. S. and Lower, W. R., *Toxicology and Industrial Health,* Vol. 5, No. 1, 1989, pp. 73–83.
[11] Ma, T.-H. in *In Vitro Toxicity Testing of Environmental Agents,* A. R. Kolber, T. K. Wong, L. D. Grand, R. S. DeWoskin, and T. J. Hughes, Eds., Plenum Press, New York, 1983, pp. 191–214.
[12] Ma, T.-H., Anderson, V., Harris, M. M., and Bare, J. L., *Environmental Mutagenesis,* Vol. 5, 1983, pp. 127–137.
[13] Ma, T.-H., Lower, W. R., Harris, F. D., Poku, J., Anderson, V. A., Harris, M. M., and Bare,

J. L. in *Short-Term Bioassays in the Analysis of Complex Environmental Mixtures III*, M. Waters, S. Sandhu, J. Lewtas, L. Claxton, N. Chernoff, and S. Nesnow, Eds., Plenum Press, New York, 1983, pp. 89–99.

[14] Ma, T.-H., Harris, M. M., Anderson, V. A., Ahmed, I., Mohammad, K., Bare, J. L., and Lin, G., *Mutation Research*, Vol. 138, 1984, pp. 157–167.

[15] Ma, T.-H. and Harris, M. M., *Hazard Assessment of Chemicals*, Vol. 4, 1985, pp. 77–105.

James M. Gentile,[1] *Peter Johnson,*[2] *and Susan Robbins*[2]

Activation of Aflatoxin B1 and Benzo(a)pyrene by Tobacco Cells in the Plant Cell/Microbe Coincubation Assay

REFERENCE: Gentile, J. M., Johnson, P., and Robbins, S., **"Activation of Aflatoxin B1 and Benzo(a)pyrene by Tobacco Cells in the Plant Cell/Microbe Coincubation Assay,"** *Plants for Toxicity Assessment: Second Volume, ASTM STP 1115,* J. W. Gorsuch, W. R. Lower, W. Wang, and M. A. Lewis, Eds., American Society for Testing and Materials, Philadelphia, 1991, pp. 318–325.

ABSTRACT: We investigated the ability of tobacco (TX1) cells in culture to activate aflatoxin B1 (AFB1) and benzo(a)pyrene (BAP) using the plant cell/microbe coincubation assay. Initial experiments with *Salmonella typhimurium* as a genetic endpoint using previously published assay conditions failed to demonstrate that either of these agents could be activated into a mutagenic form by TX1 cells. However, when modifications of the assay, including increased plant cell densities and/or increased preincubation times, were employed, both AFB1 and BAP were demonstrated to be potent plant promutagens. These data demonstrate that, for the purpose of routine screening and toxicity assessment studies focusing upon plant activation, careful attention must be paid to both the chemical or complex mixture under study as well as to the metabolic potential of the particular plant cell system employed. This will ensure a more complete understanding of the problem at hand, provide insight into the mechanisms of activation, and guard against possible false negative results.

KEY WORDS: *Salmonella,* plant activation, aflatoxin B1, benzo(a)pyrene, plant cell/microbe coincubation assay

The capability of plants to activate promutagens into genotoxic metabolites that may enter the ecosystem and human food chain takes on crucial significance with the realization of the immense diversity of xenobiotics to which plants are intentionally and unintentionally exposed.

It is well established that plants can metabolize chemicals into mutagens [1,2]. In some instances the genotoxicity of the chemical in question can be demonstrated only under the conditions of plant activation [3], while other promutagens are transformed into mutagens effectively by both plant and animal systems [4].

Several techniques have been developed to study plant activation of chemicals. *In vivo* protocols rely upon extracts from preexposed intact plants and subsequent application of these extracts to a microbial genetic indicator organism [6,7]. Although these *in vivo* protocols mimic conditions encountered in modern agriculture, problems involving microbial contamination of the plant materials, artifacts introduced into the microbial assays by plant nutrients, dosimetry problems inherent in treating intact plants, and potential modification of metabolites and concentration of samples have led to the development of alternate techniques.

---

[1] Dean, Natural Sciences and Kenneth G. Herrick Professor of Biology, Biology Department, Hope College, Holland, MI 49423.

[2] Undergraduates, Hope College, Holland, MI 49423.

Two different *in vitro* protocols have proven effective with several test agents. One protocol employs enzyme extracts from nonexposed plants in conjunction with a genetic indicator system [1] while the other requires the incubation of plant cells in culture with a genetic indicator organism [4]. This latter assay is known as the plant cell/microbe coincubation assay.

The plant cell/microbe coincubation assay is an effective assay for assessing the promutagenic potential of xenobiotics as well as for assessing mechanisms of activation [7]. Several different types of plant cell lines have been assessed for their ability to be used in this assay [8–10]; however, the total number of substrates identified as plant promutagens in this assay are limited (Table 1).

In an effort to better understand the phenomenon of plant activation as well as the overall utility of the plant cell/microbe coincubation assay, and to expand the database associated with this assay, we investigated the activation of aflatoxin B1 (AFB1) and benzo(a)pyrene (BAP) by TX1 cells.

Both AFB1 and BAP are well-established mammalian promutagens, and each has been demonstrated previously to be activated to some extent by plant systems. AFB1 was successfully activated *in vitro* in the *Salmonella* assay by cell-free homogenates from *Zea mays* seedlings [2] and by microsomal enzymes from bulbs of *Tulipa generiana* [11,12]. BAP was demonstrated to be metabolized into an oxidized form (by homogenates of vascular plants [13] as well as by whole cell suspensions cultures derived from parsley and soybean [14]). It was activated into a genotoxic form by photosynthetic algae (*Selenastrum capricornutum*) exposed to different light regimes [15], and we have recently demonstrated that BAP can be activated by *Selenastrum* in a modified preincubation assay [16].

Procedure

Chemicals—AFB1 and BAP were purchased from Sigma Chemical, Co., MO. All other chemicals were purchased from Sigma Chemical Co., while bacteriological media was purchased from Hazelton Laboratories, KS.

TABLE 1—*Chemicals identified as plant promutagens using the plant cell/microbe coincubation assay.*

Chemical	Plant Cell Type	Reference
2-aminofluorene	Carrot	*4,24*
	Cotton	*4,24*
	Tobacco	*4,7,23,24,25*
	Tradescantia	*9*
Benzo(a)pyrene	Algae	*16*
m-phenylenediamine	Carrot	*4,8*
	Cotton	*4,8*
	Tobacco	*4,5,8*
	Tradescantia	*9*
o-phenylenediamine	Tobacco	*5*
4-nitro-*o*-		
phenylenediamine[a]	Algae	*16*
	Carrot	*24*
	Cotton	*24*
	Tobacco	*5,24*

[a] The active plant promutagen is a contaminant of unknown structure associated with commercial grade 4-nitro-*o*-phenylenediamine.

Genetic Indicator Organisms—Salmonella typhimurium strain TA98 was used. This strain was originally obtained from Dr. Bruce Ames (University of California-Berkeley) and is maintained in our laboratory according to the protocols outlined by Maron and Ames [17]. Its integrity was routinely monitored using the protocol of Zeiger et al. [18], and its mutability was tested with 2-nitrosofluorene. Positive controls were run concurrently in each experiment.

Plant Cells

Long-term plant cell suspension cultures of tobacco, cell line TX1, were originally obtained from Dr. J. Widholm (University of Illinois-Urbana). This cell line is routinely maintained in our laboratory in a modified Murashige and Skoog [19] liquid culture medium (MX).

Plant Cell/Microbe Coincubation Assay

For a typical experiment, plant cells were grown to 28°C to either midlog- or stationary-phase and the cells were harvested by centrifugation, washed and resuspended in MX$^-$ medium. MX$^-$ medium is a modified Murashige and Skoog (X) liquid culture medium that lacks plant growth hormone. Fresh weights of the plant cells were adjusted to 200 mg/mL and stored on ice (<15 min) until used.

Bacterial cells were grown from a single colony isolate. Cells were grown aerobically overnight in 100-mL nutrient broth at 37°C. The bacterial suspension was centrifuged; the cell pellet was resuspended and washed twice in 100-mM potassium phosphate buffer (pH 7.4) and the cells harvested following the final wash. The bacterial pellet was resuspended in buffer to a titer of 1×10^{10} cells/mL and iced for no greater than 1 h prior to use.

For a standard coincubation assay, each reaction mixture consisted of 4.5 mL of the plant cell suspension, 0.5 mL of the bacterial cell suspension (5×10^9 cells), and a test agent (in 25-μL DMSO), which were combined in a reaction vessel and incubated in a gyratory shaker (150 rpm) at 28°C for 1 h. Negative controls consisted of plant or algal cells and bacteria without the test agent, solvent alone, and the test agent added to heat-killed plant cells coincubated with viable bacteria. Each test agent was tested alone and with Arochlor-1254-induced mammalian-hepatic S9 (Hazelton Laboratories, MD) as positive controls. After treatment time, the reaction tubes were placed on ice to quench the reaction [4]. Triplicate 0.5-mL aliquots of the reaction mixture were removed and added to 2 mL of molten top agar supplemented with 550-μM histidine and biotin. The top agar was poured onto minimal medium plates, incubated for 48 h at 37°C, and *Salmonella his$^+$* revertant colonies were scored using a New Brunswick Biotran III Automated Colony Counter (New Jersey). All experiments were conducted a minimum of three times.

Viability studies were conducted to ensure that both plant and bacterial cells survived exposure to the test agent. Plant cell viability was determined using the phenosafranin dye exclusion method of Widholm [20]. Bacterial cell viability was assessed by adding 1 mL of the cold reaction mixture to 1 mL of iced 100-mM phosphate buffer. A dilution series in phosphate buffer was performed so that approximately 500 cells were added to each of three top agar tubes and poured onto nutrient agar plates. Following 36-h incubation at 37°C, bacterial colonies were scored.

In some experiments, modifications of these standard procedures were made. These modifications are noted as appropriate in the text.

Results and Discussion

Dose-response curves were run with AFB1 (Fig. 1) and BAP (Fig. 2) and TX1 cells using standard assay conditions. Some direct-acting mutagenic activity was observed for AFB1 in the absence of TX1 cells, but no enhancement of this activity was observed when AFB1 was incubated with TX1 cells. No mutagenic activity was observed for BAP in either the absence or presence of TX1 cells. Responses for both chemicals were judged negative for cells harvested at either the midlog- or stationary-phase of growth.

Since both AFB1 and BAP had been previously identified as plant promutagens using plant tissue extracts [2,12], we sought to understand if the conditions of the plant cell/ microbe assay were such that false negative responses had occurred.

With both chemicals we sequentially evaluated the effects of: (1) varying the density of plant cells in the reaction mixture; (2) the growth phase at which plant cells were harvested; and (3) the time of coincubation (from 1 through 5 h).

Positive results were obtained with AFB1 when the density of the plant cells in the reaction mixtures was increased (Fig. 3). No positive response was observed using high densities of heat-killed plant cells coincubated with viable bacteria. A plant cell density of 300 mg/mL gave the best response. Standard assay conditions in our laboratory employ 200 mg/mL/ reaction tube. This density was chosen because greater cell densities were very thick and were difficult to use in the coincubation assay, and, with other promutagens, the increased incremental response obtained when using cell densities greater than 200 mg/mL/reaction tube was minimal. The 300 mg/mL/reaction tube cell density was difficult to handle, but with this increased cell density we found that midlog-phase cells were more responsive to AFB1 activation than were stationary-phase cells. However, prolonged incubation time (>1 h) proved ineffective because of AFB1 toxicity to the plant cells.

For BAP, no positive responses were observed following the various assay modifications indicated above; however, since BAP had been previously demonstrated to be metabolized by intact, photosynthetic algae following prolonged incubation between BAP and the algal

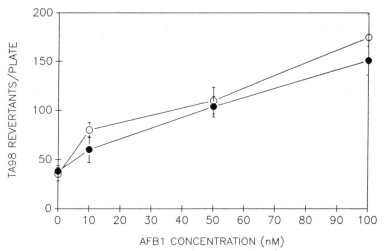

FIG. 1—*Activation of AFB1 (nM/reaction tube) in the presence (O–O) and absence (●–●) of TX1 cells (200 mg/mL reaction mixture) using the standard plant cell/microbe coincubation assay. Standard error bars, where not indicated, fall within the limits of the graphed data point.*

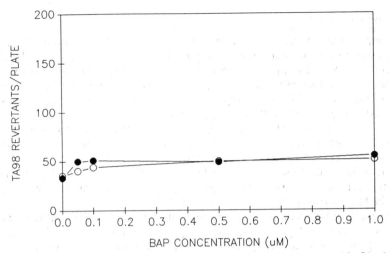

FIG. 2—*Activation of BAP (μM/reaction tube) in presence (O–O) and absence (●–●) of TX1 cells (200 mg/mL reaction mixture) using the standard plant cell/microbe coincubation assay. Standard errors bars, where not indicated, fall within the limits of the graphed data point.*

cells [15], we sought to determine if a similar prolonged incubation period between the nonphotosynthetic TX1 cells and BAP would prove effective.

For these experiments midlog-phase TX1 cells were harvested as described previously, and 200-mg/mL TX1 cells were suspended in 6.0-mL MX⁻ medium with 0.05-mL BAP (in DMSO). The reaction mixture was shaken vigorously (150 rpm) and incubated at 28°C. At time intervals between 1 and 48 h following the initiation of the preincubation reaction, 0.5 mL of an overnight culture of *S. typhimurium* was added to the reaction vials and allowed to coincubate for 1 h. Following this final coincubation phase, 0.5 mL of the reaction mixture was plated as described previously. Results from these experiments are shown in Fig. 4. BAP was activated by TX1 following a 36-h preincubation period, with increased activation

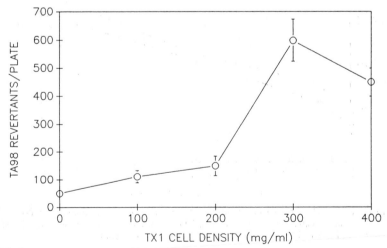

FIG. 3—*Activation of AFB1 (100 nM/reaction tube) using increasing TX1 cell densities.*

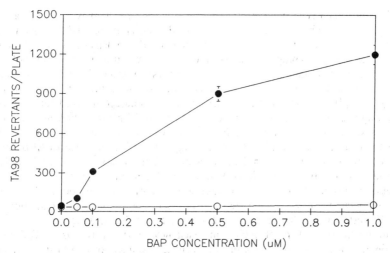

FIG. 4—*Activation of BAP (0.5 μM/reaction tube) in the presence (○-○) and absence (●-●) of TX1 cells following an extended preincubation/coincubation procedure. Standard error bars, where not indicated, fall within the limits of the graphed data point.*

observed when the preincubation period was extended to 48 h. These data are similar to those for BAP activation by *Selenastrum* where a 24-h incubation period was required before mutagenic metabolites could be isolated from the growth medium [15]. The fact that mutagenic metabolites were detected after 24 h with the algal cells (rather than the 36 to 48 h required in our experiments) could be due to either intrinsic differences in rates of cell metabolism or, more probably, due to the fact that, in the algae experiments, metabolites were isolated and concentrated from cell medium prior to genetic assessment. Using our modified preincubation/coincubation procedure, a dose-response curve was established for BAP (Fig. 5).

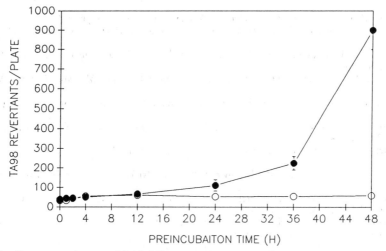

FIG. 5—*Dose responsiveness of BAP following in the presence (○-○) and absence (●-●) of TX1 cells following a 48-h preincubation/1-h coincubation procedure. Standard error bars, where not indicated, fall within the limits of the graphed data point.*

It is feasible that issues of bioavailability of substrate might play an important role in explaining why these differentially modified coincubation assays proved effective with AFB1 and BAP. However, in-depth studies on the mechanism(s) of activation must be conducted with each agent before a complete understanding of the activation process can evolve. These substrates will provide unique opportunities to understand more about the process of plant activation since each is metabolized by cytochrome P-450-dependent mechanisms in mammalian systems [21,22]. Mechanistic studies on plant promutagens have been performed on only a few agents, most notably 2-aminofluorene [23] and 4-nitro-o-phenylenediamine [24]. In both cases, perodactic oxidation rather than monooxygenase oxidation appears responsible for the activation process [5,23]. Identifying other substrates such as AFB1 and BAP as plant promutagens and resolving their mechanism(s) of activation in plant cells will provide valuable insight into the overall biotransformation of xenobiotics by plant systems.

It is also important to note that, for routine screening purposes, both AFB1 and BAP would have been classified as negative plant promutagens using routine conditions for the plant cell/microbe coincubation assay. Therefore, although this assay is a highly effective system in which to study plant activation, each new substrate that is studied should be independently judged against a variety of protocol modifications to ensure against false negative results. In addition, protocol modifications should be considered and evaluated when one looks at comparative metabolism between different types of plant cells.

Acknowledgments

This research was supported by NSF-REU Grant BB5-8712566. The authors are grateful to Glenda J. Gentile for her assistance in this research.

References

[1] Gentile, J. M. and Plewa, M. J., *Mutation Research*, Vol. 197, 1988, pp. 173–182.
[2] Plewa, M. J. and Gentile, J. M. in *Chemical Mutagens, Principles and Methods for Their Detection*, Vol. 7, F. J. DeSerres and A. Hollaender, Eds., Plenum, New York, 1982, Chapter 8, pp. 401–420.
[3] Plewa, M. J. and Gentile, J. M., *Mutation Research*, Vol. 38, 1976, pp. 287–292.
[4] Plewa, M. J., Wagner, E. D., and Gentile, J. M., *Mutation Research*, Vol. 197, 1988, pp. 207–220.
[5] Gentile, J. M., Gentile, G. J., Townsend, S., and Plewa, M. J., *Mutation Research*, Vol. 101, 1982, pp. 19–29.
[6] Plewa, M. J., Wagner, E. D., Gentile, G. J., and Gentile, J. M., *Mutation Research*, Vol. 146, 1984, pp. 233–245.
[7] Plewa, M. J., Weaver, D. L., Blair, L. C., and Gentile, J. M., *Science*, Vol. 219, 1983, pp. 1427–1429.
[8] Lhotka, M. A., Plewa, M. J., and Gentile, J. M., *Environmental and Molecular Mutagenesis*, Vol. 10, 1987, pp. 79–88.
[9] Anderson, V. A., Plewa, M. J., and Gentile, J. M., *Mutation Research*, Vol. 197, 1988, pp. 303–312.
[10] Shane, B. S. and D. J. Longstreth, *Environmental and Molecular Mutagenesis*, Vol. 15, Suppl. 1, 1990, p. 55.
[11] Higashi, K., Ikeuchi, K., and Karasaki, Y., *Bulletin of Environmental Contamination and Toxicology*, Vol. 29, 1982, pp. 505–510.
[12] Higashi, K., Ikeuchi, K., Karasaki, Y., and Obara, M., *Biochemistry Biophysics Research Communication*, Vol. 115, 1983, pp. 46–52.
[13] Higashi, K., *Mutation Research*, Vol. 197, 1988, pp. 273–288.
[14] Van der Trenck, K. T. and Sandermann, H., *Planta*, Vol. 141, 1978, pp. 245–251.
[15] Schoeny, R., Cody, T., Warshawsky, D., and Radike, M., *Mutation Research*, Vol. 197, 1988, pp. 289–302.

[16] Gentile, J.M., Lippert, M., Johnson, P., and Shafer, T., *Bulletin of Environmental Contamination and Toxicology,* 1990, *in press.*

[17] Maron, D. and Ames, B. N., *Mutation Research,* Vol. 113, 1983, pp. 173–215.

[18] Zeiger, E., Pagano, D. A., and Robertson, I, G, C., *Environmental Mutagenesis,* Vol. 3, pp. 205–209.

[19] Murashige, T. and Skoog, F., *Physiological Plant,* Vol. 15, 1962, p. 473.

[20] Widholm, J. M., *Stain Technology,* Vol. 47, 1972, pp. 189–194.

[21] Gelboin, H. V., *Physiological Reviews,* Vol. 60, 1980, pp. 1107–1166.

[22] Busby, W. F. and Wogan, G. N. in *Chemical Carcinogens,* Vol. 2, C. S. Searle, Ed., ACS Monograph, Washington, 1984, Chapter 5, pp. 945–1136.

[23] Wagner, E. D., Gentile, J. M., and Plewa, M. J., *Mutation Research,* Vol. 216, 1989, pp. 163–178.

[24] Gentile, J. M., Pluymers, D., and Plewa, M. J., *Mutation Research,* Vol. 173, 1986, pp. 181–185.

[25] Wagner, E. D., Smith, S. R., Hajek, K., and M. J. Plewa, *Environmental and Molecular Mutagenesis,* Vol. 15, Suppl. 17, 1990, p. 63.

Shahbeg S. Sandhu,[1] Jagtar S. Dhesi,[2] and Baljit S. Gill[3]

Identification of Aneuploidy-Inducing Agents Using the Wheat Seedling Assay

REFERENCE: Sandhu, S. S., Dhesi, J. S., and Gill, B. S., "**Identification of Aneuploidy-Inducing Agents Using the Wheat Seedling Assay,**" *Plants for Toxicity Assessment: Second Volume, ASTM STP 1115,* J. W. Gorsuch, W. R. Lower, W. Wang, and M. A. Lewis, Eds., American Society for Testing and Materials, Philadelphia, 1991, pp. 326–330.

ABSTRACT: A short-term assay using Chinese Spring wheat for detecting induced aneuploidy and/or clastogenicity is described. Seeds of virescent wheat (v_1v_1) are treated with a test agent. The appearance of white sectors ($v_1v_1v_1$) or green sectors (v_10) on the virescent leaves indicates the gain or loss of a chromosome or a segment of a chromosome-bearing v_1 allele. Three test chemicals, chromium trioxide (100 to 5000 µg/mL), chromium dichromate (500 to 5000 µg/mL), and cyclophosphamide (100 to 1000 µg/mL) produced a dose-related positive response for chromosome aberrations in this assay.

KEY WORDS: wheat seedling assay, aneuploidy, clastogenicity, genotoxicity

Aneuploidy (a deviation from the multiple of haploid chromosome numbers) is associated with mental retardation, spontaneous abortions, and is suspected to be a factor in carcinogenesis (see recent reviews by Dellarco et al. [1], Epstein [2], and Vig and Sandberg [3]). In spite of the evidence of the role of aneuploidy in human mortality and morbidity, relatively few short-term eukaryotic assays have been developed for evaluating the ability of environmental chemicals to disrupt normal cell division processes [4]. A number of chemicals of human concern, such as asbestos, benzene, and diethylstilbestrol, which yield nondetectable mutagenic response in most of the assays for gene mutations, have been shown to induce aneuploidy [5,6].

The most commonly used methods for detecting chromosome breaks and aneuploidy in higher eukaryotes involve cytological observations that are laborious, require highly trained personnel, and can be subjective. As an alternative, a set of new assays for detecting environmentally induced aneuploidy and chromosome breaks have been proposed that either simplify cytological observations by (*a*) quantitating the distribution of a single chromosome to the cell progeny [7–9], (*b*) identifying kinetochores in micronuclei [10–12], or by employing genetic selection [13,14]. We have recently reported the development of a bioassay that employs scoring leaf sectors in hexaploid wheat seedlings for evaluating the potential of environmental chemicals to induce chromosomal anomalies [15].

The hexaploid wheat assay offers a distinct advantage over other genetic assays using diploid species for detecting aneuploidy. In diploid species, chromosome deficiencies, mon-

---

[1] Research biologist, Genetic Toxicology Division, U.S. Environmental Protection Agency, Research Triangle Park, NC 27711.
[2] North Carolina Central University, Durham, NC 27707.
[3] Project leader, Environmental Health Research and Testing, Inc., Research Triangle Park, NC 27709.

osomy, and nullisomy may result in lethality. Hexaploid wheat (*Triticum aestivum*, $2n$ = 42) has three pairs of homologous chromosomes and can tolerate loss or gain of chromosome(s) and still reveal their effects by genetic tests. Preliminary results on a few model test agents were reported earlier [15]. In this article we report genetic response of three additional compounds for their ability to induce chromosome aberrations and/or aneuploidy.

Materials and Methods

Tester Strain

A strain of Chinese Spring wheat (*Triticum aestivum*, $2n$ = 42) was isolated by Neatby [16] and analyzed by Sears and Sears [17] and Sears [18] to have a pair of recessive alleles (v_1) on the short arm of chromosome 3B that suppress the development of green pigmentation in leaves at temperatures below 26°C. In hemizygous condition, this allele is ineffective in suppressing chlorophyll development. This was evident in monosomic analysis of this strain where monosomic line for the chromosome 3B ($v_1 0$) developed normal green pigmented leaves. Thus, when seed of this strain of genotype $v_1 v_1 V_2 V_2 V_3 V_3$ (virescent) is treated with an aneugen (an aneuploidy-producing agent) or a clastogen (a chromosome-breaking agent), the appearance of green longitudinal sectors indicates that a cell lineage has lost a chromosome (monosomy) or a segment of a chromosome bearing the v_1 allele. The appearance of a white sector is an indicator of a cell lineage that has gained a chromosome (trisomy $v_1 v_1 v_1 V_2 V_2 V_3 V_3$) or a part of chromosome 3B bearing v_1 allele. Thus the appearance of green and/or white sectors on virescent leaves are used as indices of induced aneuploidy.

Green sectors are quite easy to observe through visual inspection, while the white sectors in virescent background are difficult to score. However, at temperatures above 26°C, virescent leaves become normal green, and white sectors are easier to observe. The appearance of white and green sectors adjacent to each other is an indication of nondisjunction, whereas the appearance of isolated green sectors may arise due to loss of a chromosome induced by mechanisms other than nondisjunction.

Test Chemicals

Chromium trioxide (CrO_3, purity 98%) and potassium dichromate ($K_2Cr_2O_7$, purity 99%) were obtained from Fisher Scientific Company (Raleigh, NC). Cyclophosphamide (CP, purity 99.5%) was obtained from Sigma Scientific Co., St. Louis, MO. The test chemicals were dissolved in 1/150 M Sorensen's phosphate buffer at pH 5.00 and 7.00. Stock solutions (A and B) for Sorensen's phosphate buffer were prepared as given below:

A. 1/15 M KH_2PO_4 (9.076 g KH_2PO_4/L distilled water).

B. 1/15 M Na_2HPO_4 (11.871 g Na_2HPO_4/L distilled water).

A Sorensen's phosphate buffer at pH 5.00 at 18°C is obtained by mixing 0.07 parts of stock solution A and 9.93 parts of stock solution B and diluted 1:9 with distilled water. For pH 7, 6 parts of stock solution A were mixed with 4 parts of stock solution B and diluted 1:9 with distilled water.

Assay Protocol

Twenty seeds of the tester strain (Neatby's strain of Chinese Spring wheat) were soaked at room temperature in distilled water for 8 h before treatment. Chemical treatment was carried out on a rotary shaker for 20 h at 25°C. After treatment, seeds were thoroughly washed and transferred to a potting mixture and grown in a phytotron at 18°C in a growth chamber at Duke University, Durham, NC. Nutrients were provided by daily watering with Hoagland's solution [19]. Green sectors (in the plants grown at 18°C) and white sectors (in plants grown at 30°C) were scored by visual inspection on the second, third, and fourth leaves. The average number of sector(s) per plant was derived from the total sectors observed on all the leaves within a specific chemical treatment and dividing it by the total number of plants in that treatment [15].

Each experiment included a concurrent negative control (Sorensen's buffer). The pH of the solvent control was the same as that of the treatment solution. For evaluating statistical significance, the analysis of Lorenz [20] for Poisson distribution was applied.

Results and Discussion

The data on the induction of sectors by three test chemicals are shown in Table 1. Chromium and chromium compounds are found quite frequently at Superfund-designated sites. Genotoxicity of chromium compounds has been recently reviewed by Venitt [21] and

TABLE 1—*Induction of leaf sectors by chromium trioxide, potassium dichromate, and cyclophosphemide.*

Test Chemical $\mu g/mL$	Average Sector per Plant	Fold Increase Over the Control
Chromium trioxide (pH7)		
Solvent (pH7)		
0	0.16	1.0
100	0.12	0.8
1000	0.62[b]	3.9
2000	0.80[b]	5.0
4000	2.01[b]	12.6
5000	1.29[b]	7.8
Chromium trioxide (pH5)		
Solvent (pH5)		
0	0.06	1.0
100	0.12	2.1
1000	0.18[b]	3.1
2000	0.60[b]	10.9
5000	1.80[b]	31.3
Potassium dichromate (pH7)		
Solvent (pH7)		
0	0.13	
500	0.29[a]	2.2
1000	0.25	1.9
2000	0.39[b]	3.0
5000	2.70[b]	20.8
Cyclophosphamide (pH7)		
Solvent (pH7)		
0	0.06	1.0
100	0.06	1.0
200	0.24[b]	4.0
500	0.44[b]	7.3
1000	1.17[b]	19.5

[a] Significance at 1% level.
[b] Significance at 5% level.

Waters et al. [22]. Chromium trioxide under physiological conditions interacts with nucleic acids, proteins, and organic acids [22] and as such has a potential to cause genotoxic effects. In the wheat assay, Chromium trioxide (CrO_3) was tested at concentrations ranging from 100 to 5000 μg/mL at pH 5 and pH 7. In both hydrogen ion concentrations, CrO_3 yielded concentration-related increases, although at pH 5 there was relatively high frequency of sectors as compared to its response at pH 7. This increase at pH 5 may be due to enhanced permeability of cell membranes.

$K_2Cr_2O_7$ yielded a positive response at a concentration of 500 μg/mL. There was no further increase at the next concentration level (1000 μg/mL). However, significantly higher frequencies of sectoring were observed at treatment concentrations of 2000 and 5000 μg/mL. The hexavalent chromium salts are in general more reactive than trivalent chromium salts in inducing chromosomal abnormalities [22].

CP is used as a chemotherapeutic agent and is a known clastogen [23] and an aneugen for yeast and the germ cells of mice (see review by Waters et al. [4]). In the wheat seedling assay, it produced a concentration-dependent increase from 200 to 1000 μg/mL. In in vitro assays, CP is genetically inactive in the absence of mammalian liver microsomal activation mixture. The fact that CP yielded a positive response in the wheat assay shows that the wheat plants may have the ability to metabolically convert CP to a genetically reactive metabolite(s).

Aneuploidy may be induced by numerous mechanisms including nondisjunction, chromosome breaks, and gene mutation [24]. In our test system, we observed a greater preponderance of green sectors as compared to white sectors and twin sectors (white and green sectors adjacent to each other). The twin sectors are a reliable indicator of nondisjunction. The higher frequency of green sectors observed in this assay shows that the induced effects due to chromosome loss are most probably due to the production and loss of telocentric chromosomes.

In conclusion, our data published earlier [15], and presented in this paper as well as the evaluation of twelve additional model compounds [25,26], suggest that the wheat seedling assay may provide an inexpensive and simple tool for identifying the potential genetic hazards of environmental chemicals.

Acknowledgment

Although the research described in this article has been supported by the United States Environmental Protection Agency through contract number 68-02-4456 to Environmental Health Research and Testing, Inc., it has not been subjected to Agency review and therefore does not necessarily reflect the views of the Agency and no official endorsement should be inferred. Mention of trade names or commercial products does not constitute endorsement or recommendation for use.

We gratefully acknowledge the assistance of Shirley Milton in the preparation of this manuscript.

References

[1] Dellarco, V. L., Voytek, P. E., and Hollaender, A., Eds., *Aneuploidy: Etiology and Mechanisms,* Plenum Press, New York, 1985.
[2] Epstein, C. J., *The Consequences of Chromosome Imbalance: Principles. Mechanisms and Models,* Cambridge University Press, Cambridge, England, 1986.
[3] Vig, B. K. and Sandberg, A. A., *Aneuploidy—Part A: Incidence and Etiology,* Liss, New York, 1987.

[4] Waters, M. D., Stack, F., Mavournin, K. H., and Dellarco, V. L., "Quantitative Evaluation of Chemicals That induce Aneuploidy Using Genetic Activity Profile Method" in *Aneuploidy: Etiology and Mechanisms*, V. L. Dellarco, P. E. Voytek, and A. Hollaender, Eds., Plenum Press, New York, 1985, pp. 445–454.

[5] Barrett, J. C., Oshimura, M., Tanaka, N., and Tsutsui, T., "Genetic and Epigenetic Mechanisms of Presumed Nongenotoxic Carcinogens" in *Banbury Report 25: Nongenotoxic Mechanisms in Carcinogenesis*, B. E. Butterworth and T. J. Slaga, Eds., Cold Spring Harbor Laboratory Press, Cold Spring Harbor, NY, 1985, pp. 311–324.

[6] Waters, M. D., Bergman, H. B., and Nesnow, S., "The Genetic Toxicology of Gene-Tox Non-Carcinogens," *Mutation Research*, Vol. 205, 1988, pp. 139–182.

[7] Athwal, R. S. and Sandhu, S. S., "Use of a Human X Mouse Hybrid Cell Line to Detect Aneuploidy induced by Environmental Chemicals," *Mutation Research*, Vol. 149, 1985, pp. 73–81.

[8] Crespi, C. L., Seixas, G. M., and Penman, B. W., "The Induction of Aneuploidy by Nitrogen Mustard in a Human Lymphoblastoid Cell Line," *Mutation Research*, Vol. 190, 1987, pp. 69–76.

[9] Tenchini, M. L., Mottura, A., Velicogna, M., Pessina, M., Rainaldi, G., and De Carli, L. "Double Y as an Indicator in a Test of Mitotic Nondisjunction in Cultured Human Lymphocytes," *Mutation Research*, Vol. 121, 1983, pp. 139–146.

[10] Perry, J. M. and Perry, E. M., "Comparison of Test Methods for Aneuploidy," *Mutation Research*, Vol. 181, 1987, pp. 267–287.

[11] Oshimura, M. and Barrett, J. C., "Chemically induced Aneuploidy in Mammalian Cells: Mechanisms and Significance in Cancer," *Environmental Mutagenesis*, Vol. 8, 1986, pp. 129–159.

[12] Eastmond, D. A. and Tucker, J. D., "Identification of Aneuploidy Inducing Agents Using Cytokinesis-Block Human Lymphocytes and an Antikinetochore Antibody," *Environmental and Molecular Mutagenesis*, Vol. 13, 1989, pp. 34–43.

[13] Sandhu, S., Gudi, R., and Athwal, R. S., "A Genetic Assay for Aneuploidy: Quantitation of Chromosome Loss Using a Mouse/Human Monochromosomal Hybrid Cell Line," *Mutation Research*, Vol. 201, 1988, pp. 423–430.

[14] Gudi, R., Sandhu, S. S., and Athwal, R. S., "A Genetic Method to Quantitate induced Chromosome Breaks in Mammalian Cells: Identification of Potential Clastogenic Agents," *Mutation Research*, Vol. 225, 1989, pp. 149–156.

[15] Redei, G. P. and Sandhu, S. S., "Aneuploidy Detection with a Short-Term Hexaploid Wheat Assay," *Mutation Research Special Issue on Aneuploidy*, Vol. 201, 1988, pp. 337–348.

[16] Neatby, K. W., "A Chlorophyll Mutation in Wheat," *Journal of Heredity*, Vol. 24, 1933, pp. 159–162.

[17] Sears, L. M. S. and Sears, E. R., "The Mutants *Chlorina*-1 and Hermen's Virescent" in *Proceedings of the Third International Wheat Genetic Symposium*, Australian Academy of Science, Canbera, 1968, pp. 299–304.

[18] Sears, E. R., "Wheat Cytogenetics," *Annual Review of Genetics*, Vol. 3, 1968, pp. 451–469.

[19] Hoagland, D. R. and Arnon, D. T., "The Water Culture for Growing Plants Without Soil," *California Agricultural Experiment Station Circular*, Vol. 347, 1950, pp. 1–32.

[20] Lorenz, R. J., "Zur statistik des plaque-testes *Archiv Gesamte Virusforschung*, Vol. 12, 1962, pp. 108–137.

[21] Venitt, S., "Genetic Toxicology of Chromium and Nickel Compounds" in *Health Hazards and Biological Effects of Welding Fumes and Gases*, R. M. Stern, A. Berlin, A. C. Fletcher, and J. Jaervisalo, Eds., Excerpta Medica, Amsterdam, 1986, pp. 249–266.

[22] Waters, M. D., Stack, H. F., and Brady, A. L., "Genetic Activity Profiles of Superfund Priority 1 Chemicals," internal report, U.S. Environmental Protection Agency, Research Triangle Park, NC, 1987.

[23] Goetz, P., Malashenko, A. M., and Surkova, N., "Cyclophosphamide induced Chromosome Aberrations in Meiotic Cells of Mice," *Folia Biologica (Prague)*, Vol. 26, 1980, pp. 289–297.

[24] Bond, D. J. and Chandley, A. C., *Oxford Monographs on Medical Genetics No. II, Aneuploidy*, Oxford University Press, New York, 1983.

[25] Dhesi, J. S. and Sandhu, S. S., "Application of a Wheat Seedling Assay for Detecting Aneuploidy Induced by N-Ethyl-N-Nitrosourea and 4-Nitroquinoline-1-oxide, *Mutation Research*, in press.

[26] Sandhu, S. S., Dhesi, J. S., Gill, B. S., and Svensgaard, D., "Evaluation of 10 Chemicals for Aneuploidy in Hexaploid Wheat Assay," *Mutagenesis*, in press.

New Approaches

Julius U. Nwosu,[1] *Hilman C. Ratsch,*[2] *and Larry A. Kapustka*[2]

A Method for On-Site Evaluation of Phytotoxicity at Hazardous Waste Sites

REFERENCE: Nwosu, J. U., Ratsch, H., and Kapustka, L. A., **"A Method for On-Site Evaluation of Phytotoxicity at Hazardous Waste Sites,"** *Plants for Toxicity Assessment: Second Volume, ASTM STP 1115,* J. W. Gorsuch, W. R. Lower, W. Wang and M. A. Lewis, Eds., American Society for Testing and Materials, Philadelphia, 1991, pp. 333–340.

ABSTRACT: On-site methods using the seed germination test to evaluate the phytotoxicity of hazardous waste site soils are advantageous because samples are not removed from the site and time is saved in soil handling and preparation. Seedlings may be collected at the end of the test and taken to the laboratory for further analyses. We have used the Magenta™ GA-7 test vessels to conduct on-site the germination of cucumber, *Cucumis sativus* L. var. Hybrid Calypso, and wheat, *Triticum aestivum* L. var. Stephens. Site soil was sampled using a pre-measured cup that held a volume nominally equivalent to 100 g of soil. Twenty pregraded seeds were planted in the soil in each test container and covered with 16-mesh silica sand. Site soil was irrigated with water to approximately 45% moisture on a dry weight basis, covered, and left under a shade at the test site for five days. After five days, germinated seeds were counted and percent germination values were calculated for each seed species. Controls for the test were performed on-site, using 16 and 20-mesh silica sand. Test performance was evaluated against the companion laboratory seed germination test. Calculated percent germination values in the cucumber seeds indicated statistically significant ($\alpha = 0.05$) but biologically reasonable differences between the laboratory test and the field test. Cucumber seeds showed a greater difference, while the wheat seeds did not show any significant difference between the tests.

KEY WORDS: seed germination, phytotoxicity, site assessment and remediation, Surflan, 2-chloroacetamide, on-site evaluation, ecological assessment

The seed germination bioassay was designed as a tool to conduct direct phytotoxicity tests in the laboratory on soils from hazardous waste sites (Neubauer technique) [1,8,9,10]. We have performed numerous laboratory toxicity tests using this technique, and modifications have been made to improve the efficiency of the test. In laboratory settings, plant toxicity tests involve methods which assess both the critical developmental stages in the plant life cycle and also physiological endpoints pertinent to ecological impact in field assessments. Laboratory seed germination tests require that soil samples be removed from the site and transported to the laboratory. In the laboratory, soil samples are processed (screened and mixed) in an effort to minimize the variations that could interfere with laboratory test conditions [4]. These time-consuming laboratory manipulations, for example, screening and mixing of soils, are expensive and may compromise the integrity of the site soil and could potentially alter the sample's toxicity, biasing the toxicity assessment of the site. These manipulations also generate wastes, which then become a problem for the testing laboratory.

[1] Associate scientist, NSI Technology Services Corp., USEPA Environmental Research Laboratory, 200 SW 35th St., Corvallis, OR 97333. (Current address: P.O. Box 1673, Corvallis, OR 97339.)
[2] Plant pathologist and research ecologist, team leader, respectively, USEPA Environmental Research Laboratory, 200 SW 35th St., Corvallis, OR 97333.

An on-site seed germination test can be less time consuming, relatively inexpensive, minimize the prospects of altering the integrity of the site soil (samples are not screened and mixed), conducted under actual site conditions and no additional wastes will be generated. The main objectives of this study were to demonstrate that the laboratory seed germination test can be used for on-site phytotoxicity evaluation at hazardous waste sites to evaluate the performance of the Magenta™ GA-7 test containers (specifically their potential to minimize moisture loss while allowing adequate aeration for germination) and to test seed species in the field. Another objective was to test the applicability of the on-site test method in side-by-side comparison with the laboratory seed germination test. Toxicity data collected from on-site testing using this technique can provide field validation for data collected from laboratory tests and can also provide valuable information which may complement other hazard assessment strategies that rely heavily on terrestrial laboratory tests and chemical analysis of site samples [10]. On-site testing incorporates environmental factors that the laboratory tests control and may suggest the ecological consequences associated with these sites. This study was conducted on-site at the U.S. Environmental Protection Agency (USEPA), Environmental Research Laboratory (ERL-C), Corvallis, Oregon on a designated plot in shade-covered test containers.

Methods

Field studies were conducted at the ERL-C on a designated plot which was hand sprayed with unknown concentrations of Surflan™ (3,5-dinitro-N^4,N^4-dipropyl-sulfanilamide) [2,3]. This study was conducted about six months after the chemical was sprayed at the test plot [2,3]; field and laboratory exposures were performed concurrently using Magenta™ GA-7 (7.6 by 7.6 by 10.2-cm polycarbonate base and polypropylene lid) test containers (supplied by Magenta Corp., Chicago, IL 60641). The tests (Figs. 1 and 2 and Table 1) were performed in accordance with the ERL-C (Ecological Assessment Team) Seed Germination Standard Operating Procedure [4]. Quality assurance and quality control standard operating procedure for tests were also in accordance with the ERL-C (Ecological Assessment Team) guidelines and are summarized as follows: chemical analyses of soil samples from each sample grid point were performed to determine the concentration of the test chemical in the soil; soil pH, temperature, and moisture content were measured; germinated test seeds were defined as germinated shoots above soil surface, with a minimum requirement that 80% of the seeds in the control matrix must germinate [4]. The field study was conducted under ambient conditions at the site. The plot was divided into grids [7,12], and samples were taken at

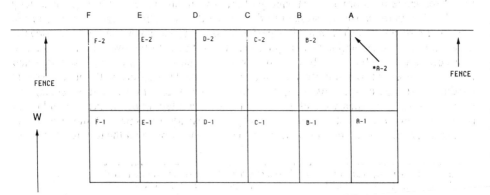

FIG. 1—*On-site test sampling locations. Samples were collected at grid intersects.*

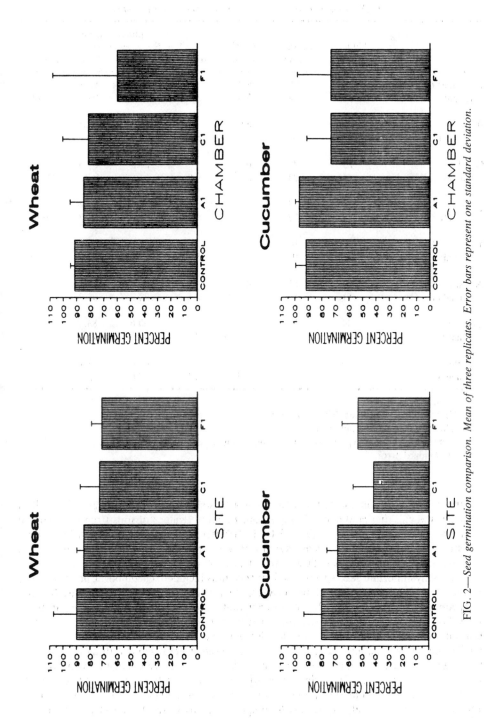

FIG. 2—*Seed germination comparison. Mean of three replicates. Error bars represent one standard deviation.*

TABLE 1—*Seed germination comparison: site versus chamber.*

	Number Germ	Mean % Germ	SE of Mean	Variance[b]	Confidence Limits Lower	Confidence Limits Upper	p-Value[c]
Cucumber control-site	48	80	5.2		69.9	90.1	
Chamber	55	92	3.5	6.3	85.1	98.9	0.06
A-1 site	41	68	6.0		56.2	79.8	
Chamber	58	97	2.2	6.9	92.7	100	<0.001[a]
C-1 site	25	42	6.4		29.5	54.5	
Chamber	44	73	5.7	9.0	61.8	84.2	<0.001[a]
F-1 site	32	53	6.4		62.0	84.2	
Chamber	44	73	5.7	8.8	61.8	84.2	0.01[a]
Wheat control-site	54	90	3.9		76.0	94.0	
Chamber	55	92	3.5	5.2	76.0	94.0	0.71
A-1 site	51	85	4.6		61.8	84.2	
Chamber	51	85	4.6	6.5	72.3	91.7	1.00
C-1 site	44	73	5.7		60.6	83.4	
Chamber	49	82	5.0	7.6	47.6	72.4	0.24
F-1 site	43	72	5.8		60.6	83.4	
Chamber	36	60	6.3	8.6	47.6	72.4	0.16

NOTE: $n = 60$.
[a] p-Value <0.05, there is significant difference.
[b] Pooled variance for site and chamber.
[c] p-value for testing the hypothesis that site and chamber yield the same percent germination.
SE = standard error.

intersects (Fig. 1) from the top soil by using a trowel to scoop up enough soil to fill a premeasured cup that held about 100 g of soil. Initial pH values of soils from each sample point were measured with a calibrated ORION™ pH meter equipped with a flat probe. Moisture content of the soils was measured with a calibrated moisture meter (Irrometer™), and the temperature of soils collected was measured with a Taylor™ thermometer. Two commercially important seed species used for this study included: cucumber, *Cucumis sativus* L. var. Hybrid Calypso (supplied by Sun Seeds, Inc., Brooks, Oregon), and wheat, *Triticum aestivum* L. var. Stephens (supplied by Oregon State University, Crop Science Department, Corvallis, Oregon) [13]. These two seed species were chosen because they are important crops in the agriculture industry. Soil samples were put into Magenta GA-7 test containers, and 20 seeds were planted in each test container. Seeds were layered with 25 g of 16-mesh silica sand and, depending on the moisture levels in the soils, were irrigated with water to about 45% moisture on a dry weight basis [6]. Test matrix for the negative control was 20-mesh silica sand, and 16-mesh silica sand was used as cover sand (supplied by Unimin Corp., Emmet, ID 83617). In all exposures, 60 pregraded seeds were planted in the test matrix (three containers per exposure, with 20 seeds planted in each test container), covered with

16-mesh sand and irrigated with water to about 45% moisture on a dry weight basis [4,6]. Positive controls were performed for each seed species (used to monitor the sensitivity of the test seeds). The positive control was prepared the same as the negative control, except that the sand was irrigated with calculated amounts of 2-chloroacetamide (supplied by Sigma Chemical Co.) solution to 45% wetness on a dry weight basis. Different test concentrations of 2-chloroacetamide were run for each of the two seed species tested [4]. Test concentrations of the 2-chloroacetamide solution used for each seed species had been established in previous experiments [4]. After the seed planting and irrigation procedures were completed, test containers were covered with lids. The laboratory exposures were completed in an environmental chamber (incubated at 24 ± 2°C and under 4300 ± 430 lx; 16-h light and 8-h dark) [17], while the field exposures were completed on-site and left in the shade. The shade was made out of a 4 by 3-ft (approximately 122 by 92-cm) piece of tarpaulin supported by four 4-ft (approximately 122-cm)-long metal frames (Fig. 3). Both tests were incubated for five days. After five days of incubation, germinated seeds (shoots visible above the soil surface) were counted and percent germination values were calculated for each seed species [4].

FIG. 3—*Shade for field exposure.*

Results

The total number of seeds that germinated and the mean percent germination ($n = 60$) were much higher in the laboratory than in the field for the cucumber seeds and were almost similar (laboratory and field) for the wheat seeds for all the sample points (A-1, C-1, and F-1) tested (Fig. 1 and Table 1). Figure 2 also indicated that the cucumber seeds showed a greater difference in germination between field and laboratory conditions, while the wheat seeds did not show any significant difference between the two test locations in all three reported sample points. Statistical comparison of the percent germination data using the Z-test (test for proportions) [5] are given in Table 1. Statistical comparison of the field and laboratory data ($\alpha = 0.05$) indicated that significant differences in germination existed between the tests (see Table 1). Although all sample points at the plot were tested, data presented in this report are from sample points that indicated the most consistency in germination.

Discussion

In the laboratory [4], test seeds were incubated in a test chamber under controlled conditions (at $24 \pm 2°C$ and 4300 ± 430 lx), while, in the field, test seeds were incubated under prevailing conditions (average weekly temperature during the test was 15.6°C). Field tests were incubated in bright sunshine (during the course of the experiment); however, under the shade the illumination was comparable to the laboratory levels (4300 ± 430 lx), based on intermittent measurements. In the laboratory conditions [4], it appeared that test seeds exhibited less sensitivity to the test chemical in comparison to their counterparts in the field. The laboratory test conditions were within the optimum conditions for the germination of the test seeds [4]. Differences in the level of sensitivity between species in the tests (field and laboratory) were anticipated since runoff and leaching of the test chemical did not present any problem in both tests. The difference in sensitivity exhibited by the cucumber seeds (field versus laboratory) could have resulted due to the fluctuating field test conditions. Furthermore, the laboratory test provided a direct comparison for the field test; the effects of the exposures of the test seeds to the test chemical under a controlled environment were established. Illumination could affect seed germination [15] and has long been identified to be important for seeds to germinate, but the extent of its influence on seed germination in the natural environment is questionable [15] and could not have contributed to the observed differences in sensitivity between the tests. Chemical analysis [high performance liquid chromatography (HPLC) analysis] of the test soils indicated the presence of the test chemical, Surflan at various levels. Also, radiolabeled studies with Surflan revealed its persistence in soil under field conditions for about one year [2,3]. The study also indicated that Surflan was readily degraded under field conditions, with 50% of the labeled material's activity remaining after one year in about 3 in. (7.62 cm) of the soil layer, with only 8% of this representing parent material [2,3]. With the data available, it is conceivable that the observed differences in germination could be a cumulative effect (effect of test chemical and environmental factors) [14,18]. Furthermore, chemicals may behave differently in the environment and their behavior are controlled by several factors. Physico-chemical interactions [11] (e.g., changes in chemical and ionic equilibrium) transform chemical compounds to form intermediates and other derivatives that could be more or less toxic than the parent materials. The toxicity exhibited by some chemicals may depend also on the bioavailability of the chemicals, their derivatives, and intermediates [11]. Environmental factors also could make the seeds more susceptible to the test chemical, which may have been responsible for the observed differences in germination between the tests. On-site evaluation of phytotoxicity utilizing the seed germination test could be a valued component in site characterization, evaluation, and remediation [16] and also could be an essential tool in toxicity evaluation at hazardous waste sites. Germination results from the positive control were not included in Table 1 because they were only used to monitor the sensitivity of the test seeds [4].

Conclusion

This study definitely demonstrated how the interactive effects of physical and chemical factors in the environment could consequently affect the outcome of biological endpoint measurements conducted in the field. Field methods represent "real world conditions" since environmental factors are taken into consideration. The fact that field methods consider other factors that are controlled by the laboratory techniques does not imply that laboratory test methods should be discarded. Laboratory methods are very important and should remain an integral aspect of the field methods since they can be conducted side by side for comparison. Observed differences in germination between the seed species (cucumber and wheat) tested in the field and laboratory are most likely explained by test conditions since both tests were exposed to comparable levels of Surflan. The lower germination rate exhibited by the cucumber seeds in the field as compared to the laboratory germination undoubtedly must have been due to the cumulative effect of the test chemical and environmental factors, such as temperature variations [14]. The biological endpoint measured in this study (seed germination) is one of the critical stages in a plant's life history and has been evaluated as a toxicity endpoint relevant to phytotoxicity assessment.

On-site evaluation methods demonstrate a distinctive approach to toxicity assessment, which could conceivably save time and money (minimizes soil handling and preparation) and still provide scientists with pertinent information about a waste site and also potentially guide scientists in making decisions for cleanup and other remedial actions.

Acknowledgments

The authors would like to thank the following for their editorial comments: Mike A. Bollman, Kevin Dewhitt, David Wilborn, Greg Linder, and Minocher Reporter. The authors are also grateful to Ted Ernst, Stuart Eide, and Lisa M. Ganio for the statistical analysis.

References

[1] Thomas, J. M. and Cline, J. E., "Modification of the Neubauer Technique to Assess Toxicity of Hazardous Chemical in Soils," *Environmental Toxicology and Chemistry*, Vol. 4, 1985, pp. 201–207.
[2] Golab, T. C., Bishop, C. E., Donoho, A. L., Manthey, J. A., and Zornes, L. L., "Behavior of ^{14}C Oryzalin in Soil and Plants," *Pesticide Biochemistry and Physiology*, Vol. 5, 1975, pp. 196–204.
[3] Probst, G. W., Golab, T., and Wright, W. L. in *Herbicides*. Vol. 1, 2nd ed., P. C. Kearney and D. D. Kaufman, Eds., Marcel Dekker, New York, 1975, p. 453.
[4] U.S. Environmental Protection Agency, "Standard Operating Procedure for Acute Toxicity Assessment of Hazardous Waste Sites," Hazardous Waste Assessment Team, Corvallis, OR, 1988, unpublished.
[5] Steele, G. D. R. and Torrie, J. H., *Principles and Procedures of Statistics: A Biometric Approach*, 2nd ed., McGraw-Hill, New York, 1980, Chapter 3.
[6] Gardner, W. H., "Water Content," *Methods of Soil Analysis*, C. A. Black et al., Eds., American Society of Agronomy, Inc., Madison, WI, 1965, Chapter 5.
[7] Schweitzer, G. E. and Santolucito, J. A., *Environmental Sampling for Hazardous Waste*, ACS Symposium Series 267, American Chemical Society, Washington, DC, 1984.

[8] Chapman, H. D. and Pratt, P. F., "Methods for the Analysis Soils, Plants, and Waters," University of California, Division of Agricultural Sciences, 1961.

[9] Vandecaveye, S. C., "Biological Methods of Determining Nutrients in Soils," *Diagnostic Techniques for Soil and Crops*, H. B. Kitchen, Ed., American Potash Institute, Washington, DC, 1948, Chapter 7.

[10] U.S. Environmental Protection Agency, National Environmental Research Center, Methods Development and Quality Assurance Research Laboratory, "Methods for Chemical Analysis of Water and Waste," EPA-600/4-79-020, EPA, Office of Technology Transfer, Washington, DC, 1979.

[11] U.S. Environmental Protection Agency, *Proceedings of the Workshop on Transport and Fate of Toxic Chemicals in the Soil*, Office of Environmental Processes and Effects Research, Office of Research and Development, Washington, DC., 1978, pp. 80–96.

[12] Keith, L. H., *Principles of Environmental Sampling*, ACS professional reference book, American Chemical Society, Washington, DC, 1988, p. 79.

[13] Fletcher, J. S., Muhitch, M. J., Vann, D. R., McFarlane, J. C., and Benenati, F. E., "PHYTOTOX Database Evaluation of Surrogate Plant Species Recommended by the U.S. Environmental Protection Agency and the Organization of Economic Cooperation and Development," a review, *Environmental Toxicology and Chemistry*, Vol. 4, 1985, pp. 523–532.

[14] Barton, L. V., "Effects of Temperature and Moisture on Viability of Stored Lettuce, Onion and Tomato Seeds," *Boyce Thompson Institute for Plant Research*, Vol. 23, 1967, pp. 285–290.

[15] Mayer, A. M. and Poljakoff-Mayber, A., *The Germination of Seeds*, 2nd ed., Pergamon Press, New York, 1963, p. 35.

[16] Athey, L. A., Thomas, J. M., Skalski, J. R., and Miller. W. E., "Role of Acute Toxicity Bioassays in the Remediation Action Process at Hazardous Waste Sites," EPA/600/8-87/004, U.S. Environmental Protection Agency, Environmental Research Laboratory, Corvallis, OR, 1987.

[17] Greene, J. C., Bartels, C. L., Warren-Hicks, W. J., Parkhurst, B. R., Linder, G. L., Peterson, S. A., and Miller, W. E., "Protocols for Short-Term Toxicity Screening of Hazardous Waste Sites," EPA/600/3-88/029, U.S. Environmental Protection Agency, Corvallis, OR, 1988.

[18] Evenari, M., "The Germination of Lettuce Seeds. I. Light, Temperature, and Coumarin as Germination Factors," *Palestinian Journal of Botany, Jerusalem*, Series 5, 1952, pp. 138–160.

Gerald E. Walsh,[1] *David E. Weber,*[1] *Tasha L. Simon,*[1]
Linda K. Brashers,[1] *and James C. Moore*[1]

Use of Marsh Plants for Toxicity Testing of Water and Sediment[2]

REFERENCE: Walsh, G. E., Weber, D. E., Simon, T. L., Brashers, L. K., and Moore, J. C., "Use of Marsh Plants for Toxicity Testing of Water and Sediment," *Plants for Toxicity Assessment: Second Volume, ASTM STP 1115,* J. W. Gorsuch, W. R. Lower, W. Wang and M. A. Lewis, Eds., American Society for Testing and Materials, Philadelphia, 1991, pp. 341–354

ABSTRACT: The freshwater wetland plants, *Echinochloa crusgalli crusgalli* and *Echinochloa crusgalli zelayensis,* and the saltmarsh plant *Spartina alterniflora* were exposed to the herbicides metolachlor and norflurazon in two types of toxicity tests: (1) seed germination and early seedling growth in water, and (2) seedling survival and growth in natural and synthetic sediments. The synthetic sediments were formulated to be similar to the natural sediments with regard to particle size distribution and organic content. The herbicides did not affect rate of germination, but significantly inhibited rate of early growth and survival and rate of growth of older seedlings in sediments. *Echinochloa* was more sensitive than *Spartina* to both herbicides. Inhibition of the growth rates of the two varieties of *E. crusgalli* was similar in natural and synthetic sediments, but inhibition of growth of *S. alterniflora* was greater in synthetic than in natural sediment. It is concluded that the species tested may be used for estimation of potential effects of toxicants on wetland plants and that synthetic sediments of known composition may be used in sediment toxicity tests.

KEY WORDS: wetland plants, *Echinochloa crusgalli, Spartina alterniflora,* metolachlor, norflurazon, germination, survival, growth, natural sediment, synthetic sediment

Freshwater and estuarine wetlands serve as sinks for waterborne pollutants [1]. These pollutants may be in the dissolved state, adsorbed to suspended particles [2], or bound to dissolved organic matter [3], but the ultimate locations of many pollutants are in interstitial water or on particles of wetland sediments [4]. Whether dissolved or adsorbed to particulates, toxic substances in sediments can be taken up by the roots of wetland plants and translocated to aerial organs, where they may inhibit growth, injure foliage, or kill the plants [5]. Also, wetland plants generally produce numerous seeds which germinate at the sediment-water interface, where they may be exposed to toxic substances in water.

Rooted plants are the dominant life forms that control the physical, chemical, and biological characteristics of wetland ecosystems. They are the major primary producers and sources of detritus, their roots and rhizomes stabilize sediment, and they provide food and habitat for animals. Chemical hazard to rooted wetland plants constitutes a threat to their ecosystems. Such hazards could occur through effects of toxic substances on seed germination, seedling growth, and survival.

[1] Research ecologist, research biologist, biological aide, biological aide, and research chemist, respectively, U.S. Environmental Protection Agency, Environmental Research Laboratory, Gulf Breeze, FL 32561.
[2] Contribution No. 694 from the Environmental Research Laboratory, Gulf Breeze, Florida. Use of trade names in this publication does not constitute endorsement by the U.S. Environmental Protection Agency.

Few studies describe effects of toxic substances in sediments on wetland plants [6], and we are not familiar with methods devoted specifically to development of synthetic sediments for toxicity testing with such plants. The research reported here was designed to (1) develop methods for exposure of freshwater and estuarine marsh plants to toxicants in water and sediment, (2) identify marsh plant species that can be used in toxicity tests, and (3) conduct toxicity tests in natural and synthetic sediments with substances known to be toxic to plants.

Use of synthetic sediments was deemed critical to evaluation of effects of toxicants in sediments. In preliminary studies, we found natural sediments to be unsuitable for toxicity tests because, while wet and without emendation, pH decreased with time and weed seeds germinated. When natural sediments were dried in air and reconstituted, pH dropped to as low as 2, and weed seeds germinated. Moreover, structure of natural sediments could not be varied experimentally; they contained unknown quantities of nutrients and perhaps toxicants, which introduced uncertain variables into the experiment. This element of uncertainty was eliminated by preparation of synthetic sediments whose structures were similar to those of natural sediments. Their characteristics and formulation methods are reported here.

The methods for toxicity testing of plants and sediments consisted of (1) a germination and early growth test in water, and (2) a seedling survival and growth test in natural and synthetic sediments. Although the natural sediments were altered by drying, they were used for comparison with synthetic sediments. Two varieties of a freshwater marsh species and one species of estuarine plant were tested with two commonly used herbicides. Effects of the herbicides in water, natural sediments, and synthetic sediments are reported, and the merit of the tests is discussed.

Materials and Methods

Plant Species

Freshwater—The common freshwater wetland plant, *Echinochloa crusgalli* (Linneaus) Palisot de Beavois (Gramineae) was used. Two varieties, *crusgalli* and *zelayensis*, were obtained as seed from Wildlife Nurseries, Oshkosh, WI, and stored dry at approximately 4°C. The varietal names were confirmed by examination of plants grown from seed, with identification according to the descriptions of Correll and Correll [7].

Estuarine—*Spartina alterniflora* Loisel (Gramineae) seeds were obtained from Environmental Concern, St. Michaels, MD. Upon receipt, the dry seeds were stored at 4°C in natural seawater diluted with deionized water to 4 parts per thousand (ppth) salinity. Species identity was confirmed from the description given by Hotchkiss [8].

Toxicity Tests

Germination and Early Growth—Seed germination and early growth tests were performed in 47-mm clear polystyrene Petri dishes with tightly capped lids (Millipore Corp., Bedford, MA). Seeds were surface-sterilized by immersion in 1% sodium hypochlorite for 20 min and rinsed in deionized water. Twelve seeds of *Echinochloa* were placed in Petri dishes with 10 mL of deionized water (freshwater control), or one of seven concentrations of herbicide. Each control and exposure concentration was prepared in triplicate. Thus, 36 seeds were exposed in the control and each treatment. Tests with *Echinochloa* were conducted for seven days under cool white fluorescent lights at approximately 35 μE/m^2/s with a diel cycle of 16-h light: 8-h darkness. With *Spartina*, eight surface-sterilized seeds were placed in 10 mL of 4-ppth diluted seawater (seawater control) and up to seven dilutions of herbicide. Controls and exposure concentrations were prepared in triplicate, so that 24 seeds of *Spartina* were used in each. The tests were conducted for 10 days under cool white fluorescent lights at approximately 35 μE/m²/s and a temperature regime of 16 h at 18 ± 1°C and 8 h at 35 ± 1°C [9]. Germinated seeds were enumerated each day. At the end of the exposure, the roots and stems were cut from the caryopses and the plant material from each Petri dish was combined and dried for 24 h at 103°C and weighed to the nearest 0.1 mg on a Mettler Model AE 103 balance.

Seedling Survival and Growth—Survival and growth tests were conducted in natural and synthetic sediments. Natural freshwater sediment was collected from a marsh near Milton, Florida, and saltmarsh sediment was collected from a marsh near Pensacola, Florida. Leaves, twigs, and other large particles were removed and the sediments dried in air at room temperature. Particle size distributions were determined by dry sieving and by settling rate in water [10]. Organic content was determined by ashing at 550°C for 24 h. Synthetic sediments were formulated to simulate the physical properties of the natural sediments (Table 1).

Synthetic sediments were formulated from washed quartz sand, silt, clay, and organic matter. Fine, medium, and coarse sands (New England Silica, Inc., South Windsor, Connecticut) were sieved to obtain the proper grain size for each synthetic sediment. Silts (average particle sizes 4.8 and 1.8 μm) and clays (average particle sizes 0.1 and 2.0 μm) were produced by Englehard Corp., Edison, NJ. Particulate organic matter was air-dried commercial peat humus (Greenleaf Products, Inc., Haines, FL) milled to an average particle size of 840 μm on a Wiley Mill.

The dried natural sediments and the synthetic sediments were reconstituted for survival and growth studies by mixing with either deionized water or 4-ppth diluted seawater at the ratio of 42 mL water: 135 g sediment. Treated sediments were prepared with water that contained dissolved herbicide. Sediments were mixed with a spatula in a glass beaker until smooth and homogeneous. Approximately 100 mL of wet sediment were added to each of three styrofoam cups, 5.5 cm high by 7.5 cm dia. Sediment pH was determined by addition of 100 g sediment to 100 mL deionized water in a glass beaker. The mixture was stirred for 1 min and allowed to settle for 1 h, at which time pH was determined with a Beckman Phi 12 pH/ISE meter. Cation exchange capacity (CEC) was determined by the ion-exchange analysis procedure [11].

Young seedlings were used in the survival and growth tests. Seeds were surface-sterilized in 1% sodium hypochlorite and set to germinate 4 days (*Echinochloa*) or 10 days (*Spartina*) before tests were to begin. *Echinochloa* seeds were germinated in deionized water at 24 ± 1°C under cool white fluorescent lights at approximately 35 μE/m²/s and a 16-h light: 8-h

TABLE 1—*Composition of natural and synthetic freshwater and saltmarsh sediments.*

Class	Particle Size, μm	% Composition, by weight Freshwater	Saltmarsh
Coarse sand	500–1500	0.6	33.6
Medium sand	250–499	9.5	58.8
Fine sand	63–249	67.4	4.9
Silt	4–62	10.3	0.6
Clay	<4	6.7	0.7
Organic	⋯	4.9	0.8
Lost during analysis	⋯	0.6	0.6

darkness cycle. *Spartina* was germinated in 4-ppth diluted seawater in a temperature regime of 16 h at 18 ± 1°C and 8 h at 35 ± 1°C with 16-h light and 8-h darkness.

Twelve seedlings were planted in each cup, triplicate cups were prepared for each control, and herbicide concentration was done in triplicate. Seedlings were planted in holes in sediment without damage to roots and with coleoptiles above the sediment. *Echinochloa* was grown for two weeks and *Spartina* for four weeks at 24 ± 1°C under cool white fluorescent lights at approximately 35 μE/m²/s on a 16-h light: 8-h darkness cycle. Twenty mL of Hoagland solution [12] were added to *Echinochloa* immediately after planting and on the 4th and 12th day, and to *Spartina* immediately after planting and on the 4th, 12th, and 20th day. At the end of the growth period, surviving seedlings were enumerated and harvested carefully by peeling the styrofoam cup from the sediment and washing sediment from the roots with deionized water. Shoots and roots were cut from the caryopses, and the plant material of each cup was dried at 103°C for 24 h and weighed to the nearest 0.1 mg.

Preparation of Herbicide Solutions

The herbicides metolachlor (2-chloro-N-(2-ethyl-6-methylphenyl)-N-(2 methoxy-1-methylethyl) acetamide, 98% pure, from Ciba-Geigy Corp., Greensboro, North Carolina) and norflurazon (4-chloro-5-methylamino-2-(3-trifluoromethylphenyl)-pyridazin-3 (2H) one, > 98% pure, Sandoz, Inc., Homestead, Fla.) were used without carrier. In seed germination and early growth tests, the highest concentration to be used was dissolved in deionized water or 4-ppth diluted seawater and diluted as needed. In seedling survival and growth tests, saturated solutions were prepared and diluted as needed to deliver selected amounts of each herbicide to the various sediment treatments.

Purity of the herbicides and all concentrations in water were confirmed by gas chromatography. Concentrations in sediments were not confirmed because percentage recovery was very low. Samples of freshwater or diluted seawater containing herbicides were extracted with solvent or mixtures of solvents and analyzed by gas chromatographs equipped with packed columns and either electron-capture or nitrogen phosphorus detectors. Average recovery of these compounds spiked into freshwater and seawater to validate analyses of test water was greater than 85% for all compounds. Depending on concentrations and sensitivity of the detector, sizes of samples extracted with solvent ranged from 2 to 20 mL. The types and amounts of solvents were different for each compound and were as follows: metolachlor (hexane, 2 to 10 mL); norflurazon (40% ethyl acetate/60% hexane [v/v] 2 to 20 mL).

Statistical Analyses

Two statistical procedures were used for analysis of germination, survival, and weight data. The mean number of germinated seeds or surviving seedlings and the average weights of seedlings per dish or cup were calculated. Comparisons of the means in each test, the lowest observed effect concentration (LOEC), and differences between each treatment within each test were computed with Tukey's Studentized Range Test [13]. Comparisons of responses in natural versus synthetic sediments were made by two-way analysis of variance (ANOVA) [13], α = 0.05.

Results and Discussion

pH and Cation Exchange Capacity

The pH of dried natural sediments was low (Table 2). Synthetic sediments were slightly basic and had no added sulfur. It is probable that oxidation of sulfide in the wetland sediments

TABLE 2—*The pH and cation exchange capacity (CEC) of freshwater and estuarine sediments used in toxicity tests with herbicides.*

	Freshwater		Estuarine	
	Natural	Synthetic	Natural	Synthetic
pH	5.8	7.5	2.9	7.4
CEC (meq/100 g)	16.6	19.0	14.1	19.0

contributed to the low pHs. Herbicidal activities at the pHs of the natural sediments are discussed below.

Cation exchange capacities of natural and synthetic sediments were similar (Table 2).

Metolachlor

Seed Germination and Early Growth—Metolachlor did not affect the germination rate of either species: the number of germinated seeds was similar in untreated controls and treated samples on each day. It did, however, suppress growth of both varieties of *E. crusgalli* and of *S. alterniflora* (Fig. 1, Table 3). The LOEC for metolachlor and seedling weights with *E. crusgalli crusgalli* and *E. crusgalli zelayensis* was 0.25 mg/L; for *S. alterniflora*, it was 0.5 mg/L. In each case, responses to concentrations at and greater than the LOEC were not statistically significantly different from each other. The highest concentration in tests with *Echinochloa* was 20 times greater than the LOEC; with *Spartina,* the highest concentration was four times greater than the LOEC. This phenomenon, in which increasing concentrations of herbicide did not reduce final seedling weight, occurred in all tests with metolachlor and norflurazon. Metolachlor is a chloroacetamide preemergence herbicide that inhibits protein [*14*] and lipid [*15*] synthesis, but does not inhibit seed germination. It is suggested that, in these tests, the seeds germinated and the seedlings grew by utilization of stored nutrient reserves, imbibition of water, and cell elongation, none of which was affected by metolachlor. However, when photoautrophic growth began, protein synthesis was inhibited and seedling weight gain was arrested at that point. Thus, average seedling weights were the same in all concentrations at and above the LOEC.

Seedling Survival and Growth—Metolachlor inhibited survival of *E. crusgalli crusgalli* and *S. alterniflora* in natural and synthetic sediments but did not affect survival of *E. crusgalli zelayensis* (Fig. 2). The LOEC for survival in metolachlor was 0.5 mg/kg in natural and synthetic sediments with *E. crusgalli crusgalli*; it was 2.5 mg/kg in synthetic sediment and 7.5 mg/kg in natural sediment with *S. alterniflora*.

Metolachlor in sediment significantly inhibited growth of seedlings (Fig. 3, Table 3). However, the effect of the herbicide with *S. alterniflora* was significantly greater in synthetic than in natural sediment. Metolachlor is stable even at pH 1 [*15*], and it is unlikely that it decomposed appreciably under the test conditions. It has been reported to be stable in loamy soil for over 64 days [*15*] and has been detected in surface and groundwaters in the United States [*15*]. Nonionic herbicides such as metolachlor adsorb to charged surfaces [*16*] such as occur on organic matter and clay in soils at pH < 4.0 [*17*]. Sposito [*18*] stated that the quantity of adsorbed matter tends to decrease as the pH increases above 4.0. It is likely that metolachlor toxicity was reduced by adsorption to particulate organic matter and clay at the pH of dried natural estuarine sediment (2.9).

Norflurazon

Seed Germination and Early Growth—Norflurazon did not affect the rate of germination of the species tested. It did reduce the rate of early growth of the two freshwater species

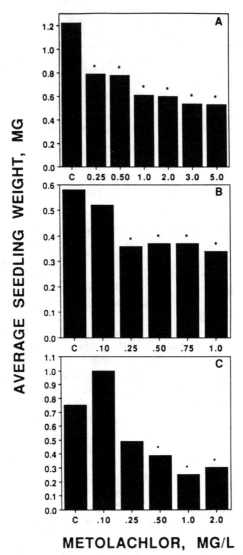

METOLACHLOR, MG/L

FIG. 1—*Average seedling weight of* Echinochloa crusgalli crusgalli (A), Echinochloa crusgalli zelayensis (B), *and* Spartina alterniflora (C) *exposed to metolachlor in water.* * = *significantly lower than control;* α = 0.05; C = control.

(Fig. 4), but the highest concentration, 1 mg/L, did not affect early growth of *Spartina* (Table 3). The LOEC for norflurazon and *Echinochloa* was 0.05 mg/L. As with metolachlor, average weights of seedlings exposed to norflurazon concentration at and above the LOEC were similar.

Norflurazon is a phenylpyridozinone herbicide that inhibits carotenoid synthesis [19], and because carotenoids protect chlorophyll from degradation by light, norflurazon treatment results in bleached seedlings (Fig. 5). Autotrophic growth of treated seedlings was arrested after initial growth by stored nutrient mobilization, imbibition of water, and cell elongation at the LOEC concentration and above, resulting in similar weights.

TABLE 3—*LOECs for growth of wetland plants exposed to herbicides in water and natural and synthetic sediments.*

		Sediment, mg/kg	
	Water, mg/L	Natural	Synthetic
Metolachlor			
E. crusgalli crusgalli	0.25	0.25	0.25
E. crusgalli zelayensis	0.25	0.10	0.25
S. alterniflora	0.50	10.0	0.50
Norflurazon			
E. crusgalli crusgalli	0.05	0.25	0.25
E. crusgalli zelayensis	0.05	0.25	0.25
S. alterniflora	>1	>2	0.25

Seedling Survival and Growth—Norflurazon reduced survival of *E. crusgalli crusgalli* in natural and synthetic sediments and of *E. crusgalli zelayensis* in natural sediments (Fig. 6). It did not affect survival of *E. crusgalli zelayensis* in synthetic sediment or *S. alterniflora* in either sediment.

The LOEC for growth for norflurazon and *Echinochloa* in both sediments and for *Spartina* in synthetic sediment was 0.25 mg/kg (Fig. 7, Table 3). As for metolachlor, the effect of norflurazon on average seedling weight of *Spartina* was significantly greater in synthetic sediment. Norflurazon is stable under acid conditions [20]. As suggested for metolachlor, it is probable that adsorption to particulate organic matter and clay at low sediment pH caused reduced toxicity.

Significance of the Research

At present, there are no standard test protocols that address the problem of potential effects of contaminated water or sediment on wetland plants. Current toxicity tests with plants utilize commercial crop species [21,22], germination on filter paper [23], or growth substrata that do not simulate natural soils [24,25]. The approach reported here demonstrates that acute exposure of seeds to toxicants in water may inhibit early growth of wetland plants and that chronic exposure of seedlings to toxicants in synthetic sediments that are similar to natural sediments may cause death or inhibit growth.

Choice of Test Species

The U.S. Environmental Protection Agency [26] described desirable attributes of organisms for use in toxicity tests conducted with benthic species. The attributes include ecological relevance, variety of endpoints (acute and chronic), all potential routes of exposure should be possible, there should be an adequate amount of tissue for analysis, and ease of organism culture and handling. Also, a plant test species should grow normally in sediments of disparate composition because natural sediments vary widely in composition. The three plants described here satisfy all of these requirements. *Echinochloa* is a widely distributed wetland genus found in North and South America, Europe, Africa, Asia, Australia, and the Pacific islands [27]; *Spartina* is often the dominant genus in many marshes of the Atlantic and Gulf Coasts of the United States, and one species, *pectinata,* is found in freshwater of the Northern United States [28]. Acute endpoints are used in the short-term germination

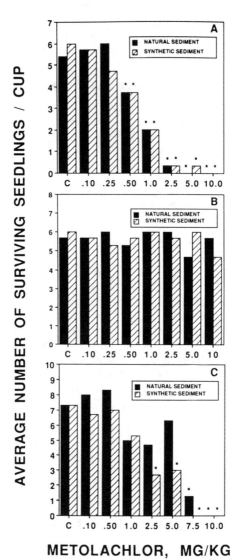

METOLACHLOR, MG/KG

FIG. 2—*Survival of* Echinochloa crusgalli crusgalli (A), Echinochloa crusgalli zelayensis (B), *and* Spartina alterniflora (C) *in natural and synthetic sediments contaminated with metolachlor.* * = *significantly lower than control;* α = *0.05; C = control.*

and early growth tests, and chronic endpoints are used in the survival and seedling growth test. Both tests provide ample tissue for analysis of uptake from water and sediment. Seeds are readily available from suppliers, can be stored in a refrigerator, and have a germination rate of approximately 90% [29]. Both species grow well in sediments of diverse composition [29]. There is also a large literature on the biology of both species. *Echinochloa crusgalli crusgalli* and *E. crusgalli zelayensis* were shown to be sensitive to industrial and municipal effluents [30]. The effluents inhibited germination, survival, and growth, and when germination and early growth tests were conducted in light and total darkness, effects of toxicants

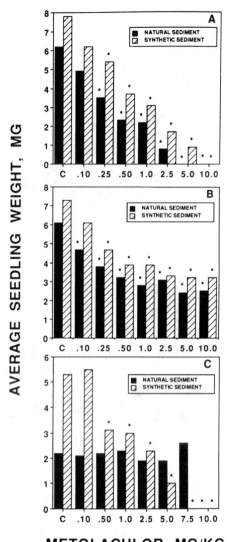

METOLACHLOR MG/KG

FIG. 3—*Average weights of seedlings of* Echinochloa crusgalli crusgalli (A), Echinochloa crusgalli zelayensis (B), *and* Spartina alterniflora (C), *exposed to metolachlor in natural and synthetic sediments.* * = *significantly lower than control;* α = *0.05; C = control.*

on imbibition of water, cell elongation, utilization of stored nutrients, and photosynthesis were identified.

Choice of Sediments

Composition and structure of sediments are probably the most important factors in substratum toxicity [*31*], and laboratory use of substances that are not similar to those in which the plant naturally grows may not provide data applicable to field conditions [*32,33*]. Grain

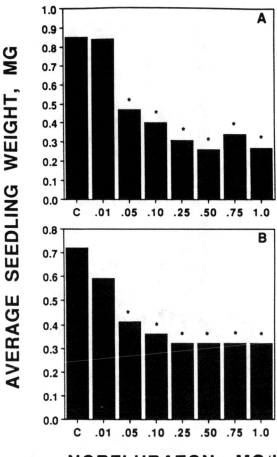

FIG. 4—*Average weights of seedling of* Echinochloa crusgalli crusgalli (A), *and* Echinochloa crusgalli zelayensis (B), *exposed to norflurazon in water.* * = *significantly lower than control;* α = 0.05; C = *control.*

size [33] and organic content [34–36] strongly influence the process of equilibrium partitioning of toxicants between sediment particles and pore water. Although natural sediments may be amended in some cases [37], they are often unsuitable for use in toxicity tests because of variation in composition between samples and data from toxicity tests with them must be normalized [32].

Standard sediments are needed for toxicity studies with plants. The standard sediment should be representative of a variety of natural sediments with regard to particle and pore sizes, chemical composition (e.g., quartz versus calcareous), organic content, and nutrient content. Bradshaw [38] gave the qualities of soil required for good plant growth: productive growth, response to fertilizers, good drainage, good water retention, and free of weeds. Others have described the principles of managing man-made soils [39] and procedures for assessment of substances suitable for growing plants [40].

Synthetic sediments can be formulated to satisfy the above requirements for plant growth

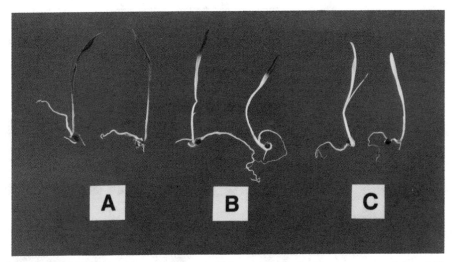

FIG. 5—*Bleaching of* Echinochloa crusgalli crusgalli *by norflurazon in water. A = control; B = 0.05 mg/L; C = 0.25 mg/L.*

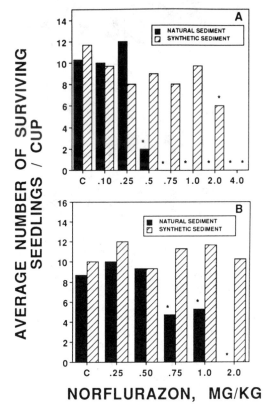

FIG. 6—*Survival of* Echinochloa crusgalli crusgalli *(A) and* Echinochloa crusgalli zelayensis *(B) in natural and synthetic sediments contaminated with norflurazon.* * = *significantly lower than control;* $\alpha = 0.05$; C = *control.*

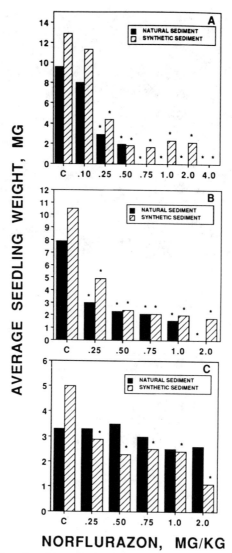

FIG. 7—*Average weights of seedlings of* Echinochloa crusgalli crusgalli (A), Echinochloa crusgalli zelayensis (B), *and* Spartina alterniflora (C), *exposed to norflurazon in sediments.* * = *significantly lower than control;* α = 0.05; C = *control.*

[29]. This report demonstrates that wetland plants responded to herbicides in synthetic sediments. In all cases, average seedling weights in synthetic sediment controls were equal to or greater than those in natural sediment controls. This indicates that the synthetic sediments were good growth media and do not contain factors that may inhibit plant growth or confound interpretation of toxicity data.

Conclusions

The wetland plants, *E. crusgalli crusgalli*, *E. crusgalli zelayensis*, and *A. alterniflora*, have been demonstrated to be useful for laboratory evaluation of herbicide toxicity in water and sediment tests. Acute (7-day) tests detect effects on early seedling growth; chronic (14- or 28-day) tests detect effects on survival and growth of older seedlings in sediment. Synthetic sediments that simulate natural sediments are of value in plant toxicity tests because they support productive growth and allow for assessment of toxic responses. Furthermore, their compositions may be held constant from test to test or may be varied in relation to experimental requirements.

References

[1] Nixon, S. W. in *Estuarine and Wetland Processes,* P. Hamilton and K. S. MacDonald, Eds., Plenum, New York, NY, 1980, pp. 437–524.
[2] Carpenter, J. H. and Huggett, R. in *Concepts in Marine Pollution Measurements,* H. H. White, Ed., University of Maryland, College Park, MD, 1984, pp. 379–403.
[3] *Aquatic Humic Substances: Influence on Fate and Treatment of Pollutants,* I. N. Suffet and P. MacCarthy, Eds., American Chemical Society, Washington, DC, 1989.
[4] Pittinger, C. A., Hand, V. C., Masters, J. A., and Davidson, L. F. in *Aquatic Toxicology and Hazard Assessment (Tenth Volume),* ASTM STP 971, N. J. Adams, G. A. Chapman, and W. G. Landis, Eds., American Society for Testing and Materials, Philadelphia, PA, 1988, pp. 138–148.
[5] Treshow, M. in *Environmental Toxicology,* J. H. Duffus and J. I. Waddington, Eds., Interim Document 13, World Health Organization, Copenhagen, 1983, pp. 168–184.
[6] Driel, W. van, Smilde, K. W., and Luit, B. van, "Comparison of the Heavy-Metal Uptake of *Cyperus esculentus* and of Agronomic Plants Grown on Contaminated Dutch Sediments," Miscellaneous Paper D-83-1, Department of the Army, U.S. Army Corps of Engineers, Washington, DC, 1985.
[7] Correll, D. S. and Correll, H. B., *Aquatic and Wetland Plants of Southwestern United States,* Water Pollution Control Research Series 16030 DNL 01/72, U.S. Environmental Protection Agency, Washington, DC, 1972.
[8] Hotchkiss, N., "Common Marsh Plants of the United States and Canada," Resource Publication No. 93, U.S. Department of the Interior, Washington, DC, 1970.
[9] Seneca, E. D., *American Journal of Botany,* Vol. 61, 1974, pp. 947–956.
[10] Folk, R. I., *Petrology of Sedimentary Rocks,* Hemphill Publishing Co., Austin, TX, 1980.
[11] U.S. Department of Agriculture, "Soil Survey Laboratory Methods and Procedures for Collecting Soil Samples," Soil Survey Investigations Report No. 1, Soil Conservation Service, Washington, DC, 1972.
[12] Hoagland, D. R. and Arnon, D. I. "The Water Culture Method for Growing Plants Without Soil," Circular 347, California Experiment Station, Berkeley, CA, 1950.
[13] *SAS User's Guide: Statistics,* 1982 ed., SAS Institute, Inc., Cary, NC, 1982.
[14] Jaworski, E. G. in *Herbicides: Chemistry, Degradation and Mode of Action,* P. C. Kearney and D. D. Kaufman, Eds., Dekker, New York, NY, 1975, Chapt. 6, pp. 349–376.
[15] U.S. Environmental Protection Agency, "Pesticide Fact Sheet, Metolachlor," Fact Sheet No. 106, Office of Pesticides and Toxic Substances, Washington, DC, 1987.
[16] Schnitzer, M. and Khan, S. U., *Soil Organic Matter,* Elsevier, New York, 1978.
[17] White, R. E., *Introduction to the Principles and Practice of Soil Science,* 2nd ed., Blackwell, Boston, MA, 1987.
[18] Sposito, G., *The Surface Chemistry of Soils,* Oxford University Press, New York, 1984.
[19] Sandman, G. and Boger, P. in *Biochemical Responses Induced by Herbicides,* D. E. Moreland, J. B. St. John, and F. D. Hess, Eds., ACS Symposium Series 181, American Chemical Society, Washington, DC, 1982, Chapt. 7, pp. 111–130.
[20] *The Pesticide Manual: a World Compendium,* 6th ed., C. R. Worthing, Ed., British Crop Protection Council, London, 1979.
[21] U.S. Environmental Protection Agency, "Seed Germination/Root Elongation Toxicity Test," EG12, Office of Toxic Substances, Office of Pesticides, Washington, DC, 1982.
[22] OECD, "Terrestrial Plants, Growth Test, OECD Guideline for Testing of Chemicals, No. 208," Organization for Economic Cooperation and Development, Paris, 1984.
[23] Wang, W., *Environmental Toxicology and Chemistry,* Vol. 6., 1987, pp. 409–414.
[24] Lee, C. R., Sturgis, T. C., and Landin, M. T., *Journal of Plant Nutrition,* Vol. 3, 1981, pp. 139–151.
[25] Azpiazu, M. N., Romero, E., and Diaz, J. M., *Water, Air, and Soil Pollution,* Vol. 28, 1986, pp. 1–26.

[26] U.S. Environmental Protection Agency, "Issues Related to the Assessment and Resolution of Problems Associated With Contaminated Sediment," Office of Water, Office of Water Regulations and Standards, Washington, DC, 1989.

[27] Holm, L. G., Plucknett, D. I., Pancho, J. W., and Herberger, J. P., *The World's Worst Weeds,* The University of Hawaii Press, Honolulu, HI, 1972.

[28] Muenscher, W. C., *Aquatic Plants of the United States,* Comstock Publishing Associates, Cornell University Press, Ithaca, NY, 1944.

[29] Walsh, G. E., Weber, D. E., Brashers, L. K., and Simon, T. L., *Environmental and Experimental Botany,* Vol. 30, 1990, pp. 391–396.

[30] Walsh, G. E., Weber, D. E., Simon, T. L. and Brashers, L. K., *Environmental Toxicology and Chemistry,* Vol. 10, 1991, pp. 517–525.

[31] LeBaron, H. M. in *Residue Reviews,* Vol. 32, *The Triazine Herbicides,* F. A. Gunther and J. D. Gunther, Eds., Springer-Verlag, New York, NY, 1970, pp. 311–353.

[32] van Straalen, N. M. and Denneman, C. A., *Ecotoxicology and Environmental Safety,* Vol. 18, 1989, pp. 241–251.

[33] Spies, R. B., *Marine Environmental Research,* Vol. 27, 1989, pp. 73–75.

[34] Stackhouse, R. A. and Benson, N. H., *Aquatic Toxicology,* Vol. 13, 1988, pp. 99–108.

[35] Stevenson, F. J., *BioScience,* Vol. 22, 1972, pp. 643–650.

[36] Swindoll, C. M. and Applehans, F. M., *Bulletin of Environmental Contamination and Toxicology,* Vol. 39, 1987, pp. 1055–1062.

[37] Barko, J. W. and Smart, R. M., *Journal of Ecology,* Vol, 71, 1983, pp. 161–175.

[38] Bradshaw, A. D., *Soil Use and Management,* Vol. 5, 1989, pp. 101–108.

[39] Stewart, V. I. and Scullion, J., *Soil Use and Management,* Vol. 5, 1989, pp. 109–116.

[40] Adams, W. A., Stewart, V. I., and Thornton, D. J., *Journal of the Sports Turf Research Institute,* Vol. 47, 1971, pp. 77–85.

Thomas Pfleeger,[1] Craig Mc Farlane,[1] Racinda Sherman,[1] and Gayle Volk[1]

A Short-Term Bioassay for Whole Plant Toxicity[2]

REFERENCE: Pfleeger, T., Mc Farlane, C., Sherman, R., and Volk, G., "**A Short-Term Bioassay for Whole Plant Toxicity,**" *Plants for Toxicity Assessment: Second Volume, ASTM STP 1115*, J. W. Gorsuch, W. R. Lower, W. Wang, and M. A. Lewis, Eds., American Society for Testing and Materials, Philadelphia, 1991, pp. 355–364.

ABSTRACT: A five-day, whole plant toxicity test was developed and evaluated. The 18 chemicals tested were primarily substituted benzenes and phenols, although representative herbicides, surfactants, and other industrial chemicals were also tested. The test yields information on root and shoot growth as a function of toxic chemical concentration in the root environment and can also be used to determine the effect on transpiration. The measurement found to be the most sensitive indicator of toxicity was total plant growth. EC_{50} (effective concentration that reduced growth 50%) values were estimated for each compound using regression analysis. This simple bioassay gives a quick response to an acute exposure and can be used to screen chemicals at various concentrations. The beginning of a unique database comparing species, concentrations, and chemical toxicity are presented, and the results are discussed in relation to other phytotoxicity data.

KEY WORDS: soybean, barley, phytotoxicity, chlorophenol, dicamba, dichloroaniline, dichlorobenzene, dichlorophenol, dichlobenil, dinitrobenzene, dinitrotoluene, sodium dodecylbenzenesulfonic acid, nitrobenzene, nitrophenol, pentachlorophenol, nitrotoluene, tributylphosphate, trichloroacetic acid, trichlorophenol, thiourea

Most phytotoxicity testing has been done to identify efficacy of herbicides. Testing by regulatory agencies is concerned not about efficacy of a herbicide's intended use but with environmental contamination and associated effects. This type of testing must also evaluate phytotoxic effects of any chemical, not just those intended specifically as herbicides. The U.S. Environmental Protection Agency (EPA) has the responsibility for regulating production and use of pesticides under the Federal Insecticide, Fungicide, and Rodenticide Act (FIFRA) and industrial chemicals under the Toxics Substances Control Act (TSCA). Both acts provide the opportunity for EPA to require phytotoxicity testing. The purpose of plant testing under these laws is to evaluate the potential for chemical damage to nontarget species and assess their environmental impact.

Currently, the majority of nonherbicidal phytotoxicity testing, regardless of objective, involves seeds or seedling bioassays [1]. Although such tests provide valuable information about imbibition, enzyme activation, cell expansion, respiration, and other parameters, they miss other effects imposed on adult plants. Since seed germination and early seedling growth depend on the energy reserves in cotyledons, they do not evaluate chemical effects imposed

---

[1] U.S. EPA Environmental Research Laboratory, 200 SW 35th St., Corvallis, OR 97333.

[2] The information in this document has been funded by the U.S. Environmental Protection Agency. It has been subjected to the Agency's peer and administrative review, and it has been approved for publication as an EPA document. Mention of trade names or commercial products does not constitute endorsement of recommendation for use.

on photosynthesis, transpiration, flower initiation, fruit development, and other important events. An additional problem of phytotoxicity testing is that different species often do not respond similarly to toxic chemicals. A single test to obtain the entire breadth of information needed to evaluate the potential hazard of a suspect chemical is clearly impossible. Rather, a suite of tests including various responses and species is required to cover this need.

Seed and seedling tests were found to be inadequate for our use because concentrations which caused no effect were sometimes associated with severe reactions in adult plants. A phytotoxicity test was needed in our own laboratory to screen chemical concentrations prior to conducting experiments on chemical uptake, translocation, and metabolism in various plant species [2]. Those experiments must be done with concentrations which do not affect the physiological parameters controlling chemical uptake. Our objectives were (1) to devise and evaluate a short exposure test (TOXSCREEN) which would determine a dose response of toxic chemicals administered to the roots of adult plants and (2) to develop a test that would be a member of a suite of tests used to assess chemical impact on the environment.

Methods

Soybean [*Glycine max* (L.) Merr cv. Fiskeby v] and barley (*Hordeum vulgre* L. cv. Klages) seeds were germinated in agricultural grade vermiculite. Three days after germination the plant root/shoot interface was wrapped with 1-cm-wide by 0.2-cm-thick closed cell polyethylene tape and supported in an 8-mm hole punched in the middle of a wide-mouth canning jar lid. The plants were grown to the desired size (28 days) in a recirculating hydroponic nursery system in a glass house [3]. Plants were selected for testing from the hydroponic nursery and transferred to the test containers. Thirty percent more plants were grown than needed to ensure that uniform plants were used. Selection was made by weighing each plant and taking the most uniform group of plants as measured by the coefficient of variation. Starting with the smallest, plants were sequentially assigned (one per treatment) to the control and each treatment. This sequence was repeated until each treatment had five plants. This procedure created the most uniform distribution, so that at the start of treatment the total amount of variation in the sample population was evenly distributed among all treatments. Each plant was placed in a 0.95-L (1-qt) glass canning jar with 900 mL of nutrient solution [4] containing the appropriate amount of test chemical. Near the bottom of each jar a 27-gauge needle was inserted through a rubber septum. A small amount (2 cm^3/min) of air was pumped into each jar to ensure an aerobic environment throughout the exposure.

Eighteen different organic chemicals were tested (Table 1) at concentrations of 0.1, 1, 10, and 100 μg mL^{-1}. The chemicals were selected to give a broad range of phytotoxicity. Chemicals were also selected on the basis of the following factors: (1) low Henry's law constant (low volatility at the water/air interface); (2) high water solubility; and (3) member of a substituted benzene or phenol series. These series were considered in an attempt to evaluate chemical structure/activity relationships (SAR). Herbicides and compounds not generally considered to be phytotoxic were added for comparison. Each treatment was replicated five times. Controls and, if necessary, carrier controls (either acetone or sodium hydroxide) created an array of 25 or 30 plants per test.

The exposure lasted five days in a controlled environment [3]. Plants were selected for bench position using a randomized block design. Lights were on a 16/8 h (on/off) cycle with a light intensity of 350 μmol m^{-2} s^{-1} at the top of the canopy. Temperature was 25/21 ± 2°C and relative humidity varied between 50 to 70%.

Chemical concentration and pH of the hydroponic solutions and total plant fresh weights were measured at the beginning and end of each five-day experiment. Root and shoot fresh weights were also measured at harvest. Transpiration was determined gravimetrically by the

TABLE 1—Codes and physical properties of test chemicals. Sources of information Refs 20–25.

Code	Chemical	Formula	Molecular Weight (MW)	Solubility, mg/L (H_2O)	Vapor Pressure, mmHg	Log, K_{ow}
CP	m-chlorophenol	C_6H_5ClO	128.6	26 000 (25°)	0.119 (25°)	2.50
DC	Dicamba	$C_8H_6Cl_2O_3$	221.1	4500 (25°)	3.41×10^{-5} (25°)	2.81
DCA	3,4-dichloroaniline	$C_6H_5Cl_2N$	162.0	205[a]	3.4×10^{-4} (25°)[b]	2.69
DCB	1,2-dichlorobenzene	$C_6H_4Cl_2$	147.0	156 (25°)	1.47 (25°)	3.38
DCP	2,4-dichlorophenol	$C_6H_4Cl_2O$	163.0	4500 (25°)	0.12 (20°)	2.75
DICB	Dichlobenil (2,6-dichlorobenzonitrile)	$C_7H_3Cl_2N$	172.0	18 (20°)	5×10^{-4} (25°)	2.74
DNB	1,3-dinitrobenzene	$C_6H_4N_2O_4$	168.1	469 (15°)	7.4×10^{-5} (25°)[b]	1.46
DNT	2,4-dinitrotoluene	$C_7H_6N_2O_4$	182.1	300 (22°)	1.1×10^{-4} (20°)	1.98
NADO	Sodium dodecylbenzenesulfonic acid	$C_{18}H_{29}SO_3Na$	348.5	No data	No data	0.96
NB	Nitrobenzene	$C_6H_5NO_2$	123.1	1900 (20°)	0.15 (20°)	1.85
NP	4-nitrophenol	$C_6H_5NO_3$	139.1	16 (25°)	8.8×10^{-5} (25°)[b]	1.96
PCP	Pentachlorophenol	C_6HCl_5O	266.4	14 (20°)	1.1×10^{-4} (20°)	5.01
PNT	p-nitrotoluene	$C_7H_7NO_2$	137.1	442 (30°)	0.22 (30°)	2.37
TBP	Tributylphosphate	$C_{12}H_{27}O_4P$	266.3	5900 (20°)†	5.1×10^{-3} (25°)[b]	3.53
TCA	Trichloroacetic acid	$C_2HCl_3O_2$	163.4	1×10^7 (25°)	0.26 (25°)[b]	4.05
TCP	2,4,6-trichlorophenol	$C_6H_3Cl_3O$	197.5	800 (25°)	8.4×10^{-3} (24°)	3.60
TECP	2,3,4,5-tetrachlorophenol	$C_6H_2Cl_4O$	231.9	No data	No data	4.21
TU	Thiourea	CH_4N_2S	76.1	91 800 (13°)	No data	-0.98

[a] Estimated value.
[b] Calculated value based on Refs 20–25.

amount of solution lost during the exposure. Solution samples were analyzed by reverse-phase high performance liquid chromatography (HPLC) using a Hewlett-Packard 1090 liquid chromatograph equipped with a photodiode array detector and a C18 column (Hewlett-Packard Hypersil ODS 5 μm, 100 by 2.1 mm). An isocratic solvent system consisting of methanol and acidified water (0.1% acetic acid, v:v) was used for all analytes except 3,4-dichloroaniline, which was analyzed using a solvent system of methanol and water modified with triethylamine (25 mM). The composition of the solvent system was adjusted for each analyte, and the flow rate was 0.3 mL/min.

Statistical analysis were performed using SAS™. Growth was the response (dependent) variable used in these statistical analyses because preliminary work showed it to be the most chemical-sensitive measure recorded. All tests were determined to have no pretreatment differences or differences between controls and carrier controls by using analysis of variance procedures (ANOVA) (alpha = 0.05). Regression analysis was used to estimate the EC_{50} (effective concentration that caused a 50% decrease in growth) and the associated standard error. EC_{50} estimates were determined using the chemical concentrations measured at the start of each experiment (with the exception of NADO, TBP, TCA, and TU, which were not measurable using HPLC, and nominal concentrations were used). Linear, spline (or segmented), and the nonlinear Weibull [5] regression models were compared to determine the best-fitting model. The linear model was included since it is the traditional method. However, when testing over large concentration ranges it did not account for as much variation in the data as either the spline or Weibull models. The Weibull model fit the data best in almost all cases as measured by coefficient of determination (R^2). When R^2s were similar (within 3%), the model with the smallest standard error was used.

Results and Discussion

An example of the growth data and the nonlinear Weibull regression used to calculate EC_{50} values is shown in Fig. 1. Calculated EC_{50} values are presented in Table 2. Similar industrial chemicals have been classified by Veith et al. [6] as being toxic to fish by general narcosis. They demonstrated that narcotic toxicity increases with increasing log K_{ow} (log of the octanol/water coefficient). For this reason scatter plots were used in an attempt to determine relations between log K_{ow} and EC_{50} values for chemicals within chemical groups. However, no consistent or useful trends were identified. One trend was apparent; within the limits of the chemicals tested, barley was more sensitive to substituted benzenes than soybean. In contrast, soybeans were more sensitive to three of the substituted phenols but about equally sensitive to the other three. In general, no overall pattern for species sensitivity was evident from the chemicals tested.

The least phytotoxic of all chemicals tested were thiourea (TU) and trichloroacetic acid (TCA). Trichloroacetic acid is used as a soil sterilant to control perennial weed grass [7]. The low toxicity probably explains its declining use. The most phytotoxic chemicals tested were the two herbicides, dichlobenil (DICB) and dicamba (DC). Dichlobenil (Table 2) had an EC_{50} value of 0.01 μg/mL. Dichlobenil is marketed as a preemergent herbicide which "is a powerful inhibitor of germination and of actively dividing meristems and acts primarily on growing points and root tips" [8]. Since the root meristems were directly exposed to the chemical in the hydroponic solution, this response corresponds to expected results. Dicamba, a broadleaf herbicide, was more toxic to soybeans than barley. Again, this corresponds to its stated use as a broadleaf herbicide [8].

The dinitro-substituted benzenes, dinitrobenzene and dinitrotoluene, were both significantly more toxic than the monosubstituted homologues (Table 2). With the exception of the barley tests with dichlorophenol (DCP), all the chlorinated phenols had similar toxicities.

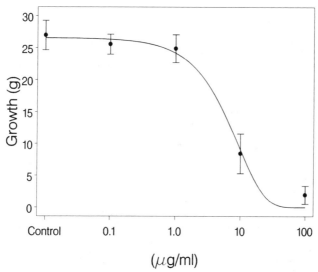

(μg/ml)

FIG. 1—*Weibull curve fitted to the response of soybean growth to various concentrations of dichlo-rophenol. The Weibull adjusted treatment mean with error bars of one standard deviation are shown for each chemical concentration. Concentrations are log transformed except for controls. Since the controls contained no chemical, they were assigned very low concentration values so they could be shown on this plot. Growth was measured using fresh weights. EC_{50} = 6.6 μg, degree of freedom (DF) = 24, R^2 = 0.95, visidual sums of squares, (RSS) = 67.6.*

Sund and Nomura [9] tested seed germination toxicity by a series of substituted phenols and reported a stronger correlation with the degree of chlorination. An interesting pattern is observed by considering the EC_{50} for substituted benzenes and phenols in relation to their molecular weights (Fig. 2). Those with molecular weights above 170 were uniformly toxic, while toxicity appeared to decrease with smaller compounds (with the exception of CP) of these chemical classes.

The last two chemicals, tributylphosphate (a plasticizer) and NADO (a surfactant), were both moderately phytotoxic, with NADO being more toxic to soybean.

Since TOXSCREEN is a new method, it was desirable to compare its results to other published tests results [10,11]. However, the comparison of phytotoxicity data among various studies is difficult because of the diversity in testing protocols, including differences between exposed tissue, exposure duration, species, age, measurement of insult, and environmental conditions. Most phytotoxicity data are for subcellular processes, seed germination, or seedlings. In making any comparison, it must be remembered that even the simplest test includes a complex series of events. For instance, a seed germination test includes effects on imbibition and respiration as it affects cell expansion and many other enzymatic and hormonal responses involved in growth. For these reasons, taxonomically similar species and tests using similar growth stages were selected when possible, but other species and tests were used when similar ones were unavailable. This makes the comparison dissimilar in many respects but still useful as a starting point for an expanding data set.

Chemical toxicity, as measured by EC_{50} concentrations, varied from 0.01 to 190 μg/mL, a range of 10^4 (Table 3). All chemicals in this test had EC_{50} concentrations that were as low or lower than other values reported in the literature (Table 3). Seed germination tests were consistently the least sensitive tests reported.

Differential species sensitivity is one of the basic principles underlying the development

TABLE 2—*NOEC range as determined by ANOVA and estimated* EC_{50} *values using the Weibull model unless noted.*

Chemical[1]	Species	NOEC, µg/mL	EC_{50}, µg/mL	Standard Error (SE), µg/mL	R^2, %
		SUBSTITUTED BENZENES			
NB	Soybean	10.0 < 100.8	76.7[s]	11.5	58
	Barley	0.9 < 4.7	41.9[s]	11.7	48
PNT	Soybean	6.8 < 81.0	51.2[s]	4.9	77
	Barley	26.3 <	>26.3		
DCB	Soybean[2]				
	Barley	1.7 < 18.8	37.5[s]	11.8	43
DCA	Soybean	0.04 < 0.9	38.0[s]	4.2	77
	Barley	4.7 < 56.1	14.3	4.4	85
DNB	Soybean	0.9 < 9.8	6.3	5.6	49
	Barley	0.1 < 1.0*	0.1	0.4	79
DNT	Soybean	1.3 < 10.1	6.4	1.0	92
	Barley	0.1 < 1.0	2.9	0.8	92
		SUBSTITUTED PHENOLS			
CP	Soybean	0.9 < 10.4	5.3[s]	0.4	97
	Barley	2.8 < 8.4	2.4	2.1	62
NP	Soybean	1.0 < 9.9	39.2[s]	2.4	93
	Barley	9.6 < 93.0	41.4[s]	5.4	86
DCP	Soybean	1.0 < 9.9	6.6	1.0	95
	Barley	4.9 < 54.2	31.3[s]	4.4	67
TCP	Soybean	0.04 < 0.9	1.0	0.2	83
	Barley	0.8 < 8.2	7.6	1.3	84
TECP	Soybean	0.01 < 0.1	0.1	0.01	88
	Barley	0.01 < 0.1	5.4[s]	1.7	28
PCP	Soybean	0.7 < 7.5	1.1	0.5	94
	Soybean	0.1 < 0.7	1.4	0.3	93
		HERBICIDES			
TCA	Soybean	1 < 10*	55.8	139.7	40
	Barley	10 < 100*	189.7	123.8	59
DICB	Soybean	<0.1	0.01	0.07	72
	Barley	<0.04	0.01	0.03	92
DC	Soybean	<0.1	0.01	0.05	92
	Barley	1.3 < 10.6	4.9	1.7	87
		OTHERS			
TU	Soybean	10 < 100*	99.0	1.4	82
	Barley	100<*	>100	No data	
TBP	Soybean	10 < 100*	34.8	5.1	96
	Barley	10 < 100*	24.1	9.1	78
NADO	Soybean	1 < 10*	25.9	3.2	97
	Barley	10 < 100*	14.2	6.2	81

[1] Chemical codes are presented in Table 1.
[2] Experiment failed.
[s] Spline model used to estimate EC_{50}.
* Nominal values used in calculation for NOEC and EC_{50}.

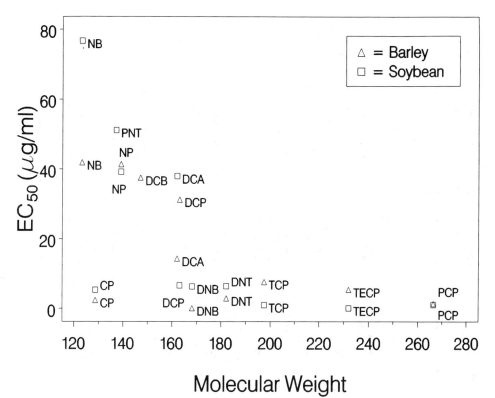

Molecular Weight

FIG. 2—*The effect of molecular weight on the EC$_{50}$'s for substituted benzenes and phenols. The chemical code is given in Table 1.*

of herbicides; however, such selectivity has not been thoroughly investigated with other chemicals. The results from this study along with data from a study using a number of different species tested with nitrobenzene [2] suggest that plant species, in general, exhibit different sensitivities to toxic organic chemicals. This finding while not surprising, has important ecological implications when considering that over 2.6 billion pounds of toxic chemicals were released to the air from manufacturing facilities in 1987 [12]. While the quantity of chemical reaching the terrestrial environment is unknown, the potential exists for the restructuring of plant communities based on the localized loading of toxicants.

Conclusions

There is a lack of information in the published literature on the response of adult plants exposed to organic chemicals other than herbicides. The TOXSCREEN test is one method that could be used to collect this type of phytotoxicity information. The test is quick, accurate, and simple. It takes only five days to complete and requires less than two work days per test. The test results are as sensitive or more sensitive than existing phytotoxicity tests. This increased sensitivity may be due to using plants that are not dependent on stored reserves for energy. The test is uncomplicated and can be accomplished with minimally trained technical help.

TABLE 3—*Comparison between TOXSCREEN and published data [10,11]. No data were found in the literature on the following chemicals tested: DCA, DNB, and DNT. Certain chemicals had very limited information (TCA, TBP, PNT, NB, and NADO) on either monocot or dicot species.*

Chemical[a]	TOXSCREEN Species	EC_{50}, µg/mL	EC_{50}, mol/L	Literature Species	Test Type	Result, mol/L
CP	Soybean	5.3	4.2E-5	Radish	Seed germ	EC_{50} 4.81E-4
	Barley	2.4	1.8E-5	Sudan grass	Seed germ	EC_{50} 1.05E-4
DC	Soybean	0.01	4.5E-8	Cucumber	Seedling	Root decrease 4.5E-6
	Barley	4.9	2.2E-5	Oat	Seedling	Shoot decrease EC_{50} 4.5 E-5
						Root decrease EC_{50} 4.5E-6
DCB	Soybean	37.5	2.6E-4	Soybean	Seedling	Growth EC_{50} 6.8E-4 < 6.8E-3
	Barley	6.6	4.0E-5	Rice	Seedling	Growth EC_{50} 6.8E-4 ≪ 6.8E-3
DCP	Soybean	31.3	1.9E-4	Radish	Seed germ	LD_{50} 3.01E-4
	Barley	0.01	5.8E-8	Sudan grass	Seed germ	LD_{50} 6.13E-4
DICB	Soybean	0.01	5.2E-8	Cucumber	Seedling	Root decrease 50% 5.81E-6
	Barley	25.9	7.4E-5	Oat	Seedling	Root decrease 50% 5.81E-6
NADO	Soybean	14.2	4.1E-5	Bean	Greenhouse	Fresh mass decrease 2.87E-4
	Barley	76.7	6.2E-4	No data		
NB	Soybean	41.9	3.4E-4	No data		
	Barley	39.2	2.8E-4	Rice	Seedling	Growth EC_{50} > 8.1E-4
NP	Soybean	41.4	3.0E-4	Radish	Seed germ	LD_{50} 3.96E-4
	Barley	1.1	4.0E-6	Sudan grass	Seed germ	LD_{50} 4.96E-4
PCP	Soybean	1.4	5.1E-6	Cucumber	Seedling	EC_{50} 2.33E-5
	Barley	51.2	3.7E-4	Sudan grass	Seedling	EC_{50} 6.76E-5
PNT	Soybean	>26.3	>1.9E-4	Rice	Seedling	Growth EC_{50} >7.29E-4 (o-NT)
	Barley	34.8	1.3E-4	No data		
TBP	Soybean	24.1	9.0E-5	Rice	Seedling	Root EC_{50} 3.8E-6 < 3.8E-5
						Shoot EC_{50} >3.8E-5
TCA	Soybean	55.8	3.4E-4	No data		
	Barley	189.7	1.2E-3	Wheat	Seedling	Root decrease 6% 6.12E-5
TCP	Soybean	29.5	4.9E-6	Radish	Seed germ	LD_{50} 2.28E-4
	Barley	7.6	3.9E-5	Sudan grass	Seed germ	LD_{50} 5.58E-4
TECP	Soybean	0.1	4.3E-7	Radish	Seed germ	LD_{50} 6.9E-5
	Barley	0.01	3.0E-7	Sudan grass	Seed germ	LC_{50} 2.89E-4
TU	Soybean	99.0	1.3E-3	Soybean	Seedling	Growth>1.3E-2
	Barley	>100	>1.3E-3	Rice	Seedling	Growth EC_{50} 1.3E-3 < 1.3E-2

[a] Chemical codes are presented in Table 1.

The results from testing the selected chemicals have led to several interesting points: (1) substituted benzenes and phenols with molecular weights above 170 have similar toxicity; (2) substituted benzenes and phenols with molecular weights below 170 have a toxicity that generally decreases with decreasing molecular weight, and; (3) chemical sensitivity of soybean and barley was within one order of magnitude on all chemicals except dicamba (DC). This general similarity between soybean and barley in phytotoxicity response probably represents a common mode of action. This type of response is referred to as narcosis in animal studies and implies a general suppression of metabolism. Other reactions are more specific and are the basis of species selective herbicides (note that dicamba is sold as a broad leaf herbicide).

Small differences in species sensitivity to toxic chemicals can result in significant alterations in plant community composition and production. Changes in plant communities have been demonstrated with both selective [13] and nonselective [14] herbicides, plant growth retardants [15], fungicides [16], and insecticides [17]. Changes in community composition associated with forest decline have been hypothesized to be the result of aerially deposited organic pollutants [18]. Furthermore, it has been suggested that pesticides in general may determine evolutionary trends not only directly through toxicity but through chromosomal aberrations and mutations [19]. Therefore, what appear as small differences in phytotoxicity in the laboratory have the potential to be ecologically significant.

Acknowledgments

We acknowledge the contributions and offer our appreciation to Kevin DeWhitt and Joel McCrady for chemical support and Henry Lee for statistical assistance.

References

[1] Rubinstein, D., Cuirle, E., Cole, H., Ercegovich, C., Weinstein, L., and Smith, J., "Test Methods for Assessing the Effects of Chemicals on Plants," EPA 560/5-75-008, U.S. EPA, Washington, DC, 1975.

[2] Mc Farlane, J. C., Pfleeger, T. G., and Fletcher, J. S., *Environmental Toxicology and Chemistry*, Vol. 9, 1990, pp. 513–520.

[3] Mc Farlane, J. C. and Pfleeger, T., *Journal of Environmental Quality*, Vol. 16, 1987, pp. 361–371.

[4] Berry, W. L., in *A Growth Chamber Manual*, R. W. Langhans, Ed., Comstock Pub. Assoc., Ithaca, NY, Chap. 7, p. 22.

[5] Rawlings, J. O. and Cure, W. W., *Crop Science*, Vol. 25, No. 5, 1985, pp. 807–814.

[6] Veith, G., Call, D. J., and Brooke, L. T., *Canadian Journal of Fisheries and Aquatic Science*, Vol. 40, 1983, pp. 743–748.

[7] *Herbicide Handbook of the Weed Science Society of America*, 5th ed., Weed Science Society of America, Champaign, IL, p. 515.

[8] *Farm Chemical Handbook '89*, C. Sim, ed., Meister Publishing Co., Willoughby, OH, 1989, p. 326.

[9] Sund, K. A. and Nomura, N., *Weed Research*, Vol. 3, 1963, pp. 35–43.

[10] Royce, C. L., Fletcher, J. S., Risser, P. R., Mc Farlane, J. C., and Benenati, F. E., *Journal of Chemical Information and Computer Sciences*, Vol. 24, 1984, pp. 7–10.

[11] Hikino, H., "Study on the Development of the Test Methods for Evaluation of the Effects of Chemicals on Plants," Chemical Research Report No. 4/1978, Office of Health Studies, Environment Agency, Tokyo, Japan, 1978.

[12] "The Toxic-Release Inventory, A National Perspective," EPA 560/4-89-005, U.S. EPA, Office of Toxic Substances, Washington, DC, June 1989.

[13] Hume, L., *Canadian Journal of Botany*, Vol. 65, 1989, pp. 2530–2536.

[14] Malone, C. R., *Journal of Ecology*, Vol. 53, 1972, pp. 507–512.

[15] Marshall, E. J. P., *Journal of Applied Ecology*, Vol. 25, 1988, pp. 619–630.

[16] Paul, N. D., Ayres, P. G., and Wyness, L. E., *Functional Ecology*, Vol. 3, 1989, pp. 759–769.

[17] Shure, D. J., *Ecology*, Vol. 52, 1971, pp. 271–279.

[18] Krahl-Urban, B., Papke, H. E., Peters, K., and Schimansky, C., "Forest Decline: Cause-Effect Research in the United States of North America and Federal Republic of Germany," Julich Nuclear Research Center, Julich, FGR, 1988, p. 137.

[19] Grant, W. F., *Symposia Biologica Hungarica*, Vol. 12, 1972, pp. 43–50.

[20] Hansch, C. and Leo, A., *Substitute Constants for Correlation Analysis in Chemistry and Biology*, Wiley, NY, 1979, p. 338.

[21] Howard, P. H., *Handbook of Environmental Fate and Exposure Data for Organic Chemicals*, Vol. I, Large Production and Priority Pollutants, Lewis Publishers, Chelsea, MI, 1989, p. 574.

[22] *The Merck Index, An Encyclopedia of Chemicals and Drugs*, 9th ed., M. Windholz, Ed., Merck and Co., Inc., NJ, 1976, p. 1313.

[23] *The Pesticide Manual, A World Compendium*, 6th ed., C. Worthington, Ed., The British Crop Council, England, 1979, p. 655.

[24] Verschueren, K., *Handbook of Environmental Data on Organic Chemicals*, 2nd ed., Van Nostrand Reinhold Co., Inc., New York, 1983, p. 1310.

[25] "Water-Related Environmental Fate of 129 Priority Pollutants," Vol. II, U.S. EPA-440/4-79-029b, 1979.

R. A. Shimabuku[1], H. C. Ratsch,[1] C. M. Wise,[1] J. U. Nwosu,[2] and L. A. Kapustka[1]

A New Plant Life-Cycle Bioassay for Assessment of the Effects of Toxic Chemicals Using Rapid Cycling *Brassica*[3]

REFERENCE: Shimabuku, R. A., Ratsch, H. C., Wise, C. M., Nwosu, J. U., and Kapustka, L. A., "**A New Plant Life-Cycle Bioassay for Assessment of the Effects of Toxic Chemicals Using Rapid Cycling *Brassica*,**" *Plants for Toxicity Assessment: Second Volume, ASTM STP 1115,* J. W Gorsuch, W. R. Lower, W. Wang and M. A. Lewis, Eds., American Society for Testing and Materials, Philadelphia, 1991, pp. 365–375.

ABSTRACT: Initial evaluation of a new plant life-cycle bioassay for the assessment of the effects of toxic chemicals is presented. The bioassay features a rapid cycling *Brassica* species that can complete its life cycle in as little as 36 days. The herbicide dalapon (2,2-dichloropropionic acid) was used as the test substance to develop the bioassay. Plants were exposed to nine levels of dalapon solution in a $0.5x$ geometric series ranging from 0 to 200 ppm (mg/L) in separate postemergence and preemergence application experiments. Harvest biomass was generally reduced by concentrations as low as 12.5 ppm. Similarly, foliar height was reduced at concentrations as low as 25 ppm. Stem diameter was reduced at concentrations above 50 ppm. Although fertile silique counts were quite variable, no siliques were formed at dosages greater than 12.5 and 50 ppm for postemergence and preemergence treatments, respectively. Mortality occurred only with the 200-ppm preemergence treatment. The higher sensitivity of the reproductive stage is of significance to the ecological success of the plant population and demonstrates the value of life-cycle bioassays.

KEY WORDS: plant, bioassay, short life cycle, *Brassica, Arabidopsis,* crucifer, dalapon, 2,2-dichloropropionic acid

Currently, the U.S. Environmental Protection Agency's (EPA) Office of Toxic Substances has approved as test guidelines only two plant bioassays to evaluate the hazard of exposure of plants to toxic and hazardous chemicals. These are the Seed Germination/Root Elongation Toxicity Test (EG-12) and the Early Seedling Growth Test (EG-13) [1]. More recently, EPA's Office of Research and Development has recommended use of seed germination and root elongation tests using lettuce as the test species for ecological assessment of hazardous waste sites [2].

Presently under review by the EPA/OTS is the *Arabidopsis thaliana* Whole Plant Life-cycle Test. This small herbaceous winter annual from the Cruciferae family is widely distributed throughout the Northern Hemisphere. It has a short life cycle capable of six, to as much as ten, life cycles a year under ideal laboratory conditions. In an earlier work, Ratsch

[1] USEPA Environmental Research Laboratory, Corvallis, OR 97333-4996.

[2] Mantech Environmental Technology, Inc., Corvallis, OR 97333.

[3] The information in this manuscript has been funded wholly by the U.S. Environmental Protection Agency. It has been subjected to the Agency's peer and administrative review, and it has been approved for publication as an EPA document. Mention of trade names or commercial products does not constitute endorsement of recommendation for use.

et al. [3] reported that for the *Arabidopsis* whole life-cycle bioassay, mature seed weight was the most sensitive endpoint, while foliar biomass was much less sensitive. We concluded from our experience with *Arabidopsis* that, although it is a good laboratory bioassay organism, it has significant limitations which may preclude its effectiveness in field situations. Its small seed size (approximately 0.12 mm in diameter) increases handling difficulty, making its use more labor intensive. The small seed and consequent small seedling may make these life stages almost impossible to track *in situ*. The siliques dehisce upon maturity, making it difficult to quantitate its most sensitive endpoint, especially under field conditions. Finally, *Arabidopsis thaliana* is a weedy species of little economic value.

The *Brassica* test species (*Brassica rapa* cv. CrGC 1-1; of the mustard family Cruciferae) selected has a seed approximately 1.3 mm in diameter that is ten times larger than *Arabidopsis* seeds. The larger seedling allowed easier handling of seeds and also facilitated making measurements and observations. The larger seed size together with an indehiscent silique type enable seed biomass measurements to be made easier under field conditions. Finally, *Brassica* is a better surrogate genus by virtue of its worldwide economic importance as a major source of food, fodder, and oil as root, leaf, or seed crops [4].

Horticultural varieties of *Brassica* usually have long reproductive cycles of six months to a year or more. Paul Williams of the University of Wisconsin was able to select for and breed rapid cycling *Brassica* strains (*B. rapa* cv. CrGC 1-1) with life cycles as low as 36 days [4]. Thus, the advantage of an accelerated research program using a short life-cycle plant such as *Arabidopsis* is shared by using a rapid cycling *Brassica*.

This paper presents the initial evaluation of another whole plant life-cycle bioassay using a different cruciferous genus, *Brassica*. Our objective was to develop a laboratory bioassay that has more potential for development into a field test.

Materials and Methods

Seeds of *Brassica rapa* L. were obtained from the Crucifer Genetics Cooperative (CrGC) of the University of Wisconsin. The cultivar used, CrGC 1-1, is a base population for the species and was selected for its short life cycle. The plants flower in 16 days, and the first seed set is produced as soon as 36 days [8]. We used first-generation progeny after increasing the original seed stock with CrGC [5] procedures.

The growth matrix used was Turface™ (AIMCOR; Deerfield, IL), an inert calcined montmorillonite soil amendment. The Turface™ used was regular grade screened through a 4.75-mm sieve (Tyler Screen Equivalent No. 4) but retained by a 1.40-mm sieve (Tyler Screen Equivalent No. 14). Preliminary studies with Turface™ and various horticultural media (including commercially available peat-based greenhouse mixes, vermiculite, perlite, silica sand, and native soil) showed that an aggregate hydroponic system using Turface™ grew plants as good as or better than the other media, especially in reference to foliar biomass and height. Also, Turface™ was the easiest to wash off the roots, allowing for better measurement of root biomass.

The test system developed was designed for simplicity and availability of the components. The pots were 16-oz opaque disposable plastic drinking cups (Dixie Cups No. A16-18, James River Corp. Norwalk, CN). Eight drain holes (nominally 2.5 mm) were punched with an awl circumscribing the bottom perimeter. A 13-mm hole was drilled in the bottom center of the cup. A wick was placed through this hole and was made from a 13-cm length of 14-mm-diameter polyester cable cord (Conso Products Co.; Union, SC). An opaque 9-oz cup (No. P509C Solo Cup Co.; Urbana, IL) was placed under each pot as a reservoir and to also catch the leachate after each watering event. Five of these pots were held in a Deepot collar (J. M. McConkey & Co. Inc.; Sumner, WA).

Each pot containing 300-g dry of sieved Turface™ was planted with two seeds placed in the crease of a 3 by 4-cm piece of regular laboratory tissue folded in half. The seed packet kept the seeds stationary and was planted 1 cm deep. A 46-cm-long stainless steel rod (2.5 mm in diameter) was placed in each pot to provide plant support. A 15-cm length of stainless steel rod was placed between the reservoir cup and pot to permit proper drainage by preventing air lock.

The test chemical was the herbicide dalapon (2,2-dichloropropionic acid as Dowpon M, 74% a.i., Dow Chemical Co., Midland, MI), which was applied in a nine concentration $0.5x$ geometric series ranging from 0 to 200 ppm. This short-term chronic test used a modified static renewal exposure system. The Turface™ in each pot held approximately 160 mL of dalapon-treated nutrient solution, and the reservoir cup contained approximately 90 mL for a total of 250 mL of solution treated with dalapon. Two separate experiments were conducted in which dalapon was applied as either postemergence or preemergence application treatment. Forty-five pots for each application treatment were arranged in a completely randomized design with five replicates for each concentration.

The plants were grown in a growth chamber under continuous illumination to relate to our other xenobiotic uptake and metabolism studies [6] and to reduce the length of the life cycle [5]. The average initial light level was 330 μmol s^{-1} m^{-2}, which at harvest was reduced to 285 μmol s^{-1} m^{-2} due to normal aging of the fluorescent lamps. The temperature was $25 \pm 2°C$ with a relative humidity range of 45 to 50%. The pots were drenched twice daily with the solution in the reservoir adjusted back to 90 mL as needed with nutrient solution [7].

Postemergence Treatment

In the postemergence application experiment, after the seed packets were planted, the pots were gently flooded with tap water to rinse out the growth media dust. They were then placed on a greenhouse mist bench and automatically misted with tap water. The greenhouse was supplied with auxiliary lights for 16 h a day, and the approximate average day/night temperature was 23/14°C.

Four days later at least one of the two seeds in each pot had emerged. The pots were flooded twice with nutrient solution and allowed to drain before being moved to the growth chamber. The plants were thinned to one plant per pot after 8 days had elapsed after sowing. The pots were treated with dalapon at 11 days after sowing, when the first true leaf began to emerge. The respective concentrations were delivered in 90 mL of distilled water by pouring it uniformly over the surface of the growth media. The 0-ppm treatment served as the control and was identical to the other treatment except it was only 90 mL of distilled water without any dalapon. Immediately after all the pots were treated, the leachate that collected in the reservoir was poured back over the surface. The solution was wicked from the reservoir cup into the growth medium by the polyester cord as the plants transpired.

Preemergence Treatment

Since the plants in both preemergence and postemergence treatments were grown for a total of 42 days and dosed at the same time, the preemergence treatment was exposed for the full 42 days, whereas the postemergence treatment (which was dosed when the plants were 11 days old) was only exposed for 31 days. Another difference was that the preemergence plants were not placed on the greenhouse mist bench. The seeded pots were flooded with tap water, drained for 20 min, and then drenched twice with nutrient solution before being moved to the growth chamber. They were then treated with the same concentrations of dalapon used in the postemergence treatment. The bottom of a standard 100-mm-diameter plastic petri dish was placed over the pot to help retain moisture. The petri dish cover was removed after the seeds emerged, and a stainless steel rod was inserted in the pot for support. The daily height measurements began at six days after sowing when the plants were of sufficient size in comparison with the postemergence treatment plants, which were measured from the dosing date.

Endpoint Measurements

Measurements of various morphological, phenological, and other endpoints were periodically taken throughout the experiment. These included foliar height, stem diameter, internode length, leaf length and width, and branching morphology. Also observed were germination, emergence, initial flowering date, survival, chlorosis, and stunting. The periodic measurement most useful was foliar height for postemergence treatment (Fig. 1) and preemergence treatment (Fig. 2). These measurements were made by gently straightening the foliage and measuring between the cotyledons and the highest point on the plant (foliage, terminal bud, or silique tip).

Under optimal conditions the life cycle of the *Brassica rapa* L. CrGC 1-1 can be completed in as few as 36 days. This was achieved in the control and the lowest dalapon treatments; however, higher sublethal doses delayed completion of the life cycle. Therefore, all plants were harvested 42 days after sowing to compensate for this delay.

Harvest measurements included foliar and root biomass, maximum foliar height, foliar area, stem diameter, number and length of axillary stems, and number of siliques and seeds. Harvest data are presented in Table 1 for the postemergence treatment and in Table 2 for the preemergence treatment. The maximum foliar height was taken after the plants were cut at the cotyledons by measuring the distance between the cotyledons and the highest point of the plant after pulling it straight. Foliar area measurements were made only for the preemergence treatment plants because the earlier postemergence harvest showed a good gradient of response to dalapon concentration. The plant was spread like a fan, then measured with a LI-COR Model 3100 Portable Area Meter™ (LI-COR; Lincoln, NE). The roots were gently washed free of Turface™ to minimize damage to the fine root system of *Brassica rapa*. The foliar and root dry weights were taken after 72 h in a 70°C oven.

Results

The results for both the postemergence and preemergence dalapon application treatments are presented both as periodic and harvest endpoint measurements. The mean daily foliar height measurements of the five replicates per treatment are illustrated in Figs. 1 and 2. In the postemergence treatment (Fig. 1), the first measurement was made on the day the plants were dosed (11 days after sowing). At 16 days, the 0 to 25-ppm treatment plants started to bolt with the first flower fully opening on approximately the 18th day. The plants were near maximum height at 28 days. During the subsequent period, numerous axillary stems formed and eventually produced more flowers on this indeterminate plant type. At the end of the experiment, the foliar heights separated into three groups. Plants treated at 1.6 to 12.5 ppm were no different from the controls, but the plants from 25-ppm treatment were approximately half the height of the controls. Plants treated with 50 to 200 ppm had little additional growth and were severely stunted. This trend was distinguishable as early as 7 days after the dalapon dose (i.e., elapsed Day 18).

Figure 2 illustrates the daily foliar height measurement for the preemergence treatment.

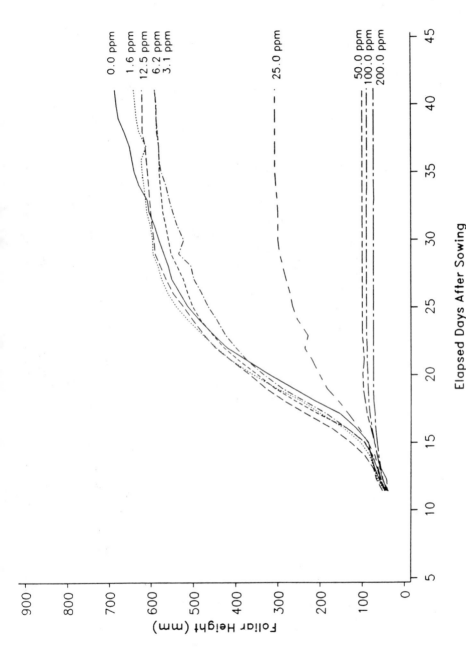

FIG. 1—*Mean daily foliar height of 11-day-old seedlings after postemergence treatment with dalapon.*

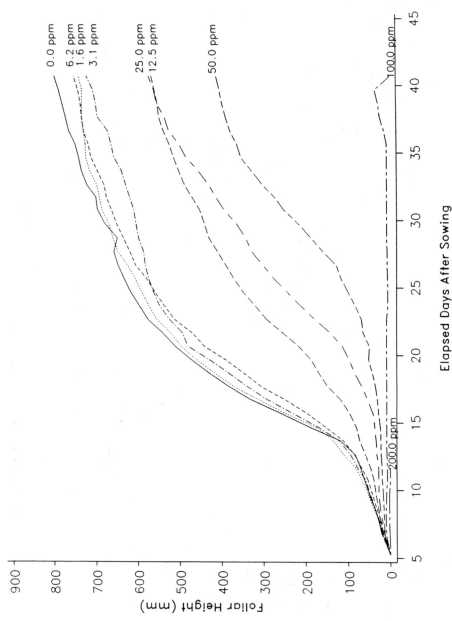

FIG. 2—*Mean daily foliar height after preemergence treatment with dalapon.*

TABLE 1—*Harvest measurements after exposure to postemergence application of dalapon.*[1]

Dalapon, ppm	Foliar Height, mm	Stem Dia, mm	Foliar Dry Wt., g	Root Dry wt., g	No. of Siliques
0	694a[5]	3.12ab[5]	7.40a[5]	0.84a[5]	14.8[6]
1.6	660a	3.88a	7.70a	0.78ab	12.6
3.1	609a	3.89a	7.18a	0.69ab	19.6
6.2	589a	3.48ab	6.23ab	0.66bc	7.2
12.5	612a	3.22ab	6.78a	0.67bc	29.4
25	330b	3.51ab	4.76b	0.53c	0
50	97c	2.64b	1.30c	0.17d	0
100	54c	0.80c	0.51c	0.06d	0
200	37c	0.88c	0.17c	0.01d	0
CV[2]	23	38	21	18	199
R-square[3]	0.90	0.58	0.91	0.93	0.26
EMS[4]	9100	1.17	1.02	0.008	12 491
LSD (0.01)[5]	164	1.19	1.74	0.15	na

[1] Mean of 5 replicates, 1 plant/replicate.
[2] Coefficient of variation of all 45 plants in measurement variable.
[3] R-square between dalapon concentration and measurement variable.
[4] Error Mean Square.
[5] Fisher's Protected Least Significant Different Value at P ≥ 0.01. Column means with a different letter are significantly different.
[6] F-test not significant difference even at P ≥ 0.05; therefore no mean separation test was applied.

TABLE 2—*Harvest measurements after exposure to preemergence application of dalapon.*[1]

Dalapon, ppm	Foliar Height, mm	Stem Dia, mm	Foliar Dry Weight, g	Root Weight, g	No. of Siliques	Foliar Area, cm²
0	797a[5]	4.16a[5]	8.64a[5]	0.79a[5]	18.4ab[6]	360a[5]
1.6	732a	3.58a	8.13a	0.87a	8.8bc	283a
3.1	723a	3.78a	7.14abc	0.65ab	24.0a	323a
6.2	743a	3.65a	7.38ab	0.63ab	18.0ab	309a
12.5	571ab	3.88a	5.54bc	0.60ab	8.8bc	249ab
25	575ab	3.81a	4.52c	0.37bc	5.2bc	247ab
50	413b	1.96b	1.60d	0.21cd	3.0c	106b
100	48c	0c	0.07d	0.06cd	0c	9c
200	0c	0c	0d	0d	0c	0c
CV[2]	28	32	34	44	117.2	43
R-square[3]	0.83	0.81	0.82	0.72	0.40	0.72
EMS[4]	20 802	0.76	2.70	0.04	126	8047
LSD (0.01)[5]	248	1.12	2.82	0.35	14.4	154

[1] Mean of 5 replicates, 1 plant/replicate.
[2] Coefficient of variation of all 45 plants in measurement variable.
[3] R-square between dalapon concentration and measurement variable.
[4] Error Mean Square.
[5] Fisher's Protected Least Significantly Different Value at $P ≥ 0.01$. Column means with a different letter are significantly different.
[6] Fisher's Protected Least Significantly Different Value at $P ≥ 0.05$. Column means with a different letter are significantly different.

The preemergence plants were placed directly in the growth chamber at the start of the experiment. Seedlings emerged after 3 days versus 4 days for the mist bench germinated postemergence treatment plants. Some of the plants (0 to 6.2 ppm) started to bolt as early as 14 days. The plants of the 0 to 6.2 ppm treatments increased in height slower than to the postemergence treatment plants of comparable levels of dalapon. The 12.5 to 50-ppm treatment plants bolted later and for a longer duration than the plants exposed to less dalapon. The 50-ppm treatment did not stunt the plants as much as was found in the postemergence treatment; however, the height of the 100-ppm treatment plants were similar to that of the postemergence treatment. The 200-ppm treatment plants germinated and emerged but did not survive past the 12th day. The height measurements separated into four groups with the first group (1.6 to 6.2 ppm) having essentially no difference in height from the controls, followed by 25% reduction for 12.5 and 25 ppm level, 50% reduction for 50 ppm, and severe stunting or death for 100 and 200-ppm treatments.

The results of the final measurements for both the postemergence and preemergence treatments are presented in Tables 1 and 2, respectively. In both of these exploratory studies several measurements were made which were erratic and had no distinguishable relationship to the treatment levels. They include the number and height of each axillary stem and seed count. The morphology of the plant made it very difficult to accurately count and measure the numerous axillary stems. The reproductive capacity can be best assessed by the silique counts even though they were more variable than other harvest data measurements.

An analysis of variance was performed on the harvest measurements to determine dalapon treatment effects. Treatment effect was measured with Fisher's Protected Least Significant Difference Mean Separation analysis in Tables 1 and 2. Most measurements reported had F-tests with significance levels at $P \geq 0.01$. The exception is silique counts, which had a significance level of $P \geq 0.05$ for preemergence treatment and was not significant for postemergence.

The postemergence treatment results in Table 1 for the maximum foliar height measurements indicate that the 25-ppm level was significantly smaller than those from the lower rates. Increasing the level of dalapon resulted in greater reduction in height. The foliar dry weight measurements show the same trend, but no statistical difference could be measured between 25 and 6.2 ppm although a significant difference was measured between 25 and 12.5 ppm. The dry root weights had a response gradient to the dalapon concentration, with 6.2 ppm being the lowest significantly different treatment from the control. Stem diameter did not appear to be as sensitive as other measures to dalapon and indicated only a positive or negative effect with only the 100 and 200-ppm treatment being significantly different from the control.

Comparison of the daily foliar height measurements (Figs. 1 and 2) between the two treatment types show that the preemergence treatment height during the most rapid growth phase lags about four days behind the postemergence treatment. Other than the difference in the stage of the life cycle of dalapon application, the preemergence treatment plants received nutrient solution from time of sowing and were grown only under constant temperature and illumination. Comparison of the harvest endpoint measurements between both treatment types indicates that the preemergence treatment plants, including the controls, are larger although they were exposed to dalapon for 42 days versus only 31 days for the postemergence treatment.

Although germination occurred, no 200-ppm preemergence treatment plants advanced beyond the seedling stage. The results indicate the same trend with the postemergence treatment experiment. For maximum foliar height, the 50-ppm treatment is the lowest dose that is significantly different from the control. For dry foliar weight, the lowest significantly different treatment was 12.5 ppm. Similarly, that level was 25 ppm for the dry root weight.

The stem diameter was taken midway between the third and fourth node on the main stem. The effect on the stem diameter at 50-ppm treatment was significantly different from the lower dosages. Plants in the 100-ppm treatment level were either too stunted or misshapen to measure and were listed as zero. The results for the foliar height measurement concur with results for stem diameter with 50 ppm being the lowest dose that was significantly different from the control.

Evidence of the floral stage sensitivity was observed during assessment of the reproductive capacity of the plants. The silique counts for the preemergence treatment had a F-test significance level of $P \geq 0.05$, and dalapon levels higher than 50- ppm treatment had no fertile siliques. The postemergence silique counts were not significant at $P \geq 0.05$. There was no reproduction at dalapon levels higher than 12.5 ppm. These findings concurred with observations of pronounced silique deformations as the dosages approached the level where reproduction ceased; however, slight deformation was noted starting as low as 3.1 ppm with curving of the siliques. As the siliques matured, they curled further and some split open to expose aborted embryos. As the dose increased from 12.5 ppm, the severity of the deformation progressed to where the silique and adjacent leaves formed succulent and stunted tissues, which had the exploded appearance of popcorn.

The length of the life cycle also appears to be impacted. At harvest, plants dosed with 3.1 ppm and lower were near completion of their life cycle as determined by the maturity of the siliques and foliar senescence. Plants receiving 12.5 ppm and higher (Fig. 2) bolted later and for a longer time than those receiving lower doses. When siliques were present on these plants at harvest, they were immature and the leaves were not senescencing.

Discussion and Conclusions

There is increasing interest in the use of rapid cycling *Brassica rapa* as a model system in studies involving genetic engineering, molecular biology, as well as the plant sciences [5]. This is the first report of its use in environmental sciences as a phytotoxicity test. This initial evaluation of a toxicological study involving rapid cycling *Brassicas* suggests that it is a promising life-cycle bioassay for assessing the effects of toxic chemicals. Our previous experience with developing the *Arabidopsis thaliana* Whole Plant Life-cycle Test [3] showed that mature seed weight was the most sensitive variable to measure with an EC_{50} to dalapon of only 4 ppm versus an EC_{50} of 141 ppm to dalapon for total biomass. The reduction in seed weight was attributed to a change in the morphology of the flower, which prevented pollination. This subtle change in the floral morphology of *Brassica rapa* was not observed; however, visual symptoms indicate that *Brassica rapa* siliques were affected down to 3.1 ppm. At levels greater than 12.5 and 50 ppm for the postemergence and preemergence treatments, respectively, no fertile siliques were formed. The sensitivity of a particular plant's reproductive system to stress imposed by toxic chemicals could be of significance to its ecological success.

The effect of dalapon on biomass was more sensitive for *Brassica rapa* than for *Arabidopsis thaliana*. The results in Tables 1 and 2 show that a significant reduction generally started at 12.5 to 25 ppm. Calculation of the EC_{50}s for postemergence foliar height, foliar dry weight, and root dry weight were 30, 32, and 33 ppm, respectively. Similar results for the preemergence treatment are 26, 24, and 50 ppm, respectively. Both treatments had considerably lower biomass EC_{50} than that for *Arabidopsis*, which was 141 ppm [3]. For corn and beans, dalapon is recommended as a preplant application at 8 to 16 lb per acre. This is approximately equivalent to the 25 and 50-ppm levels we used [8]. Thus, the recommended field rates of dalapon are capable of eliciting a significant reduction in *Brassica rapa* biomass and lowering its reproductive potential.

The harvest endpoint measurements presented in Tables 1 and 2 indicate that *Brassica rapa* responded to the concentration gradient of dalapon. Daily maximum foliar height measurements illustrated in Figs. 1 and 2 indicate a similar effect and also show that postemergence and preemergence treatment results can be determined as early as one and two weeks, respectively, after dosing.

Postemergence application of dalapon to oilseed rape (*Brassica napus* cv. Jet Neuf) at approximately 25 ppm was reported to significantly reduce height and seed yields [9]. The age of the plant was important, with the older plants being more sensitive to dalapon than the younger plants. Askew [10], using the same cultivar, found that when approximately 6.25-ppm dalapon was applied to the three- to four-leaf stage, the seed and oil yield was significantly reduced. These findings concur with our postemergence results in which no fertile siliques were formed at levels higher than 12.5 ppm.

For dalapon, germination and early seedling growth life stages (which at least initially rely on the stored resources of the seed for growth) do not seem to be as sensitive as the mature stages that rely totally on photosynthesis. A modification of a seed germination technique [2] was performed using *Brassica rapa* with a LC_{50} of 105 ppm. Our findings from the preemergence treatment showed that seeds dosed at 200 ppm did not survive. The plants which survived the 100-ppm treatment were of no ecological value to subsequent generations by virtue of their inability to reproduce.

The implication here is that the *Brassica rapa* life-cycle bioassay is more sensitive than seed germination/root elongation for dalapon because other mature stages of the life cycle can be more sensitive. To assess the effects on the life cycle, these maturer stages were measured by periodic and harvest foliar height measurements, harvest biomass, foliar area, and determination of the reproductive capacity of the plant. These measures are more likely to identify sublethal effects that have ecological significance, such as lowering of the reproductive potential to a level that might affect the continued success of subsequent generations.

Acknowledgment

We are appreciative of the valuable technical assistance of Don Stufflebeem, Somvong Tragoonrung, Apichart Vanavichit, James Alexander, Fred Senecal, and Margaret Hageman. We acknowledge the statistical advice and assistance from David Zirkle and Lisa Gaunio; computer assistance from Ted Ernst; and toxicology discussions with Gregory Linder.

References

[1] U.S. Environmental Protection Agency, "Environmental Effects Test Guidelines," EPA/560/6-82-002, Publication No. PB82-232992, Office of Pesticides and Toxic Substances, Washington, DC, 1982.

[2] Warren-Hicks, W., Parkhurst, B. R., and Baker, S. S. Jr., Eds., "Ecological Assessment of Hazardous Waste Sites: A Field and Laboratory Reference," EPA/600/3-89/013, U.S. Environmental Protection Agency, Environmental Research Laboratory—Corvallis, OR, 1989.

[3] Ratsch, H. C., Johndro, D. J., and Mc Farlane, J. C., *Environmental Toxicology and Chemistry,* Vol. 5, 1986, pp. 55–60.

[4] Williams, P. H. and Hill, C. B., *Science,* Vol. 232, 1986, pp. 1385–1389.

[5] Williams, P. H., Ed., "CrGC Resource Book," Crucifer Genetics Cooperative, University of Wisconsin, Madison, WI, 1985.

[6] Mc Farlane, C., Nolt, C., Wickliff, C., Pfleeger, T., Shimabuku, R., and McDowell, M., *Environmental Toxicology and Chemistry,* Vol. 6, 1987, pp. 847–856.

[7] Downs, R. J. and Smith, J. F., "Phytotron Procedural Manual. For Controlled Environment Research at the Southeastern Plant Environment Laboratory," Technical Bulletin 225 (revised), North Carolina State University, Raleigh, NC, 1983.

[8] Burrill, L. C. et al., "Pacific Northwest Weed Control Handbook," Oregon State University's, Washington State University's, and University of Idaho's Extension Service, Corvallis, OR & Pullman, WA, 1990.

[9] Lutman, P. J. W., *Annals of Applied Biology,* Vol. 107, 1985, pp. 515–527.

[10] Askew, M. I., *Aspects of Applied Biology,* Vol. 6, 1984, pp. 251–256.

R. K. Somashekar[1] and Siddaramaiah[1]

Use of Plants for Toxicity Assessment of Heavy Metal Rich Industrial Effluents

REFERENCE: Somashekar, R. K. and Siddaramaiah, **"Use of Plants for Toxicity Assessment of Heavy Metal Rich Industrial Effluents,"** *Plants for Toxicity Assessment: Second Volume, ASTM STP 1115,* J. W. Gorsuch, W. R. Lower, W. Wang and M. A. Lewis, Eds., American Society for Testing and Materials, Philadelphia, 1991, pp. 376–382.

ABSTRACT: The objective of this study was to evaluate phytotoxicity tests for screening and biomonitoring complex effluent samples. Seed germination tests were conducted using three effluent samples from three industrial sources. Five types of seeds were tested. Complete inhibition of germination was, however, not observed in any case. *Zea mays* and *Dolichos biflorus* were the most sensitive. There occurred a significant difference in root/shoot length and dry weight between treated and control samples at $P < 0.01$ and $P < 0.05$. Results of definitive tests indicated a linear concentration-effect relation. The study indicated that phytotoxicity tests involving higher plants have a high potential for use in the biomonitoring of industrial effluents because of simplicity and sensitivity. It is concluded that the test employed here could be suitably adopted with slight modifications for tropical conditions.

KEY WORDS: phytotoxicity tests, screening, industrial effluents, seed germination, tropical legumes, vigor index

Important criteria for selecting biomonitoring techniques for effluents are speed, sensitivity, and effectiveness [1]. Although many faunal tests are available, only a few are appropriate for effluent samples containing heavy metals and biocides. Algal toxicity tests are well developed [2], but have not been critically evaluated with respect to higher plants, especially those from terrestrial environments. The method suggested by Ratsch and Johndro [3] is an adaptation of the lettuce elongation test and has attracted little attention. Using crop seeds, they employed two methods: (1) seeds completely immersed in nutrient solution, and (2) seeds kept in contact with filter paper containing the test solution. The results showed the former method to be more sensitive than the latter for cadmium chloride ($CdCl_2$), silver nitrate ($AgNO_3$), and 2,4-dichlorophenoxy acetic acid (2-4D). In addition to detecting herbicidal toxicity, higher plants are useful in detecting toxicity from low concentrations of heavy metals. Toxicity induced by metals using the rye-grass elongation method is also demonstrated [4]. The use of common duck weed (*Lemna minor*) to test heavy metal toxicity revealed the test to be eight times more sensitive than published 96-h acute fish test data [5].

The U.S. Environmental Protection Agency [6] and the U.S. Food and Drug Administration [7] have recommended many terrestrial plants such as *Brassica olaracea, Cucumis sativus, Triticum aestivum,* and *Panicum miliaceum* for conducting phytotoxicity tests involving heavy metals. In view of this, the present study is an attempt to develop a phytoassay procedure using terrestrial cultivated plants to screen and biomonitor complex industrial effluents by employing the seed germination test.

---

[1] Department of Botany, Bangalore University, Bangalore 560 056, India.

Several local species were used to test waste waters obtained from three industries that were found to contain elevated levels of metal ions. They included some common pulses (*Dolichos biflorus, Vigna sinensis,* and *Phaseolus aureus*), mustard (*Brassica juncea*), and corn (*Zea mays*), belonging to Fabaceae, Brassicaceae, and Poaceae, respectively. The treatment was made as follows:

Industries	Plants Used
Hindustan Aeronautics Ltd. (HAL)	*Dolichos biflorus* and *Vigna sinensis*
Hindustan Machine Tools (HMT)	*Zea mays* and *Dolichos biflorus*
UB-Mec. Batteries (UB-Mec)	*Brassica juncea* and *Phaseolus aureus*

Phytotoxicity tests via seed germination and seedling growth were used since they are among the best tests available [8–11].

Materials and Methods

The effluent samples were obtained from three industries at the point of discharge from the treatment plant. Treated effluents from these industries either find their way into a nearby sewage canal (HMT, UB-Mec) or a freshwater tank (HAL). The samples collected from the industry were regarded as 100% concentrates. In order to facilitate easy evaluation, four concentrations (v/v) were prepared: 10, 25, 50, and 75% using double-distilled water as dilutant. Chemical analyses were carried out using standard methods [2]. A Perkin Elmer model 403 atomic absorption spectrophotometer was used for the estimation of heavy metals, using the described procedure [12].

Certified seeds samples were obtained from Sutton and Co., Bangalore, India, and kept at −10°C. Seeds were treated with hypochlorite solution, imbibed, and washed five times. The experimental conditions are given in Table 1. Germination counts were taken 120 h after incubation; the appearance of a stub in place of a hypocotyl was considered as a germinated seed. Subsequently, the germinated seedlings were transferred to earthen pots at the rate of ten seedlings per pot and irrigated with 150 mL of different concentrations of the effluents on alternate days. The data on seedling growth were recorded after the 30th day. Double-distilled water was used as the control and dilutant. The experiment was replicated three times.

TABLE 1—*Summary of test conditions.*

Type	Static
Temperature	25 ± 3°C
Light intensity	Cool white fluorescent in case of petri dishes and NA in case of pots
Light period	70 µE/m^2 3 s in case of petri plates and 12 h + 12 h light and dark for pots
Vessel	60 by 15-mm petri plates 10 by 10-in. (25.4 by 25.4 cm) earthen pots
Effluent volume used	5 mL for petri plates, daily; 150 mL on alternate days for pots
No. specimens per vessel	10 seeds in petri plates; 10 seedlings in pots
Total No. seed samples	100 in each case
Replicates	3 in each case
Nutrients	No extra nutrient added; only distilled water for control sets
Duration	120 h for seed germination test; 30 days for seedling growth
Saturation point	Germination and survival after transfer to pots

The wet weight was measured after freeing and washing the plant material from the soil. The dry weight was estimated after drying the material for 48 h at 105°C. The root/shoot length was measured after careful removal of plants from the soil. The vigor index (actual performance of the plant as a percentage of the expected performance calculated on the basis of percent germination in comparison with the control) was calculated by following the procedure of Vaidehi et al. [13].

A concentration-effect relation was measured to determine 50% effect concentration, IC_{50}. IC_{50} is a concentration of a toxicant/pollutant which produces a 50% reduction in the growth of plants in comparison with the control. In this study, each IC_{50} value represents an overall growth response of 100 seedlings under each category. Using a PSY omni computer running on a binomial program, IC_{50} and 95% confidence values were calculated. Duncan's new multiple range test was employed to determine any significant difference between mean values for the test groups. A 95% significance level ($P < 0.01$ and $P < 0.05$) was adopted.

Results and Discussion

Physico-chemical analyses were done for nearly 30 parameters. The quality of the effluents can be characterized as highly complex. Except those of UB-Mec, which were relatively

TABLE 2—*Physico-chemical characteristics of waste waters[a] (all values except pH and turbidity are in mg/L).*

	Industries		
Factors	HMT	HAL	UB-Mec
pH	8.3 ± 0.4	8.5 ± 0.6	2.8 ± 0.8
Total dissolved solids	14 988.5 ± 4.8	3312.7 ± 3.8	2790.0 ± 3.8
Total suspended solids	6 139.5 ± 3.4	188.3 ± 1.0	2454.9 ± 2.9
Total alkalinity	1 860.8 ± 3.1	3897.7 ± 1.7	1312.0 ± 1.9
Residual chlorine	6.7 ± 0.8	1.0 ± 0.4	0.9 ± 0.3
Hydrogen sulfide	2.7 ± 0.4	0.6 ± 0.1	0.6 ± 0.2
Dissolved oxygen	4.9 ± 0.3	5.6 ± 0.3	3.5 ± 0.4
BOD at 20°C for 5 days	8 596.0 ± 4.7	275.0 ± 2.3	3435.0 ± 6.4
Chemical oxygen demand (COD)	25 555.0 ± 9.8	492.5 ± 2.8	7400.0 ± 7.9
Chlorides as Cl	1 619.0 ± 6.4	261.7 ± 1.8	1876.0 ± 3.2
Fluorides as F	6.4 ± 1.0	1.7 ± 0.4	6.3 ± 0.1
Calcium as Ca	191.5 ± 1.8	148.7 ± 1.2	174.8 ± 1.3
Magnesium as Mg	112.5 ± 1.0	16.4 ± 0.8	95.6 ± 1.5
Sodium as Na	384.7 ± 1.8	0.5 ± 0.1	676.9 ± 2.0
Potassium as K	366.0 ± 1.6	290.2 ± 1.9	5.1 ± 0.4
Iron as Fe	82.4 ± 1.0	3.3 ± 0.3	9.1 ± 0.5
Sulfates as SO_4	3 758.6 ± 1.8	292.2 ± 1.3	1738.8 ± 1.9
Phosphates as PO_4	0.2 ± 0.1	19.0 ± 1.4	1.5 ± 0.4
Nitrates as NO_3	45.7 ± 1.6	2.2 ± 0.8	2.2 ± 0.6
Silicates as SiO_2	1.0 ± 0.4	0.2 ± 0.0	0.8 ± 0.2
Ammoniacal (N)	4.2 ± 0.7	32.2 ± 0.7	3.1 ± 0.5
Nickel as Ni	116.8 ± 1.1	22.3 ± 0.9	16.3 ± 0.8
Lead as Pb	13.6 ± 0.7	13.8 ± 0.3	282.7 ± 1.7
Zinc as Zn	22.5 ± 1.0	124.5 ± 1.3	282.4 ± 2.4
Copper as Cu	152.5 ± 1.8	143.3 ± 2.7	33.8 ± 1.8
Cobalt as Co	6.8 ± 0.7	2.3 ± 0.4	56.4 ± 1.3
Chromium as Cr	134.2 ± 0.8	146.4 ± 1.5	12.3 ± 1.0
Turbidity (NTU)	8 300.0 ± 6.8	294.7 ± 2.8	1912.0 ± 2.9

[a] Mean ± SD of 6 observations.

acidic, the rest were alkaline. Those of HMT contained more of suspended solids, and the turbidity value was 8300 NTU. These effluents also contained residual chlorine, and their biological oxygen demand and chemical oxygen demand values were high. Effluents from UB-Mec were highest for lead and zinc, those of HMT for copper, and that of HAL for chromium. A richer proportion of cobalt was also observed in the effluents of UB-Mec (Table 2).

Seed germination percentage (Tables 3-5) remained 85% or more after 120 h of incubation for all the control samples except in the case of *Z. mays*. The relative percentage of survival following germination also remained higher. Variation also occurred with respect to seed germination from 48 to 120 h. Seeds exposed to 100% effluents always had a lower percentage of germination and survival. The data obtained indicated that a 120-h incubation period is essential to allow sufficient time for all the seeds to germinate. Complete inhibition of germination was not observed in any case. The effluents from HAL suppressed the germination of *D. biflorus* more effectively than that of HMT. Of the different seed samples tested, *B. juncea* appeared to be the most resistant and *Z. mays* the most susceptible. This conclusion is based on the relative percentage of germination and survival of seedlings in all the categories, although the same species was not exposed to effluents of all the industries. Indeed, the data reflected that the effluent from UB-Mec was phytotoxic to *P. aureus,* the effluent of HAL to *D. biflorus,* and the effluent of HMT to *Z. mays.*

It is known that, during germination, root and shoot growth occurs. The relative suppression of this growth could therefore be considered as a measure of phytotoxicity, besides germination, because not all germinated seeds survive. The relative percent of seedlings that survived after germination varied in the present study with respect to the concentration of the effluents used. Many studies [4,9–11,14] have shown that the root system is more sensitive than other plant organs to environmental toxicants because the roots are in direct contact with the effluents and because they are the organs of absorption.

After exposure to the effluents of HAL, the inhibition of root growth was greater in *D.*

TABLE 3—*Effect of waste waters from HMT on germination and growth behavior of crop plants (average values of 100 seedlings at 120 h).*

Concentration, %	Germination, %	Length, cm		Inhibition, %		Weight, g		Vigor Index
		Shoot	Root	Shoot	Root	Wet	Dry	
ZEA MAYS L.								
10	70	14.97	20.65	9.54	10.69	16.0	3.46	2493.4
25	65	13.82	20.29	16.49	12.24	13.6	3.04	2217.1
50	58	11.58	7.04[b]	30.03	69.55	13.1	1.81	1079.9
75	51	6.23[b]	3.04[b]	62.35	86.85	12.2[a]	0.93[a]	472.7
100	49	5.25[b]	1.16[b]	68.27	94.98	8.8[b]	0.48[b]	314.0
Control	78	16.55	23.12	. . .	. . .	19.0	3.84	3092.7
DOLICHOS BIFLORUS L.								
10	80	18.11	13.21	19.18	17.17	8.2	2.91	2506.4
25	72	16.50	9.00	26.37	43.57	7.1	2.73	1836.0
50	67	12.48[a]	7.32[a]	44.41	54.10	6.6	2.58	1326.6
75	61	8.64[b]	6.81[b]	61.44	57.30	4.6[b]	2.40[a]	942.4
100	56	7.45[b]	5.33[b]	66.75	66.58	3.9[b]	1.97[a]	715.6
Control	87	22.41	15.95	. . .	. . .	10.4	3.04	3337.3

[a] Significantly different from control: $P < 0.05$.
[b] Significantly different from control: $P < 0.01$.

TABLE 4—*Effect of waste waters from HAL on germination and growth behavior of crop plants (average values of 100 seedlings at 120 h).*

Concentration, %	Germination, %	Length, cm		Inhibition, %		Weight, g		Vigor Index
		Shoot	Root	Shoot	Root	Wet	Dry	
DOLICHOS BIFLORUS L.								
10	87	20.47	18.11	12.70	18.11[a]	7.66	2.63	3265.9
25	80	19.14	13.86	18.37	27.01	7.10	2.49	2640.0
50	76	17.61[a]	10.52[a]	24.90	44.60	6.20	2.37	2137.8
75	69	13.51[b]	9.84[b]	42.38	48.18	5.84[a]	2.16[a]	1611.1
100	63	11.68[b]	7.62[b]	50.19	59.87	4.54[a]	1.00[b]	1215.9
Control	91	23.45	18.99	. . .	. . .	9.83	2.97	3862.0
VIGNA SINENSES L.								
10	82	34.40	14.87	13.67	1.06	33.2	5.27	4040.1
25	79	29.18	14.64	26.77	2.59	28.2	5.01	3461.7
50	71	25.13	13.36	36.93	11.11	27.2	4.88	2732.7
75	67	21.69[a]	11.81	45.57	21.42	25.4	4.26[a]	2244.5
100	60	19.38[b]	10.34	51.36	31.20	23.0[a]	3.40[a]	1783.2
Control	86	36.85	15.03	. . .	. . .	36.8	5.63	4719.6

[a] Significantly different from control: $P < 0.05$.
[b] Significantly different from control: $P < 0.01$.

biflorus than in *V. sinensis*. Nearly complete inhibition of root growth occurred in *Z. mays* exposed to HMT waste water, although shoot growth was normal. A similar observation has been made in the case of Japanese millet [1]. The root and shoot growth in *B. juncea* and *P. aureus* treated with UB-Mec effluents remained the same at 25 and 50% concentrations.

TABLE 5—*Effect of waste waters from UB-MEC on germination and growth behavior of crop plants (average values of 100 seedlings at 120 h).*

Concentration, %	Germination, %	Length, cm		Inhibition, %		Weight, g		Vigor Index
		Shoot	Root	Shoot	Root	Wet	Dry	
BRASSICA JUNCEA (L.) CZER.								
10	87	10.84	6.42	12.68	17.90	10.34	2.76	1501.6
25	80	9.04	5.25	26.68	32.86	9.68	2.19	1143.2
50	74	8.23	5.12	33.25	34.52	7.01	1.88	987.9
75	70	6.10[a]	4.97[a]	50.52	36.44	5.85[a]	1.76	774.9
100	65	5.34[a]	4.13[b]	56.69	47.18	4.38[b]	1.33[a]	615.5
Control	92	12.33	7.82	. . .	. . .	12.90	2.80	1853.8
PHASEOLUS AUREUS ROXB.								
10	80	16.76	7.38	7.76	15.75	8.78	2.03	1939.2
25	73	14.38	6.30	20.85	28.08	6.66	1.90	1509.6
50	67	10.14[a]	5.39[a]	44.19	38.47	5.18[a]	1.49	1040.5
75	63	7.49[b]	4.94[b]	58.17	43.60	4.37[b]	1.13[a]	783.0
100	58	6.34[b]	4.03[b]	65.01	53.99	3.64[b]	0.98[a]	601.4
Control	87	18.17	8.76	. . .	. . .	10.34	2.31	2342.9

[a] Significantly different from control: $P < 0.05$.
[b] Significantly different from control: $P < 0.01$.

TABLE 6—IC_{50} and 95% confidence limit for the effluents of industries calculated using the moving average method of Wang and Williams [1], expressed as % effluent concentration.

	Industries					
	HMT		HAL		UB-Mec	
Crop Plant Tested	IC_{50}	95% CL	IC_{50}	95% CL	IC_{50}	95% CL
Brassica juncea	. . .	. . .	. . .	. . .	7	10–24
Phaseolus aureus	. . .	. . .	. . .	. . .	10	16–28
Zea mays	18	19–33	. . .	. . .	. . .	. . .
Dolichos biflorus	12	17–31	8	19–30	. . .	. . .
Vigna sinensis	. . .	. . .	6	11–23	. . .	. . .

The wet and dry weights of the samples were measured separately using plants obtained from one of the replicates. In the majority of cases, there appeared to be no difference between plants treated with 10% effluents and the control as evidenced by Z. *mays* treated with 10% HMT effluent (Tables 3–5). The use of pulses for the seed germination test with respect to UB-Mec and HMT waste waters gave IC_{50} values of 10 to 12 for *P. aureus* and *D. biflorus,* respectively, at 100% effluent (Table 6). The use of Z. *mays* yielded an IC_{50} value of 18. For *D. biflorus* treated with HMT and HAL waste waters, the IC_{50} values were 12 and 8, respectively.

Conclusions

The use of higher plants for ecotoxicology testing has been a regular practice in the past. A majority of investigators have used aquatic macrophytes because they are commonly affected after the discharge of effluents into fresh water. In our locality, diluted waste waters discharged from industries were being used for growing fodder and vegetable plants. This practice is suspected to add sufficient quantities of toxic compounds to the plants being cultivated. For this reason the present investigation used only native plants. The results obtained suggest that these plants may be suitably adapted for phytotoxicity tests under tropical conditions. It is expected that many such investigations using the technique described here would increase our understanding of the hazards of using either pure or diluted effluents for agriculture. These tests can be adapted to other species as well.

It is advisable to treat a particular industrial effluent as a single unit while conducting phytotoxicity tests since plant communities of both terrestrial and aquatic environments are exposed to them in a similar way. Further, the complexity of industrial effluents in containing inorganic, organic, and trace metals in varied proportion is well documented. Following treatment, the plant systems respond to some of these agents by exhibiting symptoms. Therefore, to understand the reason factor behind symptoms that may be produced following treatment with industrial waste waters, investigations with solutions of known compound have to be conducted. In the complex effluents tested here, phytotoxicity may be due to the presence of one or more toxic substances and the resulting toxic interactions.

References

[1] Wang, W. and Williams, J. M., *Environmental Toxicology and Chemistry,* Vol. 7, 1988, pp. 645–652.

[2] APHA, AWWA, and WPCF, *Standard Methods for the Examination of Water and Waste Water,* APHA, Inc., New York, 1985.

[3] Ratsch, H. C. and Johndro, D., *Environmental Monitoring and Assessment,* Vol. 6, 1986, pp. 267–276.

[4] Wong, M. H. and Bradshaw, A. D., *New Phytologist,* Vol. 91, 1982, pp. 255–261.

[5] Wang, W., *Environmental Toxicology and Chemistry,* Vol. 5, 1986, pp. 891–896.

[6] U.S. Environmental Protection Agency, "Seed Germination/Root Elongation Toxicity Test," EG-12 Office of Toxic Substances, Washington, DC, 1982.

[7] U.S. Food and Drug Administration "Seed Germination and Root Elongation," FDA Environmental Assessment Technical Guide No. 4.06, Centre for Food Safety and Applied Nutrition and Centre for Veterinary Medicine, U.S. Department of Health and Human Services, Washington, DC, 1987.

[8] Wang, W., *Environment International,* Vol. 10, 1984, pp. i–iii.

[9] Wang, W., *Environment International,* Vol. 11, 1985, pp. 49–55.

[10] Wang, W., *Environmental Toxicology and Chemistry,* Vol. 5, 1986, pp. 891–896.

[11] Wang, W., *Environmental Toxicology Chemistry,* Vol. 6, 1987, pp. 409–414.

[12] Van Loon, J. C., *Analytical Atomic Absorption Spectrophotometry,* Academic Press, New York, 1980.

[13] Vaidehi, B. K., Jagadamba, G. V., and Lalitha, P., *Indian Botanical Reporter,* Vol. 4, No. 1, 1985, pp. 92–94.

[14] Luessem, H. and Rahman, A., *Vom Wasser,* Vol. 54, 1980, pp. 29–35.

Marjorie P. Kraus[1]

Cyanophage/Host Assay for Toxicity Assessment in Water, Wastewater, Sludges, and Composts

REFERENCE: Kraus, M. P., **"Cyanophage/Host Assay for Toxicity Assessment in Water, Wastewater, Sludges, and Composts,"** *Plants for Toxicity Assessment: Second Volume, ASTM STP 1115,* J. W. Gorsuch, W. R. Lower, W. Wang and M. A. Lewis, Eds., American Society for Testing and Materials, Philadelphia, 1991, pp. 383–391.

ABSTRACT It has been demonstrated that cyanobacterial viruses (cyanophages) arise in sewage treatment and are present in streams receiving treated sewage. While algae alone are often proposed for toxicity assessment, the method presented here uses the cyanophage/host interaction, thus bringing the assay into the realm of molecular biology. Examples of its application to water, sewage, and sludge treatment, to high nitrate, to herbicides and viricides, and to composts support its value for initial screening or first-tier toxicity assay. Its ability to deal with principles and mechanisms at the molecular level and to incorporate progress in DNA and protein methodologies adds to its significance and utility at higher levels of toxicity assessment.

KEY WORDS: aquatic toxicology, compost assay, cyanobacteria, cyanophage, herbicides, methylene blue, plant assay, toxicity assay, viricides

Comparisons of deep wells, shallow wells, tributaries, and main streams clearly demonstrate that cyanophage exists in water chiefly as a result of treated sewage [1]. Comparisons of cyanophage assay with routine samplings by the United States Geological Survey in evaluating stream conditions showed general agreement [2]; however, no firm relation between *E. coli* and cyanophage virulence was apparent [1]. Contrary to the common idea that cyanophages occur whenever large quantities of hosts are present, cyanophages were found to originate by plasmid transfer to hosts as a result of bacterial breakdown of sewage particulates [1]. The study demonstrated that bacterial (or fungal) activity occurring during sewage treatment resulted in the (plasmid) transfer of an "S" host-range, plaque-morphology characteristic to hosts during the bacterial breakdown of sewage particulates. While plasmids have been demonstrated in cyanobacteria, no specific assignment of plasmid activity has previously been made [3].

The cyanophage/host assay is simple, economical, rapid, and informative. In principle it is similar to enteric virus/tissue culture or bacteriophage/bacteria systems which provide understanding at the level of DNA interactions. The photosynthetic hosts are axenic, highly purified, genetically distinct organisms of the cyanobacterial species *Nostoc sp* and *Schizothrix calcicola*. Cyanobacteria have the advantage of a green background on agar plates, which permits added distinction among plaque types. The cultural requirements of the hosts are minimal, and their phages are distinguished by host-range and plaque-morphology differences among members of a panel of hosts. These differences reflect the genetic makeup

[1] Director, Algal Research Center, Landenberg, PA 19350.

of the organisms in their responses to environmental stress. The philosophy of the host/ virus interaction has been well outlined by Coohill [4] and Loechler et al. [5]. A more detailed discussion of the cyanophage system has been presented by Kraus [6].

A major purpose of this assay is to lessen dependence on chemical statistics and to move closer to the particular target of the toxicant.

Materials and Methods

Algal hosts and cyanophage. The general scope, practice, and application of the cyano-phage host-range, plaque-morphology assay was previously described [7]. Host-range characteristics of cyanophages used in this study are shown in Table 1. Phages and hosts are given trivial symbols: hosts are indicated by a three-letter description and cyanophages by a single letter. Their different properties are described in the table; otherwise, they are morphologically alike and their differences are concerned with viral specificity. Cyanophage S was cultivated on host CRY, Cyanophage P on TWS, A on ARN, and F on FAS. The J pattern is often found in nature but does not plaque true on any available host.

Stocks used in all experimental procedures are routinely tested and are the major controls for the setups. Any failure of this stock negates the entire experiment and results would not be reported. Internal controls are part of the experimental procedure as measurements made with and without the agents concerned. They may not be specifically labelled as "controls."

Cyanophage assay. Four cyanophage hosts were selected from mature (three-week-old) stock. Half the supernatant growth medium was poured off and wasted. Labelled flasks containing 40 mL of growth medium were set up for each sample. Two millilitres of each algal host were added to the corresponding flasks of each set followed by a 40-mL water sample. These flasks were then placed in a light box on glass shelves about 30 cm from a cool white fluorescent light. A 7-W night light warmed the box to about 30 to 32°C. Between the second and the fifth days flasks were observed for color loss, indicating lysis. Using a pasteur pipette, a drop of lysate was then spotted onto a set of previously prepared (1.5%) agar overlay plates marked into sections. The pattern of lysis determined the cyanophage type. Flasks were observed for ten days. Slow lysis was generally due to conditions other than cyanophage, such as toxicities or grazing by amebas, nematodes, etc. In such cases ten drops of chloroform were added to 3 mL of growth medium and the spot or plaque was picked up on the end of a pasteur pipette and thoroughly mixed by drawing the material back and forth about 30 times. After the chloroform had settled for at least 2 h, the

TABLE 1—*Identification of cyanophage.*[a]

	Hosts							
	Nostoc		*Schizothrix calcicola*					
Cyanophage	ARN	BET	CRY	FAS	FAV	KAP	SAF	TWS
A	+	−	−	−	−	−	−	−
S	−	−	+	+	o	+	−	+
P	−	−	+	−	+	−	−	+
F	−	−	−	+	o	−	−	−
J	−	−	−	−	+	−	−	+

NOTE: − = algal growth; o = reduced growth; + = viral lysis.

[a] Phages and hosts are given trivial symbols: hosts are indicated by a three-letter description and cyanophages by a single letter.

supernatant was spotted on a range of overlay plates. Results were recorded as confirmed viral lysis ($+$), turbid plaques (o), unconfirmed lysis ot toxicity (x), or normal algal growth ($-$).

The symbol (o) indicates turbid plaques, which are well recognized as viral entities, generally indicating the production of strains resistant to the infecting cyanophages. The fact of resistant hosts is an important phase of virology.

Nitrate. Nitrate solutions were prepared from a 100-g/L sodium nitrate ($NaNO_3$) standard by dilution with distilled water. Capped test tubes marked at the 10-mL level were set up according to protocol. One millilitre of nitrate solution sufficient to produce the desired nitrogen concentration in 10 mL was introduced into corresponding tubes, followed by 1 mL of algal culture and 0.02 mL of the designated cyanophage at about 2×10^9 plaque-forming units/mL. Tubes were placed in the light box and examined at 12, 24, and 36 h for visible lysis.

Herbicides. Bladex, Bicep, and Aatrex, herbicides in current use on cropland, were obtained from a company engaged in custom spraying. They were thick white liquids containing the usual inert additives, lime and lubricating spreaders, etc., of the trade. Lime is the main component, and data showed that when the preparation was diluted to the point where there was no longer visible cloudiness or sediment, the toxicity was below detection. In practice the preparation is added to water in the spray rig and measured as quantities of active ingredient per acre. The formulae of some active ingredients are shown in Fig. 1. Many similar formulae are continually being studied as possible additives for specific herbicide activity. Under fair trade agreements, ingredients of less than 2.2% need not be disclosed. Hence, test concentrations were determined by dilution of the commercial product.

In studying herbicide effects on the host/virus relation, the cyanophage system F/FAS was chosen because of its simplicity: cyanophage F gives turbid plaques (labelled "o") on host FAV, clear plaques on FAS, and all other hosts are immune. The herbicides were diluted with growth medium, as previously indicated, and three protocols were compared in an attempt to locate conditions of greatest activity. Protocol I pretreated hosts and phage together for 2 h to allow virus attachment to hosts. The mixture was then centrifuged 5 min at high speed in a clinical centrifuge. The supernatant was discarded and the pellet resuspended in fresh growth medium. Protocol II pretreated phage and herbicide for 2 h before adding cells. Protocol III pretreated cells and herbicide for 2 h before adding phage. Freshly prepared cyanophage F at about 2×10^9 plaque-forming units/mL was diluted to 2×10^4, and 2 mL of the desired concentrations of herbicide, cells, and phage were mixed in screw-

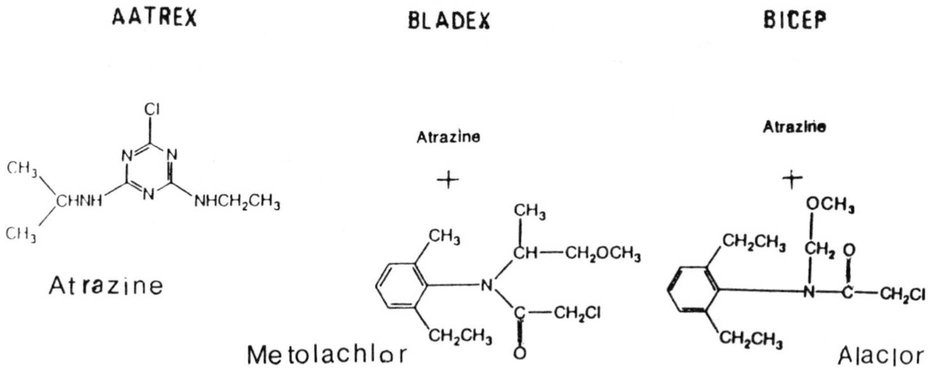

FIG. 1—*Active components of herbicides.*

capped tubes and allowed to develop in the light box. Cyanophage counts were made before the burst (4 to 6 h) and after the burst (24 h) by drop-spotting on host FAV.

Methylene blue. Methylene blue was freshly prepared from powdered stock at 100 μg/L and kept cold in the dark until needed.

Sewage treatment and water treatment. Types of sewage treatment facilities monitored include overland flow, spray irrigation, oxidation ponds, lagoons, package plants, and small and large municipal plants with many different engineering designs. A new plant using seven rotating biological contactors, monitored monthly, provided better understanding of the sequence of biological events.

Composts and sludges. Sludges were obtained from commercial enterprises, drainage ditch muds, primary and secondary sludges and pressed cake from sewage treatment, fresh and used mushroom compost, farm wastes, and from slaughter house and cannery wastes [8].

Preparation of sludges and composts. Some composts do not pulverize well due to the presence of fiber and sand. Various types of commercial homogenizers and mortar and pestle were used. Air-dried or moist weight was obtained, and the material was comminuted until it could be maneuvered by a wide bore pasteur pipette. The sample was taken up in growth medium at 1 g/100 mL.

Results

High nitrate. Nitrate concentration was investigated as a result of a fish kill thought to be caused by high nitrate leached from a fertilizer plant. Stations formerly testing cyanophage-positive were negative during the kill. Table 2 shows that at high nitrate concentration (30 mg/L nitrate nitrogen) cyanophage is not active against its normal hosts. Table 2 is also a simple illustration of the difference between the action of an agent on the host/cyanophage interplay and the action of that agent on the host alone. (It should also be noted that some negative reactions shown on Table 2 are a function of host resistance to a particular cyanophage—which emphasizes the importance and utility of a range of hosts in interpreting results.)

For purposes of comparison with Table 2, standard laboratory growth conditions contain 1.6 mg/L nitrate nitrogen. Average nitrate nitrogen in the local main stream was about 3 mg/L, and the average of about 100 local wells was about 3.7 mg/L including two farms with a nitrate nitrogen of about 10 mg/L, which is the local limiting permissible nitrate concentration. The alert reader may question the lytic activity of F/FAV on CRY since its

TABLE 2—*Effect of high nitrate on cyanophage/host interaction.*[a]

	Cyanophage							
	A/Arn				F/FAS			
Nitrate, mg/L nitrogen	Hosts							
	ARN	BET	FAV	FAS	FAV	FAS	CRY	SAF
0.3	− + +	− − *	− − *	− − *	− O O	+ + +	− + +	− − *
3	− + +	− − *	− − *	− − *	− − −	− + +	− − −	− − *
30	− − −	− − *	− − *	− − *	− − −	− − −	− − −	− − *

[a] Results are given in terms of visible lysis of the interaction as read at 12, 24, and 36 h. + = lysis; O = turbid plaques, incomplete lysis; − = no change in the suspended cells; * = resistant host.

general behavior is negative. Such disconcerting anomolies or "leakiness" do occur and as yet are not accounted for. In Row 2, at 3 mg/L nitrate nitrogen, its behavior is normal.

Herbicides. The three protocols shown in Table 3 nicely demonstrate that considerable difference exists among the three chemically similar herbicide activities. It is evident that these herbicides are involved in interactions that are complete in 2 h and hence may function in early cellular activities of penetration and takeover. Herbicides attempt to reduce cell growth by interfering with cellular processes centered mainly on injury to the photosynthetic apparatus. By varying the pretreatment conditions, the three protocols use the cyanophage to point out biomolecular specificities in each herbicide. Algae (and even cyanobacteria) are photosynthetic organisms and as test species should respond as well as other plants. While specific results will not apply equally to all plants, many major biochemical pathways will be similar, and it is likely that algae will have pathways that may be pointed out by some of the minor additives. These specificities are of vital interest to those in the herbicide trade. Evidently small differences in additives have considerable effect.

Methylene Blue. The photodynamic action of methylene blue on bacteriophage was noted by Yamamoto in 1985 [9]. Survival curves for cyanophage in sewage treated with methylene blue at doses of 1 and 2 μg/mL show strong activity (Kraus, M., unpublished). Methylene blue is an herbicide as well as a virucide. Table 4 gives data on the effect of methylene blue on two cyanophage/host systems, F/FAS and S/CRY. For the S/CRY data, methylene blue appears effective at 4 μg/mL, whereas the F/FAS data show residual activity at that concentration except in the case where the initial contact was between phage and methylene blue (Protocol II). This work needs to be repeated to discover whether this is a consistent result. The data agree with the interpretation that the action of methylene blue on phage is more rapid than the action of methylene blue on the cell and that the action of methylene blue as a virucide differs from that of methylene blue as an herbicide.

TABLE 3—*Effect of herbicides on the F/FAS cyanophage/host interaction.*[a]

Herbicide Dilution	Aatrex		Bladex		Bicep	
	Preburst	Postburst	Preburst	Postburst	Preburst	Postburst
			Protocol I			
0.000	99	+	31	+	60	+
0.001	55	+	39	+	x	k
0.02	15	4	46	+	x	k
0.01	x	k	27	+	x	k
			Protocol II			
0.000	34	+	34	+	78	+
0.001	49	+	19	+	35	+
0.02	41	1	48	x	46	+
0.01	28	1	4	k	4	+
			Protocol III			
0.000	38	99	13	+	10	+
0.001	38	67	38	+	8	9
0.02	29	69	k	k	12	k
0.01	x	99	k	k	8	k

[a] Protocol I: Pretreat phage and host with herbicide for 2 h, centrifuge, remove supernatant: resuspend pellet in fresh medium.
    II: Pretreat phage with herbicide 2 h, add cells.
    III: Pretreat cells with herbicide 2 h, add phage.
NOTE: Spotplate for preburst at 6 h; spot plate for postburst at 24 h. + = confluent lysis; x = turbid plaques; K = killing of host by herbicide; numeral = plaque count/0.03 mL.

TABLE 4—*Action of methylene blue on cyanophage.*[a]

	Host/Cyanophage System														
	F/FAS								S/CRY						
	μg/mL Methylene Blue Preburst (A) and Postburst (B) Cyanophage Count														
	0		1		2		4		0		1		2		4
	A	B	A	B	A	B	A	B	A	B	A	B	A	B	A	B
I															
FAS	34	+	19	+	8	+	1	+	–	–	–	–	–	–	–	–
FAV	–	o	–	o	–	–	–	–	–	–	–	–	–	–	–	–
FWS	–	–	–	–	–	–	–	–	11	+	3	+	–	+	–	–
CRY	–	–	–	–	–	–	–	–	11	+	9	+	5	+	–	–
KAP	–	–	–	–	–	–	–	–	11	+	8	+	–	+	–	–
KRS	–	–	–	–	–	–	–	–	–	–	–	–	–	–	–	–
SAF	–	–	–	–	–	–	–	–	–	–	–	–	–	–	–	–
II															
FAS	55	+	2	+	–	–	–	–	–	–	–	–	–	–	–	–
FAV	–	o	–	o	–	–	–	–	11	0	–	0	–	+	–	–
FWS	–	–	–	–	–	–	–	–	140	+	–	+	–	+	–	–
CRY	–	–	–	–	–	–	–	–	89	+	–	+	–	+	–	–
KAP	–	–	–	–	–	–	–	–	150	+	–	+	–	+	–	–
KRS	–	–	–	–	–	–	–	–	–	–	–	–	–	–	–	–
SAF	–	–	–	–	–	–	–	–	–	–	–	–	–	–	–	–
III															
FAS	+	+	+	+	+	+	39	–	–	–	–	–	–	–	–	–
FAV	–	o	–	o	–	o	–	–	5	o	2	o	1	o	–	–
FWS	–	–	–	–	–	–	–	–	119	o	87	+	13	+	–	–
CRY	–	–	–	–	–	–	–	–	30	+	84	+	7	+	–	–
KAP	–	–	–	–	–	–	–	–	76	+	77	+	14	+	–	–
KRS	–	–	–	–	–	–	–	–	–	–	–	–	–	–	–	–
SAF	–	–	–	–	–	–	–	–	–	–	–	–	–	–	–	–

NOTE: Numbers = number of plaques: + = lysis; o = partial lysis; – = no lysis.
[a] Protocol I: Pretreat cells and phage for 2 h with methylene blue centrifuge, discard the supernatant and resuspend in growth medium. Protocol II: Pretreat phage 2 h with methylene blue, add cells. Protocol III: Pretreat cells 2 h with methylene blue. Add phage. Preburst spot plate at 4 h; postburst spot plate at 24 h.

Future experimentation. Preliminary experimentation has shown that Bicep has a strong additive effect on methylene blue activity at concentrations below 50 μg/mL. This datum merits further study. Studies on the photosynthetic apparatus have been carried out in the past on many microalgae systems [10] but usually on impure hosts. Pure hosts and cyanophages are a necessity in producing useful data toward understanding genetic and biochemical differences among the various strains and for furthering the application of genetic and molecular biology to toxicity assessment.

Water Treatment, Sewage Treatment, Sludges, and Composts

Suburban water treatment plant. The Suburban Water Treatment Plant obtains water from both the Red Clay and the White Clay Creeks (Chester Co., PA). Flocculation treatment removes over 99.9% of the cyanophage virulence indicated in the incoming water. Ameba grazing on algae in the stream may be a factor aiding this flocculation. They are major feeders on algal hosts, and their cysts are present in high numbers. Tests show that cyanophage is still present in the sludge, which is stored in basins near the plant. Accumulated data from stream surveys show that cyanophage recovery occurs downstream from chlorinated sewage discharges. (It would be interesting to determine whether any genetic alteration is involved.) Synergistic and antagonistic effects among organisms which abound in sludge present a complex picture (Table 5).

Mushroom composts and lagoons. In the past mushroom lagoons were merely storage tanks with conventional restrictions on biological oxygen demand (BOD) and suspended solids. Current progress is due to compartmentalization. Table 6 shows data on lagooned mushroom wastes. Comparison of plates on 1-mL samples indicates that problems are more serious in the sediments. A high nitrate value (Table 2) is likely to accompany the high conductivity values (as shown in Table 6). Currently, large-scale composting operations appear to be successful in producing satisfactory products for land application. Tests on these products show them to be free from cyanophage and from undesirable fungi and protozoans, which might release immunities held by some members of the panel of hosts.

Discussion

This paper presents a variety of toxicity assessments which can be rapidly and economically carried out by a cyanophage/host approach. While it is important to gain statistically secure LD_{90} or LD_{50} values on various organisms under toxic conditions, it is also necessary to better comprehend the manner by which the organisms expire. Some risk evaluation handed out by legal means based on surrogate assay may be erroneous or unfair because of unscientific or illogical interpretation of the data. Many disease situations involve virus activity where ecological associates play a significant role to potentiate disease. Recent investigations on human disease [11] show that protozoans can seriously affect the outcome. Giardiasis is also a currently occurring disease carried in water by protozoans. Plasmids too, are common vehicles of viral transfer. In one of my early references [6], I gave data on the large numbers of what we now call plasmids found in water. Recently, in studying marine systems, it has been possible to use cyanophage assay to trace plasmid infection in human disease from

TABLE 5—*Cyanophage assay—Suburban Water Co. samples.*[a]

				Cyanophage Assay			
Date	Sample	pH	Conductivity, μmhos	Incubation	1-mL Plate	Identification	Remarks
7/27	Red clay intake	7.2	340	+ +o	500	P	
2/16	White clay intake	6.8	225	+ + +	107	P,J,S	
8/21	White clay intake	6.8	215	+ + +	254	P,J	pmp
7/27	Mixing tank	7.3	292	− − −	0		Chlorination
7/27	Fan out	7.3	286	− − −	0		Chlorination
7/27	Flocculation	6.9	267	− − −	0		Chlorination
7/27	Backwash	6.8	262	− − −	0		Chlorination
7/27	Tap	6.9	266	− − −	0		Chlorination
7/27	Sludge	7.2	342				Fungus, rs,
7/27	Supernatant			+ + +	49	A,P,S	Turbid plaques

[a] pmp = depressed plaques with amebas: A,S,P,J = cyanophages; rs = resistant host strains; + = clear plaques; 0 = turbid plaques; − = no plaques.

TABLE 6—*Cyanophage assay of concentrated lagooned mushroom wastes.*[a]

Sample	pH	Conductivity, μmhos/cm	Incubation	Cyanophage Assay, 1-mL plates	
				Sediment	Supernatant
Influent	7.7	3816	...	52 pmp	Many, turbid
Tank A	7.8	4430	...	500 pmp	22
Tank B	8.0	4140	...	300	200
Tank C	8.3	4100	...	Many, turbid	rs
Tank D	8.4	3960	...	400	Many turbid
Effluent	8.0	4118	...	Fungi	138 turbid

[a] pmp = plaques with many amebas; rs = resistant algal hosts.

saltwater samples to land-based discharges (Kraus, M. P. et al., submitted to *Applied Phycology*). The fact that an organism can *trace* a disease need not incriminate it as the *cause* of the disease. Along with population increases, there will be an increase in disease. From samplings of many, many streams across the United States, the disposal of sewage in major and minor streams remains a serious problem.

Conclusion

A series of representative studies using a cyanophage/host methodology is presented here. The first set uses cyanophage as a pollution indicator where cyanophages developed during the sewage treatment act to lyse the test culture, killing the host cells. Less violent changes involve modification or adaptation to environmental conditions where resistant host cells are developed. A third type of activity is displayed by toxins such as herbicides and methylene blue, which act specifically on cellular biochemical reactions such as phycobilin pigments and the photosynthetic apparatus. All these activities have tie-ins with the essential or modified genetic makeup and ecological condition of the test cells.

Practical problems of aquatic toxicology also deal with sewage and agricultural discharges, water treatment, and water reuse. The discovery in sewage sludge of "killing plaques" [1] (presumably a plasmid which delivers an "S" character into cyanophage hosts) has made an inroad into the enigma concerning the inception of cyanophages in sewage treatment and has led to practical use. Recent work comparing pre- and postburst counts on sludges and composts indicates that large changes between the expected and actual results appear to be related to insufficient treatment to destroy critical organisms. These observations may be the basis for a test for adequately composted sewage sludge for landscaping and agricultural use; proper treatment of the sludge should result in a material no longer carrying organisms able to alter the normal host range and plaque morphology of a test cyanophage.

Cyanophage host-range, plaque-morphology assay as a tier-one toxicity assay. Because of the molecular and genetic information available in the short span of an host/virus interaction, the opportunity exists to obtain much precise information on specific toxic events. Many important sequential biological functions take place during the 6 to 7-h infectious cycle of a virulent (and nonvirulent) cyanophage interaction. These events can be separated as a function of time [7] and can provide information on various toxicities in an economical and efficient manner.

Acknowledgments

The author is indebted to Drs. Cara Fries and Carol Litchfield, who critically read the paper and offered many useful suggestions. She also thanks an anonymous reviewer for numerous thoughtful and helpful comments and the Chester County Conservation District for herbicide samples.

References

[1] Kraus, M. P., "Value of a Cyanophage/Host System for Toxicity Assessment," in First Symposium on Use of Plants for Toxicity Assessment, ASTM, 19 Apr. 1989, Atlanta, GA (abstract).

[2] Stutman, M., "Report to the Red Clay Valley Association," Kennett Square, PA, 1983.

[3] Herdman, M., "Genomes and Genetics," *The Biology of Cyanobacteria, Botanical Monographs,* Vol. 19, University of California Press, Berkeley, CA, N. G. Carr and B. A. Whitton, Eds., 1982, pp. 263–306.

[4] Coohill, T. P., "Virus-Cell Interactions as Probes for Vacuum-UV Radiation Damage and Repair," *Photochemistry and Photobiology,* Vol. 44, 1986, pp. 359–363.

[5] Loechler, E., Straus, N., Bryant, J. L., and King, J. L., *In Vitro Toxicity Testing of Environmental Agents,* Part A, Plenum Press, New York, 1983, pp. 79–105.

[6] Kraus, M. P., "Host Range and Plaque Morphology of Bluegreen Algal Viruses," Technical Report DEL-SG-1-74, College of Marine Sciences, University of Delaware, Newark, DE, 1973, pp. 1–30.

[7] Kraus, M. P., "Cyanophage Assay as a New Concept in the Study of Environmental Toxicity," *Aquatic Toxicity and Hazard Assessment, (7th Symposium), ASTM STP 854,* R. D. Cardwell, R. Purdy, and R. C. Bahner, Eds., American Society for Testing and Materials, Philadelphia. 1985, pp. 27–41.

[8] Kraus, M. P., "Analysis of Virus Die-off in Land-Applied Industrial Wastes and Sludges," *Toxic and Hazardous Wastes,* G. S. Boardman, Ed., *Proceedings,* 18th Mid-Atlantic Industrial Waste Conference, 1986, pp. 365–375.

[9] Yamamoto, N., "Photodynamic Inactivation of Bacteriophage," *Journal of Bacteriology,* Vol. 80, 1958, p. 443.

[10] Hill E. R. and Wright, S. J. R., *Pesticide Microbiology,* Academic Press, New York, 1978, pp. 535–602.

[11] Ma, P., "Protozoa in Aquired Immune Deficiency Disease," *ATTC Quarterly,* Fall 1987, pp. 1–2 (with references).

Author Index

A

Ailstock, M. S., 267
Asfaw, A., 217

B

Backer, L. C., 309
Boston, H. L., 126
Brashers, L. K., 341
Breteler, R. J., 118
Brown, R. A., 197
Buhl, R. L., 118
Byl, T. D., 101

C

Casto, B. C., 309
Corradi, M. G., 276

D

Debus, R., 258
Dhesi, J. S., 326
Dixon, D. G., 209

E

Eirkson, C. E. III, 12

F

Farmer, D., 197
Favali, M. A., 276
Fleming, W. J., 267
Fletcher, J., 5
Fletcher, J. S., 250
Folsom, B. L., Jr., 172
Fossati, F., 276
Freemark, K. E., 77

G

Gentile, J. M., 318
Gill, B. S., 309, 326
Gobas, F. A. P. C., 178
Greenberg, B. M., 209

H

Haffner, G. D., 178
Harrass, M. C., 12
Hill, W. R., 126
Huang, X-D., 209

I

Ireland, F. A., 146, 217, 297

J

Johnson, P., 318
Judy, B. M., 146, 217, 297

K

Kapustka, L., 240
Kapustka, L. A., 333, 365
Klaine, S. J., 101
Kraus, M. P., 383
Krause, G. F., 146, 217

L

Larson, L. J., 230
Lovett-Doust, L. 178
Lower, W. R., 146, 217, 297

M

McFarlane, C., 355
MacQuarrie, P., 77

Maki, A. W., 118
Momot, J. J., 267
Moore, J. C., 341

N

Norman, C. M., 267
Nowell, L. H., 12
Nwosu, J. U., 333, 365

P

Peterson, H. G., 107
Pfleeger, T., 355
Plewa, M. J., 287
Price, R. A., 172

R

Ratsch, H., 333
Ratsch, H. C., 365
Reporter, M., 240
Rickard, P., 77
Ritter, M., 60
Robbins, S., 318
Robideaux, M., 240

S

Sandhu, S. S, 309, 326
Schroder, P., 258

Sherman, R., 355
Shimabuku, R. A., 365
Siddaramaiah, 376
Simon, T. L., 341
Smith, B. M., 41
Somashekar, R. K., 376
Stanley, R. D., 29
Stewart, A. J., 126
Stewart-Pinkham, S. M., 161
Swanson, S. M., 77
Sutton, W. W., 217

T

Tapp, J. F., 29
Thomas, M. W., 217

V

Volk, G., 355

W

Wagner, J., 240
Walsh, G. E., 341
Wang, W., 68
Weber, D. E., 341
Wickliff, C., 250
Wickster, P., 240
Windeatt, A. J., 29
Wise, C. M., 365

Subject Index

A

2-aminofluorene, 287
Acetaminophen, 287
Adducts, 297
Aflatoxin B1
Agricultural chemicals
  effect on plant growth and metabolism, 240
Agricultural soil amendment, 12
Air pollution
  lichens, 276
  toxic exposure to cadmium, 161
Alachlor, 267
Algal bioassays
  with monophenolic acids, 230
Algae
  growth medium, 218(table)
  periphytic, 126
  recommendations for test species, 77
  use in assessing toxicity, 41, 107, 126
Algae sensitivity to insecticides and herbicides, 84(table)
Analysis of covariance, 126
ANCOVA (*See* Analysis of covariance)
Aneuploidy, 326
Animal carcinogens, 297
Animal drug wastes, 12
Animals
  influence of halo-organic compounds, 258
Anthracene, 209
Anthropogenic chemicals, 240
Antimutagen, 287
Aquatic macrophytes
  pesticide toxicity, 77
  response to herbicides, 267
  toxicity of chlorinated hydrocarbons, 178
Aquatic plants, 41, 267
Aquatic systems—toxicity, 126
Aquatic test systems, 267

Aquatic toxicology, 209, 383
AQUIRE data base, 79(table)
Arabidopsis, 365
Asbestos, 326
Assay methods
  chlorophyll fluorescence, 146
Assays for gene mutation, 326
Atrazine effects on aquatic plants, 81(table), 267

B

Bacterial activity, 383
Barley, 297, 355
Benzene, 326
Benzenesulfonic acid, 355
Benzo(a)pyrene, 318
Bioassays
  algal, 230
  aneuploidy-inducing agents, 326
  cyanophage/host assay, 383
  effects of toxic chemicals, 365
  phytotoxicity, 333, 355
  seed germination, 333
  selenastrum capricornutum, 217
  terrestrial plants, 197
  tradescantia micronucleus, 309
  use of plants for toxicity assessment, 41, 126
Bioconcentration, 178
Bioindicator, 41, 101
Biomonitoring
  aquatic systems, 126
  industrial effluents, 376
Bioremediation, 41
Biotechnology, 41
Biotransformation, 250
Blocking, 60
Brassica, 365

395

C

2-chloroacetamide, 333
CO₂ (*See* Carbon dioxide)
Cadmium toxicity
  biochemical effects, 161
  hazard assessment, 240
  health effects in ecosystem, 161
  selenastrum, 107, 111–112(figs)
  toxic effects from air pollution, 161
Calculated effect, 217
Carbon dioxide
  toxic effects on marine phytoplankton,
    118
Carcinogen, 297
Carson, Rachel (author)
  Silent Spring, 1962, 5
Chemical effects on plant behavior
  experimental design, 60
Chemical hazard to wetland plants, 341
Chemical influence of plants
  on xenobiotics, 250, 287
Chemical waste sites, 309
Chemicals, test
  physical properties, 357(table)
Chinese spring wheat assay, 326
Clastogenicity, 326
Chlorella, 230
Chlorinated hydrocarbons
  toxicity in aquatic macrophytes and fish,
    178
Chlorobenzene
  lethality in fish and aquatic plants,
    178
Chlorobiphenyl(PCB), 178
Chlorophenol, 355
Chlorophyceae
  for freshwater algae tests, 77
Chlorophyll fluorescence, 146
Coincubation assay, 318
Complete amendment, 217
Compost assay, 383
Consumer protection
  U.S. Food and Drug
    Administration(FDA), 12
Contaminants
  plant uptake, 172
Copper, 101, 107
Crucifer, 365
Cultured plant cells, 287
Cyanobacteria, 383
Cyanophage/host assay, 383,
    389–390(tables)

D

2,2-dichloropropionic acid, 365
Dalapon, 365, 371(tables)
Design experiments
  effects of chemicals on plant behavior,
    60
Dicamba, 355
Dichlobenil, 355
Dichloroanaline, 355
Dichlorobenzene, 355
Dichlorophenol, 355
2,2-Dichloropropionic acid, 365
Diethyldithiocarbamate, 287
Diethylstilbestrol, 326
Difluoro-chloro-bromo-methane, 258
Dinitrobenzene, 250, 355
Dinitrotoluene, 355
DNA, 297
Drug wastes
  animal, 12
Duckweed, 68

E

Early growth, 29
Early seedling growth test, (EPA, EG-13),
    365
E₅₀, 29, 217
Echinochloa crusgalli, 341
Ecological assessment, 41, 197, 333, 365
Ecotoxicology (*See also* Toxicology) 29,
    118, 197, 240, 341
Effluent toxicity, 68, 130–131(table)
Electron microscopy, 276
Emission spectroscopy, 240
Environmental assessment
  chemical waste sites, 309
  plant bioassays to predict effects of
    chemicals, 12, 309
  plant-chemical tests, 6(table)
  use of plants for toxicity assessment, 41,
    309
  xenobiotics, 250
Environmental chemicals
  possible factor in human carcino-genesis,
    326
Environmental data profiles, 24–25(tables),
    26(table)
Environmental protection, 5, 355
Environmental toxicology, 41, 287, 309
European Community(EC), 29
Experimental design, 60

F

Factorial design, 60
FDA (*See* Food and Drug Administration)
Federal Food, Drug and Cosmetic Act
    (FFDCA), 12
FFDCA (*See* Federal Food, Drug, and
    Cosmetic Act)
Fish, 178
Fluorescence bioassays, 41
Fluorometer, 14
Food additives
    environmental assessment, 12, 23–24 and
        26(tables)
    factors for plant toxicity testing, 22(fig)
Food and Drug Administration (FDA), 12
Food chain effects, 126, 140, 287, 318
Freshwater algae species
    effect of test duration toxic response,
        81(table)
    used in pesticide toxicity tests, 80(table)
Fumigation experiments with spruce, 258

G

Genetic hazard at chemical waste sites, 309
Genetically engineered plants, 41
Genotoxic metabolites, 318
Genotoxicity of chromium compounds,
    326, 328(table)
Germination, 29, 341, 379–380(tables)
Glasshouse bioassay, 197
Glutathione-S-transferase, 258
Glyphosate, 267
Government regulations
    environment, 5
Greenhouse effect, 258
Growth, 341

H

4-hydroxybenzoic, 230
H$_2$S (*See* Hydrocarbons)
Halo-organic compounds
    influence on animals and plants, 258
Halone 1211, 258
Hazard assessment
    cadmium, 240
    effects of marine diatom on H2 and
        CO2, 118
    effects of toxic chamicals, 365
    landfill soil, 146
    polycyclic aromatic hydrocarbons, 209

    potential from manmade chemicals, 5
    terrestrial plant bioassays with pesticides,
        197, 199(table)
    toluene, 240
    xenobiotics, 250
Hazard identification, 309
Hazardous materials
    chemical waste site, 309
    evaluation of phytotoxicity, 333
    influence of plant biomass, 250
Hazardous waste sites, 333
Health effects, 161, 309
Heavy metal rich industrial effluents, 376
Herbicides
    algae sensitivity, 84(table)
    cyanophage/host assay for toxicity, 383,
        387(table)
    impact on plant growth, 250
    phytotoxicity on aquatic plants, 267
    plant toxicity, 355
    toxicity assay, 146, 152–157(tables, figs)
Humics, 230
Hydrilla verticillata
    sublethal stress, 101
Hydrocarbons, 118
Hydrophobicity, 178

I

ICPAES (*See* Inductively coupled atomic
    plasma emission spectroscopy)
Inductively coupled atomic plasma
    emission spectroscopy, 240
Industrial effluents (*See* Industrial waste)
Industrial waste
    disposal sites, 309
    effluent toxicity assessment, 68, 376
    toxicity and food chain effects, 126,
        130–131(table),
Insecticides
    algae sensitivity, 84(table)
    impact on plant growth, 250
Invertebrates, effects of contaminants, 126

K,L

Kinetics, 178

Labeling, 12
Laboratory methods, 333
Laboratory toxicity tests, 333

Landfill soil
  hazard assessment, 146
Lead, 161
Lettuce, 68
Lichens, 276
Life-cycle bioassays, 365
Light and electron microscopy, 276

M

m-phenylenediamine, 287
Macrophytes, 77
Manmade chemicals
  disposal, 5
Marine algae species
  used in pesticide toxicity tests, 86(table)
Marine diatom
  toxicity tests, 118
Marine ecotoxicology, 118
Marsh environments,
  plant uptake of contaminants, 172
Metabolism, 240, 250
Metal loading
  periphyton as indicators, 126
Metal speciation, 107
Methodology (*See* Toxicity tests and test
  methodology)
Methylene blue, 383, 388(table)
Metolachlor, 341
Monophenolic acids
  algal bioassays, 230
Municipal effluent, 68
Mutagenic effects, 297, 309

N

N-methyl-N-nitrosourea, 297
Natural sediment, 341
Nitrobenene, 355
Nitrophenol, 355
Nitrotoluene, 355
Norflurazon, 341

O

OECD (*See* Organization for Economic
  Development)
Off-shore natural gas production
  effects of ocean discharge, 118
On-site evaluation, 333
Organic pollutant, 209
Organization for Economic Development,
  29

Oust, 101
Ozone depletion, 258

P

^{32}P-postlabeling, 297
PAH (*See* Polycyclic aromatic
  hydrocarbon)
PCB, 178
Paraquat, 267
Pentachlorophenol, 355
Periphyton chlorophyll, 133(table),
  134(fig), 137(table), 138(fig)
Periphyton evaluation techniques, 126
Periphyton metal content, 142(table)
Peroxidase, 101
Peroxidase inhibitors, 287
Peroxidation, 287
Pesticide toxicity tests, 83(table),
  93(appendix tables), 312–313(tables)
Pesticides
  chemical toxicity to plants, 355
  techniques for plant bioassays, 197
  toxicity to aquatic plants, 77,
    86–89(tables)
  waste site, chemical analysis of soil,
    312–313(tables)
pH, 107, 113–115(figs)
Phenanthrene, 209
Phenolic acids, 230, 235(table)
m-phenylenediamine, 287
Photosynthesis
  chlorophyll fluorescence yield, 146
  marine diatom, 118, 121(table),
    123(table)
Photoxicity, 209
Physical properties of test chemicals,
  355(table)
PHYTOTOX data base, 8, 49
Phytotoxic effects, 355
Phytotoxicity (plant)
  evaluation at hazardous waste sites, 333
  higher plants—effluent toxicity
    assessment, 68
  pesticide effects, 197
  polycyclic aromatic hydrocarbons, 209
  soil-based plant tests, 29
  test results, 72(table)
  tests and testing, 5, 7(table), 333, 355,
    376
Pica abies, 258
Plant activation, 287, 318
Plant bioassays (*See also* Bioassays)
  cyanophage/host assay for toxicity, 383

for assessing effects of toxic chemicals, 365
for assessing uptake of contaminants, 172
to predict effects of chemicals, 12
toxicity assay, 146
with pesticides, 197
Plant cell/microbe coincubation assay, 318
Plant cell systems, 287
Plant cell suspension cultures, 240
Plant cell systems—promutagen activation, 287
Plant-chemical interactions
need for more data, 250
Plant emergence, 29
Plant growth, 355
Plant life cycle, 333
Plant metabolism, 250, 318
Plant peroxidase (POD) activity, 101
Plant promutagens, 319(table)
Plant stress, 161
Plant tests (*See also* Toxicity tests and test methods)
cadmium toxicity, 161
soil-based, 29
test results on 43 chemicals, 34–37(tables)
Plant tissue culture, 250, 254–255(tables)
Plant toxicity
tests and testing, 5, 6(table), 12
Plants
DNA adduct determination, 297
effects of chemicals on plant behavior, 60
genetically engineered, 41
influence of halo-organic compounds, 258
use for toxicity testing, 41
Pollutants, 240, 276, 341
Polycyclic aromatic hydrocarbons
Pondweed, 267
Potamogeton pectinatus, 267
Product labeling, 12
Preregistration testing, 77
Promutagens, 287, 318
Proportional amendment, 217, 224(table), 226–227(table)

Rooted plants in wetland ecosystems, 341
Ruggedness test, 217, 222–223(tables)

S

Sago pondweed, 267, 272(table)
Salmonella assays, 28, 318
Saltbush, 297
Screening, 376
Sediment, 341
Sediment sampling, 172, 173(table), 343(table)
Sediment toxicity tests, 341
Seed germination, 333, 336(table), 341, 376
Seed germination/root elongation toxicity test(EPA, EG–12), 365
Seedlings
emergence, 197, 202(table)
growth, 365
survival, 341
Selenastrum, 107, 118, 217, 228(table)
Sewage, 383
Short lifecycle, 365
Site assessment and remediation, 333
Sludges, 12, 383 (*See also* Sediment)
Sodium dodecyl-benzenesulfonic acid, 355
Soil-based plant tests, 29, 217, 309, 333
Soil toxicity, 146
Soils, 217, 220(tables), 228(table)
Solar radiation, 209
Soybean, 355
Spartina alterniflora, 341
Spectroscopy, 240
Spruce, 258
Stream toxicity assessment, 126
Sublethal stress, 101
Sulfides
toxic effects on marine photo-plankton, 118
Sulfometuron methyl, 101
Surflan, 333
Suspension cultures, 250
Superfund sites, 309
Survival, 341
Synthetic sediment, 341
Syringic, 230

R

Regulatory protocols, 107
Rice, 68

T

Terrestrial plants, 41
Terrestrial systems
effects of chemicals, 12

Test conditions
  chlorophyll fluorescence toxicity assay, 146
  selenastrum, 107
  higher plants for toxicity assessment, 69(table)
  water quality of test samples, 71(table)
Test methods (*See* Toxicity tests and test methodology)
Test species, 9(table), 77
Test systems for ecological risk assessment, 41, 240
Tests and test methodology (*See* Toxicity tests and test methodology)
Thiourea, 355
Tissue culture
  method for xenobiotics biotransformation, 250
Tobacco cells in plant cell/microbe coincubation assay, 318
Toluene, 240
Toxic chemicals
  cadmium, 161
  effect on photosynthesis efficiency, 146
  toluene, 240
Toxicity (*See also* Toxicity tests and test methodology)
  cadmiums, 161
  chlorinated hydrocarbons, 178
  chlorophyll fluorescence assay method, 146
  food-chain effects, 126
  off-shore natural gas production
    effects of waste gases on marine life, 118, 101
  pesticide effects on aquatic plants, 77
  photosynthesis short-term testing, 118
  toluene, 240
Toxicity assessment
  aflatoxin B1 and benzo(a)pyrene, 318
  cadmium, 240
  industrial effluents, 376
  periphyton communities, 126
  phytotoxicity at hazardous waste sites, 333
  toluene, 240
  use of plants, 41, 68, 309, 341
  water, wastewater, sludges and composts, 383
Toxicity tests and test methodology
  aneuploidy-inducing agents, 326
  aquatic plants—herbicide response, 267
  cadmium, 240
  chlorinated hydrocarbons, 178

electron microscopy, 276
herbicide effects on aquatic plants, 267, 341
high plants, 68
overview, 5, 9(table)
peroxidase inhibitors, 287
physical properties of test chemicals, 357(table)
phytotoxicity, 333, 360(table), 362(table)
plant tissue culture, 250
short-term photosynthesis, 118
toluene, 240
tradescantia micronucleus assay, 309
use of marsh plants, 341
using chemostat-grown green alga, 107
water and sediment, 341
x-ray microanalysis, 276
Toxicology, 118, 209, 287 (*See also* Ecotoxicology, Toxicity tests and test methodology)
Tracer fate tests
  with plant cell cultures, 240
Track-sprayer, 197
Tradescantia micronucleus assay, 309, 314–315(tables)
Transpiration, 355
Tributyl-phosphate, 355
Trichloroacetic acid, 355
Trichlorophenol, 355
Tropical legumes, 376

U

U. S. Food and Drug Administration (FDA) (*See* Food and Drug Administration)
Ultraviolet radiation, 209
UV radiation (*See* Ultraviolet radiation)
Unicellular green algae
  test species for pesticide toxicity, 77

V

Vanillic, 230, 237(table)
Vascular plants, 5
Vegetative vigor, 197
Vigor index, 376
Viricides, 383

W,X

Waste gases
  effects on marine life, 118
Waste incineration, 161

Wastewater
  cyanophage/host assay, 383
  effect on germination, 379–380(tables)
  physico-chemical characteristics,
      378(table)
  toxicity assessment, 383
Water contamination—bioassay, 217
Waterborne pollutants, 341

Wetland ecosystems, 341
Wetland plants, 341
Wetlands
  plant uptake of contaminants, 172
  protection, 267
Wheat seedling assay, 326

Xenobiotics, 240, 250, 287